Fundamentals of Nonlinear Optics

Fundamentals of Nonlinear Optics

Second Edition

Peter E. Powers
Joseph W. Haus

CRC Press
Taylor & Francis Group
Boca Raton London New York

CRC Press is an imprint of the
Taylor & Francis Group, an **informa** business

CRC Press
Taylor & Francis Group
6000 Broken Sound Parkway NW, Suite 300
Boca Raton, FL 33487-2742

International Standard Book Number-13: 978-1-4987-3683-1 (Hardback)

Library of Congress Cataloging-in-Publication Data

Names: Powers, Peter E., author. | Haus, Joseph W., 1948- author.
Title: Fundamentals of nonlinear optics / Peter E. Powers, Joseph W. Haus.
Description: Second edition. | Boca Raton, FL : CRC Press, Taylor & Francis
Group, [2017] | Includes bibliographical references and index.
Identifiers: LCCN 2016044943| ISBN 9781498736831 (hardback ; alk. paper) |
ISBN 1498736831 (hardback ; alk. paper)
Subjects: LCSH: Nonlinear optics.
Classification: LCC QC446.2 .P69 2017 | DDC 535/.2--dc23
LC record available at https://lccn.loc.gov/2016044943

Visit the Taylor & Francis Web site at
http://www.taylorandfrancis.com

and the CRC Press Web site at
http://www.crcpress.com

To

My nonlinear family: Deborah, William, Gregory, Regina, and Eric

—*PEP*

… and to mine:

Jean, Alison, Michelle, Paul, Karin, Thomas, and Monica

—*JWH*

CONTENTS

10   Nonlinear Optics Including Diffraction and Dispersion   337

11   Quantum Nonlinear Optics   381

PREFACE TO THE FIRST EDITION

The purpose of this text is to introduce students and professionals to the fundamentals of nonlinear optics. The text is designed around a one-semester course. It is written at a level accessible to graduate students and senior undergraduates who have taken a course in electro-magnetism including Maxwell's equations and electromagnetic waves. Outside of the student population, the text is also intended for scientists and engineers who are new to nonlinear optics and seek an introductory resource. The topics covered in the text encompass a broad spectrum of nonlinear phenomena from second-harmonic generation to soliton formation. The topics covered prepare students for further study in the subject matter as well as more advanced topics.

The wide use of nonlinear optical phenomena in laboratories and commercial devices requires a familiarity with the underlying physics as well as practical device considerations. The text introduces subject matter by giving the theoretical background and framework for a nonlinear effect, followed by details of how such an effect is implemented in a real system. These two aspects of nonlinear optics are complementary. The theory developed allows for a fundamental understanding of a given effect while the practical device considerations help to guide the theoretical directions.

A goal of the text is to look at different nonlinear phenomena using, where possible, the same theoretical structure. This approach helps students to master key concepts that come up repeatedly throughout nonlinear optics. It also helps students to think about how to apply what they learn to topics not covered in the text. The text starts with a chapter focused on linear phenomena important to nonlinear optics. Whether a course includes this chapter depends on the student's background. Even for students with a solid grounding in optics, however, the chapter serves as a resource to important linear phenomena and its terminology. The first nonlinear phenomena covered are the technologically important processes of second-harmonic generation, sum- and difference-frequency generation, and the electro-optic effect. These processes are covered in considerable detail at both a theoretical and practical level. The formalisms developed for these effects carry over to subsequent topics covered in the text such as four-wave mixing, self-phase modulation, Raman scattering, Brillouin scattering, and soliton formation. The text has the same bias of connecting these higher-order effects to applications as it does for the lower-order effects. To help develop both a strong theoretical foundation and a practical understanding of nonlinear optics, the text includes extensive problems at the end of each chapter. Many of the problems require some degree of plotting and simple programming. The text is not tied to a single mathematical package; instead, it gives general algorithms, which are accessible to any scientific graphical and programming package. Instructors can draw upon the problems at the end of chapters for in-class examples and for homework assignments. A solutions manual for the problems is available to instructors from the publisher.

I have many people to thank for helping with this text. My introduction to nonlinear optics came as a graduate student, and many of my insights and understanding were developed as a student of Professor Chung Tang. I am grateful for his guidance and support in my career. This text grew out of a series of lecture notes developed over several years teaching nonlinear optics. I am grateful to numerous students who gave feedback and suggestions to improve the notes and the text. Many of the examples given in the text draw upon experimental work in my laboratory and I would like to thank my graduate students for the many stimulating conversations centered on describing these effects. I would further like to thank Dr. Kenneth Schepler at the U.S. Air Force Research Laboratory, and Professors Leno Pedrotti and Joseph Haus at the University of Dayton for critiquing portions of the text. I am indebted to Professors Mark Cronin-Golomb at Tufts University, Jeff F. Young at University of British Columbia, and Marc Dignam at Queen's University, who reviewed a rough draft and provided invaluable feedback. I thank my editor, Luna Han, for her help, encouragement, and patience throughout the writing of the text. Finally, I thank my family, especially my wife Deborah, for their support while I wrote the text.

Peter E. Powers
Dayton, Ohio

PREFACE TO THE SECOND EDITION

It was a tremendous loss for the nonlinear optics community when Peter Powers passed away in June 2014. He was planning a second edition of this book to include corrections and add new material where needed. Taking on the task of writing this second edition was an awesome task, since in my view it was already the best book on the market for introducing students and researchers to the practical aspects of nonlinear optics. In this revision, it has been my intention to stay true to his vision of this book.

Looking back, my first embrace of his teaching philosophy was when I used his book to teach a course on nonlinear optics. For this second edition, I have followed his approach with the exception of some reorganization or enhancing coverage of certain topics. In particular, the treatment of Maxwell's equations in Chapter 2 is expanded to include dispersion at the outset, which is a solid foundation for the topics that follow. The chapters on parametric processes were largely left as Peter wrote them. This was his domain and he presented these topics clearly and with authority. I added problems to the chapters and made small changes where I thought it would improve the clarity of the presentation.

The main differences come in the treatment of third-order nonlinear processes, where my own expertise is applied. Material on the modulation instability was added to further motivate soliton formation. Also, the treatment of fiber optic systems was expanded with an appendix on the derivation of the slowly varying envelop equations for propagation in optical fibers. Chapter 11 on quantum nonlinear optics explores a relationship of nonlinear optics to the fundamental limitations of nature, especially where noise can be a dominant effect in the measurement process. A direct connection is made with nonlinear phenomena studied in earlier chapters, but now viewed from a quantum perspective; the prior results are translated to a quantum context and treated in a phenomenological way. The foundations of quantum mechanics, especially Einstein's objections, are examined with an emphasis on nonlinear optics through the creation of entangled photon pair states.

I want to thank the students and colleagues who have given me feedback on this book and especially Sean Kelley, who diligently read all chapters and made detailed remarks that improved the presentation. I am also grateful to Professors Imad Agha and Partha Banerjee for reading Chapter 11 and providing advice and comments. Luna Han has been a supportive and patient editor during the writing of the new edition; she gave me the opportunity to complete the task to my satisfaction. Finally, I also thank my wife Jean who put up with my mess sprawled across our kitchen table for many months.

Joseph W. Haus
Dayton, Ohio

AUTHORS

Peter E. Powers was a professor of physics and electro-optics, and the Brother Leonard A. Mann chair in the sciences at the University of Dayton. He became a member of the faculty since 1997. Before joining there, he spent 3 years at Sandia National Laboratories developing mid-IR nonlinear optical sources for remote sensing and sensitive chemical detection. Dr. Powers completed his PhD on femtosecond optical parametric oscillators at Cornell University in 1994 under the direction of Professor Chung L. Tang. Dr. Powers' research centers on nonlinear optical devices with an emphasis on optical frequency conversion. These devices have enabled applications including trace species detection, remote sensing, laser radar, and terahertz generation. He published widely in the optics literature, is a regular participant at optics conferences, and has six patents in the area of nonlinear optics. Dr. Powers was a respected member of the optics community. He chaired the SPIE symposium, *Nonlinear Frequency Generation and Conversion: Materials, Devices, and Applications*, and was a frequent journal referee. He was also a fellow of the Optical Society of America and the SPIE, and a member of the American Physical Society and Sigma Xi.

Joseph W. Haus is a professor in Departments of Electro-Optics and Photonics, Electrical and Computer Engineering, and Physics at the University of Dayton, where he joined in 1999. He served as the director of the Electro-Optics Program for 13 years; he is the director of the Ladar and Optical Communications Institute. Before joining the University of Dayton, he was a professor of physics at Rensselaer Polytechnic Institute for 15 years. He also held positions in Germany at the Universität Essen and the Kernforschungsanlage in Jülich and in Japan at the University of Tokyo, where he held the Hitachi Limited Quantum Materials chair in 1991–1992. His research has covered topics in quantum and nonlinear optics in composite materials and nonlinear fiber optics and he collaborated with Peter

Powers on optical parametric processes, including terahertz generation. Dr. Haus cofounded the International Conference on Nanophotonics, which held in China, but also has been held in Taiwan, Japan, and Brazil. He has published a book called *Fundamentals and Applications of Nanophotonics* (Woodhead Publishing, 2016) and he holds two patents in nonlinear optics. He is a fellow of the American Physical Society, Optical Society of America, and SPIE. He is a member of the IEEE.

1

INTRODUCTION

1.1   HISTORICAL BACKGROUND

You are about to embark on a study of nonlinear optics. Before we delve into the mathematical framework behind nonlinear optics, it is instructive to briefly present an historical perspective. The field of nonlinear optics encompasses a rich diversity of phenomena whose applications are growing at a seemingly exponential rate. The modern rebirth of the field began with the proposal of a laser device in 1958 by Arthur L. Schawlow and Charles H. Townes and its first operational demonstration in 1960 by Theodore Maiman. The laser was conceived as a highly monochromatic and coherent light source with emission at "optical" wavelengths, that is, at wavelengths far shorter than maser wavelengths. Indeed, the laser was initially so closely tied to the development of the maser, which generates coherent radiation at microwave frequencies, that in early publications it was referred to as the *optical maser*.

In the nineteenth century, there were two noteworthy discoveries of field-induced refractive birefringence effects. A quadratic field effect was described by the Scotsman John Kerr in 1875 and a linear field effect was reported by the German Friedrich C. A. Pockels in 1883. Today we recognize these effects as third- and second-order nonlinear effects, respectively. In both cases, these phenomena affect the phase of the wave, rather than the amplitude. In 1922, Leon N. Brillouin, a French physicist, postulated a nonlinear coupling between acoustic and optical waves in a medium, and in 1928, two Indian physicists Chandrasekhara V. Raman and Kariamanickam S. Krishnan measured an inelastic spontaneous emission effect, now commonly called the Raman effect, which was predicted in 1923 by the Austrian Adolf Smekel. Both of these effects are third-order nonlinear phenomena. Interestingly, the Russian physicist Leonid I. Mandelstam independently predicted both of these effects during that period.

Two publications of note in the history of nonlinear optics appeared in the following decade. In 1931, the German Maria Goepport-Mayer published her PhD dissertation work on two-photon absorption processes by extending concepts of quantum theory. Two-photon absorption is also a third-order nonlinear effect that corresponds to an imaginary contribution to the coefficient, whereas the Kerr effect corresponds to a real third-order contribution. The effect was observed soon after the invention of the laser as a nonlinear fluorescence phenomenon. Her work became honored by coining a molecular cross section of a Goeppert-Mayer (GM), which has the unit 10^{-50} cm^4 s/photon; it is a convenient unit in studies of nonlinear absorption in materials. The second publication is the paper by Albert Einstein, Boris Podolsky, and Nathan Rosen in 1935 (at that time at Princeton University) that purported to call into question the inadequacy of quantum mechanics. In their argument, they introduced what is now called quantum entanglement of particle pairs that obey momentum and energy conservation laws. This has been a fruitful field of research in nonlinear optics by using parametric down-conversion to split a photon into a pair of photons. This research is leading to the emergence of new fields in quantum information and quantum communications.

Within a few years after the laser was invented, there was an explosion of laser-related activities around the globe. Within a few years, nonlinear optical effects such as second-harmonic generation (SHG), sum-frequency generation (SFG), difference-frequency generation (DFG), optical rectification, two-photon absorption, and third-harmonic generation were observed for the first time at optical frequencies (see Figure 1.1 and references therein). The celebrated paper in 1961 by Peter Franken's group at the University of Michigan reported a first measurement of SHG using a quartz crystal as the nonlinear material. The published paper is also interesting for what it does not show, namely, the second-harmonic signal was interpreted by the copy editor as a smudge on the photograph and the proof of SHG signal was removed. The paper caused quite a stir because it provided new insights into optical frequency conversion at a quantum mechanical level; the new conceptualization of nonlinear optical processes was not widely appreciated at the time.

Additional studies quickly followed building a theoretical framework to describe and predict the observed nonlinear behavior. Nikolaas Bloembergen's group at Harvard University was among the early contributors (Armstrong et al., 1962; Bloembergen and Pershan, 1962; Kleinman, 1962a,b; Pershan, 1963; Ducuing and Bloembergen, 1964; Harris, 1966; Minck et al., 1966; Tang, 1966; Boyd and Kleinman, 1968; Giallorenzi and Tang, 1968). The landmark 1962 paper by Armstrong et al. presented analytical solutions of nonlinear equations and proposed new experimental strategies to achieve efficient nonlinear conversion. The invention of the laser ushered in an era of publications reporting observations of new phenomena, exploring new aspects of the field's theoretical underpinnings, and improving upon earlier results. Advances in laser and photodetection technologies have enabled studies of additional nonlinear phenomena in extended wavelength regimes and the creation of ultrashort pulses.

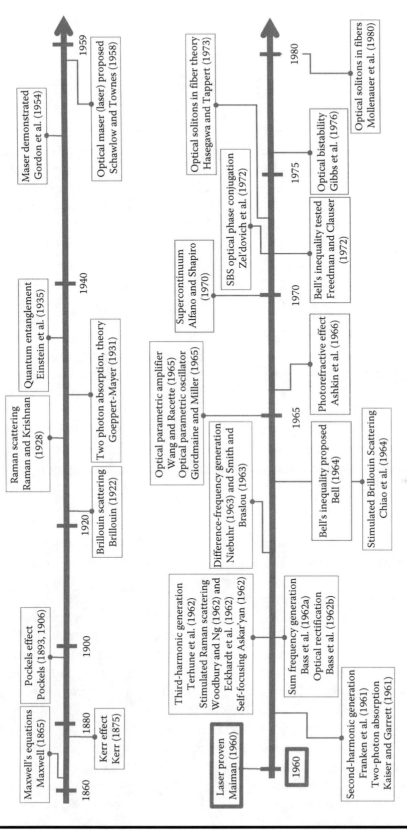

FIGURE 1.1   Timeline pointing out selected milestones in the history of nonlinear optics. *Top:* Prehistory of nonlinear optics before the laser's invention. *Bottom:* A 20-year span of rapid nonlinear optics development after the laser was demonstrated.

Behind all these developments is the high field associated with lasers. Because of their temporal and spatial coherence, lasers have a brightness, defined as $W/m^2/Sr$, which far exceeds other light sources. Early Q-switched lasers easily attained intensities of >1 MW/cm^2 by focusing, and the interaction of these high fields with materials led to observations of nonlinear behavior. The rate of discovery and development in laser science and in nonlinear optics since the early 1960s continued at a rapid pace in subsequent decades and was highly synergistic. The generation of ultrashort pulses requires nonlinear elements to mode-locked pulses in the cavity. Laser operation down to the femtosecond pulse width range is now a common tool in the lab and in commercial machining applications. The development of super-intense femtosecond lasers opened a new regime of strong-field nonlinear optics. Strong fields drive a nonlinear response in gases that can generate harmonic waves with over 100 times the energy of the pump photons; these phenomena are called high-order harmonic waves and photon wavelengths can reach into the x-ray regime. Due to the coherence of the pump laser source, the high-order harmonics also have some measure of coherence; coherence combined with the shorter wavelengths has also lead the way to breaking the femtosecond barrier and producing attosecond pulse widths.

From among the many different nonlinear phenomena reported, we mention here several of those covered in this book: spontaneous parametric fluorescence in 1967 (Akhmanov et al., 1967; Byer and Harris, 1968), optical phase conjugation in 1972 (Zel'dovich et al., 1972; Yariv, 1976), optical bistability in 1976 (Gibbs et al., 1976), and optical solitons in 1980 (Mollenauer et al., 1980).

Applications for nonlinear optical devices were readily apparent after the underlying nonlinear effects were first demonstrated. For instance, the ability to convert laser light to new frequencies adds significant flexibility to fixed frequency laser sources. Moreover, the efficiency of nonlinear frequency conversion can be high, in some cases depleting the input laser energy. It is common to find commercial lasers sold with frequency extension units to double, triple, and quadruple the fundamental frequency. Continuously tuned nonlinear sources are the result of parametric frequency conversion processes. Frequency conversion is more than a convenience—nonlinear optical techniques enable experiments that require specific frequencies unattainable by available lasers, which have limited tuning ranges.

Another class of nonlinear optics applications called nonlinear spectroscopy has established an understanding of molecular materials at the quantum level. Nonlinear absorption and scattering methods provide a means to identify trace quantities of unknown material constituents with high fidelity. Spontaneous Raman scattering (Raman and Krishnan, 1928) imparts a "fingerprint" scattered frequency signature that allows for the identification of different materials. The increased brightness of laser beams has rendered stimulated Raman scattering (SRS) signals observable and using a laser beam incident on a suitably prepared surface SRS spectroscopy has improved molecular fingerprinting sensitivity by 10–12 orders of magnitude, even enabling single-molecule detection (see Chapter 9). Nonlinear saturation spectroscopy allows for high-resolution spectroscopy by eliminating Doppler shifts in absorption features, called Lamb-dip spectroscopy (Lamb, 1964). Many laser-based spectroscopy topics are not treated in this book; the interested reader may wish to consult a survey of spectroscopic techniques for further information (Demtroder, 1996).

1.2 UNIFYING THEMES

Nonlinear effects in this book involve energy exchange between independent optical inputs, nonlinear phase shifts to optical fields, and energy exchange between optical fields and a material. For each effect, we derive a nonlinear polarization expression and insert it into a wave equation that describes the propagation of electromagnetic (EM) waves through the medium. We connect the material polarization to electric fields through the material's linear susceptibility, $\chi^{(1)}$. To simplify the discussion, we treat the polarization and electric field as scalars and introduce the relationship between the material polarization and an incident field as purely linear,

$$\underline{P} = \varepsilon_0 \chi^{(1)} \underline{E}, \tag{1.1}$$

where P is the polarization, E is the electric field, and ε_0 is the permittivity of free space. In Chapter 2, we cover linear EM effects in greater detail and emphasize the important role they play in nonlinear optics. In Chapter 2, we also derive Poynting's theorem, which shows a direct connection to energy exchange between a field and the nonlinear polarization.

When the electric field is large, as can be the case with lasers, the material properties are modified by the incident field. The first nonlinear effects we study are associated with distortions to the electron cloud surrounding atoms and molecules. Classically, the electric field displaces electrons in the medium, and when the displacement is large, nonlinear effects become apparent. This classical model is developed in Chapter 3. These electronic-based nonlinear effects are weak, and so we write the dependence as a power-series expansion in the field,

$$\underline{P} = \varepsilon_0 \chi^{(1)} \underline{E} + \varepsilon_0 \chi^{(2)} \underline{E}^2 + \varepsilon_0 \chi^{(3)} \underline{E}^3 + \cdots \tag{1.2}$$

The susceptibility is broken down into orders with $\chi^{(1)}$ corresponding to linear effects, $\chi^{(2)}$ corresponding to quadratic terms in the field, and so on. The types of processes associated with the different expansion orders are referred to as simply $\chi^{(2)}$ effects, $\chi^{(3)}$ effects, and so on. In Chapter 3, we derive expressions for the nonlinear susceptibilities based on classical models. Chapters 4 through 7 cover the detailed tensor form of $\chi^{(2)}$ and apply Maxwell's equations to study SHG, SFG, and DFG.

The same formalisms used for the earlier chapters are extended to describe $\chi^{(3)}$ effects, such as four-wave mixing, optical bistability, and optical phase conjugation, in Chapter 8. Another class of nonlinear problems involves interactions of an electric field with molecular vibrations, rotations, reorientation, and collective modes (phonons and acoustic waves). Raman scattering and Brillouin scattering are two important nonlinear optical phenomena we cover in Chapter 9, and we show that they are described as $\chi^{(3)}$ effects. Indeed, nonlinear effects are commonly used as mode-locking elements in laser cavities to help form short pulses.

The equations used in many studies of nonlinear optics are extended to incorporate dispersion and diffraction in Chapter 10. The treatment of ultrashort pulses and their linear and nonlinear effects has been widely discussed in the literature. For instance, it is important for understanding the operation of ultrashort-pulse solid-state and fiber lasers. A description of $\chi^{(2)}$ nonlinear processes with ultrashort pulses enables a deeper understanding of wavelength-tunable sources of coherent radiation; they are useful tools for spectroscopic applications. Nonlinear optics for weak fields can be used to understand the quantum nature of the EM field. Chapter 11 is an introduction to the quantum properties of the EM field and the manifestation of the quantum properties in experiments. The quantum nature has applications for quantum communication and computing, which are emerging from this important research field.

The derivation of the wave equation governing all the effects discussed in this book follows the same procedure. We find that the wave equations for different $\chi^{(2)}$ and $\chi^{(3)}$ processes have many similarities, which allow us to build up concepts that carry over to all $\chi^{(2)}$ and $\chi^{(3)}$ processes. The derivations made early in this book include many simplifying assumptions, but expose important general properties of nonlinear phenomena. The simplifying assumptions are progressively lifted in later chapters, revealing more detailed behavior of nonlinear optical systems.

In some cases, the perturbation expansion given in Equation 1.2 breaks down. For example, consider a commercially available high-energy and ultrashort-pulse laser, characterized by 10 mJ of energy in a 50-fs pulse. This laser produces pulses with peak powers of approximately 2×10^{11} W. When the output of the laser is focusable to a diffraction limited spot size, say 10 μm², the local intensity can achieve extreme values 2×10^{22} W/m². The electric field associated with the intensity are around 1×10^{12} V/m, which is larger than the Coulomb field binding the electron to the proton, approximately 1×10^{11} V/m (Problem 1.2). For these extreme field values, $\chi^{(2)}E^2$ becomes comparable to $\chi^{(1)}E$, and the expansion in Equation 1.2 breaks down. However, extreme fields are not the only route where the expansion breaks down. In resonant nonlinear phenomena, population transfer between a few quantum states is responsible for phenomena such as saturable absorption and for quantum coherence in atomic systems such as two-level atoms, where the perturbation expansion breaks down even for relatively low intensities.

The approach of this book is to treat phenomena where the series expansion given in Equation 1.2 is valid. Such phenomena cover a great deal of basic nonlinear optics and the mathematical concepts. The techniques contained in this book developed in discussing them are a good starting point for studying other, perhaps more exotic, nonlinear processes.

1.3 OVERVIEW OF NONLINEAR EFFECTS COVERED IN THIS BOOK

Although we derive classical expressions for the nonlinear susceptibilities, we also make use of quantum mechanical pictures that are centered on energy conservation to describe the phenomena. Energy-level diagrams along with schematic illustrations of photons help to describe the underlying processes quickly and intuitively. As an overview to the phenomena presented in this chapter, we present the effects schematically using the quantum mechanical picture and defer specific details of the processes to later chapters. Note that the quantum mechanical pictures given here describe the nonlinear interaction *when* it occurs, but since nonlinear effects are weak, the probability of a given interaction occurring is low. Figure 1.2 shows the different permutations of $\chi^{(2)}$ effects involving three frequencies, which correspond to three photons at the most basic level. The energy conservation statement for these interactions is given by

$$\hbar\omega_1 = \hbar\omega_2 + \hbar\omega_3, \tag{1.3}$$

where for three-wave interactions, we use the convention $\omega_1 > \omega_2 \geq \omega_3$. SFG and SHG at the most fundamental level involve two input photons that are annihilated while one photon with higher energy is created (see Figure 1.2a). DFG is the reverse process where a high-energy photon splits into two photons of lesser energy (see Figure 1.2b). In DFG, note that when the high-energy photon at ω_1 splits into a photon at ω_2 and another at the difference frequency, ω_3, the input at ω_2 is amplified by an additional photon. Hence, DFG is also called optical parametric amplification (OPA).

The same three-wave processes are described using energy-level diagrams depicted in Figure 1.3. The system starts with an electron in its ground state and is then excited to an intermediate "virtual" state. The virtual level indicates a transient state for the system, whereas real electronic levels are those where the electron can reside indefinitely.

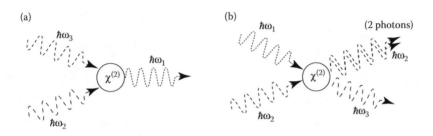

FIGURE 1.2 (a) Sum-frequency process and (b) difference-frequency process and OPA. The sinusoidal-shaped lines represent photons.

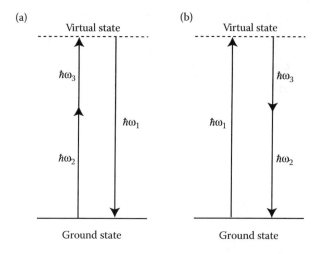

FIGURE 1.3 Energy-level diagrams for (a) a sum-frequency process and (b) a difference-frequency process and OPA.

In the case of DFG and OPA, we have two photons present at the input, one at $\hbar\omega_1$ and another at $\hbar\omega_2$ However, the energy diagram shows only the input photon at ω_1 since it is annihilated, whereas the input photon at $\hbar\omega_2$ is unaffected.

Four-wave mixing applications involve four photons at the most basic level and have photon pictures and energy-level diagrams similar to the three-wave case. For example, Figure 1.4 shows a sum-frequency process between three input lasers at ω_2, ω_3, and ω_4 defined by

$$\hbar\omega_1 = \hbar\omega_2 + \hbar\omega_3 + \hbar\omega_4. \tag{1.4}$$

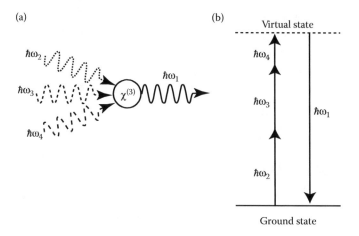

FIGURE 1.4 Illustration of a four-photon sum-frequency process: (a) photon annihilation and creation and (b) energy-level diagram.

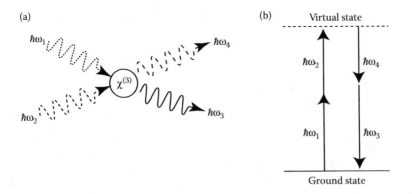

FIGURE 1.5   Illustration of a four-photon parametric process: (a) photon annihilation and creation and (b) energy-level diagram.

The sum-frequency process is not the only possible outcome for a four-wave process. Another type of process is the parametric amplifier (see Figure 1.5) defined by the energy conservation statement,

$$\hbar\omega_1 + \hbar\omega_2 = \hbar\omega_3 + \hbar\omega_4. \tag{1.5}$$

A particularly interesting case for the parametric amplifier occurs when $\omega_1 = \omega_2$ as two output frequencies are generated from a single input laser.

In the above processes, energy is exchanged between optical inputs while the material system is left in its original state. In Raman scattering and Brillouin scattering, energy is exchanged between optical inputs and the medium. For Raman scattering, an energy-level diagram gives a clear picture of the process showing a system in its ground state being excited to an intermediate virtual state and then relaxing to an excited state (see Figure 1.6). The photon emitted from the system has an energy

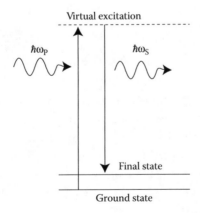

FIGURE 1.6   Illustration of Raman scattering with an energy-level diagram. The input is at the pump frequency, ω_P, and the output is at ω_S, which is called the Stokes frequency.

produced by the difference between the incident photon energy and the material excitation's energy.

The quantum mechanical pictures given above help to describe most of the phenomena covered in this book. However, a few effects, such as nonlinear phase shifts, are not easily depicted with these quantum mechanical pictures; hence, we use classical visualizations instead. The quantum and classical visualizations help to describe the basic processes, but do not give any indication as to the efficiency of the nonlinear interaction. We determine this detail and others in subsequent chapters by including the nonlinear polarization in the wave equation for EM waves.

1.4 LABELING CONVENTIONS AND TERMINOLOGY

We use specific terminology regarding sinusoidally varying time-dependent quantities. In most cases, we use complex variable representations to represent these quantities. Hence, we introduce notation to distinguish between real amplitudes and complex amplitudes. For example, let us consider the real electric field given by

$$\underline{\vec{E}} = \underline{\vec{E}_0} \cos(kz - \omega t + \phi). \tag{1.6}$$

The underscore underneath a variable here indicates that it is a real quantity. An equivalent expression written in terms of complex amplitude is

$$\underline{\vec{E}} = \frac{1}{2} \vec{A} e^{i(kz - \omega t)} + c.c., \tag{1.7}$$

where c.c. stands for "complex conjugate," and $\vec{A}$ is the complex amplitude given by

$$\vec{A} = \underline{\vec{E}_0} e^{i\phi}. \tag{1.8}$$

Note that we indicate the difference between a real amplitude, $\underline{\vec{E}}$, and a complex amplitude, $\vec{A}$, by the underscore.

A second issue arises when representing fields using complex variables. The definition of the complex amplitude is somewhat ambiguous in that the factor of 1/2 in Equation 1.7 may be absorbed into the complex amplitude. In this book, we always keep the factor of 1/2 outside of the complex amplitude, as shown in Equation 1.7. For more discussion on this topic, see Chapter 2.

A final note on terminology: In nonlinear optics literature, the radiometric term called irradiance (W/m²), which is the power per unit area through a surface, is commonly referred to as the intensity; we use the terminology of intensity in this context. The intensity units quoted in the literature are often in mixed units and appear as W/cm².

1.5 UNITS

Nonlinear interactions are described using a variety of systems of units in the literature. The two most common ones are SI and Gaussian (cgs, esu) units. Gaussian units are commonly referred to as "esu," shorthand for "electrostatic units," in the literature. Each system has its advantages, but we adopt SI for all formulae in this book. Formulae written in SI units are particularly helpful when working with experimentally measured values. It is useful to relate the SI and cgs values of certain material properties. The relevant conversion factors for this book are

$$\chi^{(1)}_{SI} = 4\pi \chi^{(1)}_{Gaussian},$$ (1.9)

$$\chi^{(2)}_{SI} = \frac{4\pi}{3 \times 10^4} \chi^{(2)}_{Gaussian},$$ (1.10)

$$\chi^{(3)}_{SI} = \frac{4\pi}{(3 \times 10^4)^2} \chi^{(3)}_{Gaussian}.$$ (1.11)

The denominator is the conversion factor from statvolts/cm to volts/m and the factor of 4π in the numerator is a notational change in the definition of the susceptibilities. The form of Maxwell's equations in Gaussian units has a factor of 4π multiplying the source terms and a factor $1/c$ multiplying the flux contributions in both Faraday's and Ampere's laws, which means that electric and magnetic fields have the same units.

PROBLEMS

1.1 This problem is an opportunity to enrich your knowledge and appreciation about historical milestones in nonlinear optics. Searching the list of Nobel prizes in physics and chemistry, choose one that uses lasers and/or nonlinear optics. Read the laureate's biography and listen or read the lecture on the Nobel website. Then write a brief description of the laureate and a summary of the technical discovery credited by the Nobel committee.

1.2 The hydrogen atom is a prototype model to compare the size of the nonlinear coefficients with an electrostatic field related to electron binding. Use classical concepts to calculate the electric field strength due to the proton's Coulomb field at the Bohr radius, the most probable value of the electron's orbital radius in the ground state. Use the electric field strength to estimate the second- and third-order nonlinear coefficients based on a unit analysis. Compare the results with second-order coefficients for inorganic compounds found in Appendix B and with third-order coefficients for several inorganic compounds you find in the literature.

REFERENCES

Akhmanov, S. A., V. V. Fadeev, R. V. Khokhlov, and O. N. Chunaev. 1967. Quantum noise in parametric light amplifiers. *JETP Letters* 6: 85–88.

Alfano, R. R. and S. L. Shapiro, 1970a. Emission in the region 4000 to 7000 Å via four-photon coupling in glass. *Physical Review Letters* 24: 584–587.

Alfano, R. R. and S. L. Shapiro, 1970b. Observation of self-phase modulation and small-scale filaments in crystals and glasses, *Physical Review Letters* 24: 592–594.

Armstrong, J. A., N. Bloembergen, J. Ducuing, and P. S. Pershan. 1962. Interactions between light waves in a nonlinear dielectric. *Physical Review* 127: 1918–1939.

Ashkin, A., G. D. Boyd, J. M. Dziedzic, R. G. Smith, A. A. Ballman, J. J. Levinstein, and K. Nassau. 1966. Optically-induced refractive index inhomogeneities in LiNbO3 and LiTaO3. *Applied Physics Letters* 9: 72–74.

Askaryan, G. A., 1962. Effects of the gradient of a strong electromagnetic beam on electrons and atoms. *Soviet Physics JETP-USSR* 15: 1088–1090.

Bass, M., P. A. Franken, A. E. Hill, C. W. Peters, and G. Weinreich. 1962a. Optical mixing. *Physical Review Letters* 8: 18.

Bass, M., P. A. Franken, J. F. Ward, and G. Weinreich. 1962b. Optical rectification. *Physical Review Letters* 9: 446–448.

Bell, J. S. 1964. On the Einstein Podolsky Rosen paradox. *Physics* 1: 195–290.

Bloembergen, N. and P. S. Pershan. 1962. Light waves at the boundary of a nonlinear media. *Physical Review* 128: 606–622.

Boyd, G. D. and D. A. Kleinman. 1968. Parametric interaction of focused Gaussian light beams. *Journal of Applied Physics* 39: 3597.

Brillouin, L. 1922. Diffusion de la Lumière et des Rayonnes X par un Corps Transparent Homogéne: Influence del'Agitation Thermique [Diffusion of light and X-rays by a transparent homogeneous medium: Influence of thermal agitation]. *Annales des Physique* 17: 88.

Byer, R. L. and S. E. Harris. 1968. Power and bandwidth of spontaneous parametric emission. *Physical Review* 168: 1064–1068.

Chiao, R. Y., C. H. Townes, and B. P. Stoicheff. 1964. Stimulated Brillouin scattering and coherent generation of intense hypersonic waves. *Physical Review Letters* 12: 592–595.

Demtroder, W. 1996. *Laser Spectroscopy*. New York, NY: Springer-Verlag.

Ducuing, J. and N. Bloembergen. 1964. Statistical fluctuations in nonlinear optical processes. *Physical Review* 133: A1493–A1502.

Eckhardt, G., R. W. Hellwarth, F. J. McClung, S. E. Schwarz, D. Weiner, and E. J. Woodbury. 1962. Stimulated Raman scattering from organic liquids. *Physical Review Letters* 9: 455–457.

Einstein, A., B. Podolsky, and N. Rosen 1935. Can quantum-mechanical description of physical reality be considered complete? *Physical Review* 47: 777–780.

Franken, P., A. E. Hill, C. W. Peters, and G. Weinreich. 1961. Generation of optical harmonics. *Physical Review Letters* 7: 118–119.

Giallorenzi, T. G. and C. L. Tang. 1968. Quantum theory of spontaneous parametric scattering of intense light. *Physical Review* 166: 225–233.

Gibbs, H. M., S. L. McCall, and T. N. C. Venkatesan. 1976. Differential gain and bistability using a sodium-filled Fabry–Perot interferometer. *Physical Review Letters* 36: 1135–1138.

Giordmaine, J. A. and R. C. Miller. 1965. Tunable coherent parametric oscillation in LiNbO$_3$ at optical frequencies. *Phyical Review Letters* 14: 973–976.

Goeppert-Mayer, M. 1931. Über Elementarakte mit zwei Quantensprüngen [About elementary actions with two quantum transitions]. *Annalen der Physik* 9: 273–294.

Gordon, J. P., H. J. Zeiger, and C. H. Townes. 1954. Molecular microwave oscillator and new hyperfine structure in microwave spectrum of NH$_3$. *Physical Review* 95: 282–284.

Harris, S. E. 1966. Proposed backward wave oscillation in the infrared. *Applied Physics Letters* 9: 114–116.

Hasegawa, A. and F. Tappert, 1973. Transmission of stationary nonlinear optical pulses in dispersive dielectric fibers. I. Anomalous dispersion. *Applied Physics Letters* 23: 142–144.

Kaiser, W. and C. G. B. Garrett. 1961. Two-photon excitation in CaF$_2$:Eu^{2+}. *Physical Review Letters* 7: 229–231.

Kerr, J. 1875. A new relation between electricity and light: Dielectrified media birefringent. *Philosophical Magazine Series 4*, 50: 337–348.

Kleinman, D. A. 1962a. Nonlinear dielectric polarization in optical media. *Physical Review* 126: 1977–1979.

Kleinman, D. A. 1962b. Theory of second harmonic generation of light. *Physical Review* 128: 1762–1775.

Lamb, W. E. 1964. Theory of an optical maser. *Physical Review* 134: A1429–A1450.

Maiman, T. H. 1960. Stimulated optical radiation in ruby. *Nature* 187: 493–494.

Maxwell, J. C. 1865. A dynamical theory of the electromagnetic field. *Philosophical Transactions of the Royal Society of London* 155: 459–512.

Minck, R. W., R. W. Terhune, and C. C. Wang. 1966. Nonlinear optics. *Applied Optics* 5: 1595–1612.

Mollenauer, L. F., R. H. Stolen, and J. P. Gordon. 1980. Experimental observation of picosecond pulse narrowing and solitons in optical fibers. *Physical Review Letters* 45: 1095–1098.

Niebuhr, K. E. 1963. Generation of laser axial mode difference frequencies in a nonlinear dielectric. *Applied Physics Letters* 2: 136–137.

Pershan, P. S. 1963. Nonlinear optical properties of solids: Energy considerations. *Physical Review* 130: 919–929.

Pockels, F. 1893. *Abhandlungen der Gesellschaft der Wissenschaften zu Gottingen [Treatises of the Society of Sciences in Gottingen]* (Vol. 39). Gottingen, Germany: Dieterichsche Verlags-Buchhandlung, p. 1.

Pockels, F. 1906. *Lehrbuch der Kristalloptik [Textbook of Crystal Optics]*. Leipzig, Germany: Teubners.

Raman, V. V. and K. S. Krishnan. 1928. A new type of secondary radiation. *Nature* 121: 501–502.

Schawlow, A. L. and C. H. Townes. 1958. Infrared and optical masers. *Physical Review* 112: 1940–1949.

Smith, A. W. and N. Braslou. 1963. Observation of an optical difference frequency. *Journal of Applied Physics* 34: 2105–2106.

Tang, C. L. 1966. Saturation and spectral characteristics of the Stokes emission in the stimulated Brillouin process. *Journal of Applied Physics* 37: 2945–2955.

Terhune, R. W., P. D. Maker, and C. M. Savage. 1962. Optical harmonic generation in calcite. *Physical Review Letters* 8: 404–406.

Wang, C. C. and G. W. Racette. 1965. Measurement of parametric gain accompanying optical difference frequency generation. *Applied Physics Letters* 6: 169–171.

Woodbury, E. J. and W. K. Ng. 1962. Ruby laser operation in the near-IR. *Proceedings of the Institute of Radio Engineers* 50: 2367 (correspondence).

Yariv, A. 1976. Three-dimensional pictorial transmission in optical fibers. *Applied Physics Letters* 28: 88–89.

Zel'dovich, B. Ya., V. I. Popovichev, V. V. Ragul'skii, and F. S. Faizullov. 1972. Connection between the wave fronts of the reflected and exciting light in stimulated Mandel'shtem–Brillouin scattering. *JETP Letters* 15: 109.

FURTHER READING

Agrawal, G. P. 2013. *Nonlinear Fiber Optics* (5th edition). New York, NY: Academic Press.

Banerjee, P. P. 2004. *Nonlinear Optics: Theory, Numerical Modeling, and Applications*. New York, NY: Marcel Dekker, Inc.

Bertolotti, M. 2015. *Masers and Lasers: An Historical Approach* (2nd edition). Boca Raton, FL: CRC Press.

Bloembergen, N. and Y. R. Shen. 1964. Quantum-theoretical comparison of nonlinear susceptibilities in parametric media, lasers, and Raman lasers. *Physical Review* 133: A37–A49.

Boyd, R. W. 2008. *Nonlinear Optics* (3rd edition). Burlington, MA: Academic Press.

Faraday, M. 1933. *Volume IV, Nov. 12, 1839—June 26, 1847* (Thomas Martin edition). London, UK: George Bell and Sons, Ltd.

He, G. S. 2014. *Nonlinear Optics and Photonics*, Oxford, U.K., Oxford University Press.

New, G. 2014. *Introduction to Nonlinear Optics*. Cambridge, UK: Cambridge University Press.

Stegeman, G. I. and Stegeman, R. A. 2012. *Nonlinear Optics: Phenomena, Materials and Devices*. Hoboken, NJ: John Wiley & Sons.

2

LINEAR OPTICS

2.1 INTRODUCTION

Linear optics plays a crucial role in determining the performance and characteristics of nonlinear optical systems. Focusing, material dispersion, and birefringence are only a few examples of linear properties that strongly influence nonlinear behavior. This chapter introduces linear properties important to nonlinear optics as well as the framework of derivations common to both linear optics and nonlinear optics. We first address the distinctions between linear and nonlinear systems.

2.1.1 LINEARITY

In linear algebra, a function, f, is said to be linear if it satisfies the relationship: $f(ax_1 + bx_2) = af(x_1) + bf(x_2)$, where a and b are constants and x_1 and x_2 are independent inputs to the function. Linearity implies that it is possible to calculate the output of the function by considering the effects of the two inputs independently, then adding the results. The same is true with linear optics: individual inputs such as different laser beams are analyzed independently of each other. It may not be obvious at first glance that a phenomenon such as refraction at an interface is linear. However, the refracted angle for a given beam is not affected by the presence of other beams. If the amplitude of the input field is doubled, then the refracted beam's amplitude doubles as well. When coherent beams of light are superposed, it seems that the beams affect each other, resulting in interesting interference patterns. However, interference is also linear since the overall field is simply the linear superposition of all the input fields. That is, it is still possible to calculate the overall output by considering one input at a time.

2.1.2 MAXWELL'S EQUATIONS

In linear optical systems, energy is not exchanged between independent optical inputs. That is, the different beams of light in a system do not "talk" to each other. In contrast, nonlinear optical systems have energy exchanges between the inputs. Understanding these nonlinear coupling phenomena is the focus of this chapter. Energy exchange distinguishes linear and nonlinear optical systems, and the vehicle for this exchange, the polarization of the optical medium, is of fundamental importance to all interactions described here. This connection is established more concretely in the form of Poynting's theorem in Section 2.1.3, which we derive from Maxwell's equations.

Maxwell's equations are derived from first principles in a large number of textbooks on electromagnetism (see Griffiths, 1999). They are the foundation for nearly all the derivations in this chapter. For macroscopic fields, Maxwell's equations are given by

$$\vec{\nabla} \cdot \underline{\vec{D}} = \rho_f , \tag{2.1}$$

$$\vec{\nabla} \times \underline{\vec{E}} = -\frac{\partial \underline{\vec{B}}}{\partial t} , \tag{2.2}$$

$$\vec{\nabla} \cdot \underline{\vec{B}} = 0 , \tag{2.3}$$

$$\vec{\nabla} \times \underline{\vec{H}} = \underline{\vec{j}}_f + \frac{\partial \underline{\vec{D}}}{\partial t} , \tag{2.4}$$

where $\underline{\vec{E}}$ is the electric field, $\underline{\vec{D}}$ is the electric displacement, $\underline{\vec{H}}$ is the magnetic field, $\underline{\vec{B}}$ is the magnetic induction, ρ_f is the free charge density, and $\underline{\vec{j}}_f$ is the free current density. An underscore underneath a variable in Equations 2.1 through 2.4 indicates that the variable is real as opposed to a complex quantity, which we encounter further in Section 2.1.6. $\underline{\vec{D}}$ and $\underline{\vec{H}}$ are defined by their respective constitutive relationships in terms of the electric field and magnetic induction:

$$\underline{\vec{D}} = \varepsilon_0 \underline{\vec{E}} + \underline{\vec{P}} , \tag{2.5}$$

$$\underline{\vec{H}} = \frac{\underline{\vec{B}}}{\mu_0} - \underline{\vec{M}} , \tag{2.6}$$

where $\underline{\vec{M}}$ is the magnetization, $\underline{\vec{P}}$ is the polarization of the medium, ε_0 is the permittivity of free space (8.854×10^{-12} F/m), and μ_0 is the magnetic permeability of free space ($4\pi \times 10^{-7}$ H/m). In this chapter, we consider nonmagnetic media by which Equation 2.6 will be written as

$$\vec{H} = \frac{\vec{B}}{\mu_0},$$ (2.7)

since most materials found in nature possess magnetic properties at optical frequencies that are the same as free space.

2.1.3 POYNTING'S THEOREM

Poynting's theorem gives us a connection between electric and magnetic fields and energy flow. Poynting's theorem is derived by looking at the work associated with the free current density. Work done by a force is defined as

$$dW = \vec{F} \cdot d\vec{x},$$ (2.8)

where $\vec{F}$ is the force and $d\vec{x}$ is the displacement. The force on a free charged particle, q_f, moving with a velocity, $\vec{v}$, is given by the Lorentz force:

$$\vec{F} = q_f \left(\vec{E} + \vec{v} \times \vec{B}\right).$$ (2.9)

In a macroscopic medium, we treat the charge as continuous and consider incremental charge dq_f occupying a volume dV, so that

$$dq_f = \rho_f \, dV.$$ (2.10)

By considering only free charges, our expression excludes work associated with bound charges in the medium. We call the work done on free charges in the medium "mechanical work."

Combining Equations 2.8 through 2.10 gives

$$dW_{\text{mech}} = (\rho_f \, dV)\vec{E} \cdot d\vec{x},$$ (2.11)

where the work has no magnetic contribution since $\left(\vec{v} \times \vec{B}\right)$ is perpendicular to $d\vec{x}$ and $\left(\vec{v} \times \vec{B}\right) \cdot d\vec{x} = 0$. We rewrite the displacement as

$$d\vec{x} = \vec{v}dt.$$ (2.12)

By recognizing that $\vec{j}_f = \rho_f \vec{v}$, Equation 2.11 becomes

$$dW_{mech} = \underline{\vec{E}} \cdot \underline{\vec{j}}_f \, dV \, dt. \tag{2.13}$$

The expression gives the mechanical work done by the fields on charges in the volume, dV, in a time interval, dt.

We obtain $\vec{j}_f$ from Equation 2.4 so that Equation 2.13 is rewritten as

$$dW_{mech} = \left(\underline{\vec{E}} \cdot \left(\vec{\nabla} \times \vec{H} \right) - \underline{\vec{E}} \cdot \frac{\partial \vec{D}}{\partial t} \right) dV \, dt. \tag{2.14}$$

Equation 2.14 is expanded with the help of the following vector identity:

$$\underline{\vec{E}} \cdot \left(\vec{\nabla} \times \vec{H} \right) = \vec{H} \cdot \left(\vec{\nabla} \times \underline{\vec{E}} \right) - \vec{\nabla} \cdot \left(\underline{\vec{E}} \times \vec{H} \right). \tag{2.15}$$

We use Equation 2.2 for $\vec{\nabla} \times \underline{\vec{E}}$ and use the definition of the Poynting vector, $\vec{S}$,

$$\underline{\vec{S}} \equiv \underline{\vec{E}} \times \vec{H}. \tag{2.16}$$

With these substitutions, Equation 2.14 becomes

$$dW_{mech} = \left[\left(-\vec{H} \cdot \frac{\partial \underline{\vec{B}}}{\partial t} - \underline{\vec{E}} \cdot \frac{\partial \vec{D}}{\partial t} \right) dV - \vec{\nabla} \cdot \underline{\vec{S}} \, dV \right] dt. \tag{2.17}$$

We use the constitutive relationships for $\vec{D}$ and $\vec{E}$ and $\vec{B}$ and $\vec{H}$ and identify the energy stored in the fields per unit volume:

$$U = \frac{1}{2} \left(\underline{\vec{E}} \cdot \underline{\vec{D}} + \mu_0 \underline{\vec{B}} \cdot \vec{H} \right). \tag{2.18}$$

Finally, we integrate over the volume in Equation 2.17 and employ the divergence theorem to obtain Poynting's theorem:

$$\int_{\substack{closed \\ surface}} \underline{\vec{S}} \cdot d\vec{a} = - \left[\frac{dW_{mech}}{dt} + \frac{d}{dt} \int U \, dV + \int \underline{\vec{E}} \cdot \frac{\partial \vec{P}}{\partial t} \, dV \right], \tag{2.19}$$

FIGURE 2.1   Laser beam enters a small volume and exits with less energy.

where $d\vec{a}$ is the vector surface element directed normal to the surface and outward from the enclosed volume, and dW_{mech} is the work done to all the charges in the volume under consideration.

Poynting's theorem as written in Equation 2.19 is an energy balance equation. The left-hand side of Equation 2.19 is the total power flow out of a closed surface. The terms on the right-hand side of the equation reveal the source of the power flow. The first expression on the right-hand side is the rate at which work is done by the fields on free charges in the volume, which typically results in joule heating of the material due to dissipative processes, such as inelastic collisions or carrier relaxation. The next term represents the rate of change in the EM energy stored in fields bounded by the surface, and the last term is the rate at which work is done by the field on the medium via the polarization. A visualization for a case where the energy entering the volume is greater than that leaving it is shown in Figure 2.1. The last term in Poynting's theorem (Equation 2.19) shows that the polarization is central to energy exchange between a medium and an electric field. The polarization of a medium plays a key role in the energy dynamics of a system and is also central in the study of nonlinear optics.

2.1.4 INTENSITY

Poynting's vector relates to energy flow, but its instantaneous value changes too quickly for detectors to follow. However, detectors do respond to the energy deposited over a given time interval. The magnitude of the time-averaged Poynting vector defines the beam's *intensity*, given by

$$I = \left| \left\langle \vec{S} \right\rangle \right| = \left| \left\langle \vec{E} \times \vec{H} \right\rangle \right|. \tag{2.20}$$

The magnitude is used as a simplification, but the time-averaged Poynting vector reveals the pattern for power flow through complex materials. The magnitude in Equation 2.20 is also called the irradiance; as mentioned in Chapter 1, nonlinear optics literature often uses the term intensity synonymously with the term irradiance. Equation 2.20 simplifies when considering monochromatic plane waves. We look at a specific example of a plane wave traveling in the z-direction in a medium with an index of refraction n; the electric field is given as

$$\vec{E} = E_0 \cos(kz - \omega t + \phi)\hat{x}, \tag{2.21}$$

where ω is the angular frequency, k is the wave vector magnitude ($k = (n\omega/c)$, the real valued refractive index is denoted as n), and ϕ is a constant phase offset. Note that we use Maxwell's equations to derive the plane-wave field given by Equation 2.21 in Section 2.3. This field and Maxwell's equation (Equation 2.2) let us solve for $\vec{B}$, resulting in

$$\underline{\vec{B}} = \frac{nE_0}{c}\cos(kz - \omega t + \phi)\hat{y}. \tag{2.22}$$

In a nonmagnetic medium, the H field is

$$\underline{\vec{H}} = \frac{nE_0}{\mu_0 c}\cos(kz - \omega t + \phi)\hat{y}. \tag{2.23}$$

Hence, the intensity is

$$I = \frac{nE_0^2}{\mu_0 c}\langle\cos^2(kz - \omega t + \phi)\rangle = \frac{nE_0^2}{2\mu_0 c}. \tag{2.24}$$

Intensity has units of W/m². Unlike the instantaneous electric field and the Poynting vector, intensity is readily measured.

2.1.5 LINEAR POLARIZATION

The connection between energy exchange and the polarization for linear optics is illustrated by looking at a class of dielectrics commonly introduced in electromagnetism courses. Correctly speaking, the dielectric constant is never constant, and in many crystalline materials, it is a tensor. The general expression for the relationship between polarization and electric field is expressed as

$$\underline{\vec{P}}(t) = \varepsilon_0 \int\limits_{-\infty}^{\infty} \underline{\underline{\tilde{\chi}}}(t - t')\cdot\vec{E}(t')dt', \tag{2.25}$$

where $\underline{\underline{\tilde{\chi}}}(t)$ is a real 3×3 matrix called the linear susceptibility tensor. The center dot indicates a multiplication between the tensor and the electric field. In component form, Equation 2.25 is written as

$$\underline{P}_i(t) = \varepsilon_0 \int\limits_{-\infty}^{\infty} \sum_{j=x,y,z} \chi_{ij}(t - t')\underline{E}_j(t')dt'. \tag{2.26}$$

The linear susceptibility tensor satisfies the physical causality principle that the polarization does not respond at time t to an electric field applied in the future; in other words, $\chi_{ij}(t) = 0, i, j = x, y, z$ for $t < 0$. To better understand the causality condition by referring to Equation 2.25, the electric field applied at a time $t' > t$ cannot physically affect the polarization at time t. Furthermore, as a consequence of the susceptibility's tensor characteristics, the polarization vector may not point in the same direction as the electric field.

2.1.6 COMPLEX REPRESENTATION OF POLARIZATION

The above approach that works with cosine and sine functions is somewhat cumbersome. A more efficient approach uses complex variables. For a refresher on complex variables, see Appendix A. We start by writing the monochromatic field (Equation 2.21) in one of the two forms,

$$\underline{\vec{E}} = \mathrm{Re}\left\{E_o e^{i(kz-\omega t+\phi)}\right\}\hat{x} = \mathrm{Re}\left\{\vec{A}e^{i(kz-\omega t)}\right\}, \tag{2.27}$$

$$\underline{\vec{E}} = \frac{1}{2}(\vec{A}e^{i(kz-\omega t)} + \text{c.c.}), \tag{2.28}$$

where c.c. is the complex conjugate. In Equations 2.27 and 2.28, the phase offset, ϕ, and the unit vector are absorbed into the complex amplitude, $\vec{A}$. It is a standard practice to encode amplitude and phase information into the complex amplitude.

The polarization is defined as

$$\underline{\vec{P}} = \mathrm{Re}\left\{\vec{P}e^{i(kz-\omega t)}\right\} = \frac{1}{2}\left(\vec{P}e^{i(kz-\omega t)} + \text{c.c.}\right). \tag{2.29}$$

The polarization is algebraically related to the electric field by applying a temporal Fourier transform. To Fourier transform the time variable, we will use the definition

$$\underline{\vec{F}}(t) = \int_{-\infty}^{\infty} \vec{F}(\omega)e^{-i\omega t}\, d\omega. \tag{2.30}$$

The Fourier transformed field $\vec{F}(\omega)$ is in general a complex function, a fact which is noted by omitting the bar under the function. The inverse Fourier transform is written as

$$\vec{F}(\omega) = \frac{1}{2\pi}\int_{-\infty}^{\infty} \underline{\vec{F}}(t)e^{i\omega t}\, dt. \tag{2.31}$$

In frequency space, Equation 2.25 is rewritten as

$$\vec{P}(\omega) = \varepsilon_0 \ddot{\chi}(\omega) \cdot \vec{E}(\omega). \tag{2.32}$$

The complex valued susceptibility tensor in frequency space does not have a bar underneath to distinguish it from the real valued time-dependent susceptibility. This notation will be used for other time-dependent variables throughout this book. The frequency-dependent susceptibility is a complex function:

$$\ddot{\chi}(\omega) = \ddot{\chi}_R(\omega) + i\ddot{\chi}_I(\omega). \tag{2.33}$$

To explore the physical processes in more depth, we consider a monochromatic plane wave traveling in the z-direction in a medium; the electric field is given by Equation 2.21. For an isotropic medium, the complex susceptibility is a diagonal tensor with the scalar function in each direction; the scalar function is denoted as

$$\chi = \chi_R + i\chi_I; \tag{2.34}$$

the reader is reminded that the susceptibility is a complex function in frequency space. The applied electric field induces a polarization in a medium that in general can include components in different directions than the incident field, and such responses are discussed in Section 2.2. Equation 2.32 is multiplied by the complex factor $\exp(i(kz - \omega t))$ and the real part is taken to construct the real polarization:

$$\underline{\vec{P}} = \mathrm{Re}\left\{\varepsilon_0 \chi \vec{A} e^{i(kz-\omega t)}\right\}. \tag{2.35}$$

In Equation 2.35, $\varepsilon_0 \chi \vec{A}$ is the complex amplitude of $\underline{\vec{P}}$ and is written as $\vec{P}$ (note absence of the underscore in the complex amplitude). That is,

$$\vec{P} = \varepsilon_0 \chi \vec{A}. \tag{2.36}$$

This expression is similar to that presented in introductory electromagnetism courses, except that in Equation 2.36, the relationship is between the *complex amplitudes* of the polarization and the electric field. Equation 2.36 clarifies the phase difference between the polarization and the electric field; it is just the phase of the complex susceptibility, χ.

For the present discussion, the polarization response can have a component in phase with the field and another component 90° out of phase (or in quadrature). For a medium with a scalar susceptibility function, the polarization is in the same direction as the incident field given in Equation 2.21; the general response is written as

$$\underline{\vec{P}} = \varepsilon_0 E_0 \hat{x}(\chi_R \cos(kz - \omega t + \phi) - \chi_I \sin(kz - \omega t + \phi)), \tag{2.37}$$

where the two susceptibilities are defined by Equation 2.34; they determine the in-phase and in-quadrature responses. Equation 2.37 yields a phase shift between the field and polarization depending on the magnitudes of χ_R and χ_I. This more general expression for the polarization is linear in field strength; however, it can lead to energy exchange with the material, called linear absorption or linear gain. The energy loss or gain is described by the energy exchange term in Poynting's theorem, which is

$$-\vec{E} \cdot \frac{d\vec{P}}{dt} = -\frac{\varepsilon_0 E_0^2 \omega}{2}(\chi_R \sin(2(kz - \omega t + \phi)) + \chi_I \cos(2(kz - \omega t + \phi)) + \chi_I). \qquad (2.38)$$

The two periodic contributions are similar to those describing the displacement of driven mass on a spring system where work is done by an external drive on a spring for part of the cycle, whereas during the other half of the cycle, the spring does work on the external source. This back-and-forth exchange averages to a net zero exchange in energy. Similarly, the harmonic terms in Equation 2.38 average to zero, leaving the constant term

$$-\frac{\varepsilon_0 E_0^2 \omega}{2} \chi_I. \qquad (2.39)$$

This constant term describes an exchange of energy from the field to the volume or vice versa depending on the sign and magnitude of χ_I. The induced polarization therefore has a connection to energy flow even in linear optics. Although energy is absorbed or lost by the beam, this energy exchange is with the medium; the beam is passing through and not with another optical input. The presence of a second beam in the same volume will not affect the amount of absorption in a *linear* medium. However, with nonlinear absorption, a second beam will affect the first one; we will deal with nonlinear absorption in Chapter 8.

2.1.7 ENERGY EXCHANGE BETWEEN A FIELD AND POLARIZATION

The phase relationship between the field and polarization plays an important role in determining the direction of energy flow as seen by examining the time-averaged energy flow term in Equation 2.19:

$$\left\langle \vec{E} \cdot \frac{d\vec{P}}{dt} \right\rangle. \qquad (2.40)$$

To work with Equation 2.40 using complex quantities, one must proceed with caution. Unfortunately, the product in Equation 2.40 is not simply the product of the complex amplitudes. However, when considering a time average of monochromatic quantities, it is possible to write Equation 2.40, in terms of complex amplitudes. Consider two monochromatic quantities written as

$$\underline{\vec{F}} = \frac{1}{2}\vec{F}e^{-i\omega t} + \text{c.c.} \tag{2.41}$$

and

$$\underline{\vec{G}} = \frac{1}{2}\vec{G}e^{-i\omega t} + \text{c.c.} \tag{2.42}$$

Again note that the underscored font represents a real quantity. The time average is written in terms of the complex amplitudes as

$$\left\langle \underline{\vec{G}} \cdot \underline{\vec{F}} \right\rangle = \frac{1}{2}\text{Re}(\vec{G} \cdot \vec{F}^*). \tag{2.43}$$

The proof of Equation 2.43 is left as an exercise at the end of this chapter (see Problem 2.3). Finding the time average in Equation 2.40 then requires the complex amplitude of the time derivative of the polarization, which, for a monochromatic field, is extracted from

$$\frac{d\underline{\vec{P}}}{dt} = \frac{d}{dt}\left(\frac{1}{2}\varepsilon_0\chi\vec{A}e^{i(kz-\omega t)} + \text{c.c.}\right) = \frac{-i}{2}\omega\varepsilon_0\chi\vec{A}e^{i(kz-\omega t)} + \text{c.c.} \tag{2.44}$$

From this equation, we see that the complex amplitude of the time derivative of the polarization is

$$-i\omega\varepsilon_0\chi\vec{A}. \tag{2.45}$$

Therefore,

$$\left\langle \underline{\vec{E}} \cdot \frac{d\underline{\vec{P}}}{dt} \right\rangle = \frac{1}{2}\varepsilon_0 \, |\vec{A}|^2 \, \omega\,\text{Re}(i\chi^*). \tag{2.46}$$

This expression directly connects the direction of power flow and the susceptibility. If χ is purely real, corresponding to a material polarization exactly in phase with the input field, then the optical beam experiences no net energy change as it traverses the medium. The linear susceptibility can be considered as a purely real quantity for many nonlinear media of interest since these media are typically transparent with little or no losses. The majority of nonlinear interactions considered in this chapter occur in transparent media. When χ is complex, the time-averaged power flow is either positive or negative depending on the sign of the imaginary part of χ.

For nonlinear interactions, Equation 2.40 also describes the energy exchange between the polarization and a field. The phase between the polarization (including nonlinear terms) and a field determines the direction of energy flow for interacting fields (if energy flows at all). The difference between the linear case and the nonlinear case is that the polarization term is expanded to include nonlinear terms as introduced in Chapter 3. This basic connection between polarization and energy flow establishes polarization as a central player in nonlinear optics.

2.2 TENSOR PROPERTIES OF MATERIALS

In addition to defining the concept of linearity, Section 2.1.6 shows that the familiar expression for induced polarization must be modified to account for phase differences between the incident field and the polarization. In an isotropic medium, the polarization points in the same direction as the incident field. As indicated in Equation 2.32, a more complete description of polarization needs to account for the fact that an input field in a given direction may induce a polarization in a different direction. For example, consider the x-component of the polarization, which, in general, can depend on all components of the electric field; hence, the complex amplitude of the polarization's x-component is represented as

$$P_x = \varepsilon_0 (\chi_{xx} A_x + \chi_{xy} A_y + \chi_{xz} A_z), \tag{2.47}$$

where χ_{xx}, χ_{xy} and χ_{xz} are complex susceptibilities of potentially different magnitudes. Note we eliminated the underscore bar for the Fourier transform functions. This notation simplifies the rest of our discussion and it is the norm when discussing the tensor properties of the susceptibility.

Any given Cartesian component, i, of the polarization's complex amplitude in Equation 2.44 is written more compactly as a summation:

$$P_i = \varepsilon_0 \sum_j \chi_{ij} A_j. \tag{2.48}$$

This summation is simplified further by using Einstein notation for which repeated indices indicate a summation over that index. In Einstein notation, Equation 2.48 becomes

$$P_i = \varepsilon_0 \chi_{ij} A_j. \tag{2.49}$$

Einstein's notation and expressions using the summation symbol are used throughout the text.

The susceptibility clearly cannot be represented by a single number; rather, as written above, it requires the specification of as many as nine different numbers. The nine numbers form the elements of a susceptibility tensor, $\ddot{\chi}$, whose elements are given by χ_{ij}. Equation 2.49 is written in terms of the susceptibility tensor as

$$\vec{P} = \varepsilon_0 \ddot{\chi} \cdot \vec{A}. \tag{2.50}$$

The next section gives the defining properties of tensors.

2.2.1 TENSORS

Many quantities such as the scalar electric potential and the electric field are tensors. The formal aspects of tensorial properties may be found elsewhere; a less formal explanation that is restricted to tensors common to optics is used in this chapter. The heart of what determines whether an object, such as the susceptibility above, is a tensor depends on how it changes under coordinate transformations. The restriction on tensors in this chapter is that only transformations that do not translate, stretch, shrink, or skew the coordinate axes are considered. Allowed transformations we will consider include rotations, reflections, inversions, and their combinations such that the basis set after transformation is still an orthogonal one. Any orthogonal transformation can be represented by a 3×3 matrix, denoted as $\ddot{R}$, whose elements are labeled as R_{ij} with i and j ranging over x, y, and z. As an example, consider a rotation about the y-axis by an angle, θ:

$$\ddot{R} = \begin{bmatrix} \cos\theta & 0 & \sin\theta \\ 0 & 1 & 0 \\ -\sin\theta & 0 & \cos\theta \end{bmatrix}. \tag{2.51}$$

For the transformation to be orthogonal, the inverse matrix is identical to its transpose, that is, $\ddot{R}^T\ddot{R} = I$, where $\ddot{R}^T$ is the transpose of $\ddot{R}$, that is, $R_{ij}^T = R_{ji}^*$, and I is the identity matrix. The orthogonality relationship is easily verified for the matrix in Equation 2.51. In component form, the orthogonality relationship is

$$(\ddot{R}^T\ddot{R})_{ij} = \delta_{ij}, \tag{2.52}$$

where δ_{ij} is the Kronecker delta. The Kronecker delta is defined by

$$\delta_{ij} = \begin{cases} 1 & i = j \\ 0 & i \neq j \end{cases}. \tag{2.53}$$

A useful relationship between the Kronecker delta and an orthogonal transformation is as follows. For a given orthogonal transformation matrix, we write out the matrix multiplication in Equation 2.52 as

$$(\ddot{R}^{\mathrm{T}})_{i\alpha}(\ddot{R})_{\alpha j} = \delta_{ij}. \tag{2.54}$$

The transposed real element is obtained by switching the order of the indices so that we may identify

$$\delta_{ij} = R_{\alpha i} R_{\alpha j}. \tag{2.55}$$

By changing the order of $\ddot{R}$ and $\ddot{R}^{\mathrm{T}}$ in Equation 2.52, we obtain

$$\delta_{ij} = R_{i\alpha} R_{j\alpha}. \tag{2.56}$$

We use Equations 2.55 and 2.56 later to show that certain quantities are tensors.

The definition of a tensor, when using the restriction of orthogonal transformations, is an object that obeys one of the following rules:

$$\text{Rank } 0: \varphi'(\vec{r}') = \varphi(\vec{r}), \tag{2.57}$$

$$\text{Rank } 1: v_i'(\vec{r}') = R_{i\alpha} v_\alpha(\vec{r}), \tag{2.58}$$

$$\text{Rank } 2: \varepsilon_{ij}'(\vec{r}') = R_{i\alpha} R_{j\beta} \varepsilon_{\alpha\beta}(\vec{r}), \tag{2.59}$$

$$\text{Rank } 3: d_{ijk}'(\vec{r}') = R_{i\alpha} R_{j\beta} R_{k\gamma} d_{\alpha\beta\gamma}(\vec{r}). \tag{2.60}$$

The primed quantities indicate the position vector and the tensor with respect to the coordinate system after the transformation. Equations 2.57 and 2.58 show that scalars and vectors can be rank 0 and rank 1 tensors, respectively, provided that they comply with the respective rule. In this chapter, if a given quantity obeys one of the above transformational rules, then it is considered a tensor. The rank 0 tensor has a value that is independent of a coordinate system. The classic example of a rank 0 tensor (i.e., scalar) is that of temperature at a given location in a room. The temperature at that point does not change when viewed with a different coordinate system. The next order tensor, rank 1, has a transformation that should be familiar because it is the same as for a vector under a rotation of the coordinate system.

Showing that a scalar or vector quantity obeys the appropriate rule above is straightforward. Going to higher ranks is less common. We now prove that the susceptibility is a rank 2 tensor. Considering Equation 2.49, it may not be obvious that χ_{ij} is a tensor. To prove that it is a rank 2 tensor, we show that it transforms according to Equation 2.59. To do so, the vector quantities are rewritten in terms of a different coordinate system:

$$P'_i = R_{i\alpha} P_\alpha, \tag{2.61}$$

$$A'_j = R_{j\beta} A_\beta. \tag{2.62}$$

Substituting these equations into Equation 2.49, rewritten as $P'_i = \varepsilon_0 \chi'_{ij} A'_j$, gives

$$R_{i\alpha} P_\alpha = \varepsilon_0 \chi'_{ij} R_{j\beta} A_\beta. \tag{2.63}$$

This equation is simplified by operating on both sides of Equation 2.63:

$$R_{i\alpha'} R_{i\alpha} P_\alpha = \varepsilon_0 R_{i\alpha'} R_{j\beta} \chi'_{ij} A_\beta. \tag{2.64}$$

Next, we identify $R_{i\alpha'} R_{i\alpha} = \delta_{\alpha\alpha'}$ (Equation 2.56) so that

$$P_\alpha = \varepsilon_0 R_{i\alpha} R_{j\beta} \chi'_{ij} A_\beta. \tag{2.65}$$

Hence, the susceptibility in the new coordinates is read from Equation 2.65 as

$$\chi_{\alpha\beta} = R_{i\alpha} R_{j\beta} \chi'_{ij}. \tag{2.66}$$

This relationship is "inverted" (see Problem 2.7), by multiplying both sides by appropriately labeled R's and using the Kronecker-delta properties giving

$$\chi'_{ij} = R_{i\alpha} R_{j\beta} \chi_{\alpha\beta}. \tag{2.67}$$

Equation 2.67 shows that the susceptibility does indeed transform as a rank 2 tensor. In Chapter 3, the tensor transformation rules greatly simplify nonlinear notation derived there. For further reading on tensor algebra, see Symon (1971).

2.3   WAVE EQUATION

We have been focusing on the central role of the polarization from a standpoint of energy transfer; indeed, much work in nonlinear optics is dedicated to determine the polarization. However, knowledge of the polarization at a given location in a material is only part of the story. It is equally important to consider the wave equation in a medium. This wave equation provides a quantitative approach to determine the evolution of optical fields as well as their effect on each other in the presence of non-linearities. The following is a derivation of the wave equation with broad simplifications and assumptions. In subsequent chapters, the wave equation is revisited and in many cases derived with less restrictive assumptions. The basic procedure is always the same and is grounded in Maxwell's equations. The derivation here shows that the wave equation brings to light some important linear optical properties of materials.

2.3.1 CONSTITUTIVE RELATIONSHIPS FOR COMPLEX AMPLITUDES

The starting point for developing constitutive relationships for the complex amplitudes is the solid foundation of Maxwell's equations given in Equations 2.1 through 2.4 and the constitutive relationships (Equations 2.5, 2.6, and 2.25). For many optics problems, the constitutive relations are written in a more compact notation. In the case of Equation 2.5, $\left(\vec{D} = \varepsilon_0\vec{E} + \vec{P}\right)$ and using Equations 2.25, a direct proportionality between the complex amplitudes of $\vec{D}$ and $\vec{E}$ is made by applying a Fourier transform defined in Equations 2.30 and 2.31 to the fields. The connection between the complex electric field amplitude and the D field is made by way of the 3×3 matrix:

$$\vec{D}(\omega) = \ddot{\varepsilon}(\omega) \cdot \vec{E}(\omega).$$

(2.68)

The matrix, $\ddot{\varepsilon}(\omega)$, is an example of a second rank tensor and it is also called the dielectric permittivity tensor. Another common way to write Equation 2.68 is to factor out the permittivity of free space giving (suppressing the dependence on ω in the notation)

$$\vec{D} = \varepsilon_0\ddot{\varepsilon}_r \cdot \vec{E},$$

(2.69)

where $\ddot{\varepsilon}_r$ is called the relative permittivity tensor. The magnetic field's complex amplitude is also rewritten in a simplified form as

$$\vec{H} = \frac{\vec{B}}{\mu}.$$

(2.70)

As previously mentioned, natural materials do not have a magnetic response at optical frequencies and $\mu = \mu_0$. However, magnetic properties have been engineered in a new class of materials called metamaterials. As metamaterials emerge from the laboratory, they make possible exciting new optical phenomena such as a negative index of refraction. Metamaterials are likely to play an important role in future nonlinear optical techniques and devices.

2.3.2 WAVE EQUATION IN HOMOGENEOUS ISOTROPIC MATERIALS

We first develop a simple wave equation appropriate for a broad class of important materials such as glasses. To do so, we make the following simplifying assumptions. These materials are homogeneous isotropic media (referred hereafter as simply isotropic), and by definition, the permittivity is independent of orientation or location. Hence, $\bar{\bar{\varepsilon}}$ is treated as a scalar and we drop the tensor notation; the relationship between $\vec{D}$ and $\vec{E}$ in Equation 2.68 is simplified to

$$\vec{D} = \varepsilon \vec{E}. \tag{2.71}$$

An equivalent expression in terms of the susceptibility defined in Equation 2.50 is

$$\varepsilon_r = (1 + \chi). \tag{2.72}$$

The electric displacement in a lossless material, where χ is real, always points in the same direction as the incident field, and its amplitude is independent of the orientation of the material.

Our first step in coming up with a wave equation is to take the curl of Equation 2.2:

$$\vec{\nabla} \times \left(\vec{\nabla} \times \vec{E} \right) = \vec{\nabla} \times \left(-\frac{\partial \vec{B}}{\partial t} \right). \tag{2.73}$$

For monochromatic complex fields, Equation 2.30 yields the following relation for the time derivatives of the time-periodic function:

$$\frac{\partial}{\partial t} \rightarrow -i\omega. \tag{2.74}$$

Equation 2.72 becomes

$$\vec{\nabla} \times (\vec{\nabla} \times \vec{E}) = \vec{\nabla} \times (i\omega \vec{B}). \tag{2.75}$$

A vector identity for the left-hand side is

$$\vec{\nabla} \times (\vec{\nabla} \times \vec{E}) = \vec{\nabla}(\vec{\nabla} \cdot \vec{E}) - \nabla^2 \vec{E}. \tag{2.76}$$

In isotropic, homogeneous materials, $\vec{\nabla} \cdot \vec{E} = 0$, which is shown as follows. In most optical applications, the material in question is not charged and so $\rho_f = 0$, hence

$$\vec{\nabla} \cdot \vec{D} = 0. \tag{2.77}$$

Since $\vec{D}$ and $\vec{E}$ are related in isotropic media by Equation 2.75, Equation 2.77 becomes

$$\vec{\nabla} \cdot \vec{D} = \vec{\nabla} \cdot \varepsilon \vec{E} = \varepsilon \vec{\nabla} \cdot \vec{E} + \vec{E} \cdot \vec{\nabla}\varepsilon. \tag{2.78}$$

In isotropic media, $\vec{\nabla}\varepsilon = 0$. Hence, $\vec{\nabla} \cdot \vec{E} = 0$ and Equation 2.73 becomes

$$\vec{\nabla} \times (\vec{\nabla} \times \vec{E}) = -\nabla^2 \vec{E}. \tag{2.79}$$

Substituting Equation 2.77 into Equation 2.75 gives

$$\nabla^2 \vec{E} = -i\omega \vec{\nabla} \times \vec{B}. \tag{2.80}$$

Using the constitutive relation (Equation 2.5) and Maxwell's equation (Equation 2.4),

$$\nabla^2 \vec{E} = -i\omega\mu \left(\vec{j}_f - i\omega\varepsilon\vec{E} \right). \tag{2.81}$$

In media with no free current density, $\vec{j}_f = 0$, resulting in the final vector wave equation:

$$\nabla^2 \vec{E} = -\mu\varepsilon\omega^2 \vec{E} = -\frac{\mu_r \varepsilon_r}{c^2} \omega^2 \vec{E}. \tag{2.82}$$

Equation 2.82 is an important form of the vector wave equation in frequency space. It dictates the properties of an EM wave in isotropic materials.

Another reasonable simplification is to restrict our considerations to linearly polarized fields, thereby turning Equation 2.82 into a scalar wave equation:

$$\nabla^2 E = -\frac{\mu_r \varepsilon_r}{c^2} \omega^2 E. \tag{2.83}$$

Many solutions to the scalar wave equation exist; the most commonly used solutions in optics are plane waves and Gaussian beams. The paraxial approximation Gaussian solution is covered in Section 2.6.1. The plane-wave solution is the simplest, and in fact plane waves form a complete set so that their superposition can be used to describe arbitrary beams. With such a superposition procedure, we can construct waves with arbitrary spatial and temporal frequency content. The field of Fourier optics develops a formalism to study optical beam propagation in the spatial domain.

The scalar plane-wave solution is

$$E = E_0 \cos(\vec{k} \cdot \vec{r} - \omega t + \phi) \qquad (2.84)$$

for a plane wave propagating in the $\hat{k}$ direction. In this equation, ω is the angular frequency of the wave and $\vec{k}$ is the wave vector. ϕ is an arbitrary phase, which is important when considering superposition of different plane waves. In complex form, Equation 2.84 is written as

$$E = \text{Re}\left\{ A e^{i(\vec{k} \cdot \vec{r} - \omega t)} \right\}. \qquad (2.85)$$

The magnitude of the wave vector is intimately tied to the frequency as is shown by inserting Equation 2.84 (or equivalently, Equation 2.85) into the wave equation. The result of the substitution is that the plane wave is a solution, provided that $\vec{k} \cdot \vec{k} = k^2 = (\mu_r \varepsilon_r / c^2) \omega^2$. Hence, k and ω are not independent. A further look at this restriction for real coefficients (μ_r, ε_r) shows that ω / k is a speed with a magnitude of $\left(c / \sqrt{\mu_r \varepsilon_r} \right)$. Indeed, this speed is the phase velocity of the wave. The index of refraction of the medium, introduced earlier, is defined as

$$n \equiv \sqrt{\varepsilon_r \mu_r}. \qquad (2.86)$$

In terms of the linear susceptibility, the index of refraction is

$$n = \sqrt{\mu_r (1 + \chi)}. \qquad (2.87)$$

Unless otherwise noted, we take $\mu_r = 1$ throughout this book.

The result of this first encounter with the wave equation is that plane waves are solutions that travel with a phase velocity of c/n. The index of refraction is a fundamental optical property of materials; the above derivation shows how its definition naturally stems from the wave equation.

2.3.3 DISPERSION

The index of refraction has an important role in nonlinear optics theory. A particularly significant property is the dependence of the index on frequency. The variation of the index with wavelength is called dispersion. An illustrative plot of the index of refraction versus wavelength is shown in Figure 2.2. The figure shows the general trend for optical frequencies: the index decreases as the wavelength becomes longer, except in resonance regions where the index change is called "anomalous" dispersion. For media that are transparent through the visible wavelength range, the anomalous

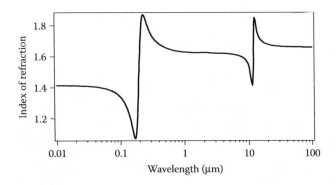

FIGURE 2.2   Illustration of dispersion in the index of refraction. The two resonant features show regions of anomalous dispersion; these regions are accompanied with high loss.

dispersion regions lie on either side of the visible spectrum. The qualitative aspects of the index dispersion curve are derived using a simple mechanical oscillator model that treats the electrons in the material as independent masses attached to springs. The classical treatment of the harmonic oscillator model is presented in Chapter 3. The functional forms of the model are used to empirically fit measured indices of refraction at several discrete wavelengths. The resultant model fit to the experimental data gives a logical function for interpolating between the measured indices. These forms of data fits are called the Sellmeier equations and are especially useful in applied nonlinear optics. A common Sellmeier form for optically transparent media given as a function of vacuum wavelength is

$$n^2(\lambda) = B_0 + \frac{\lambda^2 B_1}{\lambda^2 - \lambda_1^2} + \frac{\lambda^2 B_2}{\lambda^2 - \lambda_2^2}, \tag{2.88}$$

where B_0, B_2, B_2, λ_1, and λ_2 are fitting constants. Typically, the constants are defined for λ in units of microns. Although λ_1 and λ_2 are fitting constants, physical insight helps to provide an educated guess as to where these wavelengths should lie. As shown in Figure 2.2, the index of refraction for a typical material that is transparent in the visible has two anomalous dispersion regions or resonances: one is the ultraviolet (UV) region and the other is the infrared (IR) region. The poles of Equation 2.88 should occur at these resonances. Hence, λ_1 and λ_2 are two wavelengths outside of the visible band with one in the UV region and the other in the IR region. Such a guess helps to provide seeds to a fitting algorithm to help it converge (see Problem 2.16). Frequently, the Sellmeier form deviates from the form written in Equation 2.88; the extended Sellmeier form contains terms that account for IR and UV absorption. In general, there are many forms of the Sellmeier equations, and the particular form depends on the material in question. See Appendix B for the Sellmeier equations reported in the literature for a number of crystalline materials that have applications to nonlinear optics.

2.3.4 WAVE EQUATION IN CRYSTALS

An important feature of EM waves in crystals is that any given propagation direction has two orthogonal linear eigen polarization states, which are dubbed the *ordinary* and *extraordinary* waves, or o-waves and e-waves, respectively. For any propagation direction, one can decompose the beam into a linear superposition of an o-wave and an e-wave. The existence of these polarized waves and their properties emerges directly from the wave equation; the results are derived in this and the following sections.

The derivation of the wave equation introduced in Section 2.3.2 was greatly simplified since the dielectric tensor, $\ddot{\varepsilon}$, was treated as a scalar. Here, we lift this restriction to allow for a more general treatment that includes anisotropic media such as birefringent crystals. Deriving the vector wave equation is now more complicated, but there are still pertinent properties of $\ddot{\varepsilon}$ that somewhat simplify the process of working with the wave equation. One useful property is that in lossless media $\ddot{\varepsilon}$ is a symmetric tensor, that is $\varepsilon_{ij} = \varepsilon_{ji}$. Moreover, all of the elements of ε_{ij} are real so that the matrix is equal to its transpose, that is, $\varepsilon_{ij}^* = \varepsilon_{ji}$, making the ε_{ij} matrix Hermitian. A Hermitian matrix property is that it can be diagonalized such that all of the elements are real. The Hermitian property guarantees that a coordinate system exists where ε_{ij} is diagonal. This coordinate system is called the principal axis coordinate system. In the principal axis coordinate system, a simple relationship between $\vec{D}$ and $\vec{E}$ exists as written in the following matrix form:

$$\vec{D} = \begin{bmatrix} \varepsilon_{xx} & 0 & 0 \\ 0 & \varepsilon_{yy} & 0 \\ 0 & 0 & \varepsilon_{zz} \end{bmatrix} \vec{E}. \tag{2.89}$$

When working in the principal axis system basis, if the electric field is linearly polarized along one of the principal axes, then $\vec{D}$ and $\vec{E}$ are parallel as in the isotropic case. In most cases, the diagonal basis is the preferred coordinate system. An expression equivalent to Equation 2.86 in terms of indices of refraction is

$$\vec{D} = \varepsilon_0 \begin{bmatrix} n_x^2 & 0 & 0 \\ 0 & n_y^2 & 0 \\ 0 & 0 & n_z^2 \end{bmatrix} \vec{E}, \tag{2.90}$$

where $n_i^2 \equiv (\varepsilon_{ii}/\varepsilon_0)$ and n_x, n_y and n_z are called the principal indices. As pointed out earlier, μ_r is usually unity. For the general case where $n_x \neq n_y \neq n_z$, the crystal is called biaxial and by convention we use $n_z > n_y > n_x$ (called a positive biaxial crystal) or $n_x > n_y > n_z$ (called a negative biaxial crystal). We work mostly with positive biaxial crystals in this book. When two of the indices are equal, the crystal is called uniaxial and by convention $n_x = n_y$. This index is called the ordinary index of refraction, n_o. If

$n_z > n_o$, then the crystal is called positive uniaxial, and for the converse case, the crystal is called negative uniaxial. Uniaxial crystals and their terminology are commonly used in nonlinear optics.

Using the results from the isotopic case presented in Section 2.3.2 as a guide, solutions to the wave equation start with the assumption that a monochromatic plane-wave solution exists that has the dependence

$$e^{i(\vec{k}\cdot\vec{r}-\omega t)}. \tag{2.91}$$

We have already noted the transformation of the operators $(\partial/\partial t)$ in Equation 2.74; similarly, the operator such as $\vec{\nabla}$ acting on fields is simplified as

$$\vec{\nabla} \to i\vec{k} \tag{2.92}$$

Moreover, the wave vector is written as

$$\vec{k} = \frac{n\omega}{c}\hat{k}, \tag{2.93}$$

where the index of refraction, n, is determined by the orientation of the material and the polarization state of the electric field. The following derivation determines the specific values for n.

When operating with monochromatic plane waves, Maxwell's equations without sources are written in terms of complex amplitudes:

$$\vec{\nabla}\cdot\vec{\underline{D}}=0 \implies \hat{k}\cdot\vec{D}=0, \tag{2.94}$$

$$\vec{\nabla}\times\vec{\underline{E}}=-\frac{\partial\vec{B}}{\partial t} \implies \frac{n}{c}\hat{k}\times\vec{E}=\vec{B}, \tag{2.95}$$

$$\vec{\nabla}\cdot\vec{B}=0 \implies \hat{k}\cdot\vec{B}=0, \tag{2.96}$$

$$\vec{\nabla}\times\vec{H}=\frac{\partial\vec{\underline{D}}}{\partial t} \implies \frac{n}{c}\hat{k}\times\vec{H}=-\vec{D}, \tag{2.97}$$

where ρ_f and $\vec{j}_f$ are again assumed to vanish. The wave equation in a crystal is found using the same prescription as before, but now has an algebraic form:

$$\frac{n^2}{c^2}\hat{k}\times\left(\hat{k}\times\vec{E}\right)=-\mu_0\vec{D}. \tag{2.98}$$

Using a vector identity for the triple cross product gives

$$\frac{n^2}{c^2}\left[\left(\hat{k}\cdot\vec{E}\right)\hat{k}-\vec{E}\right]=-\mu_0\vec{D}. \tag{2.99}$$

Working with Equation 2.99 is simplest when looking at one Cartesian component at a time. The k-vector's unit vector is written as

$$\hat{k}=s_x\hat{x}+s_y\hat{y}+s_z\hat{z}. \tag{2.100}$$

With this substitution, Equation 2.99 in component form becomes

$$D_i=\frac{n^2}{\mu_0c^2}\left[E_i-\left(\hat{k}\cdot\vec{E}\right)s_i\right]. \tag{2.101}$$

Since the coordinate system is chosen to be the one where ε_{ij} is diagonal (as can always be done for ε_{ij}), it is possible to write the inverse of the dielectric matrix using Equation 2.90 as

$$\vec{E}=\frac{1}{\varepsilon_0}\begin{bmatrix}\dfrac{1}{n_x^2} & 0 & 0 \\[2mm] 0 & \dfrac{1}{n_y^2} & 0 \\[2mm] 0 & 0 & \dfrac{1}{n_z^2}\end{bmatrix}\vec{D}. \tag{2.102}$$

We extract the component E_i from Equation 2.102 and substitute it into Equation 2.101 yielding

$$D_i=\frac{n^2}{\mu_0c^2}\left[\frac{D_i}{\varepsilon_0 n_i^2}-\left(\hat{k}\cdot\vec{E}\right)s_i\right]. \tag{2.103}$$

Note that even though i is repeated in the first term in this equation, we are not using Einstein notation and no summation is carried out over the index i. Solving for D_i, we find

$$D_i=\frac{\left(\hat{k}\cdot\vec{E}\right)s_i}{\mu_0c^2\left[\left(1/n_i^2\right)-\left(1/n^2\right)\right]}. \tag{2.104}$$

Using Maxwell's Equation 2.94, $\hat{k} \cdot \vec{D} = 0$ gives

$$\frac{s_x^2}{(1/n_x^2) - (1/n^2)} + \frac{s_y^2}{(1/n_y^2) - (1/n^2)} + \frac{s_z^2}{(1/n_z^2) - (1/n^2)} = 0. \tag{2.105}$$

Equation 2.105 is called Fresnel's equation, and it must be satisfied for the original assumption of plane-wave solutions in the crystal to hold.

2.3.5 FRESNEL'S EQUATION

Fresnel's equation (Equation 2.105) gives a value for the index of refraction for any k-vector direction. In fact, there are two solutions for the index in general; they are called eigenindices.

We simplify the process of finding the values of the index by rewriting Fresnel's equation as

$$\left(\frac{1}{n_y^2} - \frac{1}{n^2} \right) \left(\frac{1}{n_z^2} - \frac{1}{n^2} \right) s_x^2 + \left(\frac{1}{n_x^2} - \frac{1}{n^2} \right) \left(\frac{1}{n_z^2} - \frac{1}{n^2} \right) s_y^2 + \left(\frac{1}{n_x^2} - \frac{1}{n^2} \right) \left(\frac{1}{n_y^2} - \frac{1}{n^2} \right) s_z^2 = 0. \tag{2.106}$$

As a first example for finding the index, consider a unit k-vector direction such as $s_x = 1$, $s_y = 0$, and $s_z = 0$. For this case, two obvious solutions for n are, by inspection, either n_z or n_y. To find the values of the index for arbitrary directions, a few constructions help us understand the behavior. The first is shown in Figure 2.3, which illustrates the intersection of Equation 2.106 with the principal planes.

2.3.6 o-WAVES AND e-WAVES

The curves in Figure 2.3 are generated by considering one principal plane at a time. For example, in the $k_x - k_y$ plane, $s_z = 0$ so that Equation 2.106 becomes

$$\left(\frac{1}{n_z^2} - \frac{1}{n^2} \right) \left[\left(\frac{1}{n_y^2} - \frac{1}{n^2} \right) s_x^2 + \left(\frac{1}{n_x^2} - \frac{1}{n^2} \right) s_y^2 \right] = 0. \tag{2.107}$$

Noting that the unit vector condition requires $s_y^2 + s_x^2 = 1$, then we have

$$\left(\frac{1}{n_z^2} - \frac{1}{n^2} \right) \left[\frac{s_x^2}{n_y^2} + \frac{s_y^2}{n_x^2} - \frac{1}{n^2} \right] = 0. \tag{2.108}$$

Hence, $n = n_z$ or n is given by the equation of an ellipse and is dependent on the particular direction of the k-vector in the $k_x - k_y$ principal plane. In the latter case,

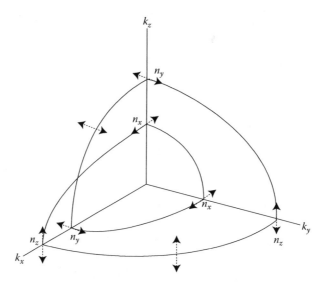

FIGURE 2.3 Intersection of Equation 2.106 with principal planes. The curves indicate the magnitude of the k-vector for a given direction and also illustrate that in general for any propagation direction there exist two magnitudes. For simplicity, the labels give the index of refraction for the given k-vector component. To find the magnitude of the k-vector, multiply by ω/c. The arrows represent the direction of the electric field.

the unit vector is written as $\hat{k} = \cos(\theta)\hat{x} + \sin(\theta)\hat{y}$, so that with respect to this k-vector direction the index becomes

$$\frac{1}{n^2(\theta)} = \frac{\cos^2(\theta)}{n_y^2} + \frac{\sin^2(\theta)}{n_x^2}. \tag{2.109}$$

A similar result is obtained for the $k_y - k_z$ and $k_x - k_z$ principal planes with two solutions: one with an index that is constant independent of the k-vector direction in the plane, called the *ordinary wave*, which has the electric field vector directed perpendicular to the principal plane of propagation, and the other with an index that changes with the k-vector direction and with a value given by the equation of an ellipse, called the *extraordinary wave*, where the electric field vector lies in the principal plane of propagation. The intersection of the curves in the $k_x - k_z$ plane corresponds to an optic axis of the crystal where both refractive indices are equal. Note that in a biaxial crystal there are two of these axes. Only one of the optic axes is illustrated in Figure 2.3 where two curves intersect. For uniaxial crystals for which two of the principal indices are the same, the optic axis is by convention taken to be the z-axis and $n_x = n_y$.

Picking a value of n that satisfies the Fresnel equation results in a valid plane-wave solution to the wave equation. However, until now, we have not considered the polarization state of the fields. First, we should note that *polarization states* such as linearly polarized and circularly polarized refer to the properties of EM waves and not to the physical polarization in a medium. Indeed, it is common to visualize the polarization state of a beam in a vacuum where there is nothing to polarize. Since a given k-vector direction in general has two solutions for n, the question we may ask is: what is the state of polarization for the corresponding index? The answer is found by taking a close look at Equation 2.104, which is used to find the directions of the D-field and, with a little more work, the direction of the E-field. Once we insert a value for the index of refraction into Equation 2.104, the polarization state is determined by the phase differences between the different components of $\vec{D}$. However, each component of $\vec{D}$, D_i, is a real number since all the quantities in Equation 2.106 are also real. Recall that earlier we determined that ε_{ij} is Hermitian, thus guaranteeing that the eigenindices are real valued. The fact that all the D_i are real shows there is no phase difference (apart from the possibility of a sign change) between the components of $\vec{D}$. Hence, $\vec{D}$ must be linearly polarized since all the components of $\vec{D}$ oscillate in phase. To get an elliptical polarization state requires that at least one of the $\vec{D}$ components oscillates out of phase with the other components. So the polarization states corresponding to the eigenindices are linear.

The directions of the eigenpolarizations are found with Equation 2.104 combined with the condition $\vec{k} \cdot \vec{D} = 0$. The directions of the electric field are indicated as double arrows in Figure 2.3. Two general observations should be noted about the directions of the fields. The first is that within a given principal plane, the two solution curves have electric fields perpendicular to each other. Moreover, the curve that is an arc of a circle in the figure, corresponding to o-waves, has $\vec{k} \cdot \vec{E} = 0$ since the electric field for this case is perpendicular to the principal plane. The e-waves correspond to the indices of the elliptical curve section, and the e-wave electric field is parallel to the principal plane. The direction of the electric field for the e-wave is surprisingly not perpendicular to the k-vector.

2.3.7 POYNTING VECTOR WALK-OFF

The fact that $\vec{k}$ and $\vec{E}$ are not perpendicular for e-waves leads to an important effect called the Poynting vector walk-off that can degrade the performance of nonlinear optical devices. This walk-off effect is illustrated by looking at the relationships between $\vec{D}, \vec{E}, \vec{H}, \vec{k}$, and $\vec{S}$, which are interrelated via Maxwell's equations. Figure 2.4

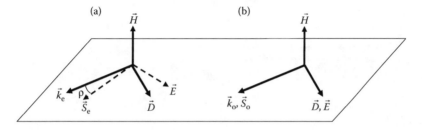

FIGURE 2.4 Directions of various components of an EM wave (a) e-polarized and (b) o-polarized in a birefringent material. Note that $\vec{k}, \vec{S}_e, \vec{E}$, and $\vec{D}$ are coplanar and that $\vec{H}$ is perpendicular to the plane. ρ is the walk-off angle.

illustrates these quantities for an e-wave and an o-wave and shows that the walk-off angle is the angle between $\vec{D}$ and $\vec{E}$ as well as between $\vec{k}$ and $\vec{S}$.

It is straightforward to show that when $\vec{D}$ and $\vec{E}$ are not parallel, then $\vec{k}$ and $\vec{S}$ are also not parallel and that the angles between $\vec{D}$ and $\vec{E}$ are the same as the angle between $\vec{k}$ and $\vec{S}$. This angle is called the walk-off angle, ρ. In the case of o-waves, $\vec{D}$ and $\vec{E}$ are parallel; hence, $\vec{k}$ and $\vec{S}$ are parallel. In nonlinear optics, it is common to have situations where multiple beams are interacting in a crystal and some of them are e-waves and others are o-waves. In these situations, it is often the case that the e-waves and o-waves will have parallel k-vectors but do not necessarily have parallel Poynting vectors. Figure 2.5 shows the case where even at normal incidence two beams emerge from a crystal when illuminated with a single beam of unpolarized light. The energy of the e-wave walks off from that of the o-wave, and the beams physically separate. Such an effect decreases the length over which the beams can interact. The walk-off angle is typically on the order of a few degrees, which becomes significant when focusing beams in a crystal and is less important for beams with large cross sections.

The walk-off angle in the case of *uniaxial* crystals is given by

$$\rho = \tan^{-1}\left(\frac{n_o^2}{n_z^2}\tan\theta\right) - \theta \qquad (2.110)$$

or equivalently

$$\tan\rho = \frac{n_e^2(\theta)\sin 2\theta}{2}\left(\frac{1}{n_z^2} - \frac{1}{n_o^2}\right), \qquad (2.111)$$

where θ is the angle between the k-vector and the z-axis. In the case of a biaxial crystal, similar expressions result in each of the principal planes.

The direction of the walk-off is determined from the sign of Equation 2.110 or 2.111, and it can also be inferred from Figure 2.3. $\vec{S}$ is, by the definition of a cross product, perpendicular to $\vec{E}$, hence it is perpendicular to the curves shown in Figure 2.3. An illustration for uniaxial crystals is shown in Figure 2.6.

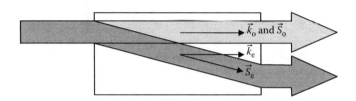

FIGURE 2.5 Poynting vector walk-off. Even at normal incidence, the energy of the e-wave will separate from the o-wave. Note that the k-vector and the Poynting vector of the o-wave are collinear and parallel to $\vec{k}_e$.

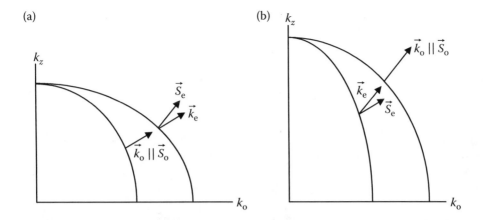

FIGURE 2.6 *k*-Vector magnitude surfaces for uniaxial crystals. The curves are the allowed magnitudes of the *k*-vector for a given propagation direction. The Poynting vector is normal to these surfaces. (a) Positive uniaxial crystal, walk-off is toward the *z*-axis (optic axis). (b) Negative uniaxial crystal, walk-off is away from the *z*-axis.

2.4 DETERMINING e-WAVES AND o-WAVES IN CRYSTALS

The constructions in Figure 2.3 contain much information, but the pertinent result is that for any direction of propagation in a crystal, there are two orthogonal linearly polarized eigenpolarizations. These polarizations form the logical basis set to describe an arbitrary polarization state. We frequently need to determine the direction of the e- and o-polarization directions for a given input field direction to a crystal. It is equally important to determine the values of the e- and o-indices associated with the e- and o-polarizations. Determining the relevant properties of the o- and e-waves is a straightforward process for which a streamlined approach applied to different crystal types is given below.

2.4.1 HOMOGENEOUS ISOTROPIC

The index of refraction is the same regardless of the direction of propagation or polarization state. For such crystals, any basis set is equivalent to any other one for linear properties. The terminology of o- and e-waves is typically not used with such crystals. Many semiconductor crystals such as GaAs fall into this category. However, even though a crystal may be homogeneous and isotropic in its linear properties, its nonlinear properties may not be isotropic.

2.4.2 UNIAXIAL CRYSTAL

As mentioned earlier, uniaxial crystals have two principal indices that are equal and one that differs from the others. The optic axis is the *z*-axis by convention

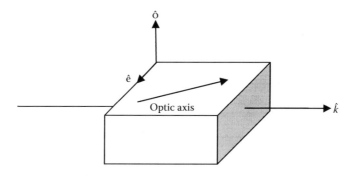

FIGURE 2.7 Uniaxial crystal with the e- and o-polarization directions indicated.

(Figure 2.6). The directions of e- and o-polarizations are found by using the following routine:

1. Draw the optic axis.
2. Draw the k-vector.
3. The o-wave polarization is in a direction perpendicular to the plane formed by the optic axis and k-vector, and the e-wave polarization is contained within the plane.

The corresponding indices of refraction are $n_o = $ constant, independent of the k-vector direction for the o-wave, and for the e-wave, the index is found using

$$\frac{1}{n_e^2(\theta)} = \frac{\cos^2\theta}{n_o^2} + \frac{\sin^2\theta}{n_z^2},$$

(2.112)

where the angle, θ, is measured with respect to the z-axis (i.e., optic axis). $n_e(\theta)$ is the extraordinary index, and it has the explicit dependence on the angle. Finally, the walk-off is found by using Equation 2.110 or 2.111 and the discussion above, thereby yielding the angle between $\vec{D}$ and $\vec{E}$.

Figure 2.7 shows a specific example where the optic axis lies at an angle to the incident beam. Depending on the input polarization, the incident beam excites both the e- and o-wave polarization states. In most applications of nonlinear optics, the input field is aligned with either the e- or o-direction so that the polarization state remains linear. Since the e- and o-waves travel with different phase velocities (i.e., their indices are different), the output polarization state differs from the input state when both e- and o-waves are excited.

2.4.3 BIAXIAL CRYSTALS

Biaxial crystals have three different principal indices. Instead of considering an arbitrary k-vector direction, it is possible to make a biaxial crystal look uniaxial by choosing to restrict the *k*-vector to one of the three principal planes (*xy*-, *xz*-, or *yz*-planes). By doing so, the same naming convention as for uniaxial crystals is applied. For example, choosing the *xz*-plane as the principal plane means that *k* lies in the *xz*-plane. For any *k*-vector direction in this plane, the polarization perpendicular to the plane (o-polarized) always points along the *y*-axis. Hence, the index of refraction for that polarization is independent of the *k*-vector's direction in the *xz*-plane. The polarization that is in the *xz*-plane sees different indices as the *k*-vector rotates in this plane; hence, it is called the e-polarization. An interesting aspect to operating in the *xz*-plane is that the crystal behaves as both positive and negative uniaxial depending on which side of the optic axis the *k*-vector is on. This arises because in this plane the ordinary index is n_y, which lies between n_z and n_x.

There are cases in which operating outside of a principal plane is desirable. In such cases, the general results still hold; namely, there are two possible eigenvalues for the index of refraction (see Fresnel's equation [Equation 2.105]), and the eigenpolarizations corresponding to these indices are linearly polarized and orthogonal to each other. The specific polarization directions and values of the indices have to be computed (Dmitriev et al., 1997).

2.5 INDEX ELLIPSOID

A second way to look at the index of refraction in crystals is through a construction known as the optical indicatrix or index ellipsoid. This construction gives a visual aid in understanding how the index of refraction changes with the propagation direction and polarization. The general idea is to compare plane waves of equal energy density traveling in different directions in a crystal. In order for the energy density to remain constant, the lengths of $\vec{D}$ and $\vec{E}$ change accordingly. In a dispersionless medium, $\vec{D}$ and $\vec{E}$ are related by $\vec{D} = \ddot{\varepsilon} \cdot \vec{E}$ and the index is related to $\ddot{\varepsilon}$, so it can be shown that a constant energy density surface maps to an index surface that is an ellipsoid. The connection is shown by starting with the energy density,

$$U_E = \frac{1}{2} \vec{E} \cdot \vec{D}.$$
(2.113)

Using Equation 2.102, this relationship is written in terms of $\vec{D}$ alone by assuming that the coordinate system for $\ddot{\varepsilon}$ is diagonal. U_E in this basis set becomes

$$U_E = \frac{1}{2\varepsilon_0} \left[\frac{D_x^2}{n_x^2} + \frac{D_y^2}{n_y^2} + \frac{D_z^2}{n_z^2} \right].$$
(2.114)

By comparing plane waves with the same energy density, U_E is a constant. So Equation 2.114 is rewritten as

$$\frac{1}{2U_E\varepsilon_0}\left[\frac{D_x^2}{n_x^2}+\frac{D_y^2}{n_y^2}+\frac{D_z^2}{n_z^2}\right]=1. \tag{2.115}$$

Now making a variable substitution of $x^2 \equiv (D_x^2 / 2U_E\varepsilon_0)$ and similarly for y^2 and z^2 results in

$$\frac{x^2}{n_x^2}+\frac{y^2}{n_y^2}+\frac{z^2}{n_z^2}=1, \tag{2.116}$$

which is the equation of an ellipsoid, and the connection between the constant energy surface and the index of refraction is complete. A given coordinate on this surface gives the magnitude of the index of refraction for a D-vector pointing in that particular direction. The ellipsoid shows a continuum of D-vector magnitudes that correspond to a continuum of polarization states. However, the ellipsoid can also be used to find the values of the index corresponding to the eigenpolarization e- and o-waves.

Figure 2.8 shows a particular k-vector. Equation 2.89 requires that $\vec{D}$ must lie in a plane perpendicular to the k-vector, so the two eigenindices lie somewhere on the

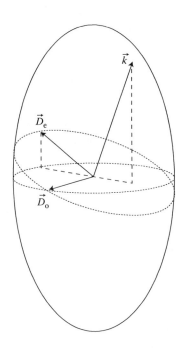

FIGURE 2.8   Index ellipsoid for a uniaxial crystal. The dotted ellipses are the intersection of the plane perpendicular to $\vec{k}$ and the ellipsoid and the equatorial plane.

intersection of this plane with the ellipsoid. This intersection curve is an ellipse and the major and minor axes correspond to the eigenindices. One of these axes lies in the "equatorial plane" of the ellipsoid and the corresponding index is the ordinary index.

The ellipsoid shows the range of values for the index of refraction. Moreover, it is a useful construct for nonlinear problems such as the electro-optic (EO) effect where the application of a static field distorts the ellipsoid. By visualizing the distortion, it is possible to observe the effect on the e- and o-polarization directions as well as the change in the extraordinary and ordinary indices.

2.6 APPLICATIONS

2.6.1 SLOWLY VARYING ENVELOPE APPROXIMATION AND GAUSSIAN BEAMS

For most nonlinear optical phenomena, the interaction strength is proportional to the intensity squared, intensity cubed, and so on. Hence, it is a common practice to focus laser beams to enhance the nonlinear effect. Moreover, beyond simple focusing, many applications require some sort of mode matching such as from a free-space laser beam to a fiber or an externally resonant cavity. This section looks at such applications from the standpoint of the canonical laser beam profile, the Gaussian beam. We introduce the slowly varying envelope equation for a linear medium and derive the basic Gaussian beam equations and then work through an example to show how to measure a beam profile to extract the Gaussian beam parameters. With those parameters in hand, it is possible to design a lens sequence to focus to a desired spot size, which can be defined by the beam radius. One common definition is to use the $1/e^2$ value of the intensity below its maximum value.

The Gaussian beam solution to the wave equation may be thought of as a modified plane wave written as

$$\underline{E}(\vec{r}_\perp, z, t) = \frac{1}{2} A \left(\vec{r}_\perp, z \right) e^{i(kz - \omega t)} + \text{c.c.}, \tag{2.117}$$

where the center of the beam propagates along the z-axis and $\vec{r}_\perp$ is the transverse distance from the z-axis. In Equation 2.116, the rapidly varying components of the field, $(kz - \omega t)$, are the same as with the plane-wave solution, but now the complex amplitude, A, is no longer a constant, that is, a plane wave. The complex amplitude is a function of the transverse coordinates, $\vec{r}_\perp$, and propagation distance, z. In other words, a beam of finite transverse size consists of many plane waves that superpose to make up the complex amplitude A. We need both $\vec{r}_\perp$ and z dependencies because a finite-sized beam diffracts such that its transverse size changes and therefore its on-axis amplitude also changes. When we consider the specific case of the Gaussian beam, the field amplitude varies in the transverse dimension as

$$|A(\vec{r}_\perp, z)| = |A'(z)| e^{-r^2/w^2},$$ (2.118)

where w defines the 1/e beam radius with respect to the field or $1/e^2$ with respect to the intensity. As the Gaussian beam propagates, the intensity maintains a Gaussian transverse cross section, but because of diffraction its size changes so that the radius is a function of position, that is $w = w(z)$, the amplitude $A'(z)$ is also a complex function of z to maintain conservation of energy flow in a lossless medium. By picking a Gaussian cross section, the wave equation then dictates other properties of the beam, like the on-axis variation $A(r = 0, z)$.

To derive the slowly varying envelope approximation (SVEA) equation, we assume that the beam consists of plane waves propagating close to the z-axis; these waves are called paraxial waves. The complex scalar field is decomposed into the following form:

$$E(\vec{r}_\perp, z, t) = \frac{1}{2} A(\vec{r}_\perp, z) e^{i(kz - \omega t)}.$$ (2.119)

$A(\vec{r}_\perp, z)$ is called the envelope function. The propagation direction is z and the transverse beam coordinates in the (x,y) plane are denoted as $\vec{r}_\perp$. Equation 2.119 is inserted into Equation 2.83 to obtain the following version of the scalar wave equation:

$$2ik \frac{\partial A}{\partial z} + \frac{\partial^2 A}{\partial z^2} + \nabla_\perp^2 A = -\left(\frac{\mu_r \varepsilon_r}{c^2} \omega^2 - k^2 \right) A,$$ (2.120)

where the transverse Laplacian operator is defined as $\nabla_\perp^2 = (\partial^2/\partial x^2) + (\partial^2/\partial y^2)$; this term describes the diffraction of the beam in the medium. The wavenumber k appearing on the right-hand side is chosen by the requirement

$$k^2 = \text{Re} \left\{ \frac{\mu_r \varepsilon_r}{c^2} \omega^2 \right\}.$$ (2.121)

The second term on the right-hand side may be neglected when $k|\partial A/\partial z| \gg |\partial^2 A/\partial z^2|$. Since $k \propto (1/\lambda)$, this is a condition that a derivative of the envelope function varies little over distances the size of the wavelength. The resulting equation is called the SVEA or the paraxial equation:

$$\boxed{\frac{\partial A}{\partial z} - \frac{i}{2k} \nabla_\perp^2 A + \frac{\alpha}{2} A = 0.}$$ (2.122)

The absorption coefficient is

$$\alpha = \frac{1}{k}\,\mathrm{Im}\left\{\frac{\mu_r \varepsilon_r}{c^2}\,\omega^2\right\}. \tag{2.123}$$

Define the Gaussian amplitude $A(\vec{r}_\perp = \vec{0}, 0) = A_0$ at $z = 0$, the solution of Equation 2.122 is (We define $r = |\vec{r}_\perp|$, the transverse distance from the z-axis)

$$A(\vec{r}, z) = A_0 e^{-\alpha z/2} \frac{w_0}{w(z)} \exp\left(\frac{-r^2}{w^2(z)}\right) \exp\left(ik\frac{r^2}{2R(z)} - i\zeta(z)\right), \tag{2.124}$$

where

$$w(z) = w_0 \sqrt{1 + \left(\frac{z}{z_R}\right)^2}, \tag{2.125}$$

$$R(z) = z\left(1 + \left(\frac{z_R}{z}\right)^2\right), \tag{2.126}$$

$$\zeta(z) = \tan^{-1}\left(\frac{z}{z_R}\right), \tag{2.127}$$

$$z_R = \frac{\pi n w_0^2}{\lambda}. \tag{2.128}$$

The quantity $w(z)$ represents the $1/e$ field amplitude radius, and w_0 is the beam waist radius, which is set at the axis position $z = 0$. $R(z)$ is the radius of curvature of the wavefront, ζ is the Guoy phase shift, and z_R is the Rayleigh range. In the absence of absorption, the power in the beam is constant with propagation distance.

A plot of the Gaussian beam radius, $w(z)$, is shown in Figure 2.9. The figure is plotted in the Rayleigh region where the beam radius is roughly constant (the beam waist has a minimum radius called w_0 and at a distance $z = z_R$, the beam radius is $\sqrt{2}w_0$), which also gives a roughly constant intensity throughout the waist region. Outside the Rayleigh range, the intensity drops off quickly as the beam radius expands and with it the strength of a given nonlinear interaction. The radius of curvature is infinite at the beam waist ($z = 0$) and negative for $z < 0$ and positive for $z > 0$.

2.6.2 GAUSSIAN BEAM PROPAGATION USING THE q-PARAMETER

Obtaining a desired spot size is achieved by combinations of lenses. Fortunately, determining the effect of a lens or other optical element on a Gaussian beam is

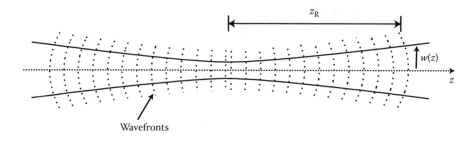

FIGURE 2.9 Cross section of a Gaussian beam along its propagation direction. The thick curves show $w(z)$ whereas the dotted curves show the wavefront curvature. z_R is the Rayleigh range, which is the distance from the beam waist to the point where $w = \sqrt{2}w_0$.

straightforward. Gaussian beam propagation with lenses is calculated using an approach similar to that used with geometric ray tracing. In ray tracing, the effects of a free-space segment, lens, or mirror are described by 2×2 matrices; to trace the ray, the matrices multiply 2×1 vectors where the first element is a transverse displacement of the ray from the symmetry axis and the second element is a ray angle relative to the same axis (Pedrotti 2007). The elements of the matrix are labeled as A, B, C, and D:

$$\text{Generic ray-tracing matrix:} \begin{bmatrix} A & B \\ C & D \end{bmatrix} \quad \text{and} \quad \text{vector:} \begin{bmatrix} x \\ \theta \end{bmatrix}. \tag{2.129}$$

An optical system consisting of multiple elements is modeled by a single ABCD matrix, which is the product of the matrices for each optical element or displacement in the optical system; this is a lumped-matrix approach. The same matrices used for the lumped-matrix approach to ray tracing are also used with Gaussian beams, except that the elements of the matrix are used to determine the change in the Gaussian beam when it passes through the optical system. The Gaussian beam is characterized by a parameter q, which is defined by[*]

$$\frac{1}{q(z)} = \frac{1}{R(z)} + i\frac{\lambda}{\pi n w^2(z)}. \tag{2.130}$$

z is, in most cases, defined with respect to the beam's focal position. The electric field's complex amplitude in Equation 2.124 can be written in terms of the Gaussian q-parameter as

$$A(r,z) = A_0 \frac{q_0}{q} \exp\left(i\frac{kr^2}{2q} \right), \tag{2.131}$$

---

[*] A derivation of the Gaussian beam equation using another convention for the plane wave, $e^{i(\omega t - kz)}$ is found in many publications. This convention differs from that found in the complex conjugate of Equation 2.120. We apply our notation given in Equation 2.119 throughout this book.

where $q_0 = q(z = 0)$. Equation 2.131 is convenient for situations where we need to separate out the complex amplitude's radial dependence. At $z = 0$, the Gaussian beam is at a focus so that it has a flat phase front and hence the radius of curvature is infinite. Therefore,

$$q_0 = -i\frac{\pi n w_0^2}{\lambda} = -i z_R.$$ (2.132)

The transformation of q for an optical system is given by the simple relationship

$$q' = \frac{Aq + B}{Cq + D},$$ (2.133)

where q' is the new Gaussian parameter and where A, B, C, and D are the matrix elements from geometric optics ray-tracing matrices, a few of which are given below:

$$\text{Free-space propagation of length, } d \Rightarrow \begin{bmatrix} 1 & d \\ 0 & 1 \end{bmatrix},$$ (2.134)

$$\text{Lens} \Rightarrow \begin{bmatrix} 1 & 0 \\ -\dfrac{1}{f} & 1 \end{bmatrix},$$ (2.135)

$$\text{Change in index (incident medium is } n_1 \text{ and second medium is } n_2) \Rightarrow \begin{bmatrix} 1 & 0 \\ 0 & \dfrac{n_1}{n_2} \end{bmatrix}.$$ (2.136)

When multiple elements exist in an optical system, the appropriate matrices are multiplied in order of the appearance of the optical element from right to left, resulting in a single ABCD matrix, which in turn allows for a calculation of the new q-parameter. Once the new q-parameter is found, it is possible to extract $w(z)$ as well as $R(z)$ using Equation 2.130.

As an example, consider a beam waist incident on the down-collimating telescope shown in Figure 2.10. The overall ABCD matrix for the sequence starting from the left lens and ending at the right lens is

$$\begin{bmatrix} A & B \\ C & D \end{bmatrix} = \begin{bmatrix} 1 & 0 \\ \dfrac{2}{f} & 1 \end{bmatrix} \begin{bmatrix} 1 & f \\ 0 & 1 \end{bmatrix} \begin{bmatrix} 1 & 0 \\ -\dfrac{1}{f} & 1 \end{bmatrix} = \begin{bmatrix} 1 & f \\ \dfrac{2}{f} & 2 \\ 0 & 2 \end{bmatrix}.$$ (2.137)

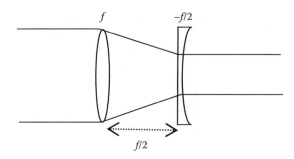

FIGURE 2.10 Down-collimating a beam based on Galileo-type telescope design.

The key to working with the Gaussian beam parameter q is the ability to determine its value at some location. Since the incident beam is a beam waist, the starting radius of curvature is infinite. Hence, we find the initial Gaussian beam parameter,

$$\frac{1}{q_0} = i\frac{\lambda}{\pi w_0^2},$$

(2.138)

where w_0 is the collimated beam radius. In terms of the Rayleigh range, Equation 2.138 is written as

$$q_0 = -iz_R.$$

(2.139)

After the lens sequence shown in Figure 2.10, the modified Gaussian beam parameter is found using Equation 2.133 and the ABCD values from Equation 2.137,

$$q' = \frac{q_0 + f}{4}.$$

(2.140)

According to Equation 2.130, the new radius of curvature and the new beam size are related to the real and imaginary parts of $1/q'$, respectively. Taking $q_0 = iz_R$, the new radius of curvature is

$$R' = \frac{f^2 + z_R^2}{4f}.$$

(2.141)

The new beam size is after some algebra,

$$w' = \frac{w_0}{2}\left(1 + \frac{f^2}{z_R^2}\right)^{1/2}.$$

(2.142)

The geometric limit is given by $z_R \rightarrow \infty$ and $w' \rightarrow w_0/2$. As the initial beam waist size decreases and f becomes comparable to z_R, the utility of the down-collimator is reduced.

2.6.3 M^2 PROPAGATION FACTOR

A further refinement is to consider beams that are not exactly Gaussian. A useful way to characterize non-Gaussian beams is through a technique that compares how the particular beam propagates compared to an ideal Gaussian beam (Ruff and Siegman, 1992). The divergence of non-Gaussian beams is always greater than Gaussian ones. A measure of the divergence is the beam width at the focal plane of the beam, times the beam width in the far field. One method of characterizing a light beam (coherent and incoherent light) is to calculate the second moment of intensity (when it exists). Using the moment definition, the beam's center-position in the x-direction is defined by

$$x_0(z) = \frac{\int xI(x,y,z)dx\,dy}{\int I(x,y,z)dx\,dy} \tag{2.143}$$

where $I(x,y,z)$ is the intensity distribution in the near and far fields. The second moment for a cross section of the beam in the x-direction is given by

$$\sigma_x^2(z) = \frac{\int (x - x_0(z))^2 I(x,y,z)dx\,dy}{\int I(x,y,z)dx\,dy} = \frac{1}{4}w_x^2(z) \tag{2.144}$$

where $w_x(z)$ defines the beam width parameter in the x-direction for a Gaussian beam. The beam's width in the y-direction, $w_y(z)$, is found in a similar fashion. For a cylindrically symmetric beam, the radial width is commonly defined. The product of initial and final widths behaves as $\sqrt{2}\sigma(0)\sqrt{2}\sigma(z) \sim \lambda z$ when z is in the far field. To show this, consider the Gaussian beam defined by Equation 2.124. The intensity may be written as

$$I(x,y,z) = I_0 e^{-2r^2/w^2(z)}. \tag{2.145}$$

The second radial moment of a Gaussian beam is

$$\sigma^2(z) = \sigma_x^2(z) + \sigma_y^2(z) = \frac{\int_0^{\infty} r^2 e^{-2r^2/w^2(z)} r\,dr}{\int_0^{\infty} e^{-2r^2/w^2(z)} r\,dr} = \frac{1}{2}w^2(z). \tag{2.146}$$

We assume that the plane $z = 0$ is the focal plane of the beam, whose width is given in Equation 2.125. Using this result, the product of the two beam widths is

$$\sigma(0)\sigma(z) = \frac{w(0)}{\sqrt{2}} \frac{w(z)}{\sqrt{2}} = w_0^2 \frac{z}{2z_R} = \frac{1}{2\pi}\lambda z. \tag{2.147}$$

To fit a non-Gaussian beam width in both the near and far field, a new beam width is defined by the function

$$w_M(z) = w_0 \sqrt{1 + M^4 \left(\frac{z}{z_R}\right)^2}, \tag{2.148}$$

With this definition in the far field, we find that for non-Gaussian beams the product in Equation 2.147 is larger by a factor M^2

$$\sigma(0)\sigma(z) = M^2 \frac{1}{2\pi}\lambda z. \tag{2.149}$$

M^2 unity reduces to the Gaussian beam equation. Beams characterized by M^2 near unity are nearly perfect Gaussian beams. The utility of M^2 comes in two regimes. In cases where M^2 is close to one, then, as noted, it is possible to consider the beam as essentially Gaussian. In regimes where M^2 is appreciably different than 1, say >3, the M^2 measurement is useful as an "energy in the bucket" approach to let the system designer know how much of the laser's energy will be contained in a certain area. For instance, about 84% of a Gaussian beam's energy is contained within a disk of radius $w_M(z)$; in other words, most of the beam's energy is captured (i.e., "in the bucket") within that area. However, it is not possible to use M^2 to infer a beam profile when M^2 is large, since there are infinitely many different beam profiles that can have the same M^2.

2.6.4 EXAMPLE OF FORMATTING A BEAM FOR SHG

Once a system designer determines how tightly to focus a beam, it is necessary to take a particular laser and map it to the desired beam size. Even if the beam size is known at the exit plane of the laser, that information is insufficient since the beam could be collimated, focusing, or diverging. The best approach is to directly measure and characterize the beam, and from these measurements, determine the optics to correctly format the beam for a given application. A particularly simple approach for beam characterization is to use a camera to measure the beam at several locations along the beam's propagation direction. Since the recorded image is a direct measurement of $I(x,y)$, it is possible to extract $w_M(z)$ at several locations along the propagation direction. Other common techniques include scanning a pinhole across the beam as well as scanning a razor blade across the beam. The razor technique is common, but it also requires making an educated guess at the beam shape. Once measurements of $w_M(z)$ are made, it is then possible to fit the measured values of $w_M(z)$ to Equation 2.144 to determine w_0, the location of w_0, and M^2. Those parameters in turn are used to determine appropriate lenses and lens spacing to obtain a desired spot size.

As a specific example, we consider the application of SHG with a high-quality ($M^2 \sim 1$) Nd:YAG laser operating at 1064 nm doubled to 532 nm in a 1-cm-long crystal. The index of refraction of the crystal is given as 1.5. (Details of how the doubling is accomplished are given in Chapter 4.) For the doubling application, high conversion efficiency is achieved when the focused beam's Rayleigh range is matched to one half the crystal length, $z_R \sim L/2$. Therefore, we have a criterion for focusing the Nd:YAG laser.

Taking the YAG laser out of its box, the first step is to characterize its beam profile. Figure 2.11 shows the results, where the location of $z = 0$ is with respect to the output plane of the laser box. From Figure 2.11, it is evident that the beam is diverging as it comes out of the laser head, so it is necessary to focus it to a smaller spot. A practical approach is to find the virtual waist from which the beam appears to originate by fitting the measured $w(z)$ to the analytic function for a Gaussian beam in Equation 2.125. Such a location may not in fact be accessible, but the properties of the beam outside of the laser behave exactly as if the beam originated from this location. For beam profile measurements shown in Figure 2.11, the virtual beam waist is located 50 cm behind the output of the laser and has a w_0 of 50 µm. It is possible to place lenses to map the beam to the appropriate size based on these two numbers.

Now that the Nd:YAG laser beam is characterized, the next step is to focus it to the right spot size. In this case, since the crystal is 1 cm and the maximum efficiency occurs when $z_R \sim L/2$, then Equation 2.120 lets us solve for the desired focused spot size, $w_0 = 33$ µm. To determine how to obtain this spot size using the laser beam parameters as measured above, we turn to the Gaussian beam parameter q.

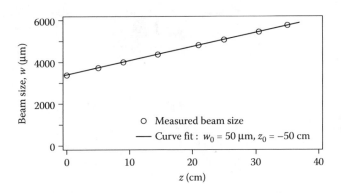

FIGURE 2.11 Measurements of $w(z)$ for an Nd:YAG laser. Notice that it is not necessary to directly measure the beam waist region in order to extract w_0.

In the case of the Nd:YAG laser, the direct measurements followed by a curve fit give the beam waist size, w_0, and its location. At the beam waist, q is particularly simple since $R = \infty$ and so q is given by

$$q_0 = -i\frac{\pi w_0^2}{\lambda}. \tag{2.150}$$

For the current example, Figure 2.11 shows $w_0 = 50\ \mu m$ and $\lambda = 1.064\ \mu m$. It is a simple matter to propagate from this virtual beam waist to some point output of the laser box; according to Equation 2.125, it is just $q_0 + d$, where d is the distance from the virtual waist to the point of interest. In the case of the second-harmonic application, the important parameter is the desired focused spot size in the crystal, which should be 33 μm. Mapping from the virtual waist to the desired waist is accomplished in a number of ways. An approach that is commonly used and analytically simple is a relay image system (Figure 2.12).

The ABCD matrix for the sequence shown in Figure 2.12 is

$$\begin{bmatrix} -\dfrac{f_2}{f_1} & 0 \\[2ex] 0 & -\dfrac{f_1}{f_2} \end{bmatrix}. \tag{2.151}$$

The ABCD matrix values from Equation 2.151 are inserted into Equation 2.133 giving the relationship between beam waists as $w_2 = (f_1/f_2)w_1$ in air. The beam waist in the crystal has the same size as in air but is located slightly farther away from the second lens; the proof is left as an exercise. Note that the focal length, f_1, must be >50 cm since it must lie outside of the laser box.

The same process used here to focus on a crystal is applied in numerous other situations. For example, when launching a laser into a fiber, the optimum launch efficiency occurs when the focused spot size matches the mode field diameter of the fiber. Another example is mode matching to a cavity. Resonators greatly enhance nonlinear processes and obtain the greatest efficiency when lasers are coupled into

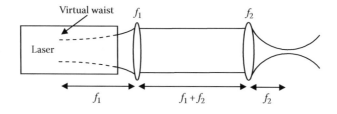

FIGURE 2.12 Relay imaging scheme to focus the beam to a desired size outside of the laser cavity.

the resonator properly by matching the focused beam waist size of the incident laser to the waist size of the resonator mode.

PROBLEMS

2.1 A monochromatic, plane wave has an electric field expressed as $\vec{E} = E_o \sin(kz - \omega t + \phi)\hat{x}$. The wave propagates in a material with a susceptibility given by $\chi = \chi_0(1 + i\sqrt{3})/2$.

    a. What is the complex amplitude of the electric field?

    b. What is the phase shift between a polarization induced by the field and the incident field?

    c. Calculate the real polarization in this medium induced by the field.

2.2 A plane wave in a vacuum has an electric field given by $\vec{E} = E_o \cos(kz - \omega t + \varphi)\hat{x}$.

    a. What is $\vec{B}$?

    b. What is the complex amplitude of $\vec{B}$?

    c. Calculate both $\vec{S}$ and $\langle \vec{S} \rangle$.

2.3 Consider two monochromatic quantities $\vec{F} = (1/2)\vec{F}e^{i\omega t} + \text{c.c.}$ and $\vec{G} = (1/2)\vec{G}e^{i\omega t} + \text{c.c.}$ Prove that the $\langle \vec{F} \cdot \vec{G} \rangle = (1/2)\text{Re}(\vec{F} \cdot \vec{G}^*)$, where $\langle \rangle$ denotes a time average.

2.4 Use Poynting's theorem and Equation 2.43 to determine the phase difference between an incident monochromatic plane-wave field and the induced polarization that leads to (a) maximum attenuation and (b) maximum gain of the incident beam. Use this result to comment on whether the polarization should lead or lag with respect to the incident field for (1) attenuation and (2) gain.

2.5 Show that absorption is linear in the sense that when two inputs at different frequencies are present in a medium that

$$\left\langle \vec{E}_{\text{total}} \cdot \frac{d\vec{P}_{\text{total}}}{dt} \right\rangle = \left\langle \vec{E}_1 \cdot \frac{d\vec{P}_1}{dt} \right\rangle + \left\langle \vec{E}_2 \cdot \frac{d\vec{P}_2}{dt} \right\rangle. \tag{2.152}$$

Use monochromatic plane waves for $\vec{E}_1(\omega_1)$ and $\vec{E}_2(\omega_2)$. The polarization's complex amplitude is related to the field via $\vec{P} = \varepsilon_0 \tilde{\chi} \vec{A}$. Assume that the susceptibility is the same for both frequencies.

2.6 Show that, when the monochromatic, complex electric field is of the form $E_0 e^{i(\vec{k} \cdot \vec{r} - \omega t)}\hat{E}$, where $\hat{E}$ is the field polarization, that the following identities are valid:

$$\vec{\nabla} \cdot \vec{E} = \text{Re}\left(i\vec{k} \cdot \vec{E}_0 e^{i(\vec{k} \cdot \vec{r} - \omega t)} \right), \tag{2.153}$$

$$\vec{\nabla} \times \vec{E} = \text{Re}\left(i\vec{k} \times \vec{E}_0 e^{i(\vec{k} \cdot \vec{r} - \omega t)} \right), \tag{2.154}$$

$$\nabla^2 \vec{E} = \mathrm{Re}\left(-(\vec{k}\cdot\vec{k})\vec{E}_0 e^{i\left(\vec{k}\cdot\vec{r}-\omega t\right)}\right), \tag{2.155}$$

$$\frac{\partial \vec{E}}{\partial t} = \mathrm{Re}\left(-i\omega\vec{E}_0 e^{i\left(\vec{k}\cdot\vec{r}-\omega t\right)}\right). \tag{2.156}$$

That is, for monochromatic plane waves, $\vec{\nabla}\to i\vec{k}$ and $\partial/\partial t \to -i\omega$.

2.7 Use the Kronecker-delta properties to prove that $\chi_{\alpha\beta} = R_{i\alpha}R_{j\beta}\chi'_{ij}$ can be "inverted" to give $\chi'_{ij} = R_{i\alpha}R_{j\beta}\chi_{\alpha\beta}$.

2.8 Use the constitutive relationship (Equation 2.5) along with Equation 2.48 to prove that $\varepsilon_{ij} = \varepsilon_0(\delta_{ij} + \chi_{ij})$.

2.9 In Chapter 3, we show that the second-order nonlinear polarization is given in component form as

$$P_i^{(2)} = \varepsilon_0\chi_{ijk}^{(2)}A_jA_k, \tag{2.157}$$

where $\chi_{ijk}^{(2)}$ is the second-order susceptibility and A_j is a component of the incident field's complex amplitude. Use Equation 2.157 to prove that $\chi_{ijk}^{(2)}$ is a rank 3 tensor.

2.10 Find the principal indices of refraction (eigenvalues) and the direction of the principal axes (eigenvectors) for the following relative dielectric tensor, $\ddot{\varepsilon}_r$:

$$\begin{bmatrix} 2.5 & 0.5 & 0 \\ 0.5 & 2.5 & 0 \\ 0 & 0 & 4 \end{bmatrix}. \tag{2.158}$$

2.11 An electric field in a material that has a dielectric tensor as given in Problem 2.10 has a complex amplitude, $\vec{E} = (\hat{x} + \hat{y} + \hat{z})(E_0/\sqrt{3})$. What is the complex amplitude of the displacement vector, $\vec{D}$? Show that $\vec{D}$ and $\vec{E}$ are not parallel, and find the angle between them.

2.12 A uniaxial crystal of indices n_o and n_z is cut so that the optic axis is perpendicular to the surface. Show that for a beam incident on the interface from air at an angle of incidence, θ_i, that the angle of refraction of the e-wave is

$$\tan\theta_e = \frac{n_z}{n_o}\frac{\sin\theta_i}{\sqrt{n_z^2 - \sin^2\theta_i}}. \tag{2.159}$$

2.13 The law of reflection in crystals is not always intuitively obvious. In a derivation similar to that for Snell's law, reflections satisfy the equation $n_i \sin\theta_i = n_r \sin\theta_r$,

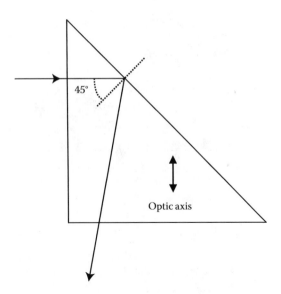

FIGURE 2.13 Illustration of a prism fabricated from a uniaxial crystal. The direction of the optic axis is shown with a double headed arrow.

where θ_i and θ_r are the incident angle and reflected angle with respect to the surface normal, respectively; n_i and n_r are the indices for the incident and reflected beams. As the following example shows, in crystals it is possible that $n_i \neq n_r$ and therefore $\theta_r \neq \theta_i$. A right prism, shown in Figure 2.13, is made out of a uniaxial birefringent crystal with the optic axis, as shown in the figure. The principal indices are $n_o = 2.2$ and $n_z = 2.1$, and the prism angles are 45° and 90°. The incident beam of light is linearly polarized in the vertical direction (parallel to the optic axis). Calculate the reflected angle with respect to the reflected surface normal (dotted line). Note that it is not 45°!

2.14 A linear optical effect that plays a role in ultrashort-pulse nonlinear optics is the group-velocity mismatch between different carrier frequencies. For example, in SHG, the fundamental and second harmonic will temporally separate due to group-velocity mismatch. The group velocity is given by $v_g = \partial\omega/\partial k$; however, it is easier to calculate inverse group velocity $v_g^{-1} = \partial k/\partial\omega$. Plot the group velocity as a function of wavelength for an o-wave traveling through β-barium borate (BBO) (see Appendix B for the Sellmeier equations). Based on the group velocity evaluated at 0.532 and 1.064 μm, over what propagation distance in the BBO will a 100-fs pulse at 0.532 μm temporally separate from a 100-fs 1.064-μm pulse? Define temporal separation when the pulses are separated by 100 fs.

2.15 Derive the wave equation in an isotropic medium for the case where j_f is not equal to zero. Assume monochromatic plane waves and use Ohm's law to relate the complex amplitudes of the free current density and the field, that is, $\vec{j_f} = \sigma\vec{A}$.

   a. Show that the resultant index of refraction is complex.

   b. Show that the imaginary part of the index leads to attenuation as the field propagates through the medium.

TABLE 2.1	Index as a Function of Wavelength
Wavelength (μm)	Index of Refraction
0.330	2.47379
0.532	2.23421
0.6328	2.20271
1.064	2.15601
1.57	2.1375
2.804	2.1029

2.16 A new transparent material has been discovered, and before nonlinear optics experiments can be performed, we need to know its dispersion curve. Measurements of the index of refraction are made at five wavelengths (using five different lasers). Table 2.1 gives the results of the measurements. Fit the above data using a two-pole Sellmeier equation,

$$n^2(\lambda) = A_0 + \frac{\lambda^2 A_1}{\lambda^2 - A_2} + \frac{\lambda^2 A_3}{\lambda^2 - A_4}, \tag{2.160}$$

where λ is kept in units of microns. The stability of the fitting routine requires reasonable guesses for the fitting parameters. A_2 and A_4 correspond to the IR and UV poles; so set the starting guesses accordingly. If we assume that the dispersion is relatively "slow," then the index should not be a strong function of wavelength. From the data, we see that it has a value centered on $n(\lambda) \sim 2.2$. Hence a good guess for A_0 is ~2.2^2. After performing the fitting routine, give the values for the fit parameters A_0 through A_4. Plot the measured and fitted index of refraction as a function of wavelength on the same graph. Check to be sure that A_2 and A_4 give poles in the IR and UV, respectively.

2.17 Use Equation 2.104 to show that the two eigenpolarizations are orthogonal. (*Hint:* Show that the D-vectors corresponding to the two eigenindices are orthogonal.)

2.18 Calculate the angle where $n_e = n_o$ in the xz-plane of a biaxial crystal. This angle is the angle of the optic axis with respect to the z-axis. Use the Sellmeier equations for potassium titanyl phosphate (KTP) to plot the angle of the optic axis with respect to the z-axis as a function of wavelength from 0.5 to 4.0 μm.

2.19 Give expressions for n_e in each of the principal planes of a biaxial crystal. In the xz- and yz-planes, assume that the k-vector propagates at an angle, θ, with respect to the z-axis. In the xy-plane, assume that the k-vector propagates at an angle, ϕ, with respect to the x-axis.

2.20 Nonlinear optical interactions typically involve multiple beams at multiple wavelengths. If the beams have different polarizations, then they may walk off each other as the following example demonstrates. Two beams of light, one at 600 nm (o-polarized) and the other at 300 nm (e-polarized), are sent through a BBO crystal with their k-vectors at an angle of 40.5° to the optic axis. The BBO crystal is 2 cm long. The beam diameters are the same and are given by "D." Determine the distance in the crystal where the two beams walk off each other. Define beam separation when the centers of the beams are displaced by one beam diameter. For what beam diameters is the full crystal length used? The Sellmeier equations for BBO are found in Appendix B.

2.21 In the following, use the Sellmeier equations for BBO found in Appendix B.
  a. Make a plot $n_e(\theta)$ from $\theta = 0$ to $\pi/2$ for both $\lambda = 1.064\ \mu m$ and $\lambda = 0.532\ \mu m$. Include on the same graph $n_o(\lambda)$ for both wavelengths.
  b. Is BBO positive or negative uniaxial?
  c. For what polarization states and for what angle does $n(1.064) = n(0.532)$? (*Hint:* This corresponds to the intersection of two of the curves in (a). This angle is the phase-matching angle for SHG.)

2.22 For nonlinear interactions in birefringent crystals, walk-off is an important factor. Plot the walk-off angle (in degrees) as a function of crystal orientation (from 0° to 90°) for the BBO crystal (see Appendix B for Sellmeier equations). Use a wavelength of 0.6 μm. The orientation of the crystal is defined as the angle between the k-vector and the optic axis of the crystal. Where is walk-off zero?

2.23 Using the geometry in Figure 2.14, prove the result for the walk-off angle in Equation 2.110. Show the equivalence of the result with Equation 2.111.

2.24 Assume that a beam waist is incident on a lens of focal length, f. The beam propagates a distance d, where it is incident on a crystal that has an index n. Provided that d is less than the focal length, prove that the focused spot size in the crystal is the same as the focused spot size without a crystal. Also assume that the initial beam waist incident on the lens is large enough that the focused spot in air is a distance of approximately f away from the lens.

2.25 Plot $w(z)$ and $1/R(z)$ for a Gaussian beam. Plotting the inverse of $R(z)$ avoids the divergence at $z = 0$ where the radius of curvature is infinite. Assume that the wavelength is 500 nm, and that $w_0 = 100\ \mu m$. Plot from $z = -10z_R$ to $z = +10z_R$ where z_R is the Rayleigh range.

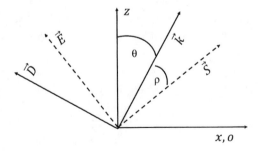

FIGURE 2.14 Geometry for determining the walk-off angle. The x and z are principal axes with the x-axis with index n_o. The wave vector, Poynting vector, electric field, and displacement field axes are labeled.

2.26 Use the result for the electric field amplitude in Equation 2.124 to calculate the intensity of a beam defined by Equation 2.20. Calculate the optical power in the beam by integrating over the transverse coordinates. Comment on the circumstances where power is conserved.

2.27 Three Gaussian laser beams are focused to have the same beam waist, $w_0 = 200$ μm, at $z = 0$ (radius of curvature $= \infty$). The wavelength of the three lasers are $\lambda_1 = 250$ nm, $\lambda_2 = 500$ nm, and $\lambda_3 = 1000$ nm. Find the ratio of the Rayleigh ranges for Laser1/Laser2, and Laser1/Laser3, that is, (z_{R1}/z_{R2}) and (z_{R1}/z_{R3}). Use the result to comment on beam overlap issues that arise for nonlinear interactions that require the overlap of beams with different wavelengths.

2.28 Consider the optical system shown in Figure 2.15. A Gaussian beam waist (w_0) is located at a distance d_1 from a lens of focal length f. A distance d_2 on the other side of the lens is a second beam waist.

a. Use the Gaussian beam parameter, q, and the ABCD matrix approach to show that the distance, d_2, is given by the expression,

$$d_2 = \frac{\left[(d_1 - f)d_1 + z_R^2\right]f}{(f - d_1)^2 + z_R^2},$$
(2.161)

where z_R is determined by the initial beam waist w_0 and is given by $z_R = \pi w_0^2 / \lambda$. (*Hint:* At a beam waist, what is the radius of curvature and therefore what should the real part of $1/q$ be?)

b. Show that the beam waist located at d_2 is given by the relationship,

$$w_{final}^2 = \left[\frac{f^2 z_R}{(f - d_1)^2 + z_R^2}\right]\frac{\lambda}{\pi}.$$
(2.162)

c. Consider a laser operating at 1.064 μm that has a beam waist of 200 μm at a location of 1 cm after the laser exit port. A certain nonlinear application requires us to focus this beam to $w = 130$ μm. At what location(s) can we put a 10-cm focal length lens to obtain the desired spot size?

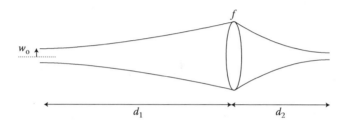

FIGURE 2.15   Transformation of a Gaussian beam using a thin lens.

2.29 Generate the solution to Equation 2.122 following the prescriptions:

a. Fourier transform the transverse coordinates using the definition,

$$A(\vec{r},z) = \iint \tilde{A}(\vec{q},z) e^{i(q_x x + q_y y)} dq_x \, dq_y ,$$ (2.163)

to write and equation for the function $\tilde{A}(\vec{q},z)$.

b. Defining the function at $z = 0$,

$$A(\vec{r},0) = A_0 \exp\left(\frac{-r^2}{w_0^2}\right),$$ (2.164)

find its Fourier transform using the inverse transform,

$$\tilde{A}(\vec{q},0) = \frac{1}{(2\pi)^2} \iint A(\vec{r},0) e^{-i(q_x x + q_y y)} dx \, dy.$$ (2.165)

c. Solve the differential equation in (a) using the amplitude $\tilde{A}(\vec{q},0)$ to fix the value at $z = 0$.

d. Take the Fourier transform of $\tilde{A}(\vec{q},z)$ to find the result in Equation 2.124.

2.30 Table 2.2 gives data for the beam radius (1/e field radius) of a laser beam as a function of position. The wavelength of the laser is 1.064 µm.

a. Plot w_M versus z, and on the same plot include the curve,

$$w_M(z) = w_0 \left[1 + M^4 \left(\frac{\lambda(z - z_0)}{\pi w_0^2} \right)^2 \right]^{1/2}.$$ (2.166)

For this part of the problem (a), set $M = 1$. For the values of w_0, and z_0, estimate them based on the data. *Note:* z_0 is the location of the minimum beam waist and w_0 is the minimum radius. Tweak the values of w_0 and z_0 to obtain a best fit for all the data. Include the best values of z_0 and w_0 on the plot.

b. Using the results of (a) as a starting point, perform a fit of the function $w(z)$ above to the data to determine best values for w_0, z_0, and M^2.

2.31 An elliptical Gaussian beam at the focal plane has a field amplitude,

$$A(\vec{r}_\perp,0) = A_0 \exp\left(-\frac{x^2}{w_{x0}^2} - \frac{y^2}{w_{y0}^2} \right).$$ (2.167)

TABLE 2.2 Beam Radius as a Function of Position	
W (µm)	z (cm)
54.4	0.125
44.5	0.135
39.5	0.145
32.5	0.155
26.2	0.165
21.9	0.175
16.3	0.185
12.0	0.195
10.6	0.205
11.3	0.215
14.1	0.225
19.1	0.235
24.0	0.245
31.1	0.255
37.5	0.265
41.7	0.275
48.1	0.285
53.0	0.295
58.0	0.305

a. Using the SVEA equation (Equation 2.122), calculate beam widths $w_x^2(z)$ and $w_y^2(z)$. Use the procedure outlined in Problem 2.29.

b. Using the result of (a), calculate the near-field and far-field beam width product for the x- and y-axes. From these results, infer the asymptotic product for the elliptical beam is Equation 2.147.

2.32 A flat-top beam has a radius a at $z = 0$.

a. Calculate the beam in the far field using the Fraunhofer diffraction formula.

b. Describe why the second moment of the intensity in the far field does not exist for this beam.

c. Use the second moment of the beam in the near field to define the beam width and the first zero of the beam in the far field to define the far-field width and calculate the product of the two widths. Compare your result with the Gaussian beam product and infer a value of M^2.

d. Using the M^2 value from (c), calculate the far-field Gaussian intensity using the width $w_M(z)$ and compare the profile with the flat-top intensity profile in the far field. Comment on the accuracy of the M^2 fit to the exact beam profile?

REFERENCES

Dmitriev, V. G., G. G. Gurzadyan, and D. N. Nikogosyan. 1997. *Handbook of Nonlinear Optical Crystals* (2nd edition). Heidelberg, Germany: Springer-Verlag.

Griffiths, D. J. 1999. *Introduction to Electrodynamics* (3rd edition). Upper Saddle River, NJ: Prentice-Hall.

Pedrotti, F. L., L. S. Pedrotti, and L. M. Pedrotti. 2007. *Introduction to Optics* (3rd edition). Upper Saddle River, NJ: Prentice-Hall.

Ruff, J. A. and A. E. Siegman. 1992. Single pulse laser beam quality measurements using a CCD camera system. *Applied Optics* 31: 4907–4909.

Symon, K. R. 1971. *Mechanics* (3rd edition). Reading, MA: Addison-Wesley.

FURTHER READING

Born, M. and E. Wolf. 1980. *Principles of Optics* (6th edition). Cambridge, UK: Pergamon Press.

Goodman, J. W. 2004. *Introduction to Fourier Optics*. Greenwood Village, CO: Roberts & Company Publishers.

Roberts, D. A. 1992. Simplified characterization of uniaxial and biaxial nonlinear optical crystals: A plea for standardization of nomenclature and conventions. *IEEE Journal of Quantum Electronics* 28: 2057–2074.

Saleh, B. E. A. and M. C. Teich. 1991. *Fundamentals of Photonics*. New York, NY: John Wiley & Sons.

Siegman, A. E. 1986. *Lasers*. Mill Valley, CA: University Science Books.

Yariv, A. and P. Yeh. 2003. *Optical Waves in Crystals, Propagation and Control of Laser Radiation*. Hoboken, NJ: John Wiley & Sons.

3

INTRODUCTION TO THE NONLINEAR SUSCEPTIBILITY

3.1 INTRODUCTION

As we discussed in Chapter 2, a polarization induced in a material by an optical field is key to understanding linear optical properties. This observation carries over to nonlinear optics, as well. Poynting's theorem, Equation 2.19, involves the total polarization, which includes both linear and nonlinear effects. When nonlinearities are included in the polarization expression, energy exchange between optical inputs at different frequencies becomes possible. For example, an input field at a frequency ω can induce a nonlinear polarization with a component at the second harmonic, 2ω. The polarization at 2ω in turn drives a field that radiates energy at the second-harmonic frequency. The nonlinear polarization is central to second-harmonic generation (SHG), and it is central to understanding all nonlinear phenomena.

In this chapter, we introduce the nonlinear polarization through a field-dependent susceptibility. We justify this connection using a classical picture based on an anharmonic oscillator model. The connection between the polarization and fields is expanded to include more possible field couplings by treating the susceptibility as a tensor. This allows, for example, couplings between a nonlinear polarization pointing in the x-direction and an incident field that points in the y-direction. The total nonlinear polarization we introduce encompasses a host of nonlinear interactions. We restrict our discussion by focusing primarily on and developing a formalism for the nonlinear susceptibility for the so-called second-order effects. A detailed discussion on higher-order effects is deferred to later chapters, but much of the same techniques used here carry over to the higher-order effects. We also defer the discussion on the connection between the nonlinear susceptibility and the wave equation to later chapters. The final section of this chapter is applications oriented. It focuses on one particular second-order process, the electro-optic (EO) effect, which describes one way to control the index of refraction through the nonlinear properties of a material.

3.1.1 NONLINEAR POLARIZATION

The polarization induced in a medium is related to an incident field through the susceptibility, χ. The susceptibility is introduced in Chapter 2 with the assumption that it is independent of the fields. A more complete description of susceptibility includes its dependence on the electric field strength. In the simplest case, we introduce the field dependence when χ and the electric field are represented by scalar real quantities. This approach lets us focus on the key nonlinear concepts while minimizing notational complexity. In subsequent sections, we cover the more general aspects of the nonlinear susceptibility. Assuming weak electric field dependence, as introduced in Chapter 1, we write the susceptibility, χ, as a power-series expansion:

$$\chi = \chi^{(1)} + \chi^{(2)}\underline{E}(t) + \chi^{(3)}\underline{E}^2(t) + \cdots. \tag{3.1}$$

Here, $\chi^{(1)}$, $\chi^{(2)}$, ... represent the various expansion orders of susceptibility, and the superscript numbers in parentheses should not be confused with exponentiation. The underscore under the scalar electric field indicates that it represents a real field rather than complex amplitude. The units of the higher-order susceptibilities are m/V for $\chi^{(2)}$, m^2/V^2 for $\chi^{(3)}$, and so on. The validity of Equation 3.1 requires that the successive orders in the expansion decrease rapidly in amplitude, as discussed in Chapter 1:

$$\left|\chi^{(1)}\right| \gg \left|\chi^{(2)}\underline{E}(t)\right| \gg \left|\chi^{(3)}\underline{E}^2(t)\right|\cdots. \tag{3.2}$$

Using the expansion in Equation 3.1 results in a real polarization:

$$\underline{P} = \varepsilon_0\chi^{(1)}\underline{E}(t) + \varepsilon_0\chi^{(2)}\underline{E}^2(t) + \varepsilon_0\chi^{(3)}\underline{E}^3(t) + \cdots. \tag{3.3}$$

By writing all the quantities in Equation 3.3 as real, we impose a specific phase relationship between the electric field and the total polarization. As noted in Chapter 2, writing χ as a purely real number indicates that an electric field induces a polarization that is in phase with it. The notation in Equation 3.3 for the first-order (linear) term is used for convenience to show the dependence of χ on field strength. For more general linear properties, the reader is referred back to Chapter 2. Assuming that the susceptibilities in the expansion are real is equivalent to assuming that the polarization response to the incident field is instantaneous. That is, the polarization depends on the electric field at the present time and not on its past history. This assumption is reasonable for many nonlinear interactions such as SHG, difference-frequency generation (DFG), and sum-frequency generation (SFG), all of which respond on a sub-fs timescale. Other nonlinear phenomena such as photorefraction require a description for the nonlinear polarization that includes past history. The instantaneous response approximation is also equivalent to ignoring dispersion in higher-order susceptibilities. We formalize this assumption later in this chapter with a discussion on Kleinman's symmetry.

The nonlinear polarization given in Equation 3.3 covers parametric interactions involving interactions between optical fields. Although Equation 3.3 is valid for many important applications, it is not a complete description of the total polarization in a material, and it is not always valid. The Raman scattering, covered in Chapter 9, is one case for which χ is expanded in terms of a molecular vibration coordinate instead of the optical field. A case for which the power-series expansion breaks down is saturable absorption.

3.1.2 PARAMETRIC PROCESSES

For processes where the expansion in Equation 3.3 is valid, nonlinear effects are illustrated by considering an incident field consisting of two linearly polarized monochromatic plane waves traveling in the z-direction. We represent the total input field as a superposition of the two scalar fields:

$$\underline{E} = [E_2 \cos(k_2 z - \omega_2 t + \phi_2) + E_3 \cos(k_3 z - \omega_3 t + \phi_3)]. \tag{3.4}$$

We use the convention that $\omega_2 > \omega_3$. Looking at the lowest-order nonlinearity in Equation 3.3, the $\chi^{(2)}$ term gives a response proportional to the field squared:

$$\underline{P}^{(2)} = \varepsilon_0 \chi^{(2)} \underline{E}^2 = \varepsilon_0 \chi^{(2)} \Big[E_2^2 \cos^2(k_2 z - \omega_2 t + \phi_2) + E_3^2 \cos^2(k_3 z - \omega_3 t + \phi_3 $$
$$+ 2E_2 E_3 \cos(k_2 z - \omega_2 t + \phi_2)\cos(k_3 z - \omega_3 t + \phi_3) \Big]. \tag{3.5}$$

Equation 3.5 is rewritten in terms of specific nonlinear mixing processes by expanding it with trigonometric identities:

$$\underline{P}^{(2)} = \varepsilon_0 \chi^{(2)} \begin{bmatrix} \dfrac{1}{2} E_2^2 \cos\left[2(k_2 z - \omega_2 t + \phi_2)\right] & \text{Second harmonic of } \omega_2 \\[2ex] + \dfrac{1}{2} E_3^2 \cos\left[2(k_3 z - \omega_3 t + \phi_3)\right] & \text{Second harmonic of } \omega_3 \\[2ex] + \dfrac{1}{2}(E_2^2 + E_3^2) & \text{Optically rectified field} \\[2ex] + E_2 E_3 \cos\left([k_2 + k_3]z - [\omega_2 + \omega_3]t + [\phi_2 + \phi_3]\right) & \text{Sum frequency of } \omega_2 \text{ and } \omega_3 \\[2ex] + E_2 E_3 \cos\left([k_2 - k_3]z - [\omega_2 - \omega_3]t + [\phi_2 - \phi_3]\right) & \text{Difference frequency} \\ & \text{between } \omega_2 \text{ and } \omega_3 \end{bmatrix}.$$

$$\tag{3.6}$$

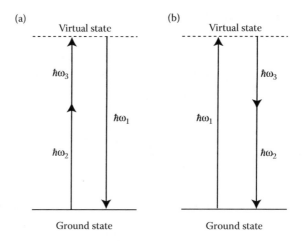

FIGURE 3.1 Visualization of the different mixing processes in terms of photon energies. (a) SFG and (b) DFG. SHG is a special case of SFG where the input photons have the same energy. Optical rectification is a special case of DFG where $\omega_1 = \omega_2$.

The expressions in Equation 3.6 show that two incident plane waves generate non-linear polarizations at the sum and difference of all frequencies involved. A visualization of the processes in Equation 3.6 with respect to photon energy is shown in Figure 3.1. With most $\chi^{(2)}$ processes, we assume that the photon energies are not resonant with any real electronic states of the material. Therefore, the excited state is a transient state, which is called a virtual state. Figure 3.2 represents another visualization in terms of photons for the processes in Equation 3.6. The photons are represented by wiggly lines in Figure 3.2. Note that in a difference-frequency process, two photons are present at the input (at ω_1 and ω_2) and three photons are present at the output: one at the difference frequency, ω_3, and two at ω_2. Hence, a difference-frequency process is also an optical parametric amplifier for ω_2.

As with linear properties of materials, the complex representation of a field is a convenient and efficient way to express nonlinear processes. In the case of nonlinear

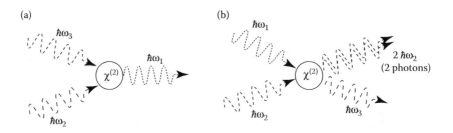

FIGURE 3.2 Visualizations of three-wave interactions. Each wiggly line represents one photon. The circle represents the nonlinear medium. (a) SFG and (b) DFG. Note that for DFG, the ω_1 photon splits into an ω_2 and an ω_3 photon; hence, the ω_2 photon is amplified in the process. So, DFG is also an optical parametric amplifier for ω_2.

interactions, which involve the square of the field, complex notation is used and the complex conjugate term is explicitly included:

$$P^{(2)} = \varepsilon_0 \chi^{(2)} \underline{E}_{tot}^2 = \varepsilon_0 \chi^{(2)} \left(\frac{1}{2} \sum_n A(\omega_n) e^{-i\omega_n t} + \text{c.c.} \right)^2 . \tag{3.7}$$

Here, $A(\omega_n)$ are the complex amplitudes for each frequency component, ω_n. The possibility of more than two fields is included by summing over the total number of distinct fields. Writing the electric field in the complex form of Equation 3.7 includes a factor of 1/2 in front of the complex amplitudes. A real field written as

$$\underline{E} = E_0 \cos(kz - \omega t + \phi) \tag{3.8}$$

is rewritten in complex form as

$$\boxed{\underline{E} = \frac{1}{2} A e^{i(kz - \omega t)} + \text{c.c.,}} \tag{3.9}$$

where

$$A = E_0 e^{i\phi}. \tag{3.10}$$

Note that some authors define the complex field amplitudes (A) differently by writing the field without the factor of 1/2:

$$\underline{E} = A' e^{i(kz - \omega t)} + \text{c.c.} \tag{3.11}$$

We observe some ambiguity in defining the complex amplitude, and so it is important to be aware of which definition is being used. This issue is especially important when fields are squared, as in Equation 3.7, and in relating intensity to the complex field amplitude. The intensity using the complex field definition in Equation 3.9 is (see also Section 2.1.4)

$$\boxed{I = \frac{n|A|^2}{2\mu_0 c},} \tag{3.12}$$

whereas using the definition in Equation 3.11 gives

$$I = \frac{2n|A'|^2}{\mu_0 c}. \tag{3.13}$$

Of course, the intensity using either definition yields the same value. This book exclusively uses the definition of the field amplitude in Equation 3.9 and the corresponding intensity as given in Equation 3.12. This definition is also commonly used in the literature, and the values given for nonlinear coefficients will correspond to this definition, as well.

3.2 CLASSICAL ORIGIN OF THE NONLINEARITY

The origin of the nonlinear susceptibility is most completely described by quantum mechanics, but many of its qualitative properties can be obtained using a classical model. In the classical model, the interaction of an incident field with a material is primarily treated as an interaction with the electrons in the host material. For dielectric media in the classical model, the medium's electrons are held in equilibrium by a potential well. The shape of the well is, to the lowest order, quadratic. However, as the incident field strength increases, the electron is pushed further from equilibrium and the potential's deviation from a pure quadratic is noticeable. Classically, an electron in a potential well experiences a restoring force given by

$$\vec{F} = -\vec{\nabla}U, \tag{3.14}$$

where U is the potential energy. Electrons held in a quadratic potential well therefore experience a linear restoring force. This restoring force is identical in form to that of a mass-on-spring harmonic oscillator. As higher-order terms in the potential are added, the restoring force for the electron is modeled as a mass on a nonlinear spring, and the system is called an anharmonic oscillator. The classical anharmonic response is a perturbation to the linear harmonic oscillator solution.

3.2.1 ONE-DIMENSIONAL LINEAR HARMONIC OSCILLATOR

In the classical model, an electron of mass m and charge -e sits in a one-dimensional quadratic potential well. For purely linear interactions, the potential energy of the electron is quadratic in its displacement, $\underline{X}$:

$$U = \frac{1}{2} m\omega_0^2 \underline{X}^2, \tag{3.15}$$

where ω_0 denotes a characteristic angular frequency. The restoring force, $\underline{F}_r$, on the electron is, from Equation 3.14,

$$\underline{F}_r = -m\omega_0^2 \underline{X}. \tag{3.16}$$

Equation 3.16 is Hooke's law with $m\omega_0^2$ acting as the effective spring constant; ω_0 is interpreted as a resonant frequency for the mass-spring system.

Since an electron in the material is not completely isolated, other forces act on it, resulting in a damping force proportional to its velocity. We lump the damping forces together as

$$\underline{F}_d = -m\Gamma \frac{d\underline{X}}{dt}, \tag{3.17}$$

where Γ is the phenomenological damping rate. When an external electric field is applied to the system, the electron experiences an additional force given by

$$\underline{F}_E = -e\underline{E}(t). \tag{3.18}$$

The forces in Equations 3.16 through 3.18 are used in Newton's second law to give

$$m\frac{d^2\underline{X}}{dt^2} = -m\Gamma \frac{d\underline{X}}{dt} - m\omega_0^2\underline{X} - e\underline{E}. \tag{3.19}$$

The electric field is assumed to be monochromatic and linearly polarized so that it is written in terms of its complex amplitude as

$$\underline{E} = \frac{1}{2}A e^{-i\omega t} + \text{c.c.} \tag{3.20}$$

The linear response to the field is an electronic displacement that oscillates at the same frequency as the driving field but possibly out of phase:

$$\underline{X} = \frac{1}{2}x e^{-i\omega t} + \text{c.c.} \tag{3.21}$$

Here, the phase shift of the displacement is incorporated in the time-independent, complex amplitude, x. Substituting Equation 3.21 into the harmonic oscillator equation, Equation 3.19, lets us solve for the complex amplitude:

$$x = -\frac{eA}{mD(\omega)}, \tag{3.22}$$

where

$$D(\omega) = \omega_0^2 - \omega^2 - i\omega\Gamma. \tag{3.23}$$

Equation 3.22 gives a displacement for the linear response. We assume that the electrons in a material are independent of each other so that the linear polarization's complex amplitude is related to the displacement by

$$P = -Nex = \frac{Ne^2 A}{mD(\omega)},$$
(3.24)

where N is the number density of "electron oscillators" in the medium, and local field corrections are ignored. The linear polarization and field complex amplitudes are also related by the expression

$$P = \varepsilon_0 \chi^{(1)} A.$$
(3.25)

The superscript in the notation $\chi^{(1)}$ indicates the first-order, or linear, susceptibility. Equating Equation 3.25 to Equation 3.24 and solving for $\chi^{(1)}$ give

$$\chi^{(1)} = \frac{Ne^2}{\varepsilon_0 mD(\omega)}.$$
(3.26)

Equation 3.26 gives the form of dispersion for the index of refraction via Equation 2.87:

$$n^2 = 1 + \chi^{(1)},$$
(3.27)

where we assume that the frequency is far from resonance so that the imaginary part of $\chi^{(1)}$ is small and thus absorption is negligible. A closer look at many Sellmeier equations, discussed in Chapter 2, reveals that the form of dispersion comes from Equations 3.26 and 3.27, with the possibility of several resonant frequencies instead of one.

3.2.2 ONE-DIMENSIONAL ANHARMONIC OSCILLATOR

In the above derivation, we assume that the electron resides in a quadratic potential well. We improve the approximation by including higher-order contributions to the potential:

$$U = \frac{1}{2} m\omega_0^2 \underline{X}^2 + \frac{1}{3} am\underline{X}^3 + \frac{1}{4} bm\underline{X}^4 + \cdots,$$
(3.28)

where "a" and "b" are the expansion coefficients for the particular shape of the potential well. In materials that exhibit inversion symmetry, called centrosymmetric materials, $U(\underline{X}) = U(-\underline{X})$, and the first-order correction must be zero, that is, $a = 0$. In the general case, we include the first two corrections to the quadratic potential well giving a restoring force of

$$\underline{F}_r = -m\omega_0^2\underline{X} - ma\underline{X}^2 - mb\underline{X}^3. \tag{3.29}$$

Newton's second law is similar to Equation 3.19 with the addition of the new correction terms:

$$\frac{d^2\underline{X}}{dt^2} + \Gamma\frac{d\underline{X}}{dt} + \omega_0^2\underline{X} + a\underline{X}^2 + b\underline{X}^3 = -\eta\frac{eE}{m}. \tag{3.30}$$

The parameter η is a dimensionless number equal to 1, and it is introduced to identify the order of the expansion in subsequent calculations. For a small displacement, a perturbation approach is taken and the displacement is written as a power-series expansion:

$$\underline{X} = \eta\underline{X}^{(1)} + \eta^2\underline{X}^{(2)} + \eta^3\underline{X}^{(3)} + \cdots. \tag{3.31}$$

By substituting Equation 3.31 into Equation 3.30, we obtain an equation with terms multiplying η, η^2, and so on. Separating out the expressions for the orders of η gives

$$\eta : \frac{d^2\underline{X}^{(1)}}{dt^2} + \Gamma\frac{d\underline{X}^{(1)}}{dt} + \omega_0^2\underline{X}^{(1)} = -\frac{eE}{m}, \tag{3.32}$$

$$\eta^2 : \frac{d^2\underline{X}^{(2)}}{dt^2} + \Gamma\frac{d\underline{X}^{(2)}}{dt} + \omega_0^2\underline{X}^{(2)} + a(\underline{X}^{(1)})^2 = 0, \tag{3.33}$$

$$\eta^3 : \frac{d^2\underline{X}^{(3)}}{dt^2} + \Gamma\frac{d\underline{X}^{(3)}}{dt} + \omega_0^2\underline{X}^{(3)} + 2a\underline{X}^{(1)}\underline{X}^{(2)} + b(\underline{X}^{(1)})^3 = 0. \tag{3.34}$$

Note that in centrosymmetric media such as glasses and gases, $a = 0$, so that the quadratic restoring force in Equation 3.30 is zero and the second-order response vanishes, $\underline{X}^{(2)} = 0$. If Equations 3.32 through 3.34 along with higher-order equations in η are all satisfied, then Equation 3.30 is also satisfied. Provided that the expansion given in Equation 3.31 is valid, then the successive orders of η indicate successively decreasing contributions to the displacement. In the perturbation approach, the lowest-order equation in η is first solved and then that result is used in the next highest order. That is, Equation 3.32 is used to find $\underline{X}^{(1)}$, which is then substituted into Equation 3.33 to find $\underline{X}^{(2)}$. $\underline{X}^{(2)}$ is used to find the next highest-order perturbation and so on. We solved the lowest-order equation, Equation 3.32, in Section 3.2.1 with a solution given by Equation 3.22. Before moving on to the next highest-order solution, we extend the linear solution to the case where the input consists of a superposition of two fields at different frequencies:

$$E = \frac{1}{2} A_1 e^{-i\omega_1 t} + \frac{1}{2} A_2 e^{-i\omega_2 t} + \text{c.c.} \tag{3.35}$$

This field drives the displacement, and to first-order in η, the total linear displacement is

$$\underline{X}^{(1)} = \frac{1}{2} x_1^{(1)} e^{-i\omega_1 t} + \frac{1}{2} x_2^{(1)} e^{-i\omega_2 t} + \text{c.c.}, \tag{3.36}$$

where $x_1^{(1)}$ and $x_2^{(1)}$ are the complex amplitudes of the linear displacements, each of which has a form given by Equation 3.22. The linear displacement appears as a term in the order η^2 equation, Equation 3.33. To find the second-order correction to the displacement, we substitute Equation 3.36 into Equation 3.33:

$$\frac{d^2 \underline{X}^{(2)}}{dt^2} + \Gamma \frac{d\underline{X}^{(2)}}{dt} + \omega_0^2 \underline{X}^{(2)} + a \left(\frac{1}{2} x_1^{(1)} e^{-i\omega_1 t} + \frac{1}{2} x_2^{(1)} e^{-i\omega_2 t} + \text{c.c.} \right)^2 = 0. \tag{3.37}$$

In the last terms, the linear displacements are squared yielding an inhomogeneous driving term with components at the sum and difference frequencies of the two input frequencies (including second harmonics and a 0-frequency or rectification term) just as in Equation 3.6. Since second-order contributions to the overall displacement are small perturbations to the linear motion, the different frequency components are considered independent of each other. For example, the presence of the second harmonic at $2\omega_1$ does not interfere with the difference-frequency term at $\omega_1 - \omega_2$. Therefore, the overall second-order displacement is a superposition of the displacements occurring at different frequencies. When the propagation of the waves is included, other factors will make it possible for one nonlinear mixing process to dominate over the others. A major discriminator is an effect called phase matching (see Chapter 4). Thus, rather than considering all possible nonlinear frequency combinations, we will be able to choose one of them.

Equation 3.37 is solved for a particular frequency from the squared term. In what follows, we work with only the difference frequency at $\omega_1 > \omega_3 = \omega_1 - \omega_2$ so that the second-order displacement is written as $\underline{X}^{(2)}(\omega_3)$. Note that we use the convention that $\omega_1 > \omega_2 \geq \omega_3$.

The second-order displacement at ω_3 is solved using Equation 3.37:

$$\frac{d^2 \underline{X}^{(2)}}{dt^2} + \Gamma \frac{d\underline{X}^{(2)}}{dt} + \omega_0^2 \underline{X}^{(2)} = -\left(a\frac{1}{2} x_1^{(1)} x_2^{(1)*} e^{-i\omega_3 t} + \text{c.c.} \right). \tag{3.38}$$

Equation 3.38 is in the form of a linear harmonic oscillator with a driving term at ω_3. The solution is found by the same procedure used to obtain the first-order solution. We take the displacement to be

$$\underline{X}^{(2)} = \frac{1}{2} x^{(2)} e^{-i\omega_3 t} + \text{c.c.} \tag{3.39}$$

and use this form in Equation 3.38. After evaluating the time derivatives and simplifying, we find

$$[-\omega_3^2 x^{(2)} - i\omega_3 \Gamma x^{(2)} + \omega_0^2 x^{(2)} + a x_1^{(1)} x_2^{(1)*}] e^{-i\omega_3 t} + \text{c.c.} = 0. \tag{3.40}$$

The linear complex amplitudes $x_1^{(1)}$ and $x_2^{(1)}$ are obtained from Equation 3.22 so that the second-order solution for the complex amplitude is

$$x^{(2)}(\omega_3) = \frac{-ae^2 A_1 A_2^*}{m^2 D(\omega_1) D^*(\omega_2) D(\omega_3)}. \tag{3.41}$$

We solve for $\chi^{(2)}$ using $x^{(2)}$ in the same way that we found $\chi^{(1)}$ using $x^{(1)}$. The complex amplitude for the second-order nonlinear polarization is

$$P^{(2)} = -Nex^{(2)}. \tag{3.42}$$

The complex amplitude, $P^{(2)}$, is found from Equation 3.7 by keeping only the difference-frequency term,

$$P^{(2)} = \varepsilon_0 \chi^{(2)} A_1 A_2^*. \tag{3.43}$$

Using Equations 3.43 and 3.42, we find

$$\chi^{(2)}(\omega_3; \omega_1, -\omega_2) = \frac{Nae^3}{\varepsilon_0 m^2 D(\omega_1) D^*(\omega_2) D(\omega_3)}. \tag{3.44}$$

The notation, $\chi^{(2)}(\omega_3; \omega_1, -\omega_2)$, indicates a dependence on three frequencies. The first frequency corresponds to nonlinear polarization, and the second two correspond to the frequencies of the two inputs. The negative sign before ω_2 indicates a complex conjugate dependence in Equation 3.44. The second-order susceptibility is related to the first-order susceptibility by noting that $D(\omega)$ is related to $\chi^{(1)}$ via Equation 3.26. With this substitution, Equation 3.44 becomes

$$\chi^{(2)}(\omega_3; \omega_1, -\omega_2) = \frac{am\varepsilon_0^2}{N^2 e^3} \chi^{(1)}(\omega_1)\chi^{(1)*}(\omega_2)\chi^{(1)}(\omega_3). \tag{3.45}$$

For cases where the frequencies are not near an absorption, $\chi^* = \chi$. Equation 3.45 shows that the classical anharmonic oscillator model gives a connection between first-order and second-order susceptibilities. Equation 3.45 is generalized to take into account the tensor nature of a material,

$$\chi_{ijk}^{(2)}(\omega_1; \omega_2, \omega_3) = \chi_{ii}^{(1)}\chi_{jj}^{(1)}\chi_{kk}^{(1)}\Delta_{ijk}. \tag{3.46}$$

Equation 3.46 is known as Miller's rule (Miller, 1964). The quantity Δ_{ijk} is called Miller's delta, and in many cases from one material to another, Δ is nearly constant. Since there is a relationship between the refractive index and $\chi^{(1)}$, Miller's rule shows the general trend that materials with a larger the index of refraction, also possess a larger the second-order susceptibility. Furthermore, by holding Δ constant, Miller's rule is used to estimate the $\chi^{(2)}$ dispersion characteristics (see Problem 3.7).

3.2.3 THIRD-ORDER EFFECTS IN CENTROSYMMETRIC MEDIA

The procedure for going to the next order, $\chi^{(3)}$, is similar to that for $\chi^{(2)}$. We substitute the second-order and first-order solutions into Equation 3.34 and then solve for $\underline{X}^{(3)}$. In noncentrosymmetric media, the second-order displacement term in Equation 3.31 dominates over the third-order term. We single out the third-order displacement by considering centrosymmetric media, where the second-order displacement is zero (the expansion coefficient, $a = 0$ in Equation 3.30). In this case, the third-order response, $\underline{X}^{(3)}$, is related to $(\underline{X}^{(1)})^3$. Note that when cubing the first-order displacement, several possible couplings of the input amplitudes may yield the same displacement frequency. For example, consider the case when two input frequencies at ω_1 and ω_2 are present and we are interested in a nonlinear displacement at the same frequency as one of the inputs, ω_1. We obtain the product

$$(\underline{X}^{(1)})^3 = \frac{1}{8}\left(x_1^{(1)}e^{-i\omega_1 t} + x_2^{(1)}e^{-i\omega_2 t} + c.c.\right)$$
$$\times \left(x_1^{(1)}x_1^{(1)*}e^{i(\omega_1 - \omega_1)t} + x_2^{(1)}x_2^{(1)*}e^{i(\omega_2 - \omega_2)t} + (x_1^{(1)})^2 e^{-i(\omega_1 + \omega_1)t}\right.$$
$$\left. + (x_2^{(1)})^2 e^{-i(\omega_2 + \omega_2)t} + 2x_1^{(1)}x_2^{(1)}e^{-i(\omega_1 + \omega_2)t} + 2x_1^{(1)}x_2^{(1)*}e^{-i(\omega_1 - \omega_2)t} + c.c.\right). \tag{3.47}$$

Of the possible output frequencies, we consider only the displacement induced at ω_1, and so we only choose to look at

$$(\underline{X}^{(1)})^3\Big|_{\omega_1} = \frac{1}{8}\left[3(x_1^{(1)})^2 x_1^{(1)*} + 6x_1^{(1)} x_2^{(1)} x_2^{(1)*}\right]e^{-i\omega_1 t} + \text{c.c.} \tag{3.48}$$

Equation 3.48 shows two types of first-order to second-order couplings that lead to a frequency of ω_1 with pairings given by

$$\begin{aligned}\omega_1 &= \omega_1 + \omega_1 - \omega_1, \\ \omega_1 &= \omega_1 + \omega_2 - \omega_2.\end{aligned} \tag{3.49}$$

Going from Equation 3.48 to $\chi^{(3)}$ follows the same procedure as $\chi^{(2)}$. We substitute Equation 3.48 into Equation 3.34:

$$\frac{d^2\underline{X}^{(3)}}{dt^2} + \Gamma\frac{d\underline{X}^{(3)}}{dt} + \omega_0^2\underline{X}^{(3)} + \left[\frac{3b}{8}(x_1^{(1)})^2 x_1^{(1)*} e^{-i\omega_1 t} + \frac{3b}{4} x_1^{(1)} x_2^{(1)} x_2^{(1)*} e^{-i\omega_1 t} + \text{c.c.}\right] = 0. \tag{3.50}$$

The third-order solution for a nonlinear displacement at ω_1 is

$$x^{(3)}(\omega_1) = -\frac{3be^3}{8m^3}\frac{A_1}{D^2(\omega_1)}\left[\frac{|A_1|^2}{D(\omega_1)D^*(\omega_1)} + 2\frac{|A_2|^2}{D(\omega_2)D^*(\omega_2)}\right]. \tag{3.51}$$

Therefore, the third-order polarization's complex amplitude is

$$P^{(3)}(\omega_1) = \frac{3Nbe^4}{8m^3}\frac{A_1}{D^2(\omega_1)}\left[\frac{|A_1|^2}{D(\omega_1)D^*(\omega_1)} + 2\frac{|A_2|^2}{D(\omega_2)D^*(\omega_2)}\right]. \tag{3.52}$$

From this expression, we see two contributions to the polarization's complex amplitude, and typically a different susceptibility is used for each term so that Equation 3.52 is rewritten as

$$P^{(3)}(\omega_1) = \frac{3\varepsilon_0}{4}\chi^{(3)}(\omega_1; \omega_1, \omega_1, -\omega_1)A_1 |A_1|^2 + \frac{3\varepsilon_0}{2}\chi^{(3)}(\omega_1; \omega_1, \omega_2, -\omega_2)A_1 |A_2|^2. \tag{3.53}$$

The factors of 3/4 and 3/2 are present to be consistent with the formalism developed in Chapter 8, where $\chi^{(3)}$ effects are covered in more detail.

3.3 DETAILS OF THE NONLINEAR SUSCEPTIBILITY, $\chi^{(2)}$

Returning to $\chi^{(2)}$ processes, the same perturbation derivation is carried out to include all the possible frequency components for the nonlinear polarization, not just at the difference frequency at ω_3. The result of such a procedure is a second-order nonlinear polarization:

$$\underline{P}^{(2)}(t) = \frac{Nea}{4}\left(\frac{e}{m}\right)^2 \sum_{n,m} \frac{A(\omega_n)A(\omega_m)}{D(\omega_n)D(\omega_m)D(\omega_n + \omega_m)} e^{-i(\omega_m + \omega_n)t}. \tag{3.54}$$

The sum is over n and m, which runs over all the frequencies present in the field, both positive and negative. The negative frequencies come from the complex conjugate terms. For example, when two lasers are present, the input frequencies present are $\pm\omega_1$ and $\pm\omega_2$. Note that when working with negative frequencies, $A(-\omega) = A^*(\omega)$. Thus, every term in the sum in Equation 3.54 has an associated complex conjugate so that the overall expression is a real quantity.

At this stage, we combine the microscopic quantities into a single coupling coefficient $\chi^{(2)}(\omega_m + \omega_n; \omega_m, \omega_n)$ specific to each frequency coupling. That is, Equation 3.54 is rewritten as

$$\underline{P}^{(2)}(t) = \frac{\varepsilon_0}{4} \sum_{n,m} \chi^{(2)}(\omega_m + \omega_n; \omega_m, \omega_n) A(\omega_m)A(\omega_n)e^{-i(\omega_m + \omega_n)t}. \tag{3.55}$$

The frequency dependence of $\chi^{(2)}(\omega_m + \omega_n; \omega_m, \omega_n)$ reflects the dependence on three frequencies. The classical model for the nonlinear susceptibility gives an explicit frequency dependence (see Equation 3.44). The convention used in this text for frequency ordering in $\chi^{(2)}(\omega_m + \omega_n; \omega_m, \omega_n)$ is that the first entry is the frequency of the nonlinear polarization followed by the frequencies of the inputs. Hence, for $\chi^{(2)}(\omega_m + \omega_n; \omega_m, \omega_n)$, the first term is the sum of the second two. This same convention carries over to higher susceptibilities, where the first frequency is that of the nonlinear polarization, and the rest correspond to the frequencies of the input fields. The factor of $1/4$ in front of the expression in Equation 3.55 makes the nonlinear polarization consistent with the power-series expansion in Equation 3.1 and the definition of the complex field amplitudes given in Equation 3.7.

3.3.1 DEGENERACY AND SUBTLETIES OF SQUARING THE FIELD

Equation 3.55 gives the total second-order nonlinear polarization under the assumption that each field is distinct. The derivations that follow from Equation 3.55 also make this assumption. It may be tempting later on to, for example, compare the second-harmonic process $2\omega_1$ to the sum-frequency process $\omega_1 + \omega_2$ in the limit $\omega_2 \rightarrow \omega_1$

and expect the result for SFG and SHG to be the same. However, such comparisons only work if we account for the degeneracy that is introduced into Equation 3.55. A simple way to see how degeneracy affects the result is to start with an electric field as in Equation 3.35 and look at its square:

$$
\underline{E}^2 = \frac{1}{4}\Big(2|A_1|^2 + 2|A_2|^2 + \Big[A_1^2 e^{-i2\omega_1 t} + A_2^2 e^{-i2\omega_2 t}
$$
$$
+ 2A_1 A_2^* e^{-i(\omega_1-\omega_2)t} + 2A_1 A_2 e^{-i(\omega_1+\omega_2)t} + \text{c.c.}\Big]\Big). \tag{3.56}
$$

As seen before, this relationship results in every possible sum and difference of the input frequencies. Let us assume that we pick the sum-frequency process. Once we decide on the sum frequency, all the other terms at other frequencies in Equation 3.56 are ignored, including $2\omega_1$ and $2\omega_2$. However, in the limit, $\omega_2 \to \omega_1$, then $2\omega_1$, $2\omega_2$, and $\omega_1 + \omega_2$ all yield the same frequency. Hence, the limiting case needs to be corrected by including all terms in Equation 3.56 that give the same SFG frequency. Another case where a limit leads to difficulties is when one of the frequencies in a sum- or difference-frequency process goes to zero. The best strategy is to avoid limiting cases altogether and stick to a specific process throughout the calculation.

3.3.2 TENSOR PROPERTIES OF SUSCEPTIBILITY

We have been treating the nonlinear susceptibility as a scalar where $\chi^{(2)}(\omega_m + \omega_n; \omega_m, \omega_n)$ connects a nonlinear polarization at $\omega_n + \omega_m$ to two inputs at ω_n and ω_m. However, as with linear optics, it is often necessary to take into account more general couplings. The most general expression connects a given Cartesian component of the nonlinear polarization to every possible frequency combination and with every possible combination of field directions:

$$
\underline{P}_i^{(2)}(t) = \frac{\varepsilon_0}{4} \sum_{m,n} \sum_{j,k} \chi_{ijk}^{(2)}(\omega_m + \omega_n; \omega_m, \omega_n) A_j(\omega_m) A_k(\omega_n) e^{-i(\omega_m+\omega_n)t}. \tag{3.57}
$$

The indices i, j, and k correspond to Cartesian components x, y, and z. The second-order susceptibility in Equation 3.57 is a Rank 3 tensor that connects each Cartesian component of the fields at the input frequencies to a given Cartesian component of the nonlinear polarization. Proving that the susceptibility transforms according to the tensor rules is left as a problem in Chapter 2 (see Problem 2.9).

Typically, Equation 3.57 simplifies considerably since the majority of situations involve linearly polarized input fields coupled to linearly polarized output fields. The nonlinear polarization can be expressed as a sum over frequencies with complex polarization amplitudes,

$$
\underline{P}_i^{(2)}(t) = \sum_{n,m} P_i^{(2)}(\omega_m + \omega_n) e^{-i(\omega_m+\omega_n)t}. \tag{3.58}
$$

Furthermore, Equation 3.58 includes every possible frequency combination, whereas in practice only a specific subset of frequencies is of interest. For these common situations, the nonlinear polarization's complex amplitude is proportional to the product of two complex field amplitudes at the input frequencies and a single effective nonlinearity. For example, the expression for the nonlinear polarization contribution to SFG simplifies to

$$P^{(2)}(\omega_{\text{sum}} = \omega_1 + \omega_2) = \varepsilon_0 \chi_{\text{eff}}^{(2)} A(\omega_1) A(\omega_2), \tag{3.59}$$

where χ_{eff} is a function of the propagation direction and depends on the polarization and propagation directions of the two linearly polarized input fields. Obtaining a nonlinear polarization that simplifies to an expression as given in Equation 3.59 is the goal of Sections 3.3.3 through 3.3.7.

3.3.3 PERMUTING THE ELECTRIC FIELDS IN THE NONLINEAR POLARIZATION

The first simplification of Equation 3.58 comes from noticing that the order of the two fields $A_j(\omega_m)$ and $A_k(\omega_n)$ should not matter. That is, it should be possible to swap the field order in Equation 3.57 and have the same nonlinear polarization. Swapping the fields gives

$$\underline{P}_i^{(2)}(t) = \frac{\varepsilon_0}{4} \sum_{\substack{m,n \\ \omega_q = \omega_m + \omega_n}} \sum_{j,k} \chi_{ijk}^{(2)}(\omega_m + \omega_n, \omega_m, \omega_n) A_k(\omega_n) A_j(\omega_m) e^{-i(\omega_m + \omega_n)t}. \tag{3.60}$$

But j, k, n, and m are all dummy variables; so, relabeling the dummy variables, n as m, m as n, j as k, and k as j, gives

$$\underline{P}_i^{(2)}(t) = \frac{\varepsilon_0}{4} \sum_{m,n} \sum_{j,k} \chi_{ikj}^{(2)}(\omega_m + \omega_n, \omega_n, \omega_m) A_j(\omega_m) A_k(\omega_n) e^{-i(\omega_m + \omega_n)t}. \tag{3.61}$$

Equating Equation 3.60 into Equation 3.61 yields

$$\chi_{ijk}^{(2)}(\omega_n + \omega_m, \omega_n, \omega_m) = \chi_{ikj}^{(2)}(\omega_m + \omega_n, \omega_m, \omega_n). \tag{3.62}$$

Equation 3.62 shows a form of permutation symmetry in $\chi^{(2)}$ since it is possible to permute j and k as long as the frequencies ω_m and ω_n are permuted with them. Note that there is little physics here, just some labeling details.

3.3.4 FULL PERMUTATION SYMMETRY IN LOSSLESS MEDIA

When the medium is lossless for all frequencies involved, then it is possible to freely permute all the i, j, and k indices as long as the frequency indices m and n are permuted with them. For example,

$$\chi_{ijk}^{(2)}(\omega_m + \omega_n; \omega_m, \omega_n) = \chi_{kij}^{(2)}(\omega_n; \omega_m + \omega_n, -\omega_m). \tag{3.63}$$

The minus sign in front of ω_m is present because the first frequency is always the sum of the second two.

The significance of full permutation symmetry is that it connects seemingly different nonlinear processes. For example, the sum-frequency process of $\omega_1 = \omega_2 + \omega_3$ has the same nonlinearity as the difference-frequency process $\omega_3 = \omega_1 - \omega_2$:

$$\chi_{ijk}^{(2)}(\omega_2 + \omega_3; \omega_2, \omega_3) = \chi_{kij}^{(2)}(\omega_3; \omega_2 + \omega_3, -\omega_2). \tag{3.64}$$

Equation 3.63 shows that running a nonlinear process "backward" gives the same nonlinearity.

Another consequence of the lossless assumption is that $\chi^{(2)}$ is a purely real quantity, although we do not prove it. Hence,

$$\chi_{ijk}^{(2)*}(\omega_2 + \omega_3; \omega_2, \omega_3) = \chi_{ijk}^{(2)}(-\omega_2 - \omega_3; -\omega_2, -\omega_3) = \chi_{ijk}^{(2)}(\omega_2 + \omega_3; \omega_2, \omega_3). \tag{3.65}$$

3.3.5 KLEINMAN'S SYMMETRY

Another permutation symmetry, called Kleinman's symmetry, arises if we neglect the dispersion of $\chi^{(2)}$. This condition is valid when the frequencies of interest are in the material's transparency region and are far from absorption resonances. Neglecting dispersion is equivalent to the assumption made at the beginning of this chapter that interactions are instantaneous. Under Kleinman's symmetry, the susceptibility's indices are permuted freely without permuting the corresponding frequencies. Therefore, Equation 3.61 simplifies to

$$\underline{P}_i^{(2)}(t) = \frac{\varepsilon_0}{4} \sum_{n,m} \sum_{j,k} \chi_{ijk}^{(2)} A_j(\omega_m) A_k(\omega_n) e^{-i(\omega_n + \omega_m)t}. \tag{3.66}$$

Note that the susceptibility is written without explicit frequency dependences.

Equation 3.66 includes nonlinear polarizations for all possible frequency mixing processes; however, we are usually only interested in one of them. Provided that the interactions are weak, the processes are treated independently of each other.

Consider the case when two frequencies at ω_1 and ω_2 are present at the input so that m and n in Equation 3.65 run over -2, -1, 1, and 2. However, we only need to consider those frequency pairings that yield the output frequency of interest. For example, in the case of DFG, we need to only retain terms with the exponential factor $e^{\pm i(\omega_1 - \omega_2)t}$.

Choosing DFG as the only process combined with Kleinman's symmetry greatly simplifies Equation 3.66. The first step in the simplification is writing out the summation over m and n, keeping only the DFG terms:

$$P_{i,\text{DFG}}^{(2)}(t) = \frac{\varepsilon_0}{4} \sum_{j,k} \chi_{ijk}^{(2)} \Big(A_j(-\omega_2)A_k(\omega_1)e^{-i(\omega_1-\omega_2)t} + A_j(-\omega_1)A_k(\omega_2)e^{i(\omega_1-\omega_2)t}$$
$$+ A_j(\omega_1)A_k(-\omega_2)e^{-i(\omega_1-\omega_2)t} + A_j(\omega_2)A_k(-\omega_1)e^{i(\omega_1-\omega_2)t} \Big) \tag{3.67}$$

Note that Equation 3.67 is identical, and j and k are interchanged. This observation allows us to rewrite the nonlinear polarization using a contracted notation.

3.3.6 CONTRACTING THE INDICES IN $\chi_{ijk}^{(2)}$

Equation 3.67 is simplified by noting that two of the terms in the brackets are the complex conjugate of the other two. Hence, Equation 3.67 is rewritten as

$$P_{i,\text{DFG}}^{(2)} = \frac{\varepsilon_0}{4} \sum_{j,k} \chi_{ijk}^{(2)} \Big(A_j(-\omega_2)A_k(\omega_1)e^{-i(\omega_1-\omega_2)t} + A_j(\omega_1)A_k(-\omega_2)e^{-i(\omega_1-\omega_2)t} + \text{c.c.} \Big). \tag{3.68}$$

Carrying out the summation over j and k explicitly allows us to see a pattern that leads to the contraction of j and k into a single index:

$$P_{i,\text{DFG}}^{(2)} = \frac{\varepsilon_0}{4} \begin{cases} \chi_{ixx}^{(2)}(A_x(-\omega_2)A_x(\omega_1) + A_x(\omega_1)A_x(-\omega_2))e^{-i(\omega_1-\omega_2)t} + \text{c.c.} \\ + \chi_{ixy}^{(2)}(A_x(-\omega_2)A_y(\omega_1) + A_x(\omega_1)A_y(-\omega_2))e^{-i(\omega_1-\omega_2)t} + \text{c.c.} \\ + \chi_{iyx}^{(2)}(A_y(-\omega_2)A_x(\omega_1) + A_y(\omega_1)A_x(-\omega_2))e^{-i(\omega_1-\omega_2)t} + \text{c.c.} \\ + \chi_{ixz}^{(2)}(A_z(-\omega_2)A_z(\omega_1) + A_x(\omega_1)A_z(-\omega_2))e^{-i(\omega_1-\omega_2)t} + \text{c.c.} \\ + \cdots. \end{cases} \tag{3.69}$$

Note that the $\chi_{ijk}^{(2)}$ and $\chi_{ikj}^{(2)}$ terms both multiply the same combination of fields; see, for example, the terms multiplying $\chi_{ixy}^{(2)}$ and $\chi_{iyx}^{(2)}$. Therefore, we combine the like expressions, for example,

$$\left(\chi_{ixy}^{(2)} + \chi_{iyx}^{(2)} \right)\left(A_x(-\omega_2)A_y(\omega_1) + A_x(\omega_1)A_y(-\omega_2) \right)e^{-i(\omega_1-\omega_2)t}. \tag{3.70}$$

Since the susceptibilities are factored out from the field amplitudes, and since Kleinman's symmetry gives $\chi^{(2)}_{ijk} = \chi^{(2)}_{ikj}$, we contract j and k into a single index given in Table 3.1.

Rewriting Equation 3.69 with the contracted notation gives

$$
P^{(2)}_{i,\text{DFG}} = \frac{\varepsilon_0}{2} \begin{cases} \chi^{(2)}_{i1} A_x(-\omega_2)A_x(\omega_1)e^{-i(\omega_1-\omega_2)t} + \text{c.c.} \\ + \chi^{(2)}_{i2} A_y(-\omega_2)A_y(\omega_1)e^{-i(\omega_1-\omega_2)t} + \text{c.c.} \\ + \chi^{(2)}_{i3} A_z(-\omega_2)A_z(\omega_1)e^{-i(\omega_1-\omega_2)t} + \text{c.c.} \\ + \chi^{(2)}_{i4}(A_y(-\omega_2)A_z(\omega_1) + A_y(\omega_1)A_z(-\omega_2))e^{-i(\omega_1-\omega_2)t} + \text{c.c.} \\ + \chi^{(2)}_{i5}(A_x(-\omega_2)A_z(\omega_1) + A_x(\omega_1)A_z(-\omega_2))e^{-i(\omega_1-\omega_2)t} + \text{c.c.} \\ + \chi^{(2)}_{i6}(A_x(-\omega_2)A_y(\omega_1) + A_x(\omega_1)A_y(-\omega_2))e^{-i(\omega_1-\omega_2)t} + \text{c.c.} \end{cases}
\tag{3.71}
$$

Each Cartesian component of the nonlinear polarization has six separate $\chi^{(2)}$ terms. A further simplification is to work with the complex amplitude of the nonlinear polarization instead of the real quantity. The real nonlinear polarization is written as

$$
P^{(2)}_{i,\text{DFG}} = \frac{1}{2} P^{(2)}_i(\omega_1 - \omega_2)e^{-i(\omega_1-\omega_2)t} + \text{c.c.}
\tag{3.72}
$$

We equate Equation 3.72 with Equation 3.71 to derive an expression for the complex amplitude:

$$
P^{(2)}_i(\omega_1 - \omega_2) = \varepsilon_0 \begin{cases} \chi^{(2)}_{i1} A_x(-\omega_2)A_x(\omega_1) \\ + \chi^{(2)}_{i2} A_y(-\omega_2)A_y(\omega_1) \\ + \chi^{(2)}_{i3} A_z(-\omega_2)A_z(\omega_1) \\ + \chi^{(2)}_{i4}(A_y(-\omega_2)A_z(\omega_1) + A_y(\omega_1)A_z(-\omega_2)) \\ + \chi^{(2)}_{i5}(A_x(-\omega_2)A_z(\omega_1) + A_x(\omega_1)A_z(-\omega_2)) \\ + \chi^{(2)}_{i6}(A_x(-\omega_2)A_y(\omega_1) + A_x(\omega_1)A_y(-\omega_2)). \end{cases}
\tag{3.73}
$$

Note that Equation 3.73 is written more compactly as a matrix multiplication,

$$P_i^{(2)}(\omega_1 - \omega_2) = \varepsilon_0 \begin{pmatrix} \chi_{i1}^{(2)} & \chi_{i2}^{(2)} & \chi_{i3}^{(2)} & \chi_{i4}^{(2)} & \chi_{i5}^{(2)} & \chi_{i6}^{(2)} \end{pmatrix} \begin{pmatrix} A_x(\omega_1)A_x^*(\omega_2) \\ A_y(\omega_1)A_y^*(\omega_2) \\ A_z(\omega_1)A_z^*(\omega_2) \\ A_y(\omega_1)A_z^*(\omega_2) + A_z(\omega_1)A_y^*(\omega_2) \\ A_x(\omega_1)A_z^*(\omega_2) + A_z(\omega_1)A_x^*(\omega_2) \\ A_x(\omega_1)A_y^*(\omega_2) + A_y(\omega_1)A_x^*(\omega_2) \end{pmatrix},$$

(3.74)

where $A(-\omega) = A^*(\omega)$ is used. The Cartesian index i runs through x, y, and z; an equivalent and interchangeable notation is to let i run through 1, 2, and 3, where 1 corresponds to x, 2 to y, and 3 to z. All three components of the nonlinear polarization's complex amplitude are now written as

$$\vec{P}^{(2)}(\omega_1 - \omega_2) = \varepsilon_0 \begin{pmatrix} \chi_{11}^{(2)} & \chi_{12}^{(2)} & \chi_{13}^{(2)} & \chi_{14}^{(2)} & \chi_{15}^{(2)} & \chi_{16}^{(2)} \\ \chi_{21}^{(2)} & \chi_{22}^{(2)} & \chi_{23}^{(2)} & \chi_{24}^{(2)} & \chi_{25}^{(2)} & \chi_{26}^{(2)} \\ \chi_{31}^{(2)} & \chi_{32}^{(2)} & \chi_{33}^{(2)} & \chi_{34}^{(2)} & \chi_{35}^{(2)} & \chi_{36}^{(2)} \end{pmatrix}$$
$$\times \begin{pmatrix} A_x(\omega_1)A_x^*(\omega_2) \\ A_y(\omega_1)A_y^*(\omega_2) \\ A_z(\omega_1)A_z^*(\omega_2) \\ A_y(\omega_1)A_z^*(\omega_2) + A_z(\omega_1)A_y^*(\omega_2) \\ A_x(\omega_1)A_z^*(\omega_2) + A_z(\omega_1)A_x^*(\omega_2) \\ A_x(\omega_1)A_y^*(\omega_2) + A_y(\omega_1)A_x^*(\omega_2) \end{pmatrix}.$$

(3.75)

We emphasize that for elements in the $\chi^{(2)}$-matrix, the contraction occurs in the second index of $\chi_{im}^{(2)}$.

In the early development of nonlinear optics, a quantity called a d-tensor was used in place of $\chi^{(2)}$. This practice has been carried over to the present, and second-order device parameters where Kleinman's symmetry applies are usually stated in terms of d instead of $\chi^{(2)}$. d has units of m/V, and in the literature, it is usually quoted in pm/V. The relationship between d and $\chi^{(2)}$ is

$$d_{ijk} = \frac{1}{2}\chi_{ijk}^{(2)}.$$

(3.76)

Again, the contracted notation is used with d_{ijk} to combine j and k into one index. The expression for the complex amplitude in Equation 3.73 is written in terms of a d-matrix as

$$\vec{P}^{(2)}(\omega_1 - \omega_2) = 2\varepsilon_0 \begin{pmatrix} d_{11} & d_{12} & d_{13} & d_{14} & d_{15} & d_{16} \\ d_{21} & d_{22} & d_{23} & d_{24} & d_{25} & d_{26} \\ d_{31} & d_{32} & d_{33} & d_{34} & d_{35} & d_{36} \end{pmatrix}$$
$$\times \begin{pmatrix} A_x(\omega_1)A_x^*(\omega_2) \\ A_y(\omega)A_y^*(\omega_2) \\ A_z(\omega_1)A_z^*(\omega_2) \\ A_y(\omega_1)A_z^*(\omega_2) + A_z(\omega_1)A_y^*(\omega_2) \\ A_x(\omega_1)A_z^*(\omega_2) + A_z(\omega_1)A_x^*(\omega_2) \\ A_x(\omega_1)A_y^*(\omega_2) + A_y(\omega_1)A_x^*(\omega_2) \end{pmatrix}. \tag{3.77}$$

The column vector in Equations 3.75 and 3.77 is specific to DFG. A different column vector is needed for SHG, SFG, and optical rectification. Those column vectors are found through a similar derivation as for DFG. The simplest case is SFG, which looks nearly identical to Equation 3.77, except that it does not have complex conjugates. The relationship for SHG is given by

$$\vec{P}^{(2)}(2\omega) = \varepsilon_0 \begin{pmatrix} d_{11} & d_{12} & d_{13} & d_{14} & d_{15} & d_{16} \\ d_{21} & d_{22} & d_{23} & d_{24} & d_{25} & d_{26} \\ d_{31} & d_{32} & d_{33} & d_{34} & d_{35} & d_{36} \end{pmatrix} \begin{pmatrix} A_x^2 \\ A_y^2 \\ A_z^2 \\ 2A_yA_z \\ 2A_xA_z \\ 2A_xA_y \end{pmatrix}, \tag{3.78}$$

where the field amplitudes correspond to the fundamental field at ω. Note that in addition to a different column vector, the overall expression does not have the factor of 2 that is present in Equation 3.77. The derivation of the column vectors for SHG and SFG is left as an exercise at the end of this chapter. We again note that one must be careful with any of the column vectors in the limit that two of the frequencies become equal or one of the frequencies goes to zero (see Section 3.3.1).

Note that the contracted notation above always involves only the j and k indices in d_{ijk}. However, under Kleinman's symmetry, we are free to permute all three indices before the contraction takes place. For example, consider d_{12}, which in its uncontracted form is d_{122}. Permuting the indices gives $d_{122} = d_{212}$, and contracting the second two indices now yields $d_{212} = d_{26}$. Therefore,

$$d_{12} = d_{122} = d_{212} = d_{26}. \tag{3.79}$$

Similar equalities are found for other elements of the d-matrix. The d-matrix under Kleinman's symmetry has only 10 independent values:

$$\begin{pmatrix} d_{11} & d_{12} & d_{13} & d_{14} & d_{15} & d_{16} \\ d_{21} & d_{22} & d_{23} & d_{24} & d_{25} & d_{26} \\ d_{31} & d_{32} & d_{33} & d_{34} & d_{35} & d_{36} \end{pmatrix} \Rightarrow \begin{pmatrix} d_{11} & d_{12} & d_{13} & d_{14} & d_{15} & d_{16} \\ d_{16} & d_{22} & d_{23} & d_{24} & d_{14} & d_{12} \\ d_{15} & d_{24} & d_{33} & d_{23} & d_{13} & d_{14} \end{pmatrix}. \qquad (3.80)$$

A further reduction in the number of independent terms comes about due to material symmetry considerations. A given crystal falls into one of 32 point group symmetries, and each of these crystal classes has an associated d-matrix (see Table 3.2). As described in Section 3.4, the crystal symmetries lead to restrictions on the d-matrix values. The zeroes found in Table 3.2 are purely due to symmetry considerations as discussed in Section 3.4.

3.3.7 EFFECTIVE NONLINEARITY AND d_{eff}

A simplifying consideration in condensing the notation relating the nonlinear polarization is to recognize that, in most cases, the interacting fields are linearly polarized. Therefore, the fields are written as either an e-wave or an o-wave for uniaxial crystals (see Chapter 2). For a given propagation direction, the e- and o-field directions are well defined in terms of the polar angles θ and ϕ of the k-vector. Figure 3.3 illustrates the relationship between the k-vector direction and the e- and o-directions for the field. The unit k-vector in spherical coordinates is given by

$$\hat{k} = \sin\theta\cos\phi\hat{x} + \sin\theta\sin\phi\hat{y} + \cos\theta\hat{z}. \qquad (3.81)$$

Referring to Figure 3.3, the corresponding unit vectors for the e- and o-polarized fields are

$$\hat{E}_e = -\cos(\theta+\rho)\cos\phi\hat{x} - \cos(\theta+\rho)\sin\phi\hat{y} + \sin(\theta+\rho)\hat{z}, \qquad (3.82)$$

$$\hat{E}_0 = \sin\phi\hat{x} - \cos\phi\hat{y}. \qquad (3.83)$$

Here, ρ is the walk-off angle. Note that it is possible to use e- and o-unit vectors that are negatives of the above expressions. Whichever sign is chosen, one must be consistent throughout a given problem to use the same convention, including carrying it over to the corresponding direction of $\vec{H}$.

The matrix multiplication in Equation 3.77 can be written such that the magnitudes of the two fields at the input frequencies are factored out of the column vector. This factoring leaves a matrix multiplication between a d-matrix and a column vector that only depends on geometric factors. After carrying out the matrix multiplication, the

TABLE 3.2 *d*-Matrices for Noncentrosymmetric Crystals

Point Group	*d*-Matrix	*r*-Matrix
1	$\begin{pmatrix} d_{11} & d_{12} & d_{13} & d_{14} & d_{15} & d_{16} \\ d_{21} & d_{22} & d_{23} & d_{24} & d_{25} & d_{26} \\ d_{31} & d_{32} & d_{33} & d_{34} & d_{35} & d_{36} \end{pmatrix}$	$\begin{pmatrix} r_{11} & r_{12} & r_{13} \\ r_{21} & r_{22} & r_{23} \\ r_{31} & r_{32} & r_{33} \\ r_{41} & r_{42} & r_{43} \\ r_{51} & r_{52} & r_{53} \\ r_{61} & r_{62} & r_{63} \end{pmatrix}$
2	$\begin{pmatrix} 0 & 0 & 0 & d_{14} & 0 & d_{16} \\ d_{21} & d_{22} & d_{23} & 0 & d_{25} & 0 \\ 0 & 0 & 0 & d_{34} & 0 & d_{36} \end{pmatrix}$	$\begin{pmatrix} 0 & r_{12} & 0 \\ 0 & r_{22} & 0 \\ 0 & r_{32} & 0 \\ r_{41} & 0 & r_{43} \\ 0 & r_{52} & 0 \\ r_{61} & 0 & r_{63} \end{pmatrix}$
M	$\begin{pmatrix} d_{11} & d_{12} & d_{13} & 0 & d_{15} & 0 \\ 0 & 0 & 0 & d_{24} & 0 & d_{26} \\ d_{31} & d_{32} & d_{33} & 0 & d_{35} & 0 \end{pmatrix}$	$\begin{pmatrix} r_{11} & 0 & r_{13} \\ r_{21} & 0 & r_{23} \\ r_{31} & 0 & r_{33} \\ 0 & r_{42} & 0 \\ r_{51} & 0 & r_{53} \\ 0 & r_{62} & 0 \end{pmatrix}$
222	$\begin{pmatrix} 0 & 0 & 0 & d_{14} & 0 & 0 \\ 0 & 0 & 0 & 0 & d_{25} & 0 \\ 0 & 0 & 0 & 0 & 0 & d_{36} \end{pmatrix}$	$\begin{pmatrix} 0 & 0 & 0 \\ 0 & 0 & 0 \\ 0 & 0 & 0 \\ r_{41} & 0 & 0 \\ 0 & r_{52} & 0 \\ 0 & 0 & r_{63} \end{pmatrix}$
mm2	$\begin{pmatrix} 0 & 0 & 0 & 0 & d_{15} & 0 \\ 0 & 0 & 0 & d_{24} & 0 & 0 \\ d_{31} & d_{32} & d_{33} & 0 & 0 & 0 \end{pmatrix}$	$\begin{pmatrix} 0 & 0 & r_{13} \\ 0 & 0 & r_{23} \\ 0 & 0 & r_{33} \\ 0 & r_{42} & 0 \\ r_{51} & 0 & 0 \\ 0 & 0 & 0 \end{pmatrix}$
3	$\begin{pmatrix} d_{11} & -d_{11} & 0 & 0 & d_{15} & -d_{22} \\ -d_{22} & d_{22} & 0 & d_{15} & 0 & -d_{11} \\ d_{15} & d_{15} & d_{33} & 0 & 0 & 0 \end{pmatrix}$	$\begin{pmatrix} r_{11} & -r_{22} & r_{13} \\ -r_{11} & r_{22} & r_{13} \\ 0 & 0 & r_{33} \\ 0 & r_{51} & 0 \\ r_{51} & 0 & 0 \\ -r_{22} & -r_{11} & 0 \end{pmatrix}$

(Continued)

TABLE 3.2 (*Continued*)   *d*-Matrices for Noncentrosymmetric Crystals

Point Group	*d*-Matrix	*r*-Matrix
3*m*	$\begin{pmatrix} 0 & 0 & 0 & 0 & d_{15} & -d_{22} \\ -d_{22} & d_{22} & 0 & d_{15} & 0 & 0 \\ d_{31} & d_{31} & d_{33} & 0 & 0 & 0 \end{pmatrix}$	$\begin{pmatrix} 0 & -r_{22} & r_{13} \\ 0 & r_{22} & r_{13} \\ 0 & 0 & r_{33} \\ 0 & r_{51} & 0 \\ r_{51} & 0 & 0 \\ -r_{22} & 0 & 0 \end{pmatrix}$
$\bar{6}$	$\begin{pmatrix} d_{11} & -d_{11} & 0 & 0 & 0 & -d_{22} \\ -d_{22} & d_{22} & 0 & 0 & 0 & -d_{11} \\ 0 & 0 & 0 & 0 & 0 & 0 \end{pmatrix}$	$\begin{pmatrix} r_{11} & -r_{22} & 0 \\ -r_{11} & r_{22} & 0 \\ 0 & 0 & 0 \\ 0 & 0 & 0 \\ 0 & 0 & 0 \\ -r_{22} & -r_{11} & 0 \end{pmatrix}$
$\bar{6}m2$	$\begin{pmatrix} 0 & 0 & 0 & 0 & 0 & -d_{22} \\ -d_{22} & d_{22} & 0 & 0 & 0 & 0 \\ 0 & 0 & 0 & 0 & 0 & 0 \end{pmatrix}$	$\begin{pmatrix} 0 & -r_{22} & 0 \\ 0 & r_{22} & 0 \\ 0 & 0 & 0 \\ 0 & 0 & 0 \\ 0 & 0 & 0 \\ -r_{22} & 0 & 0 \end{pmatrix}$
6, 4, 6*mm*, 4*mm*	$\begin{pmatrix} 0 & 0 & 0 & 0 & d_{15} & 0 \\ 0 & 0 & 0 & d_{15} & 0 & 0 \\ d_{15} & d_{15} & d_{33} & 0 & 0 & 0 \end{pmatrix}$	$\begin{pmatrix} 0 & 0 & r_{13} \\ 0 & 0 & r_{13} \\ 0 & 0 & r_{33} \\ 0 & r_{51} & 0 \\ r_{51} & 0 & 0 \\ 0 & 0 & 0 \end{pmatrix}$
$\bar{4}$	$\begin{pmatrix} 0 & 0 & 0 & d_{14} & d_{15} & 0 \\ 0 & 0 & 0 & -d_{15} & d_{14} & 0 \\ d_{15} & -d_{15} & 0 & 0 & 0 & d_{14} \end{pmatrix}$	$\begin{pmatrix} 0 & 0 & r_{13} \\ 0 & 0 & -r_{13} \\ 0 & 0 & 0 \\ r_{41} & -r_{51} & 0 \\ r_{51} & r_{41} & 0 \\ 0 & 0 & r_{63} \end{pmatrix}$
32	$\begin{pmatrix} d_{11} & -d_{11} & 0 & d_{14} & 0 & 0 \\ 0 & 0 & 0 & 0 & -d_{14} & -d_{11} \\ 0 & 0 & 0 & 0 & 0 & 0 \end{pmatrix}$	$\begin{pmatrix} r_{11} & 0 & 0 \\ -r_{11} & 0 & 0 \\ 0 & 0 & 0 \\ r_{41} & 0 & 0 \\ 0 & -r_{41} & 0 \\ 0 & -r_{11} & 0 \end{pmatrix}$

(*Continued*)

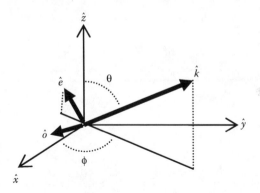

FIGURE 3.3 k-Vector direction and associated o- and e-wave directions in a uniaxial crystal. The optic axis is along the z-axis.

resultant combination of d-matrix elements and geometric factors is written as a single effective nonlinearity, $d_{\text{eff},i}$ for each Cartesian component of the nonlinear polarization,

$$P_i^{(2)}(\omega_1 - \omega_2) = 2\varepsilon_0 d_{\text{eff},i}(\theta, \phi) A_1(\omega_1) A_2^*(\omega_2). \tag{3.84}$$

The final step in deriving a compact expression for $P^{(2)}$ is to recognize that all we are interested in is the nonlinear polarization that couples to a propagating mode, that is, an e-wave or an o-wave. Hence, the nonlinear polarization given in Equation

3.84 is projected onto either the e- or o-direction. The result of the projection is that the nonlinear polarization (either e or o) is proportional to the product of the two field amplitudes at the input frequencies and a single effective nonlinearity:

$$P_0^{(2)}(\omega_1 - \omega_2) = 2\varepsilon_0 d_{\text{eff},0}(\theta, \varphi) A_1(\omega_1) A_2^*(\omega_2), \tag{3.85}$$

$$P_e^{(2)}(\omega_1 - \omega_2) = 2\varepsilon_0 d_{\text{eff},e}(\theta, \varphi) A_1(\omega_1) A_2^*(\omega_2). \tag{3.86}$$

d_{eff} for arbitrary propagation directions is tabulated for all point groups. Note that 11 groups are centrosymmetric, and hence, $d = 0$ for those classes (see Section 3.4). Tabulated formulae for d_{eff} are given in Table 3.3.

TABLE 3.3 d_{eff} for Crystals When Kleinman's Symmetry is Valid

Point Group		Interactions with Two o-Waves and One e-Wave	Interactions with Two e-Waves and One o-Wave
$\bar{4}2m$		$d_{36} \sin\theta \sin 2\phi$	$d_{36} \sin 2\theta \cos 2\phi$
$3m$		$d_{15} \sin\theta - d_{22} \cos\theta \sin 3\phi$	$d_{22} \cos^2\theta \cos 3\phi$
$4, 4mm, 6, 6mm$		$d_{15} \sin\theta$	0
4		$-(d_{14} \sin 2\phi + d_{15} \cos 2\phi) \sin\theta$	$(d_{14} \cos 2\phi - d_{15} \sin 2\phi) \sin 2\theta$
3		$(d_{11} \cos 3\phi - d_{22} \sin 3\phi) \cos\theta$ $+ d_{15} \sin\theta$	$(d_{11} \sin 3\phi + d_{22} \cos 3\phi) \cos^2\theta$
32		$d_{11} \cos\theta \cos 3\phi$	$d_{11} \cos^2\theta \sin 3\phi$
6		$(d_{11} \cos 3\phi - d_{22} \sin 3\phi) \cos\theta$	$(d_{11} \sin 3\phi + d_{22} \cos 3\phi) \cos^2\theta$
$\bar{6}m2$		$-d_{22} \cos\theta \sin 3\phi$	$d_{22} \cos^2\theta \cos 3\phi$
$422, 622$		0	0
2	XY	$d_{23} \cos\phi$	$d_{25} \sin 2\phi$
	YZ	$d_{21} \cos\theta$	$d_{25} \sin 2\theta$
	XZ	0	$d_{25} \sin 2\theta + d_{21} \cos^2\theta + d_{23}$ $\sin^2\phi$
m	XY	$d_{13} \sin\phi$	$d_{15} \sin^2\phi + d_{24} \cos^2\phi$
	YZ	$d_{15} \sin\theta$	$d_{12} \cos^2\theta + d_{13} \sin^2\theta$
	XZ	$-d_{12} \cos\theta + d_{24} \sin\theta$	$d_{12} \sin 2\theta$
$mm2$	XY	0	$-d_{15} \sin^2\phi - d_{24} \cos^2\phi$
	YZ	$-d_{15} \sin\theta$	0
	XZ	$-d_{24} \sin\theta$	0
222	XY	0	$-d_{14} \sin 2\phi$
	YZ	0	$-d_{14} \sin 2\theta$
	XZ	0	$d_{14} \sin 2\theta$
Centrosymmetric		0	0

Source: Zernike, F. and J. E. Midwinter. *Applied Nonlinear Optics.* 1973. New York, NY: John Wiley & Sons. Copyright Wiley-VCH Verlag GmbH & Co. KGaA. Reproduced with permission; With kind permission from Springer Science+Business Media: Dmitriev, V. G. et al. 1997. *Handbook of Nonlinear Optical Crystals.* Heidelberg: Springer-Verlag; Sutherland, R. L. 2003. *Handbook of Nonlinear Optics* (2nd edition). New York, NY: Marcel Dekker, Inc. With permission.

Note: Walk-off is neglected in the expression, but it can be included by replacing θ with $\theta + \rho$, where ρ is the walk-off angle. For biaxial crystals, $(n_z > n_y > n_x)$, and d_{eff} is given for the principal planes.

3.3.8 EXAMPLE CALCULATION OF d_{eff}

In practice, deriving the effective nonlinearity for a given point group is straightforward. As a specific example, consider DFG in BBO between 400 and 600 nm used to generate a nonlinear polarization at 1200 nm. BBO is a negative uniaxial class 3m crystal, and its d-matrix is given in Table 3.2. For reasons that are covered in Chapter 4, an efficient output (phase matched) at the 1200 nm difference frequency is obtained when the 600-nm beam is an o-wave, the 400-nm beam is an e-wave, and the 1200-nm beam is an o-wave. A further important piece of information is that the propagation of the beams' k-vectors lies in the x–z-plane. According to the prescription in Chapter 2, Section 2.4.2, for a k-vector in the x–z-plane at an angle θ with respect to the z-axis,

$$\hat{o} = -\hat{y}, \tag{3.87}$$

$$\hat{e} = -\cos(\theta + \rho)\hat{x} + \sin(\theta + \rho)\hat{z}, \tag{3.88}$$

where ρ is the walk-off angle. We include the walk-off angle because the column vector in Equation 3.77 is in terms of the electric field, which subtends an angle ρ to the electric displacement vector and the Poynting vector subtends the same angle with respect to the wave vector. Hence, the electric field amplitudes are written as

$$\vec{A}_{600} = A_2\hat{o} = -A_2\hat{y}, \tag{3.89}$$

$$\vec{A}_{400} = A_1\hat{e} = A_1[-\cos(\theta + \rho)\hat{x} + \sin(\theta + \rho)\hat{z}]. \tag{3.90}$$

The field components from Equations 3.89 and 3.90 are substituted into the column vector in Equation 3.77:

$$\vec{P} = 2\varepsilon_0 \begin{pmatrix} 0 & 0 & 0 & 0 & d_{31} & -d_{22} \\ -d_{22} & d_{22} & 0 & d_{31} & 0 & 0 \\ d_{31} & d_{31} & d_{33} & 0 & 0 & 0 \end{pmatrix} \begin{pmatrix} 0 \\ 0 \\ 0 \\ -A_1A_2^*\sin(\theta+\rho) \\ 0 \\ A_1A_2^*\cos(\theta+\rho) \end{pmatrix}$$

$$= -2\varepsilon_0 \begin{pmatrix} d_{22}A_1A_2^*\cos(\theta+\rho) \\ d_{31}A_1A_2^*\sin(\theta+\rho) \\ 0 \end{pmatrix}. \tag{3.91}$$

Thus, a nonlinear polarization is induced in the x- and y-directions. Recall that in this example, the output DFG beam is an o-wave for the output to be efficient (phase matched). To determine the effective nonlinearity, the nonlinear polarization is projected onto the o-wave direction, and thus, the nonlinear polarization "drives" the output o-field:

$$P_0 = \vec{P}_0 \cdot \hat{o} = 2\varepsilon_0 \begin{pmatrix} 0 & -1 & 0 \end{pmatrix} \begin{pmatrix} -d_{22}A_1 A_2^* \cos(\theta + \rho) \\ -d_{31}A_1 A_2^* \sin(\theta + \rho) \\ 0 \end{pmatrix}$$

$$= 2\varepsilon_0 [d_{31}\sin(\theta + \rho)]A_1 A_2^*. \tag{3.92}$$

The effective nonlinear coefficient for this interaction is $d_{31} \sin(\theta + \rho)$, which agrees with the result given in Table 3.3.

3.4 CONNECTION BETWEEN CRYSTAL SYMMETRY AND THE d-MATRIX

Further reduction in the number of independent elements in the d-matrix comes about due to specific crystal symmetries. Any crystal falls into one of the 32 "point groups." Point groups define an object by the type of transformations that leave it invariant. Such symmetry operations depend on the shape of the basic unit cell of the crystal. A general unit cell is shown in Figure 3.4, and the unit cell dimensions along with the different point groups are given in Table 3.4, arranged according to

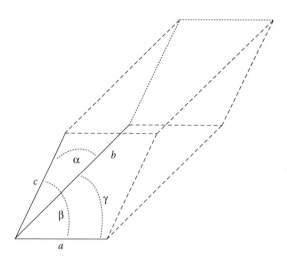

FIGURE 3.4   The unit cell of a crystal. Various crystal types are defined by the angles α, β, and γ and the relative lengths a, b, and c. α is the angle between b and c, β is the angle between a and c, and γ is the angle between a and b.

TABLE 3.4	Point Group Properties		
Crystal System	**Unit Cell Properties**	**Point Groups**	**Centrosymmetric**
Triclinic	$\alpha \neq \beta \neq \gamma \neq 90°$ $a \neq b \neq c$	1	$\bar{1}$
Monoclinic	$\alpha = \gamma = 90°$ $a \neq b \neq c$	$2, m$	$2/m$
Orthorhombic	$\alpha = \beta = \gamma = 90°$ $a \neq b \neq c$	$222, mm2$	mmm
Tetragonal	$\alpha = \beta = \gamma = 90°$ $a = b \neq c$	$4, \bar{4}, 422, 4mm, \bar{4}2m$	$4/m, \dfrac{4}{m}mm$
Trigonal (rhombohedral)	$\alpha \neq \beta \neq \gamma \neq 90°$ $a = b = c$	$3, 32, 3m$	$\bar{3}, \bar{3}m$
Hexagonal	$\alpha = \beta = 90°\ \gamma = 120°$ $a \neq c$	$6, 6/m, 622, 6mm, -\bar{6}2m$	$\bar{6}, \dfrac{6}{m}mm$
Cubic	$\alpha = \beta = \gamma = 90°$ $a = b = c$	$23, 432, \bar{4}3m$	$m\bar{3}, m\bar{3}m$

Note: Refer to Figure 3.4 for an illustration of angles in the unit cell.

crystal class. The point group labeling in Table 3.4 is according to the Hermann–Mauguin notation where the numbers refer to an *n*-fold rotational symmetry, the letter *m* refers to mirror symmetry, and a bar over a number indicates an inversion. For more information regarding crystal structures and the associated symmetries, the reader is referred to Koster (1963). For our purposes, it is enough to know that a given point group has certain symmetry operations that leave the crystal invariant. For example, the 3m class has a threefold symmetry about a vertical axis and three mirror planes (see Figure 3.5). The other point groups contain a different set of symmetry operations.

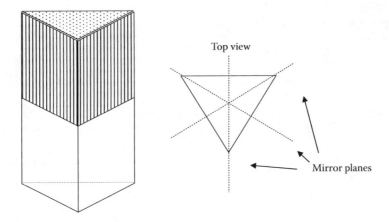

FIGURE 3.5 An object that has the symmetries of class 3m: three mirror planes and threefold rotational symmetry about the vertical axis.

3.4.1 CENTROSYMMETRIC CRYSTALS

The reduction in the number of independent d-matrix elements by symmetry operations comes from the tensor nature of the nonlinear susceptibility. Since the nonlinear susceptibility is a tensor, it transforms like a tensor. If that transformation happens to be a symmetry operation for the material, then the susceptibility before and after the transformation must be the same. We first show one of the most powerful examples of this procedure by using it to prove that all the d-matrix elements are zero for centrosymmetric media. A centrosymmetric medium is symmetric under inversion. Examples of centrosymmetric media include gases and liquids. A crystalline example is silicon.

The rules for tensor transformation are given in Section 2.2, where a given transformation is represented by an R-matrix. The R-matrix for inversion is given by

$$
\begin{bmatrix}
-1 & 0 & 0 \\
0 & -1 & 0 \\
0 & 0 & -1
\end{bmatrix},
$$
$$
x \rightarrow -x,
$$
$$
y \rightarrow -y,
$$
$$
z \rightarrow -z.
$$
(3.93)

The elements of this R-matrix can be written as

$$
R_{ij} = -\delta_{ij},
$$
(3.94)

where δ_{ij} is Kronecker's delta (see Equation 2.53). Applying the inversion operation to centrosymmetric media must leave a particular property of interest unchanged after the transformation. Let us first look at $\chi_{ij}^{(1)}$ for centrosymmetric media. The first-order susceptibility transforms like a Rank 2 tensor; hence, under inversion,

$$
\chi_{ij}^{(1)'} = R_{i\alpha} R_{j\beta} \chi_{\alpha\beta} = (-\delta_{i\alpha})(-\delta_{j\beta})\chi_{\alpha\beta}^{(1)}.
$$
(3.95)

Recall that a summation is performed for repeated indices (Einstein's notation). In Equation 3.95, the only terms in the sum that survive are those where $\alpha = i$ and $\beta = j$, and so it gives

$$
\chi_{ij}^{(1)'} = \chi_{ij}^{(1)}.
$$
(3.96)

So, the linear properties of centrosymmetric media remain unchanged under an inversion.

Following the same process as above, the second-order susceptibility transforms under an inversion according to

$$\chi_{ijk}^{(2)'} = R_{i\alpha}R_{j\beta}R_{k\gamma}\chi_{\alpha\beta\gamma}^{(2)} = (-\delta_{i\alpha})(-\delta_{j\beta})(-\delta_{k\gamma})\chi_{\alpha\beta\gamma}^{(2)} = -\chi_{ijk}^{(2)}. \tag{3.97}$$

Hence, the sign of the susceptibility changes when the crystal is inverted. However, if the material happens to be centrosymmetric, then the susceptibility must be the same after the transformation. The only way for $\chi_{ijk}^{(2)} = -\chi_{ijk}^{(2)}$ to hold is if $\chi_{ijk}^{(2)} = 0$. Hence, there are no second-order processes in centrosymmetric media. Following this procedure to higher orders shows that all even orders of $\chi^{(even)}$ are 0 for centrosymmetric media. For odd orders, the inversion operation leaves $\chi^{(odd)}$ unchanged so that the inversion symmetry condition is satisfied.

3.4.2 EXAMPLE CALCULATION OF *d*-MATRIX FOR 3m CRYSTALS

A similar process to reduce the number of elements in the *d*-matrix may be carried out for crystals with other symmetries. For example, the crystal class 3m has a threefold rotational symmetry about an axis as well as three mirror symmetry planes (see Figure 3.5). These symmetries result in a *d*-matrix given by

$$\begin{pmatrix} 0 & 0 & 0 & 0 & d_{31} & -d_{22} \\ -d_{22} & d_{22} & 0 & d_{31} & 0 & 0 \\ d_{31} & d_{31} & d_{33} & 0 & 0 & 0 \end{pmatrix}. \tag{3.98}$$

The zeroes in the *d*-matrix in Equation 3.98 come about because of the crystal's mirror symmetries. For example, consider the symmetry for a mirror reflection through the *y*–*z*-plane, which has an *R*-matrix given by

$$\begin{bmatrix} -1 & 0 & 0 \\ 0 & 1 & 0 \\ 0 & 0 & 1 \end{bmatrix}, \tag{3.99}$$

$$x \to -x,$$
$$y \to y,$$
$$z \to z.$$

The *d*-tensor transforms according to

$$d_{ijk}' = R_{i\alpha}R_{j\beta}R_{k\gamma}d_{\alpha\beta\gamma}. \tag{3.100}$$

Since the *R*-matrix in Equation 3.100 has only diagonal elements,

$$d'_{ijk} = (\pm\delta_{i\alpha})(\pm\delta_{j\beta})(\pm\delta_{k\gamma})d_{\alpha\beta\gamma} = \pm d_{ijk}. \tag{3.101}$$

The ± depends on the particular diagonal *R*-matrix element: + for the 22 and 33 elements, and − for the 11 elements. Hence, any *d*-tensor element with an odd number of 1's is zero, for example, $d_{111} = 0$ and $d_{122} = 0$. This mirror symmetry leads to all but two of the zeroes (d_{23} and d_{34}) in Equation 3.98.

Examples with inversion symmetry and reflection symmetry consist of diagonal *R*-matrices simplifying the summation in the tensor transformation equation. An example where the sum does not collapse is the rotational symmetry of the 3m class. The 3m class has 120°, 240°, and −120° rotational symmetry about the axis shown in Figure 3.5. The *R*-matrix for the 120° rotation is

$$\begin{bmatrix} -\dfrac{1}{2} & \dfrac{\sqrt{3}}{2} & 0 \\ -\dfrac{\sqrt{3}}{2} & -\dfrac{1}{2} & 0 \\ 0 & 0 & 1 \end{bmatrix}. \tag{3.102}$$

The transformation of the *d*-tensor under a 120° rotation operation follows the same rules as before, except that now the number of terms in the summation is larger. For example, we look at d_{211},

$$d'_{211} = R_{2\alpha}R_{1\beta}R_{1\gamma}d_{\alpha\beta\gamma}, \tag{3.103}$$

where the repeated indices over α, β, and γ indicate a summation over these indices. Given the specific *R*-matrix in Equation 3.102, the elements $R_{\alpha 2}$, $R_{\beta 1}$, and $R_{\gamma 1}$ are nonzero only for α, β, γ = 1 or 2. The summation in Equation 3.103 is expanded, keeping only nonzero *R*-matrix terms,

$$d'_{211} = R_{21}R_{11}R_{11}d_{111} + R_{12}R_{11}R_{21}d_{112} + R_{12}R_{21}R_{11}d_{121} + R_{21}R_{12}R_{12}d_{122}$$
$$+ R_{22}R_{11}R_{11}d_{211} + R_{22}R_{11}R_{12}d_{212} + R_{22}R_{12}R_{11}d_{221} + R_{22}R_{12}R_{12}d_{222}. \tag{3.104}$$

The summation in Equation 3.104 is further simplified by recognizing that several of the terms are zero due to the previous mirror symmetry: d_{111}, d_{122}, d_{212}, and d_{221} are all zero. Now, we substitute in the particular *R*-matrix values from Equation 3.102,

$$d'_{211} = \frac{3}{8}d_{112} + \frac{3}{8}d_{121} - \frac{1}{8}d_{211} - \frac{3}{8}d_{222}. \tag{3.105}$$

Kleinman's symmetry simplifies Equation 3.105 to

$$d'_{21} = \frac{5}{8}d_{21} - \frac{3}{8}d_{22}. \tag{3.106}$$

Since the rotation transformation is also a symmetry operation for 3m crystals, $d'_{21} = d_{21}$, and hence,

$$d_{21} = -d_{22}. \tag{3.107}$$

This same rotational symmetry operation along with $d_{133} = 0$ is used to show that $d_{233} = 0$ for 3m crystals (see Problem 3.16). A similar procedure is carried out for each of the 32 point groups giving the d-matrices shown in Table 3.2. Note that the process of using crystal symmetry extends to higher-order susceptibilities and to nonparametric nonlinear processes.

3.5 ELECTRO-OPTIC EFFECT

The derivation of a nonlinear polarization given in the previous sections is typically used for most second-order nonlinear effects. However, the so-called linear EO effect, also known as the Pockels' effect, uses a different formalism and a different terminology. The EO effect is the mixing of a static or slowly varying field with an optical field, which results in a change in a material's index of refraction. The index of refraction is related to ε_{ij}, so it is natural to look at perturbations in ε_{ij} due to an applied static field. The unperturbed expression for the displacement vector's complex amplitude is

$$D_i = \varepsilon_0 \varepsilon_{ij} A_j, \tag{3.108}$$

where A_j are the components of the complex amplitude for the incident optical field. We note that the polarization in the constitutive relationship, $\vec{D} = \varepsilon_0 \vec{E} + \vec{P}$, can be split into linear and nonlinear components,

$$\vec{D} = \varepsilon_0 \vec{E} + \vec{P}^{(1)} + \vec{P}^{(2)} + \cdots. \tag{3.109}$$

Therefore, we write the complex amplitude of the displacement vector as a linear term plus the nonlinear contribution,

$$D_i = \varepsilon_0 \varepsilon_{ij} A_j + P_i^{(2)}. \tag{3.110}$$

The nonlinear polarization results from mixing of the optical field with the static field. The nonlinear polarization is found by writing out the sum in Equation 3.66:

$$\underline{P}_{i,\text{Pockels}}^{(2)}(t) = \frac{\varepsilon_0}{4} \sum_{j,k} \sum_{\substack{n=-\omega,0,0,\omega \\ m=-\omega,0,0,\omega}} \chi_{ijk}^{(2)} A_j(\omega_m) A_k(\omega_n) e^{-i(\omega_n + \omega_m)t}$$

$$= \frac{\varepsilon_0}{4} \sum_{j,k} \chi_{ijk}^{(2)} \left[2A_j(\omega) A_k(0) + 2A_j(0) A_k(\omega) \right] e^{-i\omega t} + \text{c.c.}, \qquad (3.111)$$

where only frequency terms at ω are kept. Note that n and m run through $-\omega$, 0, 0, ω.[*] Double counting of zero frequency accounts for the assumption that the fields are written in the form

$$\underline{E} = \frac{1}{2} A(\omega) e^{-i\omega t} + \frac{1}{2} A(-\omega) e^{i\omega t}. \qquad (3.112)$$

When the field is static ($\omega = 0$), we maintain the same form as in Equation 3.112 by double counting the 0-frequency in the sum found in Equation 3.111.

We assume Kleinman's symmetry so that $\chi_{ijk} = \chi_{ikj}$, and hence, Equation 3.111 simplifies to

$$\underline{P}_{i,\text{Pockels}}^{(2)}(t) = 2\varepsilon_0 \sum_{j,k} d_{ijk} A_j(\omega) A_k(0) e^{-i\omega t} + \text{c.c.}, \qquad (3.113)$$

where the substitution of $\chi_{ijk}^{(2)} = 2d_{ijk}$ is made. To simplify the following calculations, we distinguish between the optical field and the static field by writing the static field, $A(0)$, as E.

With the new labeling, Equation 3.113 becomes

$$\underline{P}_i^{(2)}(\omega + 0) = 2\varepsilon_0 d_{ijk} A_j E_k e^{-i\omega t} + \text{c.c.}, \qquad (3.114)$$

---

[*] The treatment is much simpler if we take $\chi^{(2)}$ as a scalar:

$$\underline{P}^{(2)} = \varepsilon_0 \chi^{(2)} \left(\frac{1}{2} A(\omega) e^{-i\omega t} + \frac{1}{2} A(-\omega) e^{i\omega t} + E_{\text{static}} \right)^2$$

Performing the square and collecting terms at $e^{-i\omega t}$ give the complex amplitude,

$$P^{(2)} = 2\varepsilon_0 \chi^{(2)} A(\omega) E_{\text{static}}$$

where we use the Einstein notation. The complex amplitude of Equation 3.114 is

$$P_i^{(2)}(\omega + 0) = 4\varepsilon_0 d_{ijk} A_j E_k.$$ (3.115)

Substituting Equation 3.115 into Equation 3.110 gives

$$D_i = \varepsilon_0 (\varepsilon_{ij} + 4d_{ijk}E_k)A_j.$$ (3.116)

Equation 3.116 shows that the permittivity tensor is perturbed by $4d_{ijk}E_k$. In general, if ε_{ij} starts out in a diagonal basis, the application of the static field results in new principal axes and new principal indices. It is possible to proceed to find the new principal axes and indices using the nonlinear susceptibility notation, but the more common approach is to start with the inverse of Equation 3.116 and a notation that corresponds to the inverse. In what follows, the new notation is introduced, but we also show the correspondence between the two approaches and notations in Section 3.5.5.

3.5.1 EO EFFECTS AND THE r-MATRIX

With no applied static field, the inverse of Equation 3.116 is

$$A_i = \frac{1}{\varepsilon_0} B_{ij} D_j,$$ (3.117)

where B_{ij} is called the impermeability tensor. The logical basis set is the principal axis system such that B_{ij} is a diagonal matrix:

$$\ddot{B} = \begin{bmatrix} \dfrac{1}{\varepsilon_{xx}} & 0 & 0 \\ 0 & \dfrac{1}{\varepsilon_{yy}} & 0 \\ 0 & 0 & \dfrac{1}{\varepsilon_{zz}} \end{bmatrix} = \begin{bmatrix} \dfrac{1}{n_x^2} & 0 & 0 \\ 0 & \dfrac{1}{n_y^2} & 0 \\ 0 & 0 & \dfrac{1}{n_z^2} \end{bmatrix}.$$ (3.118)

When a voltage is applied to the crystal, the B_{ij} elements change:

$$B_{ij}' = B_{ij} + \Delta B_{ij}.$$ (3.119)

By calculating ΔB_{ij}, it is possible to find new principal indices and new principal axes for the perturbed crystal by diagonalizing the new B-matrix. In the prevalent terminology of the EO effect, the elements of ΔB_{ij} are related to the applied field through an expansion of B_{ij} in the applied static field,

$$\Delta B_{ij} = r_{ijk} E_k + s_{ijkl} E_k E_l \dots, \tag{3.120}$$

where r_{ijk} is the EO tensor and s_{ijkl} is the quadratic EO tensor. This expansion parallels that of the susceptibility in Equation 3.1. For the current discussion, only the effects to first-order in the static field are considered.

As with the contraction of d_{ijk} to d_{im}, it is possible to use a contracted notation for r_{ijk} and ΔB_{ij}. The main contraction for the EO effect comes from noting that, since B_{ij} is the inverse of ε_{ij}, and since $\varepsilon_{ij} = \varepsilon_{ji}$, it follows that $B_{ij} = B_{ji}$. This connection makes it possible to combine the indices i and j into a single index, m, which runs from 1 to 6 and uses the same substitutions as in Table 3.1. Using the contracted notation, Equation 3.120 becomes, to first-order in the static field,

$$\Delta B_m = r_{mk} E_k. \tag{3.121}$$

This contracted notation allows a simple matrix multiplication between a 6×3 matrix (called the r-matrix) and a 3×1 column vector for the applied DC field:

$$\Delta \vec{B} = \begin{bmatrix} r_{11} & r_{12} & r_{13} \\ r_{21} & r_{22} & r_{23} \\ r_{31} & r_{32} & r_{33} \\ r_{41} & r_{42} & r_{43} \\ r_{51} & r_{52} & r_{53} \\ r_{61} & r_{62} & r_{63} \end{bmatrix} \begin{bmatrix} E_x \\ E_y \\ E_z \end{bmatrix}. \tag{3.122}$$

The r-matrices for different point groups are given in Table 3.2. The form of the matrices follows from the same symmetry principals that shaped the d-matrices. For example, Kleinman's symmetry allows the free permutation of all the indices in r_{ijk}. Once we obtain the ΔB column vector from Equation 3.122, it is a straightforward exercise to determine the change in the index and the new principal axes. In matrix form, and using the contracted indices for ΔB_{ij}, the new B_{ij} matrix is

$$\ddot{B}' = \begin{bmatrix} \dfrac{1}{n_x^2} & 0 & 0 \\ 0 & \dfrac{1}{n_y^2} & 0 \\ 0 & 0 & \dfrac{1}{n_z^2} \end{bmatrix} + \begin{bmatrix} \Delta B_1 & \Delta B_6 & \Delta B_5 \\ \Delta B_6 & \Delta B_2 & \Delta B_4 \\ \Delta B_5 & \Delta B_4 & \Delta B_3 \end{bmatrix}, \tag{3.123}$$

where the ΔB_m elements in the perturbation matrix correspond to the resultant column vector in Equation 3.122. The location of a given ΔB_m in the $\Delta \ddot{B}$ matrix makes sense if we "undo" the contraction. For example, $\Delta B_5 = \Delta B_{13} = \Delta B_{31}$ which is the 13 or 31 location in the matrix. To find the new indices and principal axes, we diagonalize the overall B' matrix in Equation 3.123.

3.5.2 EXAMPLE CALCULATION OF EO EFFECT IN KH$_2$DPO$_4$

As an example, consider applying a static field along the z-axis of a KH$_2$DPO$_4$ (KDP) crystal. KDP is uniaxial with $n_x = n_y = n_o$. KDP has a $\overline{4}2m$ symmetry, and this information along with the applied field direction gives

$$\Delta \vec{B} = \begin{bmatrix} 0 & 0 & 0 \\ 0 & 0 & 0 \\ 0 & 0 & 0 \\ r_{41} & 0 & 0 \\ 0 & r_{41} & 0 \\ 0 & 0 & r_{63} \end{bmatrix} \begin{bmatrix} 0 \\ 0 \\ E_z \end{bmatrix} = \begin{bmatrix} 0 \\ 0 \\ 0 \\ 0 \\ 0 \\ r_{63}E_z \end{bmatrix}. \tag{3.124}$$

Hence, the new impermeability matrix is, according to Equation 3.123,

$$\begin{bmatrix} \dfrac{1}{n_x^2} & r_{63}E_z & 0 \\ r_{63}E_z & \dfrac{1}{n_y^2} & 0 \\ 0 & 0 & \dfrac{1}{n_z^2} \end{bmatrix}. \tag{3.125}$$

Using standard diagonalization techniques, the eigenvalues of Equation 3.125 are

$$\lambda_{1,2} = \frac{1}{n_o^2} \pm r_{63}E_z,$$

$$\lambda_3 = \frac{1}{n_z^2}. \tag{3.126}$$

The new principal indices are found from the inverse of the eigenvalues. That is,

$$\frac{1}{n_{1,2}^2} = \frac{1}{n_o^2}\left(1 \pm n_o^2 r_{63}E_z\right), \tag{3.127}$$

and $n_3 = n_z$. From Equation 3.127, we obtain

$$n_{1,2} = n_o \left(1 \pm n_o^2 r_{63} E_z \right)^{-1/2}.$$
(3.128)

For small nonlinearities, we may expand the square root to give

$$n_{1,2} = n_o \mp \frac{n_o^3 r_{63} E_z}{2}.$$
(3.129)

The corresponding principal axes with respect to the original axes are

$$\frac{1}{\sqrt{2}}\begin{pmatrix} 1 \\ 1 \\ 0 \end{pmatrix}, \quad \frac{1}{\sqrt{2}}\begin{pmatrix} 1 \\ -1 \\ 0 \end{pmatrix}, \quad \begin{pmatrix} 0 \\ 0 \\ 1 \end{pmatrix},$$
(3.130)

where

$$\begin{pmatrix} 1 \\ 0 \\ 0 \end{pmatrix}, \quad \begin{pmatrix} 0 \\ 1 \\ 0 \end{pmatrix}, \quad \begin{pmatrix} 0 \\ 0 \\ 1 \end{pmatrix},$$
(3.131)

are the unperturbed principal axes.

In the previous example, the KDP crystal starts out uniaxial, but as the field is applied, it becomes biaxial. The new principal axes are rotated about the z-axis by 45°. This large rotation may seem surprising considering that the axes shift dramatically even for the application of a weak field. However, recall that in this example, the crystal started out uniaxial, where the x- and y-axes are degenerate. We can visualize the process through the index ellipsoid (see Figure 2.8). Before the field is applied, the ellipsoid is rotationally invariant about the z-axis. When we apply the field, it begins to bulge along the lines $y = x$ and $y = -x$.

3.5.3 EO WAVE PLATES

The EO effect has several practical applications. One is a variable wave plate. Continuing with the example of Section 3.5.2, we see that if an incident optical field is propagating along the z-axis with a field polarized along the original x-axis, then it excites two eigenpolarizations along the (110) and (1 − 10) axes. These eigenpolarizations have different indices, and hence, the polarization state after exiting the crystal is modified. The phase difference between the eigenpolarizations is

$$\Delta\phi = \frac{2\pi(n_1 - n_2)L}{\lambda},$$
(3.132)

where n_1 and n_2 are the eigenindices for the two eigenpolarizations, L is the propagation length, and λ is the vacuum wavelength. The phase shift in Equation 3.132 is rewritten in terms of the applied field and the r-coefficient as

$$\Delta\phi = \frac{2\pi n_o^3 r_{41} E_z L}{\lambda}. \tag{3.133}$$

If the beam propagation is parallel to the applied field and if the field is uniform over the crystal length, then $E_z = V/L$ and Equation 3.133 is rewritten in terms of the applied voltage:

$$\Delta\phi_{\text{Longitudinal}} = \frac{2\pi n_o^3 r_{41} V}{\lambda}. \tag{3.134}$$

This type of geometry is called a longitudinal configuration. If the direction of propagation of the beam is perpendicular to the applied field (transverse), then the phase shift given by Equation 3.133 is

$$\Delta\phi_{\text{Transverse}} = \frac{2\pi n_o^3 r_{41} V L}{\lambda L_{\text{transverse}}}, \tag{3.135}$$

where we have made the substitution, $E_z = V/L_{\text{transverse}}$. The voltage at which the phase shift in Equation 3.134 or 3.135 is $\pi/2$ is called the quarter-wave voltage. For KDP in the longitudinal configuration, the quarter-wave voltage for a 1.064 μm beam is roughly 7000 V. The transverse EO configuration can bring the quarter-wave voltage down significantly (see Problem 3.19).

The EO effect results in a programmable wave plate where the degree of phase shift is proportional to the applied voltage. Polarizing the input beam along either of the eigenpolarization directions and then applying a time-varying voltage result in a phase-modulated beam.

3.5.4 EO SAMPLING: TERAHERTZ DETECTION

The EO effect relies on the application of a static or slowly varying field that modifies the material's refractive index. Our assumption in previous sections is that the field is directly applied to the material via an electrical contact structure. Another possibility is that the field comes from an EM wave passing through the medium. Consider EM waves with long wavelengths such that appreciable portions of the crystal have the same electric field strength at a given instant in time. Within this region, the applied field induces an index change that we measure with a separate probe beam, thus allowing a direct measurement of the long-wavelength field. As a specific example, we consider terahertz (THz) bursts generated using femtosecond lasers. The THz

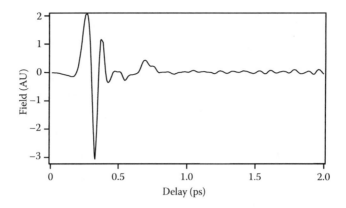

FIGURE 3.6   THz waveform measured using EO sampling.

portion of the EM spectrum (~0.3–10 THz) has promising applications for imaging through a variety of materials such as plastics.

One way to generate a THz field from a femtosecond laser is to optically rectify an ultrashort pulse in a $\chi^{(2)}$ crystal; however, we do not go into the details of generating the field here (Wu et al., 1996). An example of such a burst is shown in Figure 3.6.

Measurement of the THz field is possible if we use a probe that has a much shorter duration than a THz field's cycle duration. EO sampling experiments use a 100 fs or shorter pulse, which satisfies this condition. By overlapping the probe pulse with a portion of the THz waveform in the time domain, the probe beam sees a birefringence induced by that portion of the THz field (see Figure 3.7). Changing the relative delay between the probe pulse and the THz burst makes it possible to sample different parts of the THz waveform. We assume that the group velocity of the probe and THz fields are the same so that the relative position of the probe with respect to the THz waveform is fixed as they travel through the EO crystal. In a real system, group-velocity mismatch between the probe and THz waveforms leads to a smoothing of the measured THz waveform.

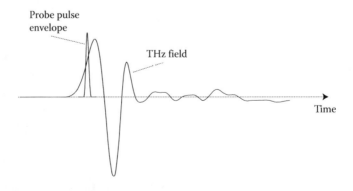

FIGURE 3.7   Probe pulse overlaps with a portion of the THz waveform in the EO crystal.

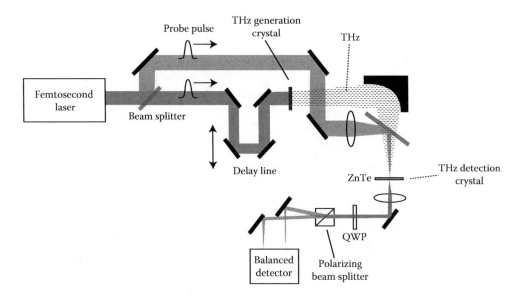

FIGURE 3.8   Schematic of THz generation and detection setup. A femtosecond laser is used to generate a THz burst, and it is used to subsequently probe the THz field. QWP, quarter-wave plate.

The THz field is measured by looking at the change in polarization state to the probe pulse due to the induced birefringence in the EO crystal. A particularly effective measurement uses an EO crystal such as ZnTe as the detection medium (Wu et al., 1996). ZnTe is not birefringent; so, in the absence of an applied field, the polarization state of the probe beam is unaffected, which allows us to use a null-background detection approach. Such an approach is shown in Figure 3.8 along with the THz generation setup based on optical rectification in a nonlinear crystal. Synchronization between the probe and THz field is obtained because both are generated using pulses from the same femtosecond laser.

THz detection works by overlapping the probe beam and THz beam in the ZnTe detection crystal. After passing through the ZnTe crystal, the probe beam is sent through a quarter-wave plate (QWP). The QWP is oriented such that the probe beam's output polarization is circular when no THz field is present. When the THz field is present, the output polarization state changes from circular to elliptical. Following the QWP is an anisotropic, polarizing beam splitter, which angularly separates orthogonal linear polarizations. When the THz field is absent, the circular polarization of the probe beam yields two separated outputs from the transmissive, anisotropic polarizer that have equal strengths. When the THz field is applied, the probe beam's polarization state goes from circular to elliptical resulting in outputs from the polarizer that have slightly different intensities. The two outputs from the polarizing beam splitter are sent to a balanced detector where the two signal currents are subtracted. When no THz field is present, the output of the balanced detector is zero, and when the THz field is applied, the detector output changes. Because the signal is nulled when no THz wave is present, we can increase the gain of the balanced detector system to measure small changes to the polarization state. In this way, small field strengths are detectable in the THz field.

A more analytical description is as follows. When an electric field is applied along the z-axis of the ZnTe crystal, we can show (see Problem 3.24) that $n_x' = n_o + \Delta n$ and $n_z' = n_o$, where n_o is the unperturbed index and where

$$\Delta n = \frac{n_o^3 r_{41} E_{THz}}{2}. \tag{3.136}$$

Going into the ZnTe crystal, the probe beam is linearly polarized and is oriented such that it propagates in the y-direction and has its polarization direction aligned at 45° to the z-axis. Its input polarization state is described by

$$\hat{E}_{Probe} = \frac{1}{\sqrt{2}} (\hat{x} + \hat{z}). \tag{3.137}$$

After emerging from the crystal, its polarization is modified:

$$\hat{E}_{Probe} = \frac{1}{\sqrt{2}} (\hat{x} + e^{-i\Gamma} \hat{z}), \tag{3.138}$$

where Γ is the phase shift between the x- and z-components of the probe's field, which for a crystal length of L is given by

$$\Gamma = \frac{\pi n_o^3 r_{14} E_{THz}}{\lambda_{probe}} L. \tag{3.139}$$

The probe beam is then transmitted through the QWP so that a 90° phase shift is introduced between the x- and z-components of the field, yielding

$$\hat{E}_{Probe} = \frac{1}{\sqrt{2}} (\hat{x} + i e^{-i\Gamma} \hat{z}). \tag{3.140}$$

The beam then passes through the polarizing beam splitter, which is oriented such that its transmitting axes are 45° to the $\hat{x}$ and $\hat{z}$ axes of Equation 3.140. The polarizer splits the probe beam into two, and in terms of the incident-normalized input one output from the polarizer is

$$E_1 = \frac{1}{2\sqrt{2}} (1 + i e^{-i\Gamma}). \tag{3.141}$$

The second output is

$$E_2 = \frac{1}{2\sqrt{2}}(-1 + ie^{-i\Gamma}). \tag{3.142}$$

These two fields are incident on two detectors, whose output currents are proportional to the squared magnitude of the incident fields

$$i_1 = i_0(1 + \sin\Gamma) \tag{3.143}$$

and

$$i_2 = i_0(1 - \sin\Gamma), \tag{3.144}$$

where i_0 is a constant with units of amperes with a magnitude dependent on the detector response, which we assume is the same for the two detectors. The balanced detector subtracts the two currents giving a final signal proportional to

$$\text{signal} \propto (i_1 - i_2) = 2i_0 \sin\Gamma. \tag{3.145}$$

Since the EO effect is weak, Γ is a small number, and we take $\sin\Gamma \sim \Gamma$. Using this result and Equation 3.139 gives a signal proportional to the THz electric field,

$$\text{signal} \propto E_{\text{THz}}. \tag{3.146}$$

Note that the direct proportionality is to the real THz field, and so the signal goes positive and negative when the THz field goes positive and negative. In other words, we obtain a signal that follows the real THz field.

3.5.5 CONNECTION BETWEEN *d* AND *r*

The formalism based on the r-tensor centered on calculating changes in birefringence is used in a wealth of applications ranging from Q-switch Pockels cells to frequency modulation. However, this linear EO effect is a second-order nonlinear process, and hence, a connection between the r- and d-tensor is possible. The connection between the r-tensor approach and a standard nonlinear polarization approach is to recognize from Equations 3.116 and 3.117 that

$$B_{ij} + \Delta B_{ij} = (\varepsilon_{ij} + 4d_{ijk}E_k)^{-1}. \tag{3.147}$$

Connecting *d*-coefficients to the *r*-coefficients starts with calculating the inverse of the perturbed permittivity tensor in Equation 3.147. The perturbed permittivity matrix is rewritten as

$$
\begin{bmatrix}
\varepsilon_{11} + \delta_1 & \delta_6 & \delta_5 \\
\delta_6 & \varepsilon_{22} + \delta_2 & \delta_4 \\
\delta_5 & \delta_4 & \varepsilon_{33} + \delta_3
\end{bmatrix},
\tag{3.148}
$$

where the δ's are the perturbations to the originally diagonal matrix. The perturbations are given by

$$
\delta_{ij} = 4d_{ijk}E_k.
\tag{3.149}
$$

The *ij* combination is contracted using the same substitutions as in Table 3.1. Writing out the δ's gives

$$
\begin{aligned}
\delta_1 &= \delta_{11} = 4d_{11k}E_k = 4(d_{11}E_x + d_{16}E_y + d_{15}E_z), \\
\delta_2 &= \delta_{22} = 4(d_{26}E_x + d_{22}E_y + d_{24}E_z), \\
\delta_3 &= \delta_{33} = 4(d_{35}E_x + d_{34}E_y + d_{33}E_z), \\
\delta_4 &= \delta_{23} = 4(d_{25}E_x + d_{24}E_y + d_{23}E_z), \\
\delta_5 &= \delta_{13} = 4(d_{15}E_x + d_{14}E_y + d_{13}E_z), \\
\delta_6 &= \delta_{12} = 4(d_{16}E_x + d_{12}E_y + d_{14}E_z).
\end{aligned}
\tag{3.150}
$$

Now, our task is to find the inverse of Equation 3.148. One can use standard linear algebra techniques (Strang, 2006). The inverse of $\bar{\bar{\varepsilon}}$ is the adjugate of $\bar{\bar{\varepsilon}}$ divided by the determinant of $\bar{\bar{\varepsilon}}$. The adjugate is the transpose of the cofactor matrix of $\bar{\bar{\varepsilon}}$. The cofactor matrix, to first-order in δ, is

$$
\begin{bmatrix}
\varepsilon_{22}\varepsilon_{33} + \delta_2\varepsilon_{33} + \delta_3\varepsilon_{22} & -\delta_6\varepsilon_{33} & -\delta_5\varepsilon_{22} \\
-\delta_6\varepsilon_{33} & \varepsilon_{11}\varepsilon_{33} + \delta_1\varepsilon_{33} + \delta_3\varepsilon_{11} & -\delta_4\varepsilon_{11} \\
-\delta_5\varepsilon_{22} & -\delta_4\varepsilon_{11} & \varepsilon_{22}\varepsilon_{11} + \delta_2\varepsilon_{11} + \delta_1\varepsilon_{22}
\end{bmatrix}.
\tag{3.151}
$$

The adjugate of $\bar{\bar{\varepsilon}}$ is the same matrix as Equation 3.151 since the matrix is symmetric. The determinant of Equation 3.148, to first-order in δ, is

$$
\det\bar{\bar{\varepsilon}} = \varepsilon_{11}\varepsilon_{22}\varepsilon_{33}\left[1 + \frac{\delta_1}{\varepsilon_{11}} + \frac{\delta_2}{\varepsilon_{22}} + \frac{\delta_3}{\varepsilon_{33}}\right].
\tag{3.152}
$$

The inverse of Equation 3.152 is, to order δ,

$$\frac{1}{\det \ddot{\varepsilon}} = \frac{1}{\varepsilon_{11}\varepsilon_{22}\varepsilon_{33}}\left[1 - \frac{\delta_1}{\varepsilon_{11}} - \frac{\delta_2}{\varepsilon_{22}} - \frac{\delta_3}{\varepsilon_{33}}\right]. \tag{3.153}$$

The inverse of $\ddot{\varepsilon}$ is the matrix in Equation 3.151 multiplied by Equation 3.153. Again, to first-order in δ, the multiplication gives the following matrix:

$$\ddot{B} = \frac{1}{\ddot{\varepsilon}} = \begin{bmatrix} \dfrac{1}{\varepsilon_{11}} - \dfrac{\delta_1}{\varepsilon_{11}^2} & -\dfrac{\delta_6}{\varepsilon_{11}\varepsilon_{22}} & -\dfrac{\delta_5}{\varepsilon_{11}\varepsilon_{33}} \\[2ex] -\dfrac{\delta_6}{\varepsilon_{11}\varepsilon_{22}} & \dfrac{1}{\varepsilon_{22}} - \dfrac{\delta_2}{\varepsilon_{22}^2} & -\dfrac{\delta_4}{\varepsilon_{22}\varepsilon_{33}} \\[2ex] -\dfrac{\delta_5}{\varepsilon_{11}\varepsilon_{33}} & -\dfrac{\delta_4}{\varepsilon_{22}\varepsilon_{33}} & \dfrac{1}{\varepsilon_{33}} - \dfrac{\delta_3}{\varepsilon_{33}^2} \end{bmatrix}. \tag{3.154}$$

Assuming that the unperturbed basis is diagonal with $1/\varepsilon_{11}$, $1/\varepsilon_{22}$, and $1/\varepsilon_{33}$ as the diagonal entries, then the perturbation matrix is split out of Equation 3.154:

$$-\begin{bmatrix} \dfrac{\delta_1}{\varepsilon_{11}^2} & \dfrac{\delta_6}{\varepsilon_{11}\varepsilon_{22}} & \dfrac{\delta_5}{\varepsilon_{11}\varepsilon_{33}} \\[2ex] \dfrac{\delta_6}{\varepsilon_{11}\varepsilon_{22}} & \dfrac{\delta_2}{\varepsilon_{22}^2} & \dfrac{\delta_4}{\varepsilon_{22}\varepsilon_{33}} \\[2ex] \dfrac{\delta_5}{\varepsilon_{11}\varepsilon_{33}} & \dfrac{\delta_4}{\varepsilon_{22}\varepsilon_{33}} & \dfrac{\delta_3}{\varepsilon_{33}^2} \end{bmatrix} = \begin{bmatrix} \Delta B_1 & \Delta B_6 & \Delta B_5 \\ \Delta B_6 & \Delta B_2 & \Delta B_4 \\ \Delta B_5 & \Delta B_4 & \Delta B_3 \end{bmatrix}. \tag{3.155}$$

By including the expressions for δ_m from Equation 3.150, it can be shown that the ΔB_m in Equation 3.155 can be written as a column vector, $\Delta\vec{B}$, which results from the multiplication of a 6×3 matrix of d-coefficients and a 3×1 vector of the applied DC field,

$$\Delta\vec{B} = -4 \begin{bmatrix} \dfrac{d_{11}}{n_x^4} & \dfrac{d_{21}}{n_x^4} & \dfrac{d_{31}}{n_x^4} \\[2ex] \dfrac{d_{12}}{n_y^4} & \dfrac{d_{22}}{n_y^4} & \dfrac{d_{32}}{n_y^4} \\[2ex] \dfrac{d_{13}}{n_z^4} & \dfrac{d_{23}}{n_z^4} & \dfrac{d_{33}}{n_z^4} \\[2ex] \dfrac{d_{14}}{n_y^2 n_z^2} & \dfrac{d_{24}}{n_y^2 n_z^2} & \dfrac{d_{34}}{n_y^2 n_z^2} \\[2ex] \dfrac{d_{15}}{n_x^2 n_z^2} & \dfrac{d_{25}}{n_x^2 n_z^2} & \dfrac{d_{35}}{n_x^2 n_z^2} \\[2ex] \dfrac{d_{16}}{n_x^2 n_y^2} & \dfrac{d_{26}}{n_x^2 n_y^2} & \dfrac{d_{36}}{n_x^2 n_y^2} \end{bmatrix} \begin{bmatrix} E_x \\ E_y \\ E_z \end{bmatrix}, \tag{3.156}$$

where Kleinman's symmetry is used to make substitutions such as $d_{26} = d_{12}$. Furthermore, the linear EO effect is given by Equation 3.122 and is reproduced here:

$$\Delta \vec{B} = \begin{bmatrix} r_{11} & r_{12} & r_{13} \\ r_{21} & r_{22} & r_{23} \\ r_{31} & r_{32} & r_{33} \\ r_{41} & r_{42} & r_{43} \\ r_{51} & r_{52} & r_{53} \\ r_{61} & r_{62} & r_{63} \end{bmatrix} \begin{bmatrix} E_x \\ E_y \\ E_z \end{bmatrix}. \tag{3.157}$$

Relating Equation 3.156 to Equation 3.157 gives the connection between d and r,

$$r_{mi} = -4 \frac{d_{im}}{n_m^4}, \tag{3.158}$$

where m runs from 1 to 6 (standard contraction notation), i runs from 1 to 3, and

$$n_m^4 = n_i^2 n_j^2. \tag{3.159}$$

With i and j determined from the contracted notation, for example, $i = 1$, $j = 2$ for $m = 6$.

When relating d and r, Miller's rule should be used to estimate the dispersion between DC and optical frequencies.

PROBLEMS

3.1 Three fields are incident on a nonlinear crystal at frequencies ω_2, ω_3, and ω_4 with $\omega_2 > \omega_3 > \omega_4$. What are the possible output frequencies for (a) a $\chi^{(2)}$ process and (b) a $\chi^{(3)}$ process?

3.2 Estimate the power level in a focused beam required to make the expansion in Equation 3.1 break down. Assume that the intensity is given by $P/(0.01 \text{ cm}^2)$, where P is the laser power, that $d = 10$ pm/V, and that the index of refraction is 1.6.

3.3 Faraday's effect: The classical harmonic oscillator equation of motion with a static magnetic field is

$$m \frac{d^2 \vec{r}}{dt^2} = -m\omega_0^2 \vec{r} - e\vec{E} - e \frac{d\vec{r}}{dt} \times (B\hat{z}), \tag{3.160}$$

where damping is ignored and $\vec{B}$ is applied along the z-axis. For an incident beam with an electric field of the form $\vec{E}e^{-i\omega t}$, we assume that the position vector is also of the form $\vec{r}e^{-i\omega t}$.

a. Write out Equation 3.160 in terms of complex amplitudes for each Cartesian component. Show that the resultant equations can be written as

$$
\begin{bmatrix}
a & ib & 0 \\
-ib & a & 0 \\
0 & 0 & a
\end{bmatrix}
\vec{r} = -\frac{e}{m}\vec{E},
\tag{3.161}
$$

where a and b are constants dependent on parameters given in the problem.

b. Use Equation 3.161 to solve for $\vec{r}$, and use the polarization definition $\vec{P} = -Ne\vec{r} = \varepsilon_0\, \ddot{\chi} \cdot \vec{E}$ to find the susceptibility matrix.

c. Show that the eigenvectors of the susceptibility matrix are right and left circularly polarized.

3.4 A laser at frequency ω is incident on a nonlinear material. Use the anharmonic oscillator approach to calculate $\chi^{(2)}(2\omega; \omega, \omega)$ in terms of the parameters given in Equation 3.30. Assume that the incident field is linearly polarized and is written as

$$
\underline{E} = \frac{1}{2}Ae^{i(kz-\omega t)} + \text{c.c.}
\tag{3.162}
$$

3.5 Two lasers at frequencies ω_2 and ω_3 are incident on a centrosymmetric nonlinear material. Use the anharmonic oscillator approach to calculate $\chi^{(3)}(\omega_1; \omega_2, \omega_3, \omega_3)$, where $\omega_1 = \omega_2 + 2\omega_3$. Solve for $\chi^{(3)}(\omega_1; \omega_2, \omega_3, \omega_3)$ in terms of the parameters given in Equation 3.30. Assume that the relationship between the polarization's complex amplitude and the input field amplitudes is written as

$$
P^{(3)} = \frac{3\varepsilon_0}{4}\chi^{(3)}(\omega_2 + 2\omega_3; \omega_2, \omega_3, \omega_3)A_2A_3^2.
\tag{3.163}
$$

3.6 A certain crystal has an index of refraction of approximately 1.6 and $\chi^{(2)}$ equal to 4.4 pm/V. A different material is discovered to have an index of 2.2. Use Miller's rule to estimate $\chi^{(2)}$ in the new material.

3.7 $\chi^{(2)}$ is measured for frequency doubling 1.064 to 0.532 µm to be 1.2 pm/V. The index of refraction is 1.654 for both wavelengths. Use Miller's rule to estimate $\chi^{(2)}$ for SHG of 1.6–0.8 µm, given that the index of refraction for this interaction is 1.646.

3.8 For lossless interactions, justify the Kleinman's symmetry condition, $\chi^{(2)}(\omega_3; \omega_1, -\omega_2) \approx \chi^{(2)}(\omega_2; \omega_1, -\omega_3) \approx \chi^{(2)}(\omega_1; \omega_2, \omega_3)$ by using the classical expressions for $\chi^{(2)}$. The classical expression for $\chi^{(2)}(\omega_3; \omega_1, -\omega_2)$ is given in Equation 3.45.

3.9 Show that the matrix elements given in Equation 3.80 are equated using Kleinman's symmetry.

3.10 The nonlinear polarization for DFG is written in the form of a d-matrix multiplied by a column vector as in Equation 3.77. Derive nonlinear polarization expressions for (a) SFG and (b) SHG that are also in the same form, that is, a d-matrix multiplied by the appropriate column vector. For the second-harmonic case, assume that one field is present at the input given by Equation 3.162.

3.11 Two beams at ω_P and ω_S are incident on a KDP crystal. The designations P and S stand for "pump" and "signal," respectively. KDP is uniaxial, and it has a symmetry class of $\overline{4}2m$. The fields' k-vectors are traveling in the y–z-plane at an angle of θ to the z-axis. The pump field is e-polarized, and the signal field is o-polarized.
   a. Write out the electric field of the pump and the signal in Cartesian coordinates. You may assume that the fields have a magnitude of E_P and E_S, respectively.
   b. Find the second-order polarization's complex amplitude for a difference-frequency process $(\omega_P - \omega_S)$.
   c. Find the projection of this polarization onto the o-axis and e-axis to find d_{eff} for each interaction.

3.12 Verify that a DFG process between an e-polarized wave and an o-polarized wave in a class $\overline{4}$ crystal has d_{eff} as given in Table 3.3. Verify the result for both cases where the DFG is either e- or o-polarized. (See Problem 3.11 for an example of a step-by-step procedure.)

3.13 Calculate $d_{eff}(\theta, \phi = 0)$ for a DFG process in Lithium niobate where ω_1, ω_2, and ω_3 are all e-polarized. θ is the angle between the k-vector and the z-axis. Lithium niobate is uniaxial and has a symmetry class of 3m.

3.14 For the crystal symmetry class 222 (orthorhombic 222), the group symmetry operations are the identity operation, and $180°$ rotations about the x-, y-, and z-axes.
   a. Use Kleinman's symmetry to show $d_{14} = d_{25} = d_{36}$.
   b. Write down the R-matrix for each of the above four transformations that leave the crystal invariant, for example, Identity operation:

$$R_{\text{Identity}} = \begin{bmatrix} 1 & 0 & 0 \\ 0 & 1 & 0 \\ 0 & 0 & 1 \end{bmatrix}.$$

   c. Use the tensor transformation rules along with the symmetry operations to show that $d_{11} = d_{22} = d_{33} = 0$. Note, you may obtain, for example, $d_{11} = d_{11}$ instead of $d_{11} = -d_{11}$ for *one* of the R-matrices. This means that d_{11} is a nonzero element for *that* symmetry operation. It will be the case that it must be zero for one or more of the other symmetry operations.
   d. Show that d_{14} is a nonzero element under all the symmetry operations.

3.15 The crystal class $\bar{4}2m$ has the following symmetry properties (along with the identity operation) represented as R-matrices:

$$R_{S4} = \begin{bmatrix} 0 & 1 & 0 \\ -1 & 0 & 0 \\ 0 & 0 & -1 \end{bmatrix}, \quad R_{S4^{-1}} = \begin{bmatrix} 0 & -1 & 0 \\ 1 & 0 & 0 \\ 0 & 0 & -1 \end{bmatrix},$$

$$R_{\sigma d1} = \begin{bmatrix} 0 & 1 & 0 \\ 1 & 0 & 0 \\ 0 & 0 & 1 \end{bmatrix}, \quad R_{\sigma d2} = \begin{bmatrix} 0 & -1 & 0 \\ -1 & 0 & 0 \\ 0 & 0 & 1 \end{bmatrix}.$$

a. What are the transformations represented by each matrix? That is, is it a rotation by a certain angle plus reflection, and so on?

b. Show that d_{11} and $d_{21} = 0$ for crystals with this symmetry.

c. Use one of the symmetry operations to show that $d_{14} = d_{25}$ (do not use Kleinman's symmetry to equate the two).

3.16 In class 3m crystals, $d_{13} = 0$. Use this d-value plus the 120° rotational symmetry about the z-axis (see R-matrix below) to show that $d_{23} = 0$.

$$R = \begin{bmatrix} -\dfrac{1}{2} & \dfrac{\sqrt{3}}{2} & 0 \\ -\dfrac{\sqrt{3}}{2} & -\dfrac{1}{2} & 0 \\ 0 & 0 & 1 \end{bmatrix}.$$

3.17 Waveguide EO modulators in lithium niobate are commonly used for telecommunications amplitude and phase modulators. A waveguide is fabricated on the surface of a lithium niobate crystal, and two electrodes are placed on either side of the waveguide (see Figure 3.9). The separation between the electrodes is 10 μm. Light passing through the waveguide propagates in the x-direction

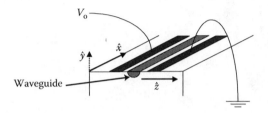

FIGURE 3.9 Lithium niobate waveguide phase modulator. Light propagates in the waveguide along the x-direction. The light is polarized along the y-direction, and the field is applied in the z-direction.

and is polarized in the z-direction. The applied static electric field is along the z-direction, and the waveguide is 4 cm long. Lithium niobate is uniaxial with $n_o = 2.211$ and $n_z = 2.139$ at 1.55 µm, and it is a class 3m crystal with $r_{33} = 30$ pm/V.

    a. Show that the output polarization state remains linear when an electric field is applied to the electrode. (Hint: show that the principal axis directions are unchanged by the application of the field.)

    b. Calculate V_π, which is the voltage that gives rise to a π phase shift as compared to when no field is applied.

3.18 Use Kleinman's permutation symmetry to show that Equation 3.111 simplifies to Equation 3.113.

3.19 Equations 3.134 and 3.135 give expressions for the longitudinal and transverse EO phase shifts in KDP. Compare the voltage required to attain a given phase shift for the transverse configuration to the longitudinal configuration. Under what condition does the transverse configuration attain a given phase shift with a lower voltage than the longitudinal configuration? What is the practical limit on reducing this voltage?

3.20 An EO modulator using two identical class 3m crystals is shown in Figure 3.10. The second crystal is rotated 90° about the y-axis with respect to the first crystal. The incident laser is polarized at 45° with respect to the x- and z-axes, and it propagates in the y-direction through both crystals. A static electric field is applied along $+z$-direction in the first crystal and along the $-z$-direction in the second crystal. The length in the y-direction is L_y and the length in the z-direction is L_z.

    a. Show that the output polarization state is linear when no voltage is applied to the crystals.

    b. Calculate the quarter-wave voltage in terms of the given parameters and in terms of the r-coefficients for a 3m crystal.

3.21 An electric field is applied to a hexagonal 222 crystal along the y-axis. Assuming that the crystal starts out nonbirefringent with $n_x = n_y = n_z = n_o$,

    a. What are the new principal indices?

    b. What are the new principal axes?

    c. If the electric field applied to the crystal is modulated, what should the incident polarization direction be to make a phase modulator? (Assume propagation along the y-axis.)

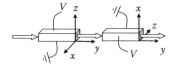

FIGURE 3.10 Two-crystal EO modulator. Note that the crystals are rotated 90° with respect to the y-axis.

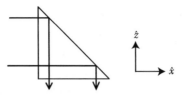

FIGURE 3.11   EO beam steering.

3.22 Beam steering with the EO effect. A class $\bar{6}m2$, nonbirefringent crystal (index n_o) is cut into a prism as shown in Figure 3.11. The perpendicular sides of the prism have length L.

    a. A voltage, V, is applied along the y-axis of the crystal. The positive potential is on the $+y$ surface, and the ground is on the $-y$ surface of the prism. The length of the prism in the y-direction is L_y. If the input is polarized along the z-axis, what is the phase shift from one side of the beam to the other at the output.

    b. Convert the phase shift into the number of wavelengths (be sure to use wavelength in the medium and not in vacuum). Hence, one side of the beam is advanced with respect to the other by this number of wavelengths. Use this information to calculate the beam deflection as a function of voltage. You can also do this problem using the approach given in Problem 2.13.

3.23 A variation on the beam deflector in Problem 3.22 is to cascade multiple steering elements. A particularly efficient design is one that uses a ferroelectric material such as lithium niobate. In these crystals, it is possible to engineer the crystal domain structure (see Chapter 5 for more details). A beam steering example is shown in Figure 3.12 where prism-shaped regions with the z-axis pointing up are adjacent to prism regions with the z-axis pointing down. Note that the arrows in Figure 3.12a indicate the direction of the z-axis. The top and bottom of the crystal are metal plated so that an electric field can be applied along the $\pm z$-axis. Referring now to Figure 3.12b, a laser with a width of L_y polarized in the z-direction passes through the device traveling in the x-direction. The laser

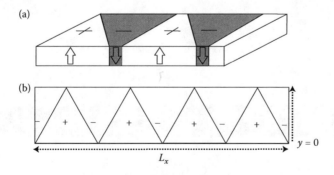

FIGURE 3.12   EO beam steering with a poled crystal: (a) three-dimensional view and (b) top view.

wavelength in vacuum is λ_{vac}. The crystal is lithium niobate (uniaxial, class 3m) with a thickness of L_z in the z-direction.

a. Solve for the index of refraction that the laser sees in the + and − regions as a function of applied voltage.

b. As the laser propagates through the structure, the top ($y = L_y$) and bottom ($y = 0$) of the laser beam experience different phase shifts. Calculate the optical phase difference between $y = 0$ and $y = L_y$.

c. Convert the phase shift into the number of wavelengths (be sure to use wavelength in the medium and not in vacuum). Hence, one side of the beam is advanced with respect to the other by this number of wavelengths. Use this information to calculate the beam deflection as a function of voltage.

d. Show that your result in (c) does not depend on the number of prisms in the structure, provided that the beam width is equal to L_y.

3.24 ZnTe is a class $\bar{4}3m$ crystal with $n_x = n_y = n_z = n_o$. Show that when a field, E_{THz}, is applied along the crystal's z-axis that n_x changes from n_o to $n_o + \Delta n$, where Δn is given by Equation 3.136, and show that n_z does not change for this applied field.

3.25 Calculate each matrix element in Equation 3.156 to obtain ΔB_1 through ΔB_6. Use these results to show that the column vector for $\Delta \ddot{B}$ can be written as in Equation 3.157.

REFERENCES

Dmitriev, V. G., G. G. Gurzadyan, and D. N. Nikogosyan. 1997. *Handbook of Nonlinear Optical Crystals*. Heidelberg: Springer-Verlag.

Koster, G. F. 1963. *Properties of the Thirty-Two Point Groups*. Cambridge, MA: MIT Press.

Miller, R. C. 1964. Optical second harmonic generation in piezoelectric crystals. *Applied Physics Letters* 5: 17–19.

Strang, G. 2006. *Linear Algebra and Its Applications* (4th edition). Belmont, CA: Thomson Brooks/Cole.

Sutherland, R. L. 2003. *Handbook of Nonlinear Optics* (2nd edition). New York, NY: Marcel Dekker, Inc.

Wu, Q., M. Litz, and X.-C. Zhang. 1996. Broadband detection capability of ZnTe electro-optic field detectors. *Applied Physics Letters* 68: 2924–2926.

Zernike, F. and J. E. Midwinter. 1973. *Applied Nonlinear Optics*. New York, NY: John Wiley & Sons.

FURTHER READING

Armstrong, J. A., N. Bloembergen, J. Ducuing, and P. S. Pershan. 1962. Interactions between light waves in a nonlinear dielectric. *Physical Review* 127: 1918–1939.

Boyd, G. D. and D. A. Kleinman. 1968. Parametric interaction of focused Gaussian light beams. *Journal of Applied Physics* 39: 3597.

Boyd, R. W. 2008. *Nonlinear Optics*. Boston, MA: Academic Press.

Kleinman, D. A. 1962a. Theory of second harmonic generation of light. *Physical Review* 128: 1761–1775.

Kleinman, D. A. 1962b. Nonlinear dielectric polarization in optical media. *Physical Review* 126: 1977–1979.

Maker, P. D., R. W. Terhune, M. Nisenoff, and C. M. Savage. 1962. Effects of dispersion and focusing on the production of optical harmonics. *Physical Review Letters* 8: 21–22.

Pershan, P. S. 1963. Nonlinear optical properties of solids: Energy considerations. *Physical Review* 130: 919–929.

Shen, Y. R. 1984. *The Principles of Nonlinear Optics*. New York, NY: John Wiley & Sons.

4

THREE-WAVE PROCESSES IN THE SMALL-SIGNAL REGIME

4.1 INTRODUCTION TO THE WAVE EQUATION FOR THREE FIELDS

In Chapter 3, we discussed the classical origins and means of determining the nonlinear polarization for various $\chi^{(2)}$ processes. A more complete description of nonlinear interactions includes the transformation of fields in the nonlinear medium. Maxwell's equations, including the nonlinear material response, provide such a description. The solution to the wave equation that follows from Maxwell's equations depends on the details of the specific system, and the possible number of scenarios is quite large. In this chapter, we consider simplified systems that shed light on some general trends in nonlinear optics. We first derive a wave equation for a low-efficiency (small-signal) three-wave interaction consisting of collinear plane waves. This interaction is characterized by two input beams and a generated output beam. In the small-signal regime, the amplitudes of the inputs do not change appreciably during the nonlinear interaction. The solution for the output in this small-signal regime introduces the important concept of phase matching, which determines how efficient the process can be. Designing an experiment also requires knowledge about the tolerances, for instance, the laser beams' divergence, bandwidth, and alignment; so the final sections of this chapter show that phase matching also allows for a description of the system's tolerance to changes in temperature, frequency, and other parameters.

4.1.1 WAVE EQUATION FOR A THREE-WAVE PROCESS

As in Chapter 2, we begin the derivation of the wave equation by taking the curl of Maxwell's equation (Equation 2.2),

$$\vec{\nabla} \times (\vec{\nabla} \times \vec{E}) = -\frac{\partial \vec{\nabla} \times \vec{B}}{\partial t} = -\frac{\partial \mu \vec{\nabla} \times \vec{H}}{\partial t} = -\mu \frac{\partial^2 \vec{D}}{\partial t^2}. \tag{4.1}$$

The left-hand side of this equation is

$$\vec{\nabla} \times \left(\vec{\nabla} \times \underline{\vec{E}} \right) = \vec{\nabla} \left(\vec{\nabla} \cdot \underline{\vec{E}} \right) - \nabla^2 \underline{\vec{E}}. \tag{4.2}$$

In Chapter 2, we took two approaches for evaluating the right-hand side of Equation 4.2. First, we assumed that the material was isotropic and homogeneous such that

$$\vec{\nabla} \cdot \underline{\vec{E}} = 0. \tag{4.3}$$

In the second approach, we applied Equation 4.2 to the more general case of birefringent crystals as a starting point for a derivation of Fresnel's equation. In this chapter, we include birefringence by considering linearly polarized beams in the crystal to be either e-waves or o-waves. However, we assume that the birefringence is weak such that $\vec{D}$ and $\vec{E}$ are essentially parallel even for e-waves. Hence, $\vec{\nabla} \cdot \vec{D} = 0$ and $\vec{\nabla} \cdot \vec{E} = 0$ for o-waves, and $\vec{\nabla} \cdot \vec{E} \approx 0$ for e-waves. Further, we assume that the medium is nonmagnetic so that $\mu = \mu_0$. With these assumptions, the wave equation has the same form as in the case of an isotropic medium:

$$\nabla^2 \underline{\vec{E}} = \mu_0 \frac{\partial^2 \vec{D}}{\partial t^2}. \tag{4.4}$$

However, instead of writing $\vec{D} = \varepsilon \vec{E}$, as done in Chapter 2, we use the defining constitutive relationship

$$\vec{D} = \varepsilon_0 \underline{\vec{E}} + \underline{\vec{P}}. \tag{4.5}$$

Upon substituting Equation 4.5 into Equation 4.4, we find

$$\nabla^2 \underline{\vec{E}} = \mu_0 \varepsilon_0 \frac{\partial^2 \underline{\vec{E}}}{\partial t^2} + \mu_0 \frac{\partial^2 \underline{\vec{P}}}{\partial t^2}. \tag{4.6}$$

Equation 4.6 is a wave equation with the polarization as a source term. At this point, the linear and nonlinear wave equations part company; we expand the polarization to include nonlinear effects. The polarization is split into linear and nonlinear parts,

$$\vec{\underline{P}} = \vec{\underline{P}}^{L} + \vec{\underline{P}}^{NL},$$ (4.7)

so that the wave equation (Equation 4.6) becomes

$$\nabla^2 \vec{\underline{E}} = \mu\varepsilon_0 \frac{\partial^2 \vec{\underline{E}}}{\partial t^2} + \mu \frac{\partial^2 \vec{\underline{P}}^{L}}{\partial t^2} + \mu \frac{\partial^2 \vec{\underline{P}}^{NL}}{\partial t^2}.$$ (4.8)

We assume that the electric field consists of monochromatic fields and hence write

$$\vec{\underline{E}} = \frac{1}{2}\sum_{m}\left[\vec{A}'_{m}e^{-i\omega_m t} + \text{c.c.}\right],$$ (4.9)

where the polarization direction of the mth field corresponds to ω_m and is either an e-wave or an o-wave. We write the complex amplitude as $\vec{A}'$ here, and later we switch to $\vec{A}$ when we factor out the rapidly varying spatial component of the field.

As shown in Chapter 3, the nonlinear polarization induced by the field in Equation 4.9 can be at frequencies not present in the input. In this chapter, we consider only $\chi^{(2)}$ processes and hence the nonlinear polarization contains terms at all the possible sum- and difference-frequency combinations of $\{\omega_m\}$ (see Equation 3.6). However, we need not consider all these nonlinear frequency terms at once. We pick the one that is of interest for a particular application and ignore the others. The justification for this procedure is the same as in Section 3.2.2; since the nonlinear interaction is weak, the different processes do not appreciably affect each other. We see in Section 4.1.3 that another effect, phase matching, provides further justification for picking one process.

Of the possible nonlinear processes, we consider difference-frequency generation (DFG) (or equivalently optical parametric amplification [OPA], see Figure 4.1) between ω_1 and ω_2, such that

$$\omega_3 = \omega_1 - \omega_2.$$ (4.10)

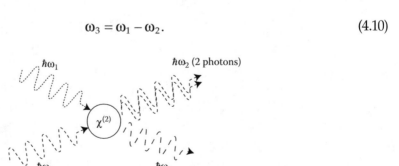

FIGURE 4.1 Difference-frequency process showing that a high-energy photon at ω_1 splits into two photons at ω_2 and ω_3. The input at ω_2 is amplified in the process.

The convention for this book is

$$\omega_1 > \omega_2 \geq \omega_3. \tag{4.11}$$

The nonlinear polarization for this difference-frequency process is written in complex form as

$$\underline{\vec{P}}^{(2)}(t) = \frac{1}{2}\vec{P}^{(2)}(\omega_3)e^{-i\omega_3 t} + \text{c.c.} \tag{4.12}$$

The nonlinear polarization oscillates at frequency ω_3 and in turn it drives a field at the same frequency. We look now at the wave equation for the field at ω_3. The linear polarization at this frequency is

$$\underline{\vec{P}}^{\text{L}} = \frac{1}{2}\varepsilon_0\chi^{(1)}\vec{A}_3'e^{-i\omega_3 t} + \text{c.c.}, \tag{4.13}$$

where we assume a lossless medium so that the linear susceptibility $\chi^{(1)}$ is real. Substituting Equations 4.12 and 4.13 into Equation 4.8 leads to a wave equation in terms of complex amplitudes:

$$\nabla^2\vec{A}_3' + \frac{\omega_3^2 n_3^2}{c^2}\vec{A}_3' = -\omega_3^2\mu_0\vec{P}^{(2)}. \tag{4.14}$$

Here, we let $(\partial/\partial t) \to -i\omega_3$, since the complex amplitudes are independent of time for monochromatic fields, and we set $1 + \chi^{(1)} = n^2$. Equation 4.14 is recast as a scalar equation since the driven field is linearly polarized (e-wave or o-wave):

$$\nabla^2 A_3' + \frac{\omega_3^2 n_3^2}{c^2}A_3' = -\omega_3^2\mu_0 P^{(2)}. \tag{4.15}$$

4.1.2 SLOWLY VARYING ENVELOPE APPROXIMATION EXTENDED

The fields corresponding to collinear plane waves have no transverse variation; and so we let the propagation direction be $\hat{z}$ and therefore $\nabla^2 \to (\partial/\partial z^2)$. One of the key assumptions in the derivation of Equation 4.15 is the power-series expansion of the polarization. For this expansion to be valid, the nonlinear terms must be a small correction to the linear polarization. So it is expected that the growth or attenuation of a field will occur gradually, over many optical cycles. This assumption, called the slowly varying envelope approximation (SVEA), was discussed in Section 2.6.1 and derived for the case when the nonlinear polarization appearing on the right-hand side of Equation 4.15 vanished. Substituting the results from Equation 2.120, neglecting diffraction and absorption, into the left-hand side of Equation 4.15 gives a first-order differential equation for the amplitude A_3,

$$2ik_3 \frac{dA_3}{dz} e^{ik_3z} = -\omega_3^2 \mu_0 P^{(2)}(\omega_3).$$ (4.16)

The wave vector amplitude is given by the dispersion relation

$$k_3^2 = \frac{\omega_3^2 n_3^2}{c^2},$$ (4.17)

Note that the same SVEA derivation may be carried out for higher-order nonlinearities with the only difference being in the nonlinear polarization term. In general, the wave equation for a plane wave at a given frequency (ω, $k = \omega n(\omega)/c$) in the SVEA is

$$\boxed{2ik \frac{dA}{dz} e^{ikz} = -\omega^2 \mu_0 P^{(NL)}.}$$ (4.18)

The expression for $P^{(NL)}$ depends on the order of the nonlinear process and the fields. This expression is used here and for third-order processes later in this book.

In the case of second-order processes, we use the results of Section 3.3 to obtain the nonlinear polarization's complex amplitude,

$$P^{(2)} = 2\varepsilon_0 d_{eff} A_1 A_2^* e^{i(k_1-k_2)z}.$$ (4.19)

Equation 4.19 is substituted into Equation 4.18 resulting in a first-order differential equation for A_3

$$2ik_3 \frac{dA_3}{dz} e^{ik_3z} = -\frac{2\omega_3^2}{c^2} d_{eff} A_1 A_2^* e^{i(k_1-k_2)z}.$$ (4.20)

Equation 4.20 shows that the fields A_1, A_2, and A_3 are coupled to each other. To determine the evolution of A_3, we need to know the evolution of A_1 and A_2. The same procedure for deriving Equation 4.20 is applied to fields at ω_1 and ω_2 to give

$$2ik_2 \frac{dA_2}{dz} e^{ik_2z} = -\frac{2\omega_2^2}{c^2} d_{eff} A_1 A_3^* e^{i(k_1-k_3)z},$$ (4.21)

$$2ik_1 \frac{dA_1}{dz} e^{ik_1 z} = -\frac{2\omega_1^2}{c^2} d_{\text{eff}} A_2 A_3^* e^{i(k_2+k_3)z}.$$ (4.22)

Equations 4.20 through 4.22 are known as the coupled amplitude equations for three-wave interactions, and they are the starting point for understanding a great deal about nonlinear processes.

4.1.3 INTRODUCTION TO PHASE MATCHING

We focus our attention first on Equation 4.20, but the following discussion can be paralleled for Equations 4.21 and 4.22. Equation 4.20 relates the change in the field amplitude of A_3 to A_1 and A_2 via the nonlinearity. The rapidly varying exponential term, $e^{ik_3 z}$, on the left-hand side of Equation 4.20 has a wavelength of $2\pi/k_3$ and the term on the right-hand side has a wavelength of $2\pi/(k_1 - k_2)$. One can think of the term on the right as relating to the spatial wavelength of the driving polarization and the term on the left as relating to the optical wavelength of the field being driven. Poynting's theorem (see Section 2.1.3) shows that the direction of energy flow depends on the sign of the time-averaged rate of work done by the fields on the medium through the polarization in Equation 2.19:

$$\left\langle \vec{E} \cdot \frac{\partial \vec{P}}{\partial t} \right\rangle.$$ (4.23)

If the spatial wavelength of the polarization and the field are not the same, then the phase relationship between the field and the polarization continuously changes through the crystal. This continuous phase change results in energy flowing from the polarization to the field in parts of the crystal and then back again later. Figure 4.2 shows such a phase slippage between the phase of the field and the

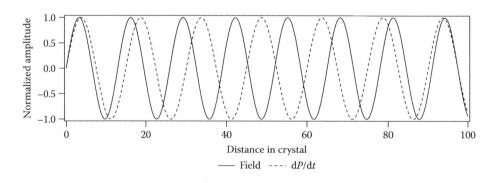

FIGURE 4.2 Snapshot in time showing that the generated field slips in phase relative to a driving polarization.

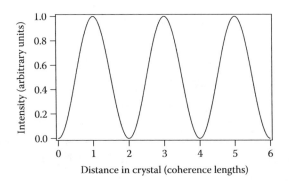

FIGURE 4.3   Time-averaged intensity as a function of position in a crystal.

time derivative of the polarization. The length over which the energy flow is from the polarization into the field is called the coherence length. If the spatial wavelength of the polarization and the field it is driving are the same, then the coherence length becomes infinite and the process is called *phase matched*. For Equation 4.20, phase matching occurs when $k_3 = k_1 - k_2$. A more convenient and conventional way to write out the phase-matching condition for a three-wave interaction is through the definition

$$\Delta k = k_1 - k_2 - k_3. \tag{4.24}$$

When $\Delta k = 0$, the process is phase matched.

The time-averaged intensity of the generated wave is shown in Figure 4.3, which shows that the intensity periodically builds up for a coherence length and then decays for the next coherence length in the crystal.

Typically, the coherence length in a crystal is too short for an appreciable buildup of the newly generated field. The nonlinear interactions occur, but in general, the efficiency is negligible unless something is done to obtain long coherence lengths. Phase-matching schemes that achieve these long lengths are first addressed in Section 4.2.

4.1.4 FIRST SOLUTION TO THE COUPLED AMPLITUDE EQUATIONS

Equation 4.20 and similar equations for A_2 and A_1 are simplified to the following three coupled amplitude equations:

$$\frac{dA_1}{dz} = i\frac{\omega_1}{n_1 c}d_{\text{eff}}A_3 A_2 e^{-i\Delta kz}, \tag{4.25}$$

$$\frac{dA_2}{dz} = i\frac{\omega_2}{n_2 c}d_{\text{eff}}A_1 A_3^* e^{i\Delta kz}, \tag{4.26}$$

$$\frac{dA_3}{dz} = i\frac{\omega_3}{n_3 c} d_{eff} A_1 A_2^* e^{i\Delta k z}. \tag{4.27}$$

Note that each one of these equations refers to a specific three-wave interaction. Equation 4.25 is a sum-frequency process between ω_2 and ω_3; Equation 4.26 is difference-frequency mixing between ω_1 and ω_3; Equation 4.27 is difference-frequency mixing between ω_1 and ω_2. Also note that all three equations have the same Δk, so that when one is phase matched, so are they all. In general, we need to consider the coupling of all three processes, but before looking at the strong coupling limit, we first solve some special cases that illustrate important nonlinear phenomena.

Consider first, a difference-frequency situation where a plane wave at ω_1 and another at ω_2 are incident on a nonlinear crystal phase matched for difference frequency to ω_3. The conversion efficiency is assumed to be small enough that the effects on the input fields are negligible. Hence, the field amplitudes A_1 and A_2 are considered constants and Equations 4.26 and 4.27 are eliminated. We are left with only Equation 4.25, which is directly integrated:

$$A_3 = i\frac{2\omega_3}{n_3 c} d_{eff} A_1 A_2^* \int_0^L e^{i\Delta k z} dz = i\frac{2\omega_3}{\Delta k n_3 c} d_{eff} A_1 A_2^* e^{i\Delta k L/2} \sin\left(\frac{\Delta k L}{2}\right). \tag{4.28}$$

The corresponding real field is

$$E_3(z,t) = \frac{2\omega_3}{\Delta k n_3 c} d_{eff} |A_1| |A_2| \sin\left(\frac{\Delta k z}{2}\right) \sin((k_3 + \Delta k)z - \omega_3 t + \phi), \tag{4.29}$$

where ϕ is a constant that comes from the phases of A_1 and A_2. Figure 4.4 is a snapshot in time of the field when $\Delta k \neq 0$ showing the beat-like behavior described in Section 4.1.3.

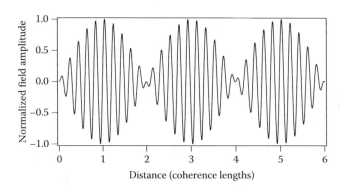

FIGURE 4.4 Plot of Equation 4.33 showing a snapshot of the generated field at ω_3 when $\Delta k \neq 0$. The amplitude has been normalized to 1.

The time-averaged intensity for monochromatic plane waves is given by Equation 2.24 and written with a complex field amplitude as

$$I = \frac{n}{2\mu_0 c} |A|^2 .$$

(4.30)

The intensity of the difference frequency is therefore obtained using the field amplitude in Equation 4.28 giving

$$I_3 = \frac{2\omega_3^2}{\mu_0 n_3 c^3} d_{\text{eff}}^2 \frac{|A_1|^2 |A_2|^2}{\Delta k^2} \sin^2\left(\frac{\Delta kz}{2}\right).$$

(4.31)

The function plotted in Figure 4.3 is a normalized form of Equation 4.31. A more convenient form of Equation 4.31 uses Equation 4.30 to express $|A_1|^2$ and $|A_2|^2$ in terms of intensity:

$$I_3 = \frac{8\mu_0 \omega_3^2}{n_1 n_2 n_3 c} d_{\text{eff}}^2 I_1 I_2 \frac{\sin^2(\Delta kz/2)}{\Delta k^2}.$$

(4.32)

The expressions for the field and the intensity in Equations 4.31 and 4.32, respectively, show a periodic beat with a periodicity and amplitude dependent on Δk. From Equation 4.31, the periodicity is $2\pi/\Delta k$, and hence the coherence length is

$$L_c = \frac{\pi}{\Delta k}.$$

(4.33)

The coherence length becomes infinite in the limit that $\Delta k \to 0$.

For a crystal of length, L, another way to write Equation 4.32 is in terms of the sinc function,

$$I_3 = \frac{2\mu_0 \omega_3^2 L^2}{n_1 n_2 n_3 c} d_{\text{eff}}^2 I_1 I_2 \text{sinc}^2\left(\frac{\Delta kL}{2}\right),$$

(4.34)

where, for the purposes of this book, we define the sinc function as

$$\text{sinc}(x) \equiv \frac{\sin(x)}{x}.$$

(4.35)

Equation 4.34 shows that the maximum conversion efficiency to the difference-frequency field occurs when $\Delta k \to 0$, since $\text{sinc}^2 \to 1$. The condition $\Delta k = 0$ is a best-case conversion efficiency scenario for a given set of plane-wave inputs. The phase mismatch parameter, Δk, plays a central role in determining the general characteristics of the nonlinear process. It governs whether or not the maximum efficiency is achieved, and it also dictates tolerances to system variables such as temperature and wavelength tuning. These dependencies are the subject of Section 4.3.

Expressions for sum-frequency generation (SFG) and second-harmonic generation (SHG) are derived in a similar fashion as for DFG in arriving at Equation 4.34. In the case of SFG, the two inputs are at ω_2 and ω_3. We assume that they are undepleted, allowing us to integrate Equation 4.25 and then convert to intensities with the result

$$I_1 = \frac{2\mu_0\omega_1^2L^2}{n_1n_2n_3c}d_{\text{eff}}^2I_2I_3\,\text{sinc}^2\left(\frac{\Delta kL}{2}\right),\tag{4.36}$$

where Δk is the same as in Equation 4.24.

For SHG, we apply Equation 4.18 with a nonlinear polarization at the second-harmonic frequency given by (see Section 3.3.6 and Equation 3.78)

$$P^{(2)}(2\omega) = \varepsilon_0 d_{\text{eff}}A_{\text{F}}^2 e^{i2k_{\text{F}}z},\tag{4.37}$$

where A_{F} is the complex amplitude of the fundamental input. The nonlinear polarization at the fundamental frequency is

$$P^{(2)}(\omega) = 2\varepsilon_0 d_{\text{eff}}A_{\text{SHG}}A_{\text{F}}^* e^{i(k_{\text{SHG}}-k_{\text{F}})z},\tag{4.38}$$

where A_{SHG} is the complex amplitude of the second harmonic. Inserting the nonlinear polarization terms into Equation 4.18 yields

$$\frac{dA_{\text{SHG}}}{dz} = i\frac{\omega_{\text{SHG}}}{2n_{\text{SHG}}c}d_{\text{eff}}A_{\text{F}}^2 e^{-i\Delta kz},\tag{4.39}$$

$$\frac{dA_{\text{F}}}{dz} = i\frac{\omega_{\text{F}}}{n_{\text{F}}c}d_{\text{eff}}A_{\text{SHG}}A_{\text{F}}^* e^{i\Delta kz},\tag{4.40}$$

For this second-harmonic case, Δk is defined as

$$\Delta k = k_{\text{SHG}} - 2k_{\text{F}}.\tag{4.41}$$

The difference between second harmonic and degenerate SFG is in how we define the inputs; recall the discussion in Section 3.3.1. Assuming that the input at the fundamental is undepleted we integrate Equation 4.39 and convert to intensities, resulting in

$$I_{SHG} = \frac{2\mu_0 \omega_F^2 L^2}{n_{SHG} n_F^2 c} d_{eff}^2 I_F^2 \, \text{sinc}^2\left(\frac{\Delta k L}{2}\right). \tag{4.42}$$

All the three-wave mixing processes, DFG, SFG, and SHG, reflected in Equations 4.34, 4.36, and 4.42, show similar characteristics. They all have the same dependence on Δk, they all depend on a product of intensities, and they all grow quadratically with interaction length L. In the case of SHG, the intensity dependence is particularly important since the output second harmonic grows quadratically with the fundamental intensity.

4.1.5 *k*-VECTOR PICTURE

Until this point, we have assumed collinear beams traveling along the z-axis. More generally, the waves may travel in different directions, and in this case, the phase mismatch condition becomes a vector relationship,

$$\Delta \vec{k} = \vec{k}_1 - \vec{k}_2 - \vec{k}_3. \tag{4.43}$$

Maximum efficiency for plane-wave interactions occurs when $\Delta \vec{k} = 0$. For both collinear and noncollinear cases, the *k*-vector mismatch is visualized with the help of a vector diagram, as shown in Figure 4.5.

The diagrams in Figure 4.5 make an assumption regarding the direction of the generated wave. In most three-wave mixing scenarios, only two of the *k*-vector directions are known since these are the inputs to the crystal. The assumption made in Figure 4.5 is that the *k*-vector associated with the generated wave is in a direction that minimizes $\Delta \vec{k}$. However, this ambiguity is an issue only when $\Delta \vec{k} \neq 0$.

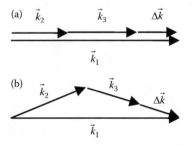

FIGURE 4.5   *k*-Vector diagrams for (a) collinear and (b) noncollinear interactions.

4.2 BIREFRINGENT PHASE MATCHING

4.2.1 BIREFRINGENT PHASE-MATCHING TYPES

Phase matching is more the exception than the rule for three-wave interactions. To examine why, consider a collinear three-wave interaction in a crystal that exhibits normal dispersion. In such a crystal with $\omega_1 > \omega_2 \geq \omega_3$, $n_1 > n_2 \geq n_3$. The phase-matching condition is

$$\Delta k = \frac{n_1 \omega_1}{c} - \frac{n_2 \omega_2}{c} - \frac{n_3 \omega_3}{c}. \tag{4.44}$$

All three-wave interactions are governed by the energy conservation condition

$$\omega_1 = \omega_2 + \omega_3. \tag{4.45}$$

Hence, we can rewrite Equation 4.44 as

$$\Delta k = \frac{1}{c}[n_1(\omega_2 + \omega_3) - n_2\omega_2 - n_3\omega_3]. \tag{4.46}$$

Rearranging Equation 4.46 gives

$$\Delta k = \frac{1}{c}[\omega_2(n_1 - n_2) + \omega_3(n_1 - n_3)]. \tag{4.47}$$

Since according to normal dispersion $n_1 > n_2$ and $n_1 > n_3$, the overall expression is always a positive quantity so that $\Delta k > 0$. Note that the index differences are never zero for a three-wave interaction and hence $\Delta k > 0$ is a strict inequality. Therefore, the process will not phase match. Furthermore, Equation 4.47 shows that Δk is always positive for normally dispersive media, and thus

$$\Delta k = k_1 - k_2 - k_3 > 0. \tag{4.48}$$

An equivalent statement is

$$k_1 > k_2 + k_3. \tag{4.49}$$

So the k-vector diagram in such cases always has a pump k-vector (k_1) that is larger than the sum of the other two. Referring to Figure 4.5, this condition implies that a noncollinear geometry will not help to make $\Delta k = 0$, in fact it makes Δk larger.

Although the current discussion is limited to isotropic media, it also applies to interactions in a birefringent medium where all the fields are e-waves or all the fields are o-waves. Note that if normal dispersion is not valid (e.g., when an absorption band lies between the frequencies), then the above argument does not hold and it may be possible to phase match the nonlinear interactions between waves (see Problem 4.3).

In birefringent materials, since the different k-vectors can be either o- or e-polarized, phase matching can sometimes be achieved. For e-polarized waves, the k-vector length is fine-tuned by choosing different crystal propagation directions. That is, the index of refraction for an e-wave, and hence the length of the e-polarized k-vector, is a function of propagation direction. Since we just proved that for normal dispersion $k_1 > k_2 + k_3$, it is clear that a strategy for phase matching should use a combination of polarizations such that the length of k_1 shrinks while k_2 and/or k_3 are lengthened.

In uniaxial crystals, specific combinations of polarizations must be used depending on whether the crystal is positive or negative uniaxial. In positive uniaxial crystals, $n_e > n_o$ and e-polarized waves have a larger k-vector magnitude than o-polarized waves. To follow the strategy of shrinking k_1 and lengthening k_2 and k_3, we see that in a positive uniaxial crystal k_1 should be o-polarized while k_2 and k_3 should be e-polarized. It turns out that it is possible that both k_2 and k_3 do not have to be e-polarized, one can be an e-wave while the other is an o-wave. Note that if we try to make k_1 e-polarized, then its k-vector length is as large as possible, thereby exacerbating the phase-matching problem. A similar argument is made for negative uniaxial crystals. Since $n_e < n_o$ in negative uniaxial crystals, e-polarized k-vectors have smaller magnitudes than o-polarized ones. Hence, k_1 should be an e-wave and both k_2 and k_3 should be either o-waves or a mixture of e- and o-waves.

A convenient shorthand for birefringent phase-matching interactions is to state the polarization of each wave. For example, the positive uniaxial interaction just discussed is depicted as

$$o \rightarrow e + e. \tag{4.50}$$

In this book, Equation 4.50 uses the following convention for frequency order:

$$\omega_1 \rightarrow \omega_2 + \omega_3. \tag{4.51}$$

Hence, k_1 is o-polarized and k_2 and k_3 are e-polarized. The arrow can point in the other direction to indicate a sum-frequency process as

$$e + e \rightarrow o. \tag{4.52}$$

The convention used in this book for frequency order is

$$\omega_3 + \omega_2 \rightarrow \omega_1. \tag{4.53}$$

The different combinations of polarizations are neatly categorized into two types of phase matching: Type I and Type II. These classifications are given below for positive and negative uniaxial crystals.

Type I phase matching

$$o \rightarrow e + e \quad \text{positive uniaxial}$$
$$e \rightarrow o + o \quad \text{negative uniaxial}$$

(4.54)

Type II phase matching

$$o \rightarrow e + o \quad \text{or} \quad o \rightarrow o + e \quad \text{positive uniaxial}$$
$$e \rightarrow e + o \quad \text{or} \quad e \rightarrow o + e \quad \text{negative uniaxial}$$

(4.55)

The characteristic of Type I interactions is that the polarization corresponding to the highest frequency is orthogonal to both polarizations of the other two frequencies. For Type II interactions, the lowest two frequencies are orthogonally polarized.

Phase matching in biaxial crystals follows the same strategy as that in uniaxial crystals. A simple approach to working with biaxial crystals is to restrict the k-vectors to one of the three principal planes illustrated in Figure 2.3. For a given plane, a k-vector with a polarization in the plane is called e-polarized and one with a polarization perpendicular to the plane is called o-polarized. Note that, as the k-vector's propagation direction is varied within the plane, the e-polarized k-vector magnitude changes while the o-polarized one does not. So, for a given plane of operation, the biaxial crystal follows the same rules as a uniaxial crystal. To visualize operation in each of the planes, we refer the reader to Figure 2.3. For positive biaxial crystals where $n_z > n_y > n_x$, when operating in the xy-plane, the crystal looks like a negative uniaxial crystal, and when operating in the yz-plane, the crystal looks like a positive uniaxial crystal. Operation in the xz-plane is a special case, since in this plane the crystal can look either negative or positive uniaxial. If the angle of propagation with respect to the z-axis in the xz-plane is less than the angle between the z-axis and the optic axis, the crystal looks like a negative uniaxial crystal; otherwise, it looks like a positive uniaxial crystal (see Problem 2.18).

Choosing the correct polarization combination is not enough to obtain $\Delta k = 0$; we also need to choose the appropriate propagation direction. For example, the phase-matching equation for an o $\rightarrow$ e + e interaction is given by

$$\Delta k = 2\pi \left[\frac{n_{1o}(\lambda_1)}{\lambda_1} - \frac{n_{2e}(\lambda_2, \theta)}{\lambda_2} - \frac{n_{3e}(\lambda_3, \theta)}{\lambda_3} \right] = 0.$$

(4.56)

Δk is written here in terms of wavelength since the Sellmeier equations are typically dependent on vacuum wavelength and not frequency. In general, a given propagation direction will not give $\Delta k = 0$. Instead, we need to solve Equation 4.56 for a specific angle where $\Delta k = 0$. The following example guides us through this calculation.

4.2.2 EXAMPLE: PHASE-MATCHING PROBLEM

When considering a particular three-wave process, the first consideration, even before phase matching, is which nonlinear crystal to use. Numerous crystals are available and the optimum choice may depend on factors independent of phase matching. Transparency range, damage threshold, hygroscopicity, durability, and cost are just a few considerations. Once a desirable crystal has been identified, the next step is to determine whether or not it supports phase matching for a given process. It needs to be emphasized that the birefringent phase-matching technique does not guarantee that every crystal can phase match every process.

Let us consider difference-frequency mixing between 780 and 1064 nm to generate 2922 nm in the nonlinear crystal, lithium niobate. Lithium niobate is a class 3 m, negative uniaxial crystal that has Type I and Type II phase-matching polarization combinations as given in Equations 4.54 and 4.55. If both interaction types have a phase-matching angle, the one with the higher d_{eff} is usually picked.

Let us first consider Type I e $\rightarrow$ o + o interaction such that Δk is given by

$$\Delta k = 2\pi \left(\frac{n_e(\lambda_1, \theta)}{\lambda_1} - \frac{n_o(\lambda_2)}{\lambda_2} - \frac{n_o(\lambda_3)}{\lambda_3} \right). \tag{4.57}$$

The index of refraction is obtained from Sellmeier equations for lithium niobate (see Appendix B). In this instance, it is possible to come up with an analytic expression for the phase-matching angle where $\Delta k = 0$. The first step is to substitute the extraordinary index (Equation 2.109) into Equation 4.57 and then solve for

$$\sin^2\theta = \frac{n_{z1}^2 \left[n_{x1}^2 - \lambda_1^2((n_{x2}/\lambda_2) + (n_{x3}/\lambda_3))^2 \right]}{\lambda_1^2((n_{x2}/\lambda_2) + (n_{x3}/\lambda_3))^2 \left(n_{x1}^2 - n_{z1}^2 \right)}, \tag{4.58}$$

where $n_{x1} = n_x(\lambda_1)$, $n_{x2} = n_x(\lambda_2)$, and so on. For lithium niobate, $n_x = n_o$. By using the Sellmeier equations for lithium niobate at the appropriate wavelengths, we solve for the angle ($T = 25°C$) and obtain $\theta \sim 47°$. The same calculation can be carried out for Type II interaction, but an analytic expression for the angle is not possible (see Problem 4.8). In such situations, the solution is found graphically or by using a root-finding algorithm. Although a numerical solution is not as satisfying as an analytic result, note that all phase-matching problems require appropriate Sellmeier equations, which involve some computer programming.

The graphical approach to finding the phase-matching angle is to plot Δk versus angle and see where it crosses zero. The Type II interaction in lithium niobate has two possibilities: $e \rightarrow o + e$ or $e \rightarrow e + o$.

For the $e \rightarrow o + e$ case

$$\Delta k = 2\pi \left(\frac{n_e(\lambda_1, \theta)}{\lambda_1} - \frac{n_o(\lambda_2)}{\lambda_2} - \frac{n_e(\lambda_3, \theta)}{\lambda_3} \right). \tag{4.59}$$

For the $e \rightarrow e + o$ case

$$\Delta k = 2\pi \left(\frac{n_e(\lambda_1, \theta)}{\lambda_1} - \frac{n_e(\lambda_2, \theta)}{\lambda_2} - \frac{n_o(\lambda_3)}{\lambda_3} \right). \tag{4.60}$$

One or both of these may phase match, and so it is best to consider both cases. The only free parameter in these equations is the angle, since in a difference-frequency process, two of the input wavelengths are prescribed and the third wavelength is dictated by energy conservation,

$$\frac{1}{\lambda_3} = \frac{1}{\lambda_1} - \frac{1}{\lambda_2}. \tag{4.61}$$

The e-wave's index of refraction is directly substituted into the above equations, and as an example, Equation 4.59 gives

$$\Delta k(\lambda_1, \lambda_2, \theta) = 2\pi \left(\frac{1}{\lambda_1} \left[\frac{\cos^2 \theta}{n_{x1}^2} + \frac{\sin^2 \theta}{n_{z1}^2} \right]^{-1/2} - \frac{n_{x2}}{\lambda_2} - \frac{1}{\lambda_3} \left[\frac{\cos^2 \theta}{n_{x3}^2} + \frac{\sin^2 \theta}{n_{z3}^2} \right]^{-1/2} \right). \tag{4.62}$$

Although this expression may seem messy, it is easy to program. All that is required are functions for $n_x(\lambda)$ and $n_z(\lambda)$ obtained from the Sellmeier equations (see Appendix B for lithium niobate). Given the two input wavelengths, Δk is plotted as a function of θ and where that curve crosses zero is the phase-matching angle. Figure 4.6 shows a plot of Δk for both $e \rightarrow o + e$ and $e \rightarrow e + o$, showing that only the $e \rightarrow o + e$ interaction phase matches at $\sim 56°$.

Another common approach to finding the zero-crossing is to use a root-finding algorithm that is typically part of a mathematical software package (see Appendix C). The graphical approach along with a root-finding algorithm is a useful combination. In some cases, the Δk function may pass through zero more than once, which may cause the root finder to crash or return a false negative. By looking at the plot, it should be clear immediately whether a solution or multiple solutions exist.

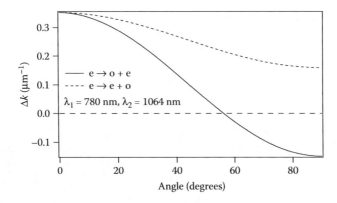

FIGURE 4.6 Type II phase matching in lithium niobate for a difference-frequency process between two inputs at 780 and 1064 nm. The angle where $\Delta k = 0$ is the phase-matched angle.

For lithium niobate and our current example, we have the result that both Type I (e → o + o) phase matching and Type II (e → o + e) phase matching are possible. The main discriminator between the two is d_{eff}. Given that lithium niobate is a class 3 m negative uniaxial crystal, d_{eff} is read off Table 3.3.

d_{eff} for the Type I interaction is

$$d_{\text{eff}} = d_{15} \sin\theta - d_{22} \cos\theta \sin 3\phi. \tag{4.63}$$

d_{eff} for the Type II interaction is

$$d_{\text{eff}} = d_{22} \cos^2\theta \cos 3\phi. \tag{4.64}$$

For uniaxial crystals, the azimuthal angle, ϕ, has no effect on the index of refraction and hence it does not affect the phase matching. So we choose its value to maximize d_{eff}. The last step is plugging in for the phase-matching angles and the d-coefficients (given in Appendix B), $d_{\text{eff}} = -4.3$ pm/V for the Type I interaction and 0.58 pm/V for the Type II interaction.

In most cases, the interaction with the larger d_{eff} is selected. The other discriminator is the Poynting vector walk-off. If one of the interactions happens to occur at $\theta = 0°$ or $\theta = 90°$, then the walk-off is zero. This process is called noncritical phase matching and is discussed further in Section 4.4.3. In some cases, the noncritical phase matching with a lower d_{eff} is chosen over a critically phase-matched interaction with a larger d_{eff}. The reason is that much longer effective crystal lengths are possible since all the interacting beams remain overlapped throughout the crystal. In our current example using lithium niobate, both interaction types lead to critically phase-matched situations with the Poynting vector walk-off angles that are comparable to each other. So, for our example, based on d_{eff} and walk-off, the Type I interaction wins. Many of these calculations are possible using freely available software. One of the most powerful is called SNLO (Smith, 2015).

4.2.3 PHASE-MATCHING SHG

A particularly simple case of Type I phase matching is that for SHG. SHG is given by

$$\omega_F + \omega_F = \omega_{SHG}, \tag{4.65}$$

where ω_F is the fundamental frequency and ω_{SHG} is the second harmonic. For a Type I interaction, both the fundamental inputs have the same polarization so that the equation for Δk is

$$\Delta k = k_{SHG} - 2k_F. \tag{4.66}$$

This relationship is phase matched when $k_{SHG} = 2k_F$. We can take this expression a step further by noting that $k = n\omega/c$, so that phase matching occurs when

$$n_F = n_{SHG}. \tag{4.67}$$

Finding a solution is visualized by plotting both indices as a function of angle, as shown for a negative uniaxial crystal in Figure 4.7. Phase-matching equations for Type I interactions are given by Equations 4.66 and 4.67.

Index matching for SHG is a special case of phase matching for which the fundamental frequency and the second harmonic travel at the same phase velocity. Phase matching is not always equivalent to phase-velocity matching, or group-velocity matching for that matter. Consider the case of Type II SHG where the two fundamental inputs have different polarizations. In this scenario, phase matching occurs when the phase velocities of the second harmonic and the two fundamental inputs are all different. Other three-wave mixing processes such as difference-frequency mixing and SFG, in general, have $\Delta k = 0$ when the phase and group velocities of the

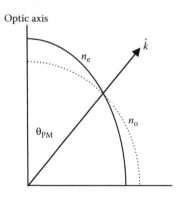

FIGURE 4.7 k-Vector magnitude plot showing the phase-matching condition for Type I SHG in a negative uniaxial crystal. For simplicity, the labels give the index of refraction instead of k. Phase matching occurs when $n_e(\omega_F) = n_o(\omega_{SHG})$.

three waves are different (see, e.g., Problem 4.6). If we want to consider phase matching in terms of phase-velocity matching, then it is the phase velocity of the nonlinear polarization wave that is matched to the field it is driving.

4.3 TUNING CURVES AND PHASE-MATCHING TOLERANCES

The previous example focused on finding a fixed phase-matching point. However, most systems have flexibility such that one of many parameters may change. For example, one of the lasers may be tunable or the crystal temperature may change. It is desirable to see, or predict, how the phase matching changes with a given parameter. A tuning curve provides this information. As for calculating a fixed phase-matching angle, many approaches are used to find a tuning curve. The first approach is to use an analytic expression for phase-matching angles, as in Equation 4.62 for the previous example. To continue that example, consider that the 780 nm wavelength is the output of a tunable Ti:sapphire laser. This laser tunes from roughly 700 to 900 nm; hence, it would be helpful for a user of this laser to know what range of phase-matching angles is needed for a difference frequency with the fixed-frequency 1064-nm laser. Since the tunable laser corresponds to the highest frequency, this type of tuning is referred to as pump tuning. For Type I phase matching in lithium niobate, a pump-tuning curve is generated by plotting angle versus pump wavelength using the analytic expression of Equation 4.58, which is shown in Figure 4.8. An interesting characteristic of this example is evident near 800 nm. For a broad range of wavelengths, the phase-matching angle does not change appreciably. This aspect is useful if the Ti:sapphire laser is mode-locked and generates femtosecond pulses, since phase matching covering the wide bandwidth of the femtosecond pulses can be accommodated. More details on such tolerances are covered in the next section.

The previous example is more the exception than the rule when it comes to generating a tuning curve. In most cases, an analytic expression does not exist for the

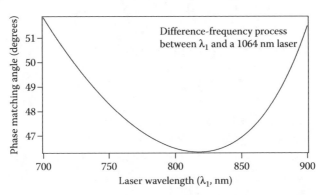

FIGURE 4.8 Example tuning curve for a Type I interaction in lithium niobate. In this case, the pump wavelength is tuned. Since the second input is fixed at 1064 nm, the output at λ_3 tunes with the pump.

phase-matching angle. Let us consider the same set of wavelengths using Type II phase matching in lithium niobate and revisit pump tuning. As mentioned earlier, if we try to come up with an analytic expression, we discover quickly that such a solution is intractable. So we must implement numerical approaches. The first technique we present uses a root-finding algorithm for each pump wavelength to solve for an angle where $\Delta k = 0$, thus generating a tuning curve similar to that shown in Figure 4.8. This technique requires a looping program to step through the pump wavelength range of interest. A typical "brute-force" program might look like the following set of pseudo code:

```
Program get_angle

Inputs: λstart, λend, Δλ, λ2
i = 0
LOOP
deltak_array=Δk(λStart, λ2, θ) || comment: The array should
|| span θ = 0 to π/2
answer_array[i]=Find_zero_crossing(deltak_array)
|| comment: there should only be one zero-crossing, and
|| this value is put into "answer_array[i]"
λStart = λStart +Δλ
i = i + 1
WHILE (λStart < λend)
```

The function $\Delta k(\lambda_{\text{Start}}, \lambda_2, \theta)$ would look like Equation 4.62 depending on the interaction type. Internal to this function would be a calculation of the third wavelength λ_3 based on energy conservation (Equation 4.61). The function Find_zero_crossing or its equivalent is available with most mathematical packages, or may be programmed. It works by interpolating between the two adjacent points in the array that have opposite signs (i.e., between the two points that span zero). Other more accurate approaches do away with creating a Δk array (e.g., numerical recipes for zero-finding algorithms) (see Appendix C).

Another numerical approach is to solve the problem graphically using a contour plot. The graphical approach keys on the characteristic that defines the tuning curve; namely, that at each point in the tuning curve, $\Delta k = 0$. The way to generate a phase-matching curve is to create a two-dimensional Δk array and pick out the $\Delta k = 0$ contour using a mathematical package. Implementing this scheme first requires that an $N \times M$ array be properly scaled or indexed, where N and M are the number of points for each dimension of the array. For the pump-tuning example, a good starting point would be an array with elements a_{ij} scaled such that $i = 0$ gives $\lambda_1 = 700$ nm and $i = N$ gives $\lambda_1 = 900$ nm. Similarly, $j = 0$ would give $\theta = 0$, and $j = N$ would give $\theta = 90°$. Each element of the array now corresponds to a particular λ_1, θ pairing, and the array element is assigned the value $\Delta k(\lambda_1, \theta)$, where λ_2 is held fixed and λ_3 is calculated from energy conservation. Once the array is populated with values for Δk, a contour plot is

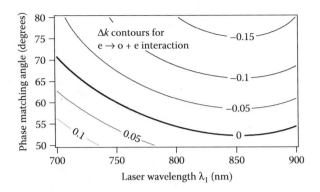

FIGURE 4.9   Contour plot for a Type II difference-frequency mixing process between λ_1 and a fixed 1064-nm laser. The zero contour is the tuning curve. The units of Δk in this plot are in μm^{-1}.

generated using built-in routines that come with most mathematical packages. For the Type II interaction discussed here, the resulting contour plot is shown in Figure 4.9.

4.3.1 PHASE-MATCHING BANDWIDTH AND ANGULAR ACCEPTANCE

Although finding $\Delta k = 0$ is a typical goal for phase matching, when Δk varies away from zero, it gives a great deal of information regarding tolerances. The first solution (nondepleted fields A_1 and A_2) to the coupled amplitude equations (Equation 4.34) shows that the intensity for a difference-frequency process is proportional to $\text{sinc}^2(\Delta kL/2)$. SHG and SFG have the same dependency in the limit of low conversion efficiency. Since the $\text{sinc}^2(\Delta kL/2)$ function ranges from 0 at its minimum to 1 at its maximum, it is a relative efficiency term that allows us to define a phase-matching efficiency function

$$\eta = \frac{I(\Delta k)}{I(\Delta k = 0)} = \text{sinc}^2\left(\frac{\Delta kL}{2}\right),\tag{4.68}$$

where I is the intensity of the newly generated frequency. If Δk is not 0, the overall performance is degraded by a factor of $\text{sinc}^2(\Delta kL/2)$. The phase-matching efficiency function is a quantitative guide to how much a parameter is allowed to change. The range over which a parameter changes while the phase-matching efficiency function, η, is greater than ~1/2 defines the bandwidth for that particular parameter.

The bandwidth is visualized by plotting $\text{sinc}^2(\Delta kL/2)$ as a function of the parameter of interest. Consider a difference-frequency process where the pump laser wavelength drifts while all other parameters such as the wavelength of the other input laser, temperature, and crystal angle remain fixed. To be more specific, we consider the Type I interaction in lithium niobate with a crystal length of 2 cm. The pump laser is tunable about 780 nm, while the second input at 1064 nm remains fixed. The effect

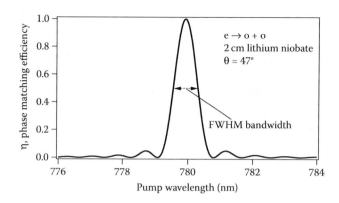

FIGURE 4.10 Phase-matching efficiency function, $\eta = \text{sinc}^2(\Delta k L/2)$, plotted versus pump wavelength λ_1.

of the drift is visualized in Figure 4.10 by plotting $\text{sinc}^2(\Delta k L/2)$ versus the pump wavelength.

Figure 4.10 shows that if the pump laser drifts by even 1 nm, the efficiency drops dramatically. The bandwidth over which the process is relatively efficient (50% points) is called the phase-matching bandwidth or acceptance bandwidth.

In the previous example, the shape of the phase-matching efficiency function is a canonical sinc^2 shape. This is not always the case. Using the same setup discussed above but simply rotating the crystal to 46.35° causes the Type I interaction to shift to a center wavelength of ~818 nm with a phase-matching bandwidth close to 15 nm (see Figure 4.11) and with a different shape.

The same process of plotting $\text{sinc}^2(\Delta k L/2)$ may be carried out for other system parameters to determine other phase-matching bandwidths. Two important tolerances are angle and temperature. The phase-matching efficiency plotted as a function of angle gives the angular acceptance of a crystal. That is, there is a certain angular range over which either the crystal can be rotated or the incident beam's angle can change while maintaining a high relative phase-matching efficiency. When focusing

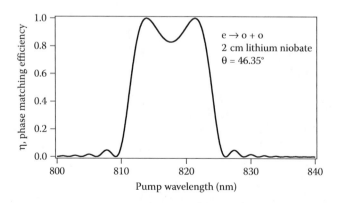

FIGURE 4.11 Phase-matching efficiency, $\text{sinc}^2(\Delta k L/2)$, plotted versus pump wavelength λ_1.

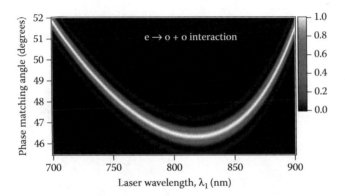

FIGURE 4.12   Plot of sinc²($\Delta k L/2$) for a Type I interaction in lithium niobate. A 500-μm crystal length was chosen to exaggerate the bandwidth. A vertical slice out of the image gives the angular acceptance and a horizontal slice gives the phase-matching bandwidth.

a laser, we must pay attention to the angular acceptance to ensure that the range of angles present in the beam will phase match.

A general note is that all bandwidths increase as the crystal length is decreased, which is a direct consequence of the sinc²($\Delta k L/2$) dependence. The sinc² function falls to 0.5 when $\Delta k L = 0.886\pi$. So the maximum Δk that can change before the process drops below this 50% relative efficiency is $0.886\pi/L$. Hence, shorter crystals allow for a larger variation in Δk, thus attaining a larger DFG wavelength bandwidth.

Another graphical technique that lends itself to visualizing bandwidths and tolerances is a two-dimensional plot of sinc²($\Delta k L/2$) displayed against two variables. Figure 4.12 shows an image plot of sinc²($\Delta k L/2$) as a function of angle and pump wavelength for a Type I interaction in lithium niobate. The second input to the crystal is fixed at 1064 nm and the crystal length is chosen to be 0.5 mm to exaggerate the bandwidth. The $\Delta k = 0$ tuning curve is easily seen, and by taking cross sections out of the plot we obtain bandwidths. For example, Figures 4.10 and 4.11 are obtained by taking horizontal slices out of Figure 4.12 at different phase-matching angles.

4.4   TAYLOR SERIES EXPANSION TECHNIQUES FOR DETERMINING BANDWIDTH

Another approach for determining bandwidths is based on a Taylor series expansion of Δk. This technique gives new insights into what determines the bandwidth. Δk depends on the wavelengths in the mixing process and on many other parameters through the index of refraction. That is,

$$\Delta k = 2\pi \left[\frac{n_1(\lambda_1, \theta, T, \ldots)}{\lambda_1} - \frac{n_2(\lambda_2, \theta, T, \ldots)}{\lambda_2} - \frac{n_3(\lambda_3, \theta, T, \ldots)}{\lambda_3} \right]. \tag{4.69}$$

A Taylor series expansion in one of these parameters, X, is

$$\Delta k(X + \Delta X) = \Delta k_\mathrm{o} + \frac{\partial \Delta k}{\partial X} \Delta X + \cdots. \qquad (4.70)$$

X is typically wavelength, temperature, or angle. We are interested in how Δk varies away from a phase-matched point, $\Delta k_\mathrm{o} = 0$; hence, Equation 4.70 becomes

$$\Delta k(X + \Delta X) = \frac{\partial \Delta k}{\partial X} \Delta X + \cdots. \qquad (4.71)$$

The bandwidth is determined by how much of a change in X is tolerated before the efficiency of the process drops below an acceptable level. The change, ΔX, causes a change in Δk, which feeds into the phase-matching efficiency function, $\mathrm{sinc}^2(\Delta kL/2)$. The minimum value for which the phase-matching efficiency is "acceptable" is a subjective determination. It is impossible to come up with a criterion to satisfy every situation; hence, we use the following range that gives the full-width-at-half-maximum (FWHM) for the sinc² function, that is

$$\left| \frac{\Delta kL}{2} \right| \le \frac{0.886\pi}{2}, \qquad (4.72)$$

which ensures that

$$\mathrm{sinc}^2\left(\frac{\Delta kL}{2} \right) \ge 0.5. \qquad (4.73)$$

So for any $|\Delta k| < (0.886\pi/L)$, the efficiency is at least 50% of the maximum achievable. This inequality is used in Equation 4.71 to obtain

$$|\Delta X| \le \frac{0.886\pi}{L\,|\partial \Delta k/\partial X|}. \qquad (4.74)$$

The full bandwidth corresponds to twice the expression in Equation 4.74 to account for going from one 50% point to the other:

$$\Delta X_{\mathrm{FWHM}} = 2\frac{0.886\pi}{L\,|\partial \Delta k/\partial X|}. \qquad (4.75)$$

We see that the crystal length always plays a role in determining bandwidths. From Equation 4.75, it is clear that the longer the interaction length, the narrower the bandwidth.

4.4.1 TEMPERATURE BANDWIDTH

Temperature bandwidth refers to the range of temperatures that allow for efficient conversion. Starting with Equation 4.71 using a temperature variation gives

$$\Delta k(T + \Delta T) = \frac{\partial \Delta k}{\partial T} \Delta T + \cdots . \tag{4.76}$$

According to Equation 4.74 with $X = T$, the expression for the extremal temperature variations is

$$\Delta T = \pm \frac{0.886\pi}{L \, |\partial \Delta k / \partial T|} . \tag{4.77}$$

To evaluate the partial derivative, Δk is expanded:

$$\frac{\partial \Delta k}{\partial T} = \frac{\partial}{\partial T} [k_1 - k_2 - k_3] . \tag{4.78}$$

Substituting $k = (2\pi n / \lambda)$ for each k gives

$$\frac{\partial \Delta k}{\partial T} = 2\pi \left[\frac{1}{\lambda_1} \frac{\partial n_1}{\partial T} - \frac{1}{\lambda_2} \frac{\partial n_2}{\partial T} - \frac{1}{\lambda_3} \frac{\partial n_3}{\partial T} \right] . \tag{4.79}$$

The expression in the brackets is evaluated using the Sellmeier equations for the crystal in question. To simplify the expression, let

$$Y \equiv \left[\frac{1}{\lambda_1} \frac{\partial n_1}{\partial T} - \frac{1}{\lambda_2} \frac{\partial n_2}{\partial T} - \frac{1}{\lambda_3} \frac{\partial n_3}{\partial T} \right] . \tag{4.80}$$

The FWHM temperature-tuning bandwidth given by Equation 4.77 becomes

$$\Delta T_{\mathrm{FWHM}} = \frac{0.886}{L \, |Y|} . \tag{4.81}$$

The temperature-tuning bandwidth becomes large when $|Y|$ approaches zero, as occurs for a Type I interaction near degeneracy (where $\lambda_2 = \lambda_3$). In this case, we obtain the bandwidth by going to second-order in the Taylor series expansion, or by plotting the $\mathrm{sinc}^2(\Delta k L / 2)$ function.

4.4.2 PHASE-MATCHING BANDWIDTH AND ACCEPTANCE BANDWIDTH

The next example looks at changing one or more of the wavelengths involved in the process. In this case, energy conservation places constraints on wavelength changes. For a three-wave interaction, $\omega_1 = \omega_2 + \omega_3$, and hence

$$\Delta\omega_1 = \Delta\omega_2 + \Delta\omega_3. \tag{4.82}$$

At this point, more information about the properties of the three sources providing these frequencies is required. A common scenario is a mixing process between two lasers where one of the inputs has a fixed frequency while the other input is tunable. The fixed-frequency input has $\Delta\omega \approx 0$ so that when the other input changes frequency, the output frequency will tune to satisfy energy conservation. As a specific example, consider difference-frequency mixing when ω_1 is fixed and ω_2 tunes. Therefore, $\Delta\omega_1 = 0$ and from Equation 4.86

$$\Delta\omega_3 = -\Delta\omega_2. \tag{4.83}$$

Now it is possible to calculate how far ω_2 tunes while maintaining a high relative efficiency. This range is called the phase-matching bandwidth or acceptance bandwidth and according to Equation 4.75 with $X = \omega_2$, it is given by

$$\Delta\omega_{2,\text{FWHM}} = 2\frac{0.886\pi}{L\,|\partial\Delta k/\partial\omega_2|}. \tag{4.84}$$

To evaluate the partial derivative, Δk is expanded:

$$\frac{\partial\Delta k}{\partial\omega_2} = \frac{\partial}{\partial\omega_2}[k_1 - k_2 - k_3]. \tag{4.85}$$

Since the input at ω_1 is fixed in frequency, its k-vector is independent of ω_2 and $(\partial k_1/\partial\omega_2) = 0$. Equation 4.83 is used to evaluate the partial derivative:

$$\frac{\partial k_3}{\partial\omega_2} = -\frac{\partial k_3}{\partial\omega_3}. \tag{4.86}$$

Now Equation 4.85 becomes

$$\frac{\partial\Delta k}{\partial\omega_2} = -\frac{\partial k_2}{\partial\omega_2} + \frac{\partial k_3}{\partial\omega_3} = v_{g3}^{-1} - v_{g2}^{-1}, \tag{4.87}$$

where v_{g2}^{-1} and v_{g3}^{-1} are the inverse group velocities corresponding to ω_2 and ω_3, respectively. Therefore, the bandwidth of the process is

$$\Delta\omega_{2,\text{FWHM}} = 2\frac{0.886\pi}{\left(v_{g3}^{-1} - v_{g2}^{-1}\right)L}. \tag{4.88}$$

We see that if the inverse group velocities are nearly the same, then the bandwidth of the process is large. So in addition to estimating the bandwidth, we also gain an insight into what determines the bandwidth. Specifically, we see that group-velocity matching results in large bandwidths. When the group velocities are equal, an accurate determination of the bandwidth requires going to second-order in the Taylor series expansion, or it can be determined by plotting the sinc$^2(\Delta kL/2)$ function. Note that group-velocity matching is not required for phase matching. Only for special cases will phase matching and group-velocity matching be achieved for the same process.

Group-velocity matching for ultrashort-pulse applications is especially important because a large phase-matching bandwidth is required to accommodate the broad bandwidth of the ultrashort pulses. We also note that for a nonlinear interaction to occur, the pulses must remain on top of each other temporally. If the group-velocity mismatch is small, then the interacting pulses stay overlapped, and from our previous analysis, the bandwidth is large. The two go together: in order to support short pulses, we require the pulses to overlap and we require a large bandwidth. In most cases, phase matching is not accompanied by group-velocity matching so that applications requiring a large phase-matching bandwidth do so by using short crystal lengths.

Since Sellmeier equations are typically written in terms of vacuum wavelength and not frequency, we look at the phase-matching bandwidth in terms of partial derivative with respect to wavelength. The partials with respect to ω are transformed into partials with respect to λ using the chain rule:

$$\frac{\partial k}{\partial \omega} = \frac{\partial k}{\partial \lambda}\frac{\partial \lambda}{\partial \omega} = 2\pi\left(\frac{1}{\lambda}\frac{\partial n}{\partial \lambda} - \frac{n}{\lambda^2}\right)\left(\frac{-\lambda^2}{2\pi c}\right) = \frac{n}{c}\left(1 - \frac{\lambda}{n}\frac{\partial n}{\partial \lambda}\right). \tag{4.89}$$

Hence, Equation 4.87 becomes

$$\frac{\partial \Delta k}{\partial \omega_2} = \frac{1}{c}\left([n_3 - n_2] + \left[\lambda_2\frac{\partial n_2}{\partial \lambda_2} - \lambda_3\frac{\partial n_3}{\partial \lambda_3}\right]\right). \tag{4.90}$$

The bandwidth is approximated using the range $|\Delta k| < (0.886\pi/L)$ so that

$$\Delta\omega_{2,\text{FWHM}} = 2\frac{0.886\pi}{(1/c)([n_3 - n_2] + [\lambda_2(\partial n_2/\partial \lambda_2) - \lambda_3(\partial n_3/\partial \lambda_3)])L}. \tag{4.91}$$

A similar approach is followed when ω_1 is allowed to tune while keeping ω_2 fixed (see Problems 4.23 and 4.26 at the end of this chapter for the related exercises).

4.4.3 ANGULAR ACCEPTANCE AND NONCRITICAL PHASE MATCHING

The sensitivity of a nonlinear process to angular deviations gives the "angular bandwidth," more commonly known as angular acceptance. This calculation is important because many nonlinear interactions depend on the intensity of the various interacting beams. Due to this intensity dependence, we may be tempted to focus as tightly as possible. A limiting factor to focusing is this angular acceptance. To calculate the angular acceptance, we go again to the Taylor series expansion:

$$\Delta k(\theta + \Delta\theta) = \frac{\partial \Delta k}{\partial \theta} \Delta\theta + \cdots. \tag{4.92}$$

According to Equation 4.75 with $X = \theta$, the FWHM angular acceptance is

$$\Delta\theta_{\mathrm{FWHM}} = 2 \frac{0.886\pi}{L\,|\partial \Delta k/\partial \theta|}. \tag{4.93}$$

The only angular dependence comes in the extraordinary index of refraction. Taking the derivative gives

$$\frac{\partial \Delta k}{\partial \theta} = 2\pi \left(\frac{1}{\lambda_1} \frac{\partial n_1}{\partial \theta} - \frac{1}{\lambda_2} \frac{\partial n_2}{\partial \theta} - \frac{1}{\lambda_3} \frac{\partial n_3}{\partial \theta} \right). \tag{4.94}$$

So depending on the type of wave (e-wave or o-wave), the derivatives in Equation 4.94 are evaluated, allowing us to solve for the angular acceptance. For example, a Type I difference-frequency process in a positive uniaxial crystal (o $\rightarrow$ e + e) gives

$$\frac{\partial \Delta k}{\partial \theta} = -2\pi \left(\frac{1}{\lambda_2} \frac{\partial n_2}{\partial \theta} + \frac{1}{\lambda_3} \frac{\partial n_3}{\partial \theta} \right). \tag{4.95}$$

The FWHM bandwidth for this process is then

$$\Delta\theta_{\mathrm{FWHM}} = \frac{0.886}{L\,|((1/\lambda_2)(\partial n_2/\partial \theta) + (1/\lambda_3)(\partial n_3/\partial \theta))|}. \tag{4.96}$$

Interestingly, for any combination of polarizations that phase match at $0°$ or $90°$, Equation 4.94 goes to 0 since at these angles

$$\frac{\partial n_e}{\partial \theta} = 0. \tag{4.97}$$

Here, the acceptance angle is large, and to calculate it, we must include second-order terms from the Taylor series in Equation 4.96, or we can plot the sinc^2 function as a function of angle. In these large-acceptance regimes, the phase matching is termed "noncritically phase matched." For all other regimes, the phase matching is called "critically phase matched."

4.5 NONCOLLINEAR PHASE MATCHING

We introduced the possibility of noncollinear phase matching in Figure 4.5. Now we revisit this idea with a goal of determining $\Delta k = 0$ and gaining an understanding of bandwidths for the noncollinear case. Consider a difference-frequency mixing process between a pump and signal beam aligned at an angle ψ, as shown in Figure 4.13. The angles θ_1 and θ_2 define the two wave vectors relative to the optic axis. This is especially important when one or both waves are polarized with a component along the extraordinary axis. Likewise, the extraordinary refractive index of the $\vec{k}_3$ wave subtends the angle $\theta_3 = \theta_1 + \alpha$ with respect to the optic axis.

The vectors lie in one plane and can be decomposed into the following equations. The direction of $\vec{k}_3$ is assumed to be in the direction that minimizes $\Delta \vec{k}$. The wave vectors for the direction of the pump wave satisfy

$$\Delta k_z = \Delta k \cos\alpha = k_1 - k_2 \cos\psi - k_3 \cos\alpha, \tag{4.98}$$

and for the orthogonal direction (i.e., along the x-axis), the relation is

$$\Delta k_x = \Delta k \sin\alpha = k_2 \sin\psi - k_3 \sin\alpha. \tag{4.99}$$

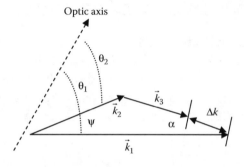

FIGURE 4.13 k-Vector diagram for a noncollinear interaction in a uniaxial crystal.

At the outset, we assume that we know the pump and signal directions since these are controlled experimentally; therefore, we also know the angle $\psi = \theta_1 - \theta_2$. If $\vec{k}_3$ is an e-wave, then α is needed in order to determine the magnitude of $\vec{k}_3$. The expressions in Equations 4.98 and 4.99 can be algebraically combined to solve for the angle α

$$\tan \alpha = \frac{(k_2/k_1)\sin \psi}{1-(k_2/k_1)\cos \psi}. \tag{4.100}$$

α is written in terms of the known input parameters, namely, ψ and the magnitudes of $\vec{k}_1$ and $\vec{k}_2$. With α, we have all the information needed to find the magnitude of $\vec{k}_3$ and hence $\Delta \vec{k}$. Further manipulation of Equations 4.98 and 4.99 gives the magnitude of $\Delta \vec{k}$,

$$\Delta k = (k_1^2 + k_2^2 - 2k_1 k_2 \cos \psi)^{1/2} - k_3. \tag{4.101}$$

Phase matching is then determined by adjusting θ_1 and θ_2 such that $\Delta k = 0$. The solution is found following one of the procedures we developed for the collinear case such as plotting against a variable to locate where $\Delta k = 0$.

Noncollinear phase matching gives an extra degree of freedom that can, for example, help compensate for the Poynting vector walk-off. In any birefringent phase-matching situation, at least one of the waves is e-polarized, and for critically phase-matched situations, the e-wave Poynting vector always walks off the o-wave direction if their k-vectors are collinear. For instance, consider a Type I interaction in a positive uniaxial crystal where $\vec{k}_2$ is e-polarized and $\vec{k}_1$ is o-polarized. In positive uniaxial crystals, the Poynting vector of e-waves walks toward the optic axis at an angle of ρ with respect to the k-vector. With a noncollinear phase-matching configuration with ψ set equal to ρ and with $\vec{k}_1$ arranged such that it is parallel to the Poynting vector associated with $\vec{k}_2$, then the energies of the two inputs stay on top of each other throughout the crystal. Admittedly, the third wave at ω_3 steers off these two. Even so, the longer overlap of the two inputs yields better overall conversion efficiency.

4.5.1 OFF-AXIS PROPAGATION SVEA EQUATIONS

The presence of waves propagating in directions that are different from one another requires a refinement of SVEA derivation. The noncollinear angles are treated only as small deviations of the beams from the z-axis, which still plays a central role in the envelope approximation. The envelope functions of the wave fields in Equation 2.119 are generalized for each field (subscripts: $\sigma = 1, 2,$ and 3 and $\vec{r} = x\hat{x} + y\hat{y} + z\hat{z} = \vec{r}_\perp + z\hat{z}$):

$$A'_\sigma (\vec{r}_\perp, z, t) = A_\sigma (\vec{r}_\perp, z) e^{i(\vec{k}_\sigma \cdot \vec{r} - \omega_\sigma t)}. \tag{4.102}$$

The Laplacian operation on the fields gives

$$\nabla^2 E_\sigma\left(\vec{r}_\perp, z, t\right) = \frac{1}{2}\left(\nabla^2 A_\sigma + 2ik_{\sigma z}\frac{\partial A_\sigma}{\partial z} + 2ik_{\sigma x}\frac{\partial A_\sigma}{\partial x} + 2ik_{\sigma y}\frac{\partial A_\sigma}{\partial y} - k_\sigma^2 A_\sigma\right)e^{i(\vec{k}_\sigma \cdot \vec{r} - \omega_\sigma t)}. \qquad (4.103)$$

Following the steps of the previous derivation, the wavenumber is defined as $k_\sigma = n_\sigma \omega_\sigma/c$ and the second-order derivative with respect to z is neglected. The transverse coordinate (x,y) contributions in the Laplacian may be retained to describe diffraction effects. Equation 4.103 is inserted into the scalar wave equation (Equation 4.15) and each wave's amplitude satisfies the SVEA extended to noncollinear wave interactions:

$$2ik_{\sigma z}\frac{\partial A_\sigma}{\partial z} + 2ik_{\sigma x}\frac{\partial A_\sigma}{\partial x} + 2ik_{\sigma y}\frac{\partial A_\sigma}{\partial y} = -\omega_\sigma^2 \mu_0 P^{(2)}(\omega_\sigma)e^{-i\vec{k}_\sigma \cdot \vec{r}}. \qquad (4.104)$$

The nonlinear polarizations for three-wave mixing are expressed as

$$\begin{aligned}
P^{(2)}(\omega_1) &= 2\varepsilon_0 d_{\mathrm{eff}} A_2 A_3 e^{i(\vec{k}_2 + \vec{k}_3)\cdot \vec{r}}, \\
P^{(2)}(\omega_2) &= 2\varepsilon_0 d_{\mathrm{eff}} A_1 A_3^* e^{i(\vec{k}_1 - \vec{k}_3)\cdot \vec{r}}, \\
P^{(2)}(\omega_3) &= 2\varepsilon_0 d_{\mathrm{eff}} A_1 A_2^* e^{i(\vec{k}_1 - \vec{k}_2)\cdot \vec{r}}.
\end{aligned} \qquad (4.105)$$

Inserting these results into Equation 4.104, we have the noncollinear version of the SVEA equations, (define $\Delta \vec{k} = \vec{k}_1 - \vec{k}_2 - \vec{k}_3$),

$$\begin{aligned}
\cos\theta_{1z}\frac{\partial A_1}{\partial z} + \cos\theta_{1x}\frac{\partial A_1}{\partial x} + \cos\theta_{1y}\frac{\partial A_1}{\partial y} &= \frac{i\omega_1^2}{k_1 c^2}d_{\mathrm{eff}} A_2 A_3 e^{-i\Delta \vec{k}\cdot \vec{r}}, \\
\cos\theta_{2z}\frac{\partial A_2}{\partial z} + \cos\theta_{2x}\frac{\partial A_2}{\partial x} + \cos\theta_{2y}\frac{\partial A_2}{\partial y} &= \frac{i\omega_2^2}{k_2 c^2}d_{\mathrm{eff}} A_1 A_3^* e^{i\Delta \vec{k}\cdot \vec{r}}, \\
\cos\theta_{3z}\frac{\partial A_3}{\partial z} + \cos\theta_{3x}\frac{\partial A_3}{\partial x} + \cos\theta_{3y}\frac{\partial A_3}{\partial y} &= \frac{i\omega_3^2}{k_3 c^2}d_{\mathrm{eff}} A_1 A_2^* e^{i\Delta \vec{k}\cdot \vec{r}}.
\end{aligned} \qquad (4.106)$$

These equations reduce to Equations 4.20 through 4.22 when the waves are collinear. The cosine angles between the wave vectors and the Cartesian axes are used. The coefficients of the x and y derivatives are assumed to be small, that is, angles close to $\pi/2$ and the coefficient of the z derivative is close to unity. Each wave propagates along an axis that is defined by the direction of its wave vector. The differential equation can be parameterized by a variable, such that $\ell_\sigma \hat{k}_\sigma = x\hat{x} + y\hat{y} + z\hat{z}$

and the unit vector is $\hat{k}_\sigma = \cos\theta_{\sigma x}\hat{x} + \cos\theta_{\sigma y}\hat{y} + \cos\theta_{\sigma z}\hat{z}$. Using these relations, the total derivative of the parameter is

$$\frac{d}{d\ell_\sigma} = \cos\theta_{\sigma x}\frac{\partial}{\partial x} + \cos\theta_{\sigma y}\frac{\partial}{\partial y} + \cos\theta_{\sigma z}\frac{\partial}{\partial z}. \tag{4.107}$$

Equation 4.106 can be expressed as

$$\frac{dA_1}{d\ell_1} = \frac{i\omega_1^2}{k_1 c^2}d_{eff}A_2 A_3 e^{-i\Delta\vec{k}\cdot\hat{k}_1\ell_1},$$

$$\frac{dA_2}{d\ell_2} = \frac{i\omega_2^2}{k_2 c^2}d_{eff}A_1 A_3^* e^{i\Delta\vec{k}\cdot\hat{k}_2\ell_2}, \tag{4.108}$$

$$\frac{dA_3}{d\ell_3} = \frac{i\omega_3^2}{k_3 c^2}d_{eff}A_1 A_2^* e^{i\Delta\vec{k}\cdot\hat{k}_3\ell_3}.$$

4.5.2 NONCOLLINEAR APPLICATION

A common application of noncollinear phase matching is to accommodate broadband conversion processes. This application is particularly important with ultrashort femtosecond frequency conversion that requires a large phase-matching bandwidth. Calculating bandwidths for the noncollinear case is similar to the collinear one. The derivation of the phase-matching efficiency for the collinear situation follows directly from the coupled amplitude equations by assuming that the two inputs to the crystal are constant. This assumption allows us to directly integrate the remaining amplitude equation, resulting in the $\text{sinc}^2(\Delta k L/2)$ dependence. For a three-wave noncollinear process where the two inputs are also approximately constants, we are led to a similar situation. If we consider the difference frequency between ω_1 and ω_2 to generate ω_3, then the phase-matching efficiency for the noncollinear case is derived from Equation 4.108. The solution with nondepleted pump and signal for a medium with thickness L is

$$A_3(L) = \frac{i\omega_3^2}{k_3 c^2}d_{eff}A_1 A_2^* \frac{\left(e^{i\Delta\vec{k}\cdot\hat{k}_3 L} - 1\right)}{i\Delta\vec{k}\cdot\hat{k}_3} \tag{4.109}$$

From this result, the phase-matching efficiency function is expressed as

$$\eta = \text{sinc}^2\left(\frac{L}{2}\Delta\vec{k}\cdot\hat{k}_3\right). \tag{4.110}$$

The various bandwidths are calculated based on how $\Delta\vec{k}\cdot\hat{k}_3$ changes with a given parameter. For the purposes of this discussion, we look at the phase-matching bandwidth where ω_1 is fixed while ω_2 tunes. The variation of $\Delta\vec{k}\cdot\hat{k}_3$ with ω_2 is

$$\frac{\partial \Delta \vec{k} \cdot \hat{k}_3}{\partial \omega_2} = \frac{\partial \left(\vec{k}_1 - \vec{k}_2 - \vec{k}_3 \right) \cdot \hat{k}_3}{\partial \omega_2} = -\frac{\partial k_2}{\partial \omega_2} \cos \Omega + \frac{\partial k_3}{\partial \omega}, \qquad (4.111)$$

where Ω is the angle between $\vec{k}_2$ and $\vec{k}_3$, and where the energy conservation condition $\Delta \omega_3 = -\Delta \omega_2$ is used since $\Delta \omega_1 = 0$. The Taylor series expansion about the set point $\Delta \vec{k} \cdot \hat{k}_3 = 0$ gives

$$\Delta \vec{k} \cdot \hat{k}_3 (\omega_2 + \Delta \omega_2) = \frac{\partial \Delta \vec{k} \cdot \hat{k}_3}{\partial \omega_2} \Delta \omega_2 = \left[-\frac{\partial k_2}{\partial \omega_2} \cos \Omega + \frac{\partial k_3}{\partial \omega_3} \right] \Delta \omega_2. \qquad (4.112)$$

Using the same criterion from the Taylor series expansions in Section 4.4.2, namely, that the maximum variation of $\Delta \vec{k} \cdot \hat{k}_3 L = 0.886\pi$, gives the FWHM bandwidth as

$$\Delta \omega_{2,\text{FWHM}} = 2 \frac{0.886\pi}{L \left| -(\partial k_2 / \partial \omega_2) \cos \Omega + (\partial k_3 / \partial \omega_3) \right|}. \qquad (4.113)$$

An advantage of the noncollinear case is that it gives some freedom to adjust the denominator of this expression to make it small or zero while still being phase matched. As with the collinear case, when the denominator goes to zero, we should include higher-order terms in the Taylor series expansion. Since $(\partial k / \partial \omega) = v_g^{-1}$, the condition for large bandwidth can be cast in terms of group velocities as

$$v_{g2} = \cos \Omega v_{g3}. \qquad (4.114)$$

This equation shows that matching the group velocities along the direction of $\vec{k}_2$ results in a large bandwidth. Such a condition is not guaranteed since $\Delta k = 0$ must also be satisfied simultaneously. One successful example, shown in Figure 4.14, is a Type I interaction in BBO (e $\rightarrow$ o + o) with one input (the pump) at $\lambda_1 = 405$ nm (e-polarized) at an angle of 30.6°. Note that this angle changes slightly if we use a different set of the Sellmeier equations. The second input (the signal) has a wavelength of λ_2 and is at an angle of θ_2, both of which are indicated in the plot. Figure 4.14 shows that phase matching for this particular pump angle ($\theta_1 = 30.6°$) has two noncollinear phase-matching solutions. This outcome is not surprising given that k_2 and k_3 are both o-waves so that the geometry where k_2 and k_3 are reflected about k_3 in Figure 4.13 also gives $\Delta k = 0$.

The phase-matching bandwidth is obtained by taking a vertical slice out of Figure 4.14. The bandwidth for a specific angle of $\theta_2 = 34.2°$ is shown in Figure 4.15. Note that we could obtain the same bandwidth by picking θ from the other solution band. However, the higher-angle solution is a better choice because the Poynting vector walk-off of the pump is approximately 4° away from the optic axis. Since the k-vector of the pump is at $\theta_1 = 30.6$, its Poynting vector would be at $\theta_1 \sim 34.6°$, which is essentially collinear with k_2. For this choice, the energy of the pump and input signal

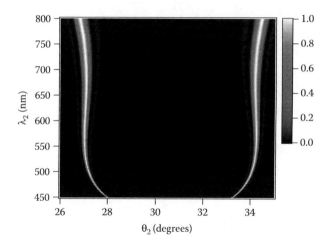

FIGURE 4.14   Plot of $\text{sinc}^2((L/2)\Delta\vec{k}\cdot\hat{k}_3)$ with a fixed input at $\lambda_1 = 405$ nm at an angle of $30.6°$ as a function of the second input's wavelength and angle (λ_2 and θ_2). $L = 1.0$ mm.

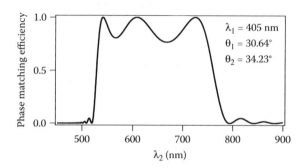

FIGURE 4.15   Noncollinear phase matching can result in a large bandwidth as shown here. The crystal is a 1-mm BBO crystal, the pump wavelength is 405 nm oriented at $30.6°$, and the signal is oriented at $34.2°$.

overlap throughout the crystal. Thus, phase matching, broad bandwidth, and walk-off compensation are all achieved in this example.

A final note on the bandwidth calculations is that they are all based on a plane-wave model. Clearly, this is not the common laboratory situation. However, these calculations give a good understanding of the qualitative aspects of phase matching as well as providing close estimates of the actual bandwidths.

PROBLEMS

In all problems note that temperature is 24°C unless otherwise specified.

4.1 In Section 4.1, we derived one of the three coupled amplitude equations for DFG between an input at ω_1 and a second input at ω_2. Make an argument justifying that Equation 4.21 is obtained from Equation 4.20 by a variable substitution. Parallel the derivation of Equation 4.20 to obtain Equation 4.22 for SFG.

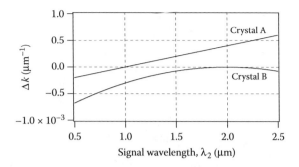

FIGURE 4.16   Comparison of Δk as a function of signal wavelength (λ_2) for two crystals.

4.2 Figure 4.16 shows a plot of Δk versus signal wavelength (λ_2) for a DFG process when the pump (λ_1) is held fixed. Two crystals, Crystal A and Crystal B, are displayed.
   a. For Crystal A and Crystal B, what is the signal wavelength where phase matching occurs?
   b. Which crystal has the larger phase-matching bandwidth for this DFG process and why?

4.3 Figure 4.17 shows the index of refraction as a function of wavelength for the nonlinear crystal, 4-N,N-dimethylamino-4'-N'-methyl-stilbazoliumtosylate (DAST), along its a-axis. Between 4 and 100 µm are several absorption bands such that the index is higher at 100 µm than at 4 µm, thus violating the normal dispersion criterion that index decreases with increasing wavelength. In this problem, we show that phase matching is possible when all three wavelengths have the same polarization.
   a. The index of refraction for frequencies polarized along the a-axis in DAST is given below and is broken into three regions. Plot the index of refraction from 1 to 4 µm and from 100 to 1000 µm and make sure that you get a plot like that shown in Figure 4.17.
   b. Assume a DFG process where all three frequencies are polarized along the crystal's a-axis and where the highest-frequency input is $\lambda_1 = 1.55$ µm. Calculate the λ_2 and λ_3 pair that phase matches. (*Hint*: Plot Δk as a function of wavelength

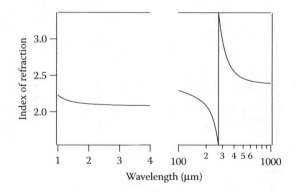

FIGURE 4.17   Index of refraction for the nonlinear crystal DAST along the a-axis.

from 1.56 to 1.57 μm and find where $\Delta k = 0$. Remember that when two wavelengths are chosen, the third is determined from energy conservation.)

c. Convert λ_3 (the longest wavelength) into a frequency in units of THz.

The index of refraction equations for DAST along its a-axis (wavelength entered in microns) is

$$1 < \lambda < 4\,\mu m \quad n(\lambda) = \sqrt{2.675 + \frac{1.64266}{1-(0.286514/\lambda^2)}},$$

$$100 < \lambda < 273\,\mu m \quad n(\lambda) = -1041.2\exp\left(\frac{-2109.7}{\lambda}\right) - 1.162\exp\left(\frac{-330.651}{\lambda}\right) + 2.3352,$$

$$\lambda > 273\,\mu m \quad n(\lambda) = 0.00517\exp\left(\frac{1434.93}{\lambda}\right) + 2.37.$$

4.4 The coupled amplitude equations for SHG are given by Equations 4.39 and 4.40. When the input (pump) is undepleted, the equations become decoupled, allowing us to directly integrate Equation 4.39. Verify that the result of the integration leads to the intensity given in Equation 4.42.

4.5 Show that the phase-matching angle for a Type I difference-frequency process, $e \rightarrow o + o$, or a Type I sum-frequency process, $o + o \rightarrow e$, is given by Equation 4.58.

4.6 Consider a DFG process in lithium niobate between a laser operating at $\lambda_1 = 0.532\,\mu m$ and a second one at $\lambda_2 = 0.600\,\mu m$.

a. What is the difference-frequency wavelength?

b. Use the Sellmeier equations in Appendix B and Equation 4.58 to find the phase-matching angle for a Type I, $e \rightarrow o + o$, interaction.

c. For the phase-matching angle found in (b), calculate the phase velocity for the three wavelengths. Does phase matching in this problem correspond to phase-velocity matching?

d. Calculate the group velocity for the three wavelengths using the angle found in (b). Does phase matching in this problem correspond to group-velocity matching?

4.7 Consider a DFG process in lithium niobate between $\lambda_1 = 0.532$ and $\lambda_2 = 0.800\,\mu m$ ($e \rightarrow o + o$).

a. What is the difference-frequency wavelength?

b. Use the Sellmeier equations in Appendix B to plot Δk as a function of angle. What is the phase-matching angle?

c. Compare the numerical result from (b) to the angle found using the analytic expression in Equation 4.58.

4.8 Consider a Type I SFG process, $e + e \rightarrow o$ in the xz-plane of a biaxial crystal. The e-wave has an index given by

$$\left(\frac{\cos^2\theta}{n_x^2} + \frac{\sin^2\theta}{n_z^2}\right)^{-1/2}$$

and the o-wave index is n_y. (The results of this problem also apply to a positive uniaxial crystal where $n_x = n_y$.)

a. Show that the phase-matching condition is given by

$$\frac{n_{x2}}{\lambda_2}\left(1+\left(\frac{n_{x2}^2}{n_{z2}^2}-1\right)\sin^2\theta\right)^{-1/2}+\frac{n_{x3}}{\lambda_3}\left(1+\left(\frac{n_{x3}^2}{n_{z3}^2}-1\right)\sin^2\theta\right)^{-1/2}-\frac{n_{y1}}{\lambda_1}=0, \quad (4.115)$$

where $n_{x2} = n_x(\lambda_2)$, and so on.

b. Equation 4.115 does not allow for an analytic expression for θ. However, for small birefringence, show that

$$\sin^2\theta \approx \frac{2((n_{y1}/\lambda_1)-(n_{x2}/\lambda_2)-(n_{x3}/\lambda_3))}{(n_{x2}/\lambda_2)\left(1-\left(n_{x2}^2/n_{z2}^2\right)\right)+(n_{x3}/\lambda_3)\left(1-\left(n_{x3}^2/n_{z3}^2\right)\right)}. \quad (4.116)$$

(*Hint*: For small birefringence, assume $|((n_x^2/n_z^2)-1)\sin^2\theta|\ll 1$.)

c. Calculate the phase-matching angle in KTP when $\lambda_1 = 0.800\,\mu m$ and $\lambda_2 = 1.50\,\mu m$ by plotting Equation 4.115 as a function of angle to find the zero-crossing point. Alternatively, use a zero-finding algorithm to find the phase-matching angle.

d. Use the approximated phase-matching angle formula in Equation 4.116 and compare the resulting angle to that found in (c).

4.9 Consider a sum-frequency process, $1.588 + 0.800\,\mu m \rightarrow 0.532\,\mu m$ in the xz-plane of a KTP crystal. In the following, use the Sellmeier equations in Appendix B.

a. Plot Δk as a function of θ, where θ is the angle between the k-vectors and the z-axis of the KTP crystal in the xz-plane for a Type I interaction $(e + e \rightarrow o)$. What is the phase-matching angle?

b. Repeat (a) for a Type II interaction $(o + e \rightarrow o)$ where the $0.800\,\mu m$ input is e-polarized.

c. Find d_{eff} for the two interactions in (a) and (b). Which interaction type is preferable based on d_{eff}? KTP is a class $mm2$ crystal.

4.10 Two laser beams are incident on a nonlinear crystal where they are optimally phase matched for SFG. The beams are loosely focused, so that they essentially look like plane waves. Calculate output intensity assuming the two input wavelengths are $\lambda_3 = 1.6\,\mu m$ and $\lambda_2 = 1.064\,\mu m$, the index of refraction for all of the wavelengths is 1.5, $d_{eff} = 1.5\,pm/V$, $L = 1\,cm$, and both beams have the same input intensity of $5\,kW/cm^2$. Assume low conversion efficiency such that the two input beams remain essentially constant.

4.11 Show that the analytic expression for the phase-matching angle for a Type I SHG phase-matching process in positive uniaxial crystals $(e + e \rightarrow o)$ is given by

$$\sin^2 \theta = \frac{n_z^2(\omega)}{n_o^2(2\omega)} \frac{\left(n_o^2(2\omega) - n_o^2(\omega)\right)}{\left(n_z^2(\omega) - n_o^2(\omega)\right)}. \tag{4.117}$$

4.12 Show that the analytic expression for the phase-matching angle for a Type I SHG phase-matching process in negative uniaxial crystals $(o + o \rightarrow e)$ is given by

$$\sin^2 \theta = \frac{n_z^2(2\omega)}{n_o^2(\omega)} \frac{\left(n_o^2(2\omega) - n_o^2(\omega)\right)}{\left(n_o^2(2\omega) - n_z^2(2\omega)\right)}. \tag{4.118}$$

4.13 Describe how to accomplish Type II phase matching for SHG using a single, linearly polarized pump laser without using a beam splitter.

4.14 Derive an analytic expression for the phase-matching angle given a Type II SHG interaction in a positive uniaxial crystal $(o + e \rightarrow o)$.

4.15 Consider frequency doubling a tunable Ti:sapphire laser, which operates from ~0.7 to 0.9 μm. Use the analytic expression for Type I phase-matching SHG in Equation 4.118 to plot the phase-matching angle as a function of fundamental wavelength in BBO (see the Sellmeier equations in Appendix B). Plot d_{eff} and the walk-off angle over the same range of angles and comment on the device performance over the tuning range of the Ti:sapphire laser. Assume that ϕ is adjusted to maximize d_{eff}.

4.16 Consider a second-harmonic process in BBO for $1.064 + 1.064$ μm $\rightarrow 0.532$ μm $(o + o \rightarrow e)$.
   a. Use the Sellmeier equations in Appendix B to plot Δk as a function of θ (angle between the k-vector and the optic axis). What is the phase-matching angle?
   b. Calculate the phase-matching angle using the analytic expression given in Equation 4.118 and compare it with your numerical calculation in (a).

4.17 Consider a Type II DFG process, $e \rightarrow o + e$, in lithium niobate where $\lambda_1 = 1.064$ μm (e-polarized) and $\lambda_2 = 1.5$ μm (o-polarized). (Ignore walk-off effects.)
   a. What is the difference-frequency wavelength, λ_3?
   b. Use the Sellmeier equations in Appendix B to solve for Δk as a function of angle. Plot Δk versus angle to find where $\Delta k = 0$, the phase-matched point.

4.18 A Type II DFG process, $e \rightarrow o + e$, in lithium niobate phase matches at $\theta \sim 59°$ when $\lambda_1 = 0.980$ μm (e-wave) and $\lambda_2 = 1.5$ μm (o-wave).
   a. Plot $\text{sinc}^2(\Delta kL/2)$ versus angle about the phase-matching angle for a 1-cm crystal.
   b. Find the angular acceptance as defined by the FWHM of the curve plotted in (a).

4.19 A Type II DFG process, $e \rightarrow o + e$, in lithium niobate phase matches at $\theta \sim 59°$ when $\lambda_1 = 0.980$ μm (e-wave) and $\lambda_2 = 1.5$ μm (o-wave).
   a. Plot $\text{sinc}^2(\Delta kL/2)$ as a function of temperature about 20°C for a 1-cm crystal.
   b. Use the FWHM of the curve in (a) to determine the temperature bandwidth.

4.20 For a Type I interaction, o → e + e or e → o + o, the phase-matching bandwidth becomes large at degeneracy. Degeneracy is the point where the two waves with the same polarization also have the same frequency. Assuming that ω_1 is fixed, we see that the Taylor series expansion technique to first-order (Equation 4.88) gives a bandwidth that blows up.
   a. Derive an expression for the bandwidth that takes the Taylor series to second-order.
   b. Calculate $\Delta\lambda_2$ for lithium niobate for e → o + o about degeneracy when the pump wavelength is 1.0 μm. Plot $sinc^2(\Delta kL/2)$ as a function of λ_2 and compare the FWHM with the analytic result.

4.21 Find the phase-matching angle for both Type I and Type II SHG in the crystal, BBO, when the fundamental wavelength is 800 nm. Compare the two interactions based on d_{eff} and walk-off angles. Which interaction appears to be the better choice for doubling?

4.22 A Type II DFG process, e → o + e, in lithium niobate phase matches at $\theta \sim 59°$ when $\lambda_1 = 0.980$ μm (e-wave) and $\lambda_2 = 1.5$ μm (o-wave). The crystal length is 1 cm.
   a. Plot $sinc^2(\Delta kL/2)$ versus λ_2 assuming that λ_1 remains fixed and λ_2 tunes about 1.5 μm (remember to recalculate λ_3 as λ_2 tunes).
   b. Use the FWHM of the curve in (a) to determine the phase-matching bandwidth.

4.23 A Type II DFG process, e → o + e, in lithium niobate phase matches at $\theta = 57.6°$ when $\lambda_1 = 1.05$ μm (e-wave) and $\lambda_2 = 1.5$ μm (o-wave). The crystal length is 1 cm.
   a. Plot $sinc^2(\Delta kL/2)$ versus λ_1 assuming that λ_2 remains fixed and λ_1 tunes from 0.9 to 1.1 μm (remember to recalculate λ_3 as λ_1 tunes).
   b. Plot $sinc^2(\Delta kL/2)$ under the same circumstances as (a) for the cases when θ is changed by +0.1° and −0.1°.

4.24 Calculate the angular acceptance for Problem 4.18 using the Taylor series expansion technique and compare the calculated angular acceptance with the plotted angular acceptance in Problem 4.18.

4.25 Calculate the temperature bandwidth for Problem 4.19 using the Taylor series expansion technique and compare the calculated value with the plotted FWHM found in Problem 4.19.

4.26 Use the Taylor series approach to calculate the pump-tuning bandwidth ($\Delta\omega_1$) when (a) ω_2 is held fixed and (b) ω_3 is held fixed. Under what condition will the pump wavelength tuning bandwidth be large?

4.27 Figure 4.18 shows an intensity plot of $sinc^2(\Delta kL/2)$ plotted versus signal wavelength (λ_2) and phase-matching angle for a Type I (e → o + o) DFG process in BBO. The pump (λ_1) is fixed at 0.500 μm. Use this plot to estimate the angular acceptance and the phase-matching bandwidth centered at
   a. $\lambda_2 = 1500$ nm.
   b. $\lambda_2 = 1000$ nm.

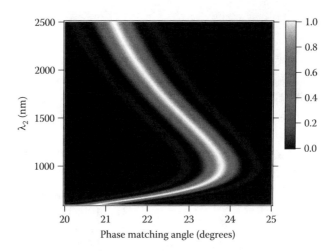

FIGURE 4.18 Intensity plot of sinc$^2(\Delta k L/2)$ plotted against phase-matching angle and signal wavelength (λ_2).

4.28 Consider a Type II DFG interaction in the xz-plane of KTP (o → e + o) where the pump laser is fixed at $\lambda_1 = 0.800\,\mu m$ (o-wave). Create a two-dimensional array where each point in the array corresponds to a unique θ, λ_2 pairing. Scale the array such that the x-values range from 35° to 90° and the y-values range from 1.0 to 4.5 μm.

    a. Assign to each element of the array $\Delta k(\theta, \lambda_2)$ assuming that λ_2 is an e-wave and λ_3 is an o-wave. Display a contour plot of the array to locate the $\Delta k = 0$ contour. Note that since λ_1 is fixed, λ_3 is dictated by energy conservation for a particular λ_2.

    b. Repeat (a) but now assume that λ_2 is an o-wave and λ_3 is an e-wave. The $\Delta k = 0$ contours from (a) and (b) make up the tuning curve for KTP.

4.29 Consider a Type II DFG interaction in the xz-plane of KTP (o → e + o) where the pump laser is fixed at $\lambda_1 = 0.800\,\mu m$ (e-wave). Assuming that λ_2 is an e-wave and λ_3 is an o-wave, write a program that generates a tuning curve by sequentially incrementing λ_2 over the range of 1.0–3.5 μm. For each λ_2, use a zero-finding algorithm to find the phase-matching angle where $\Delta k = 0$ (see Appendix C). Plot the resultant tuning curve, λ_2 and λ_3 versus 0, on the same graph.

4.30 A given process has an angular acceptance of $\Delta\theta$ for a crystal length, L. If the crystal length is increased by a factor of 5, how is the bandwidth affected?

4.31 Noncollinear phase matching is a way of compensating for the Poynting vector walk-off as well as a way to increase the bandwidth of a parametric amplifier. Assume a DFG interaction, e → o + o, in BBO with $\lambda_1 = 0.405\,\mu m$ e-polarized (let us call it the pump).

    a. Which way does the pump beam walk—toward or away from the optic axis?

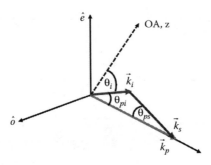

FIGURE 4.19 Type II phase-matching geometry for a BBO crystal. The optic axis (OA) lies in the plane defined by the vectors $\vec{k}_p, \hat{e}$.

b. What noncollinear angle gives $\Delta k = 0$ while simultaneously satisfying noncollinear phase matching and an angle between the pump and the idler equal to the walk-off angle. The signal (λ_2) is 0.600 µm.

4.32 BBO crystal is used for DFG. The Sellmeier equations for refractive indices are found in Appendix B. In the following, assume Type II phase matching e $\to$ o + e. The geometry is shown in Figure 4.19 and all angles are in the ($\hat{e}, \vec{k}_p$) plane.

a. Using pump wavelength 325 nm and degenerate signal and idler, write a program to find the phase-matching angle, θ_p, for collinear phase matching (answer: 54.489°), the walk-off angle ρ, and d_{eff}.

b. For noncollinear phase matching, the pump wave is kept at a fixed angle, θ_p with respect to the optic axis at the angle found in (a). The angle of the signal and idler wave vectors with respect to the pump wave direction is θ_{ps} and θ_{pi}, respectively. Using the vector decomposition of the wave vectors, show that phase matching requires

$$k_p(\theta_p) = k_s \cos(\theta_{ps}) + k_i(\theta_i)\cos(\theta_{pi}),$$
$$k_s \cos(\theta_{ps}) = -k_i(\theta_i)\cos(\theta_{pi})$$

Show that the phase-matching conditions in (b) can be written as

$$F = k_p^2(\theta_p) - k_s^2 - 2k_p(\theta_p)k_i(\theta_i)\cos(\theta_{pi}) + k_i^2(\theta_i) = 0.$$

c. The angle that the idler wave vector $\vec{k}_i$ subtends with the optic axis is $\theta_i = \theta_p - \theta_{pi}$ (see Figure 4.19). Using the result from (b), numerically find a noncollinear phase-matching angle ($F = 0$). Also calculate the walk-off angle for the idler wave ρ.

d. Use the program to numerically find the two phase-matching angles for nondegenerate signal and idler cases for signal wavelengths between 650 and 600 nm. Remember that the idler is the longest wavelength. You can do this graphically and tabulate the idler phase angle (θ_{pi}) data.

REFERENCE

Smith, A. V. 2015. *SNLO Nonlinear Optics Code*. AS-Photonics, Albuquerque, NM (available from A. V. Smith). http://www.as-photonics.com/snlo (Accessed January 30, 2017.)

FURTHER READING

Armstrong, J. A., N. Bloembergen, J. Ducuing, and P. S. Pershan. 1962. Interactions between light waves in a nonlinear dielectric. *Physical Review* 127: 1918–1939.

Boyd, R. W. 2008. *Nonlinear Optics*. Boston, MA: Academic Press.

Butcher, P. N. and D. Cotter. 1990. *The Elements of Nonlinear Optics*. Cambridge, UK: Cambridge University Press.

Giordmaine, G. D. 1962. Mixing of light beams in crystals. *Physical Review Letters* 8: 19–20.

He, G. S. and S. H. Lui. 1999. *Physics of Nonlinear Optics*. Singapore: World Scientific Publishing Co. Pte. Ltd.

Kleinman, D. A. 1962. Theory of second harmonic generation of light. *Physical Review* 128: 1761–1775.

Minck, R. W., R. W. Terhune, and C. C. Wang. 1966. Nonlinear optics. *Applied Optics* 5: 1595–1612.

Sauter, E. G. 1996. *Nonlinear Optics*. New York, NY: John Wiley & Sons.

Shen, Y. R. 1984. *The Principles of Nonlinear Optics*. New York, NY: John Wiley & Sons.

Sutherland, R. L. 2003. *Handbook of Nonlinear Optics* (2nd edition). New York, NY: Marcel Dekker, Inc.

Zernike, F. and J. E. Midwinter. 1973. *Applied Nonlinear Optics*. New York, NY: John Wiley & Sons.

5

QUASI-PHASE MATCHING

5.1   INTRODUCTION TO QUASI-PHASE MATCHING

This chapter presents a strategy for phase-matching waves that does not require crystal birefringence. The technique, called quasi-phase matching (QPM), was proposed even before birefringent phase matching (Armstrong et al., 1962), but due to technical difficulties in fabricating samples, QPM was not achievable at the time. The more direct method of birefringent phase matching was the first to be widely implemented. Both techniques are now found in a wide range of applications.

Quasi-phase-matched techniques periodically reset the phase of the nonlinear polarization to maintain a coherent buildup of the nonlinear interaction. The most common mechanism for rephasing is via the crystal's nonlinearity, $\chi^{(2)}$. A change to $\chi^{(2)}$ results in a change to the nonlinear polarization, which is the control we seek. A simple way to introduce such a change in $\chi^{(2)}$ is to physically reorient the crystal as a function of propagation. For example, a crystal can be sliced into thin slabs and then rearranged with each segment rotated 180° with respect to the previous one. However, in practice, stacking crystal plates is difficult and introduces scattering losses, periodicity errors, and other problems. Stacking crystals and other more efficient fabrication techniques are discussed after we review the basic QPM analysis.

5.2   LINEAR AND NONLINEAR MATERIAL CONSIDERATIONS

As mentioned above, the idea behind QPM is to reorient the crystal. Referring to Figures 4.3 and 4.4, we see that the field and intensity build up over one coherence length and then decay over the next coherence length. The reason for this periodic behavior is that the phase relationship between the nonlinear polarization and the field it drives is not fixed. So the energy flows from the nonlinear polarization into the field for one coherence length, reverses for the next one, and the pattern repeats. With QPM, the crystal is reoriented periodically to correct for the phase-slip between the nonlinear polarization and the field it generates. For special reorientations, the sign of the nonlinearity, $\chi^{(2)}$, is flipped; hence, the nonlinear polarization is also flipped. The periodic flipping lets the intensity grow continuously, as shown in Figure 5.1. Note that QPM is possible in any $\chi^{(2)}$ material, even *nonbirefringent* ones.

A physical reorientation of a crystal, in general, changes the linear properties as seen by a beam of light propagating through the crystal. A periodic structure results in a diffraction grating, which complicates the nonlinear mixing process. However, certain operations, such as an inversion, leave the linear properties of the crystal unchanged but lead to a sign reversal in the nonlinearity. We can understand transformations to the linear and nonlinear properties by looking at changes to the permittivity tensor, ε_{ij}, and the second-order nonlinear susceptibility, d_{ijk}. Both properties transform according to the tensor rules introduced in Chapter 2.

Let us consider a crystal in its principal axis coordinate system rotated 180° about its x-axis. The R-matrix for this transformation is

$$\begin{bmatrix} 1 & 0 & 0 \\ 0 & -1 & 0 \\ 0 & 0 & -1 \end{bmatrix} \tag{5.1}$$

Pursuant to the rules introduced in Section 2.2, ε_{ij} transforms as

$$\varepsilon'_{ij} = R_{i\alpha}R_{j\beta}\varepsilon_{\alpha\beta}, \tag{5.2}$$

where the repeated indices indicate a summation according to the Einstein notation. $\varepsilon_{\alpha\beta}$ represents the crystal before the transformation, and since $\varepsilon_{\alpha\beta}$ starts out diagonal

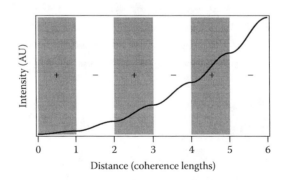

FIGURE 5.1 Intensity builds up in a quasi-phase-matched crystal by periodically inverting the sign of the nonlinearity. The different shading represents the + and − sign of the nonlinearity, $\chi^{(2)}$.

in the principal axis coordinate system, the only nonzero contributions to the sum are $\alpha = \beta$ so that

$$\varepsilon'_{ij} = R_{i\alpha}R_{j\alpha}\varepsilon_{\alpha\alpha}. \tag{5.3}$$

The only nonzero terms in the sum over α are $i = \alpha$ and $j = \alpha$. So, $i = j = \alpha$ and Equation 5.3 becomes

$$\varepsilon'_{ii} = R_{ii}R_{ii}\varepsilon_{ii}. \tag{5.4}$$

Since each of the diagonal elements of R is either $+1$ or -1,

$$\varepsilon'_{ii} = (\pm 1)^2 \varepsilon_{ii} = \varepsilon_{ii}. \tag{5.5}$$

The permittivity, and hence the index of refraction, remains unchanged for a $180°$ rotation about the x-axis. Therefore, a structure where the crystal is periodically rotated in this way has the same linear properties as a single segment of the crystal.

The changes in the nonlinear properties follow the same procedure. Since d_{ijk} is a tensor, it transforms according to

$$d'_{ijk} = R_{i\alpha}R_{j\beta}R_{k\gamma}d_{\alpha\beta\gamma}. \tag{5.6}$$

R is the same operation given in Equation 5.1, which is diagonal so that

$$d'_{ijk} = R_{ii}R_{jj}R_{kk}d_{ijk}. \tag{5.7}$$

The sign of d_{ijk} after the transformation depends on the specific combination of ijk. For example, consider the xxx, yyy, and zzz combinations. For xxx, $R_{xx} = +1$; so $d'_{xxx} = d_{xxx}$. For yyy, $R_{yyy} = -1$; so $d'_{yyy} = -d_{yyy}$. Similarly, $d'_{zzz} = -d_{zzz}$. Clearly, only certain elements of the d-tensor are inverted. The key to coupling to a d-coefficient that changes signs is to use the right combination of field polarizations. For example, according to Equation 3.77, a difference-frequency process with two input fields polarized along the z-axis gives a nonlinear polarization of

$$P_i^{(2)} = 2\varepsilon_0 d_{izz}A_z(\omega_1)A_z^*(\omega_2)e^{i(k_1-k_2)z}. \tag{5.8}$$

Provided that d_{zzz} is not zero, a nonlinear polarization is induced in the z-direction, which drives a field that propagates collinearly with the inputs. Since this interaction couples to d_{zzz}, the polarization changes signs upon a $180°$ rotation of the crystal about the x-axis. The $180°$ rotation yields the desired result: a nonlinearity that changes signs while leaving the linear properties unaltered.

5.3 QPM WITH PERIODIC STRUCTURES

A detailed understanding of QPM is found by an analysis of the wave equation. The wave equation for QPM processes is derived in the same way as for the birefringent case, resulting in coupled amplitude equations exactly as given in Chapter 4, with the exception that now we allow the nonlinearity to be a function of propagation distance, $d \rightarrow d(z)$. We consider again the case of a mixing process between two inputs that are essentially constant, which decouples the amplitude equations (Equations 4.29 through 4.31). The remaining equation (Equation 4.31) for difference frequency between ω_1 and ω_2 is reproduced here:

$$\frac{dA_3}{dz} = i \frac{\omega_3}{n_3 c} A_1 A_2^* d(z) e^{i\Delta k z}. \tag{5.9}$$

This equation is directly integrated:

$$A_3 = i \frac{\omega_3}{n_3 c} A_1 A_2^* \int_0^L d(z) e^{i\Delta k z} dz. \tag{5.10}$$

Note that the distinction between Equation 5.10 and the birefringent case is that now $d(z)$ must be included in the integral.

We begin with a crystal that has a structure as discussed above, that is, a stacking of alternating oriented crystal slabs such that the sign of d flips from one domain to the next. $d(z)$ then has the periodic form shown in Figure 5.2 with a period Λ. Such a periodic structure lends itself naturally to a Fourier series representation, and for the square-wave structure, the series is well known:

$$d(z) = d_0 \sum_{m=-\infty}^{\infty} \frac{2}{\pi m} \sin(m\pi D) e^{-i k_m z}, \tag{5.11}$$

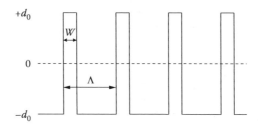

FIGURE 5.2 d-Coefficient is plotted as a function of distance in the crystal. The structure shown here can have an arbitrary duty cycle of + regions to − regions depending on the ratio of W/Λ.

where k_m is given by

$$k_m = \frac{2\pi}{\Lambda} m. \tag{5.12}$$

D is the duty cycle defined by

$$D \equiv \frac{W}{\Lambda}, \tag{5.13}$$

where W and Λ are defined in Figure 5.2. The d-coefficient of the nonlinear material is d_0. To simplify the following algebra, Equation 5.11 is rewritten as

$$d(z) = d_0 \sum_m C_m e^{-ik_m z}, \tag{5.14}$$

where

$$C_m = \frac{2}{m\pi} \sin(m\pi D). \tag{5.15}$$

Normally, the duty cycle is 50%, making $C_m = 0$ for all even orders and $C_m = (-1)^{(m-1)/2}(2/m\pi)$ for all odd orders. Now insert Equation 5.14 into Equation 5.10:

$$A_3 = i \frac{\omega_3}{n_3 c} A_1 A_2^* d_0 \int_0^L \sum_m C_m e^{-ik_m z} e^{i\Delta k z} dz. \tag{5.16}$$

The summation may be taken outside of the integral to give

$$A_3 = i \frac{\omega_3}{n_3 c} A_1 A_2^* d_0 \sum_m C_m \int_0^L e^{i(\Delta k - k_m)z} dz. \tag{5.17}$$

The integral is the same as the birefringent case in Chapter 4, except that now Δk is modified. The parallel between QPM and birefringent phase matching becomes stronger by making the substitution

$$\Delta k_m' = \Delta k - k_m. \tag{5.18}$$

Performing the integration in Equation 5.17 gives

$$A_3 = i \frac{\omega_3}{n_3 c} A_1 A_2^* d_0 L \sum_m C_m e^{i\Delta k_m' L/2} \operatorname{sinc}\left(\frac{\Delta k_m' L}{2}\right). \tag{5.19}$$

The summation may be eliminated by noting that $\Delta k_m' = 0$ is only possible for one particular value of m. For example, substituting Equation 5.12 into Equation 5.18 gives

$$\Delta k_m' = \Delta k - \frac{2\pi}{\Lambda} m = 0. \tag{5.20}$$

When $\Delta k_m' = 0$ for a particular value of m, all other values of m yield Δk_m, which is a multiple of $2\pi/\Lambda$. Hence, the value of $\operatorname{sinc}(\Delta k_m' L/2)$ is small except for the particular value of m that satisfies $\Delta k_m' = 0$. By neglecting the small contributions, the summation is eliminated from Equation 5.19 so that

$$A_3 = i \frac{\omega_3}{n_3 c} A_1 A_2^* d_0 L C_m e^{i\Delta k_m' L/2} \operatorname{sinc}(\Delta k_m' L/2). \tag{5.21}$$

Equation 5.21 leads to the same characteristics discussed for birefringent phase matching with $\Delta k_m'$ playing the same role as Δk. Specifically, $\Delta k_m' = 0$ leads to efficient generation, and a knowledge of how $\Delta k_m'$ varies away from zero leads to an understanding of various bandwidths.

Equation 5.21 is equivalent to

$$A_{3,\text{QPM}} = C_m A_{3,\text{birefringent}}. \tag{5.22}$$

where $A_{3,\text{birefringent}}$ is the field amplitude for a perfectly phase-matched crystal with nonlinearity d_0. The apparent cost of going to a QPM structure is a reduction in the field amplitude by C_m and C_m^2 for the intensity. For a first-order QPM process ($m = 1$) with a 50% duty cycle, $C_1 = 2/\pi$ and

$$A_{3,\text{QPM}} = \frac{2}{\pi} A_{3,\text{birefringent}}. \tag{5.23}$$

Equation 5.23 shows a reduction in value if both the birefringent process and the QPM process access *the same* element of the d-tensor. QPM is actually much more flexible because we can use any crystal orientation and polarization combination. The ability to choose arbitrary combinations of polarizations opens up the possibility of accessing a crystal's largest nonlinearity even if it is a combination that does not allow for birefringent phase matching. For example, in lithium niobate, the largest *d*-tensor element, d_{33}, is roughly 10 times larger than elements accessible to birefringent phase matching. Coupling to the d_{33} element requires an e → e + e interaction, as deduced from Equations 5.8 and 3.77. Such polarization combinations are now possible with QPM. Moreover, we can choose propagation in a noncritical direction where the Poynting vector walk-off is zero. Hence, all the interacting waves remain on top of each other throughout the crystal, which allows us to use much longer crystal lengths. The quasi-phase-matched interaction can access a higher nonlinearity while also allowing longer crystal lengths when compared with birefringent phase matching.

The periodicity required for a given quasi-phase-matched order is found by setting $\Delta k'_m = 0$, in Equation 5.20 giving,

$$\Lambda = \frac{2\pi}{\Delta k} m. \tag{5.24}$$

Equation 5.24 shows that a first-order QPM ($m = 1$) has a shorter periodicity than the higher orders. In the case of conversion processes with wavelengths in the visible to UV range, the coherence lengths are short and it is not always possible to fabricate a first-order structure. A higher-order QPM process ($m > 1$) increases the periodicity, and so it is possible to accommodate essentially any conversion process. The price we pay is, according to Equation 5.15, that the effective nonlinearity is reduced by a factor of m when compared with a first-order QPM process. Hence, the conversion efficiency of higher-order, phase-matched gratings is reduced by a factor $1/m^2$ below the first-order phase-matching process. The QPM process theoretically allows any mixing process within the transparency range of the crystal to quasi-phase match as long as we are able to engineer the correct periodic structure. QPM contrasts with birefringent phase matching where phase matching is dependent solely on the material's properties.

5.4 QPM CALCULATION: AN EXAMPLE

The details of QPM are best illustrated with a specific example. Consider difference-frequency generation (DFG) in lithium niobate between two lasers with wavelengths at 1.064 and 1.550 μm. Lithium niobate is a class 3m negative uniaxial crystal. The QPM approach initially ignores phase matching; instead, it focuses on determining optimal polarizations and an optimal propagation direction in the crystal. First, we would like to access the largest nonlinearity in lithium niobate, which is d_{33}. A nonlinear polarization couples to d_{33} when the two inputs and the nonlinear polarization point in the z-direction, as shown in Equation 5.8. Hence, the inputs should propagate in the x–y-plane with their fields along the z-direction. The difference-frequency output is collinear with the inputs and its field also points in the z-direction.

The second consideration is nulling the Poynting vector walk-off. The walk-off is zero when the k-vectors are at $\theta = 0°$ and $90°$ to the optic axis of lithium niobate. By choosing $\theta = 90°$, the walk-off is zero; at the same time, we access d_{33}. So the best configuration is propagation in the x–y-plane with the fields polarized along the z-axis. The next step is to calculate the required periodic structure. For a first-order QPM interaction ($m = 1$), the periodicity is given by

$$\Lambda = \frac{2\pi}{\Delta k} = \frac{1}{(n_{z1}/\lambda_1) - (n_{z2}/\lambda_2) - (n_{z3}/\lambda_3)},$$

(5.25)

where $n_{z1} = n_z(\lambda_1)$ and so on. These indices may be found from Sellmeier's equations for lithium niobate (see Appendix B). Substituting in the values for the indices and the specific wavelengths gives a periodicity of $\Lambda \sim 30$ μm.

5.5 FOURIER TRANSFORM TREATMENT OF QPM

Let us recall again Equation 5.10 and evaluate this integral in a slightly different way. First, we make the variable substitution

$$d(z) = d_0 g(z).$$

(5.26)

where $g(z)$ is a geometrical function given by

$$g(z) = \text{sign}(d).$$

(5.27)

That is, $g(z)$ is a function that alternates from $+1$ to -1 depending on the sign of $d(z)$. Now the integral in Equation 5.10 is written as

$$A_3(L) = i\frac{\omega_3}{n_3 c} A_1 A_2^* d_0 \int_{-\infty}^{\infty} \text{rect}\left(\frac{z}{L}\right) g(z) e^{i\Delta k z} dz,$$

(5.28)

where $\text{rect}(z/L)$ is a window function equal to 1 within the crystal and equal to 0 outside the crystal. Formally, the rect function is defined by

$$\text{rect}(x) = \begin{cases} 1 & 0 < |x| < 1 \\ \dfrac{1}{2} & x = 0 \quad \text{or} \quad x = 1 \\ 0 & \text{otherwise} \end{cases}$$

(5.29)

As a further simplification, we define $h(z) = \text{rect}(z/L)g(z)$ so that Equation 5.28 becomes

$$A_3(L) = i\frac{\omega_3}{n_3 c} A_1 A_2^* d_0 \int_{-\infty}^{\infty} h(z)e^{i\Delta k z}\,dz. \tag{5.30}$$

$h(z)$ is a purely geometric function determined by the length of the crystal and the periodic structure. Equation 5.30 contains the Fourier transform of $h(z)$ into Δk space. That is,

$$\tilde{H}(\Delta k) = \int_{-\infty}^{\infty} h(z)e^{i\Delta k z}\,dz, \tag{5.31}$$

where Δk in this context is just a variable. Substituting Equation 5.31 into Equation 5.30 gives

$$A_3(L) = \kappa d_0 \tilde{H}(\Delta k), \tag{5.32}$$

where

$$\kappa = i\frac{\omega_2}{n_3 c} A_1 A_2^*. \tag{5.33}$$

So the amplitude, A_3, depends on the material properties and input field amplitudes via κ and d_0, and it depends on the geometry of the QPM structure through $\tilde{H}(\Delta k)$. Perhaps a more useful quantity than Equation 5.32 is a comparison with a perfectly phase-matched process. The perfectly phase-matched process is one where $\Delta k = 0$ while accessing the same nonlinearity, d_0. For the perfectly phase-matched condition, the field amplitude is

$$A_{3,\text{perfect}}(L) = \kappa d_0 L. \tag{5.34}$$

Now we define relative efficiency as

$$\eta_{\text{QPM}} = \frac{I_{3,\text{QPM}}}{I_{3,\text{perfect}}} = \left|\frac{\tilde{H}(\Delta k)}{L}\right|^2. \tag{5.35}$$

η_{QPM} lets us ask the question of how a particular QPM process compares to one that is perfectly phase matched. Of course, this comparison is not quite fair since it may not be possible for a birefringently phase-matched process to access the same d_0. We will see that η_{QPM} also helps to visualize several other aspects of QPM.

As a first example, consider the QPM structure we have been discussing thus far, a 50% duty cycle structure varying from $+d_0$ to $-d_0$ with a periodicity of Λ. Hence, $g(z)$ has the same periodicity and varies from $+1$ to -1. The combination of $g(z)$ and the crystal length gives $h(z)$. An example with $\Lambda = 100$ μm in a 1-mm crystal is shown in Figure 5.3.

η_{QPM} is obtained from the Fourier transform of $h(z)$ and the crystal length. Fourier transforms are a standard part of most mathematical packages in the form of a fast Fourier transform (FFT). The way to perform the FFT is to create an array of values, in this case $h(z)$ shown above, and then apply the FFT to the array. Usually, the FFT is accomplished in a given mathematical package by a single statement such as output = FFT(h_array). Even though the FFT is easy to implement, a few details are important to understand. The first is that the FFT assumes a Fourier transform written in terms of frequency and not angular frequency. So the FFT gives a result that would be

$$\tilde{H}(f) = \int_{-\infty}^{\infty} h(z)e^{i2\pi fz}dz. \tag{5.36}$$

Note the difference in the exponent when comparing Equation 5.36 with Equation 5.31. So we need to remember that after performing the FFT, we need to convert from frequency to Δk using $f = \Delta k/2\pi$.

A second issue is that an FFT assumes the data being analyzed are a single period of a periodic function. Strategies to come up with an FFT that is more representative of the analytic Fourier transform include padding the array with zeroes at the beginning and end of the function and choosing an appropriate point spacing. A final issue with FFT algorithms is that they generate an output with the frequencies ordered in a nonintuitive way specific to the mathematical package. Hence, it is important to study the details of a given mathematical package's FFT operation.

To come up with η_{QPM} for the structure shown in Figure 5.3, we follow Equation 5.35 and take the magnitude squared of the FFT of $h(z)$ divided by L^2. η_{QPM} for the

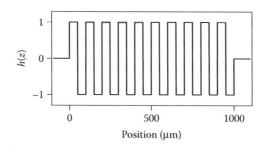

FIGURE 5.3 $h(z)$ shows the QPM structure in geometric terms.

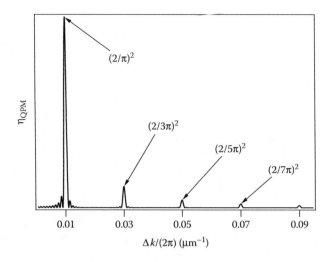

FIGURE 5.4 η_{QPM} (QPM efficiency) for a single-periodicity crystal.

particular structure shown in Figure 5.3 is plotted in Figure 5.4. η_{QPM} shown in Figure 5.4 is determined entirely by the engineered structure in the crystal. Now it is possible to consider any three-wave mixing process in this crystal and predict its efficiency relative to a perfectly phase-matched process. For example, consider second-harmonic generation (SHG). To determine its relative efficiency, we first calculate Δk based on the Sellmeier equations of the crystal alone without regard to the periodic structure. That is, $\Delta k_{SHG} = k_{2\omega} - 2k_{\omega}$. For the sake of the present discussion, let us say that $\Delta k_{SHG}/2\pi = 0.06\ \mu m^{-1}$. Next, we look up η_{QPM} for the periodic structure, which for the structure shown in Figure 5.3 is given in Figure 5.4. We observe that for the SHG process, η_{QPM} corresponding to this Δk_{SHG} is essentially zero. Now let us say that we consider a difference-frequency process, and based on the Sellmeier equations it gives $\Delta k_{DFG}/2\pi = 0.01\ \mu m^{-1}$. The η_{QPM} for this particular Δk matches with the first-order QPM peak, and so the DFG interaction will be relatively efficient. Once the η_{QPM} is determined from the QPM structure, it is treated as a lookup table that applies to any crystal with the same QPM structure. The procedure to determine the relative efficiency of a given process is to first determine Δk in the bulk crystal and then look up η_{QPM} for that particular Δk.

The Fourier transform approach, and with it η_{QPM}, helps when working with arbitrary QPM structures. Consider a chirped grating where the periodicity changes continuously. Understanding the chirped structure from a standpoint of a QPM grating vector is difficult because the periodicity is not well defined. However, from the Fourier transform point of view, all we need to know is Δk, after which we look up the value $\tilde{H}(\Delta k)$ or η_{QPM}. A comparison of a single-periodicity crystal such as that shown in Figure 5.3 and a chirped crystal with the same length is shown in Figure 5.5. The figure shows that the chirped structure gives a broader range over which Δk yields an appreciable efficiency. The trade-off is that the magnitude of the efficiency is much lower when compared with the single-periodicity crystal.

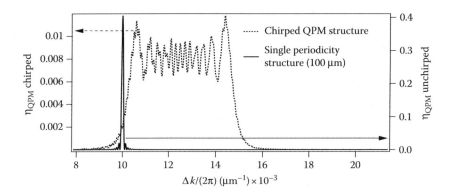

FIGURE 5.5 η_{QPM} for a single-periodicity crystal as well as a crystal with a continuously vary-ing periodicity (chirped). Both were calculated using the same crystal length. Note that the chirped scale is on the left and the unchirped scale on the right.

5.6 TOLERANCES

The behavior of a QPM device is similar to birefringent phase-matched ones. The phase-matching bandwidth, angular acceptance, and so on are calculated based on the changes to sinc($\Delta k'L/2$). All the techniques presented in Chapter 4 carry over here. One difference is that with quasi-phase-matched structures, we have an additional dependency, that of the periodic structure. Once the periodic structure is fabricated, it is fixed. Inevitably, the fabrication process is imperfect and the structure deviates from the ideal one such as shown in Figure 5.2. The Fourier approach presented above is a useful way to analyze the effect of imperfections.

As a first example, consider a periodic structure where the duty cycle varies slightly from one period to the next. Although the duty cycle varies, for the present calculation we assume that the periods are registered to a perfectly periodic lattice. The effects of duty cycle variation are modeled by first constructing an array for $h(z)$ based on the geometric pattern, including the variations. Next, we Fourier transform the $h(z)$ array to obtain $\bar{H}(\Delta k)$ and obtain η_{QPM} from Equation 5.35. Figure 5.6 shows η_{QPM} for a crystal with 1000 periods for both a perfect 50% duty cycle structure and one that has a 50% ± 30% random duty cycle variation. As shown in Figure 5.6, η_{QPM} is not strongly influenced by duty cycle variation. Moderate duty cycle variations do not present a significant problem as long as the domains are registered to a perfectly periodic lattice.

A more serious fabrication problem is a misregistration of the periodicity. Consider the example of stacking a series of oriented crystal plates. If in the process of stacking the crystals, one of the slabs does not have a good contact and leaves a slight space, then the different parts of the crystal have the same periodicity but are phase shifted with respect to each other. If such breaks from a perfectly periodic lattice occur throughout the crystal at random intervals, then the overall efficiency is degraded significantly. Figure 5.7 shows a comparison between a perfect structure and one

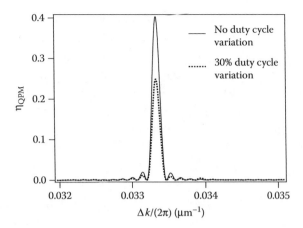

FIGURE 5.6 η_{QPM} for a quasi-phase-matched crystal with 1000 periods (30 μm each) with and without duty cycle variation.

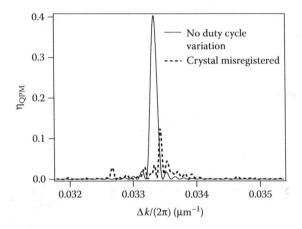

FIGURE 5.7 Comparison between a crystal with 1000 domains that have a perfect QPM structure and one with 1% of the regions randomly misregistered.

with 10 locations where a defect occurs to break the registration. The locations of the 10 misregistration sites were chosen randomly and the total number of periods is 1000. A comparison of Figure 5.6 with Figure 5.7 shows that the efficiency is much more adversely affected by misregistration than by duty cycle variation.

5.7 FABRICATING QUASI-PHASE-MATCHED STRUCTURES

When we introduced QPM, we assumed the dependence of d(z) on position came about by stacking rotated plates. The stacked-plate technique is most successful when the crystal plates are diffusion bonded to each other. Diffusion bonding consists of carefully polishing and cleaning the plates, stacking them, and then placing the stack under pressure at an elevated temperature. Although simple to visualize, stacking crystals involves several difficulties: the most detrimental of which are scattering losses and crystal-plate thickness variations. Thickness variation leads to misregistration and, as shown in Figure 5.7, misregistration quickly degrades the performance. In most cases, the stacked crystal approach uses a higher-order QPM process because it is difficult to fabricate and work with plates that are thin enough (~15 μm!) for the first-order QPM, thereby further reducing the efficiency. A final issue is that the total number of plates is usually small, on the order of 100.

Originally, the QPM technique was proposed in a crystal where the beam encounters multiple total internal reflections. With the appropriate angle, the phase shifts on reflection result in QPM. However, the total number of quasi-phase-matched periods is limited by the number of total internal reflections, which is small. A revolution in QPM sample fabrication occurred in the early 1990s when electric-field poling of ferroelectric crystals was applied to fabricating QPM structures. Ferroelectric crystals have a spontaneous material polarization that flips when a strong electric field of the opposite polarity is applied. The electric-field strength where the dipoles flip their polarity is called the coercive field. Lithium niobate and lithium tantalate are the most widely used ferroelectric crystals, but others such as potassium titanyl phosphate (KTP) have been poled successfully too. The strategy for poling these crystals is to lay down a periodic conductive contact on one surface; the applied field strength under the contacts is stronger than in the neighboring regions, which gives the surface electric field the desired periodicity (see Figure 5.8). As the voltage is increased to drive the electric field above the coercive field, then everywhere under the contacts the dipole domains are reversed while the remaining regions are unchanged. Once the domains are poled, the reversals are permanent yielding a periodically poled crystal. An example of a poled crystal is shown in Figure 5.9, which shows a photograph of the top surface of a periodically poled lithium niobate crystal. The most common method

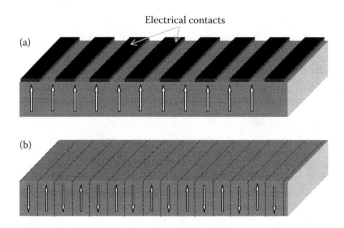

FIGURE 5.8 Ferroelectric crystal (a) before and (b) after the electric-field poling. Arrows indicate the crystal orientation. A voltage is applied to the contacts (dark areas) in (a) resulting in a reversal of the orientation. After poling, the contacts are removed since the poling is permanent. The dotted lines in (b) are the domain boundaries.

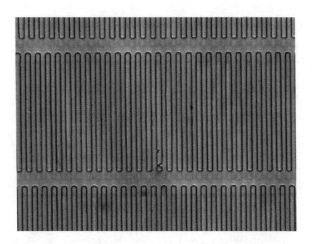

FIGURE 5.9 Photograph of a portion of a multiperiodicity, periodically poled lithium niobate crystal. The regions bounded by dark lines are −z oriented and the other regions are +z oriented.

of laying down the pattern is via photolithography, which borrows from semiconductor industry technology to achieve highly precise and accurate periodic structures. High-precision mask photolithography virtually eliminates misregistration issues with the placement of the periodic contacts. Furthermore, the photolithographic technique allows for essentially arbitrary patterning by changing the mask design, such as the multiperiodicity structure shown in Figure 5.9 and the chirped grating described in Section 5.5.

The main limitation of electric-field poling is the crystal thickness. Since the quasi-phase-matched period is typically 10–100 μm, a thick crystal leads to a large aspect ratio, that is, the crystal thickness divided by the smallest feature size. As the aspect ratio increases, the poling quality decreases and can result in an increase in duty cycle random variations and even by domains running together. The maximum achievable aspect ratio is crystal dependent, but most are poled with a 1-mm thickness or less. Using thicker crystals can result in nonvertical displacement of the domain wall boundaries, which will lead to misregistration of the period inside the crystal. As poling technology improves and as crystal stoichiometry and properties with doping (MgO) improve, even thicker crystals are becoming possible to pole.

Unfortunately, not all nonlinear crystals are ferroelectric and so the electric-field poling technique is reserved to a small class of crystals. Another successful fabrication technique is templated growth. Here, a highly precise template with the periodic structure is used to seed subsequent crystal growth. The template is fabricated using a variety of techniques, but they all start with a lithographically defined structure. The details of making the template are too vast to include here and so the reader is referred to the bibliography for further reading. The template is typically less than 1-μm thick and is not used itself for nonlinear frequency mixing. After the template is prepared, it is placed in a chamber where material is deposited and a thicker crystal is grown with a structure determined by the template. The growth preserves the template orientation with thicknesses up to 1000 μm or more. Templated growth has had good success in semiconductors such as GaAs, GaP, and ZnSe. Periodically oriented GaAs grown using this technique is an excellent example of the power of QPM since GaAs is a not birefringent crystal. Hence, devices can take advantage of the GaAs's large nonlinearity, $d_{14} \sim 90$ pm/V, and broad IR transparency to achieve low-threshold mid-wave IR frequency conversion.

PROBLEMS

The Sellmeier equations for the crystals mentioned in the following problems are listed in Appendix B.

5.1 Consider QPM in lithium niobate for the case when $\hat{k}$ is along the x-axis and all fields are polarized along the z-axis. Hence, the index of refraction for each wavelength is $n_z(\lambda, T)$.

 a. For a difference-frequency interaction where $\omega_1 > \omega_2 > \omega_3$, plot the first-order QPM period versus λ_2, given that $\lambda_1 = 1.064$ μm and the crystal temperature is 24°C. Plot over the range λ_2 equals 1.3–5 μm.

 b. For a fixed grating period of 28.75 μm, find the phase-matched wavelength pair, λ_2 and λ_3, given that $\lambda_1 = 1.064$ mm.

5.2 Calculate the first-order and third-order QPM periodicities for a second-harmonic (e → e + e) process in lithium tantalate with $\hat{k}$ along the x-axis. The fundamental wavelength is 1.064 μm.

5.3 Calculate the first-order QPM periodicity for a sum-frequency process (e + e → e) process in lithium niobate when $\hat{k}$ is along the x-axis, given $\lambda_2 = 1.064$ μm and $\lambda_3 = 1.550$ μm.

5.4 Calculate the first-order QPM periodicity for a backward DFG process in lithium niobate, given $\lambda_1 = 1.064$ μm and $\lambda_2 = 1.55$ μm. In this process, the k-vectors corresponding to λ_1 and λ_2 point in the $+\hat{x}$-direction while the k-vector corresponding to λ_3 points in the $-\hat{x}$-direction. (*Hint*: Rederive an expression for $\Delta k'$.) Typically, it is difficult to fabricate QPM structures shorter than 1 μm. Do you think a first-order QPM structure for the backward DFG is practical?

5.5 A periodically poled KTP crystal has a 38-μm periodicity. Given $\lambda_1 = 1.064$ μm, solve for the pair λ_2 and λ_3 that satisfies QPM for DFG. $\hat{k}$ is along the x-axis and all fields are polarized in the z-direction. (*Hint*: Plot $\Delta k'$ as a function of λ_2 and find the zero crossing.)

5.6 To set up a zero-finding algorithm, we start by looking at the problem graphically and then we duplicate the graphical procedure algorithmically. Consider a first-order quasi-phase-matched difference-frequency process in lithium niobate with a 30-μm periodicity. The interaction is e → e + e and $\hat{k}$ lies along the x-axis.

 a. Write a program to calculate $\Delta k'$ as a function of λ_1 and λ_2 (λ_3 is determined from energy conservation). Given $\lambda_1 = 1.1$ μm, plot $\Delta k'$ as a function of λ_2

from $\lambda_2 = \lambda_{start}$ to $\lambda_2 = 4.5\,\mu m$. λ_{start} is the wavelength that gives a difference frequency at the IR transparency cutoff of lithium niobate at 4.5 μm.

b. Show that the two zero crossings in the plot from (a) correspond to a difference-frequency pairing.

c. Duplicate the results of (a) by using a zero-finding algorithm (see Appendix C) to find the λ_2 that makes $\Delta k' = 0$. Use the plot in (a) to guide the wavelength range over which the zero-finding algorithm should search. To avoid the problem of multiple roots and wavelengths outside of the crystal's transparency range, the zero-finding algorithm should search from λ_{start} to degeneracy ($\lambda_2 = 2\lambda_1$).

5.7 (See also Problem 5.9) Use the $\Delta k = 0$ root-solver program written in Problem 5.6 to create a temperature-tuning curve for a DFG interaction in periodically poled lithium niobate crystal with a 28-μm periodicity and $\lambda_1 = 1.064\,\mu m$. The temperature-tuning curve is a plot of λ_2 as a function of T. Include λ_3 as a function of T on the same plot. Verify that both λ_2 and λ_3 tune toward degeneracy ($2\lambda_1$) as the temperature rises.*

5.8 (See also Problem 5.10) Use the $\Delta k' = 0$ root-solver program written in Problem 5.6 to create a pump-tuning curve (λ_1 tunes) for DFG in periodically poled lithium niobate crystal with a 28-μm periodicity at room temperature. The pump-tuning curve is a plot of λ_2 as a function of λ_1. Include λ_3 as a function of λ_1 on the same plot.

5.9 (See also Problem 5.7) Consider a DFG interaction in periodically poled lithium niobate crystal with a 28-μm periodicity and $\lambda_1 = 1.064\,\mu m$. Create a temperature-tuning curve by the method of contours introduced in Section 4.3. Create an array scaled along its x-axis from 1.3 to 1.7 μm and from 20°C to 200°C along its y-axis. Populate the array with values of $\Delta k'$, and then plot the contour where $\Delta k' = 0$, which is the tuning curve. Note that λ_1 is held fixed and that λ_3 is calculated based on energy conservation.

5.10 (See also Problem 5.8) Consider a DFG interaction in periodically poled lithium niobate crystal with a 28-μm periodicity and $\lambda_1 = 1.064\,\mu m$. Create a pump-tuning curve by the method of contours introduced in Section 4.3. Create an array with λ_1 along its x-axis scaled from 1.0 to 1.4 μm and λ_2 along the y-axis scaled from 3.5 to 5.0 μm. Populate the two-dimensional array with values of $\Delta k'(\lambda_1, \lambda_2)$, then plot the contour where $\Delta k' = 0$, which is the pump-tuning curve. Note that temperature is held fixed and that λ_3 is calculated based on energy conservation.

---

* An efficient way to generate the tuning curve is to write the zero-finding algorithm as a function, $f(\lambda_1, \Lambda, T)$. Given specific values for λ_1, Λ, and T, the function returns λ_2 that makes $\Delta k = 0$. With this function, a temperature tuning curve is generated by temptune(x) = $f(1.064, 28, x)$, where x is a variable representing the temperature. A pump tuning curve is generated by pumptune(x) = $f(x, 28, 24)$, where x is a variable representing the pump wavelength. Similarly, we can create a tuning curve for the grating periodicity.

5.11 A periodically poled lithium niobate crystal is 5 cm long and has a 28-μm periodicity. Given $\lambda_1 = 1.064$ μm, find λ_2 and λ_3 for a difference-frequency process at 100°C. Assume an e → e + e interaction with the k-vectors propagating along the x-axis. Find the solution by either plotting $\Delta k'$ for a first-order QPM interaction as a function of λ_2 or by using the zero-finding algorithm developed in Problem 5.6.

   a. Plot sinc²$(\Delta k'L/2)$ as a function of temperature for a temperature range from 90°C to 110°C while keeping λ_1 and λ_2 fixed (and hence λ_3 is fixed by energy conservation). What is the temperature-tuning bandwidth?

   b. Plot sinc²$(\Delta k'L/2)$ as a function of λ_2 over a range from $\lambda_{20} \pm 10$ nm, where λ_{20} is the value of λ_2 where $\Delta k' = 0$. In this calculation, λ_1 and temperature are held fixed and λ_3 is calculated using energy conservation. What is the phase-matching bandwidth?

   c. Plot sinc²$(\Delta k'L/2)$ as a function of λ_1 over a range 1.064 μm ± 10 nm. In this calculation, λ_2 and temperature are held fixed and λ_3 is calculated using energy conservation. What is the pump-tuning bandwidth?

5.12 Why is an e → e + e QPM interaction preferable to an o → o + o QPM interaction in lithium niobate? Assume $\hat{k}$ is along the x-axis.

5.13 Derive an expression for the first-order QPM period for an e → o + e interaction (either DFG or SFG) in KTP when $\hat{k}$ is along the x-axis and $\hat{o}$ is along the z-axis. Does this interaction couple to a nonzero nonlinearity (i.e., is d_{eff} nonzero)?

5.14 Prove that d_{eff} for a first-order QPM interaction is maximized for a structure that has a 50% duty cycle.

5.15 Prove that d_{eff} for a third-order QPM period is maximized for either $D = 1/2$ or $D = 1/6$. Sketch the growth of the intensity as a function of coherence lengths to explain why the two duty cycles give the same result. Explain why, from a standpoint of QPM fabrication, $D = 1/2$ is preferable to $D = 1/6$.

5.16 If it were possible to create a QPM structure where $d(z) = d_0 \sin(k_g z)$, what would be the effective QPM nonlinearity of the structure? Is it better or worse than the square-wave structure shown in Figure 5.3?

5.17 It is possible to cascade two processes within a single QPM structure. Consider a DFG process with ω_1 and ω_2 being present at the input with the difference frequency at ω_3. With the correct QPM structure, it is possible to have a second cascaded DFG process between ω_2 and ω_3 that amplifies ω_3 and generates a new difference frequency at ω_4. This cascade makes possible an overall more efficient DFG at ω_3. The process is e → e + e and the propagation direction is along the x-axis. Use energy conservation to derive an expression for ω_4 in terms of the two inputs, ω_1 and ω_2. Convert the result into an expression for λ_4 in terms of λ_1 and λ_2.

   a. In the simultaneous process, we must satisfy $\Delta k'_{process1} = 0$ and $\Delta k'_{process2} = 0$ with the same periodic structure; therefore, $k_g = k_1 - k_2 - k_3$ and $k_g = k_2 - k_3 - k_4$. Equate these two equations for k_g to come up with a function in the form $f(\lambda_1, \lambda_2) = 0$. Note that $n(\lambda_4)$ can be written in terms of λ_1 and λ_2.

b. Plot $f(\lambda_1, \lambda_2)$ as a function of λ_2 with $\lambda_1 = 1.064\ \mu m$ for lithium niobate. Plot over the range from 1.10 to 1.75 μm at a temperature of 24°C to find the zero crossing, which is the solution.

c. Given $\lambda_1 = 1.064\ \mu m$ and the solution for λ_2 in (c), calculate λ_3, λ_4, and the grating periodicity Λ_g for the cascaded structure.

5.18 GaAs is a nonbirefringent crystal, but with an appropriate structure, it is possible to quasi-phase-match three-wave mixing processes. Calculate the first-order QPM periodicity for a difference-frequency process with $\lambda_1 = 2.5\ \mu m$ and $\lambda_2 = 4\ \mu m$. Note that the orientation of the QPM periodicity does not matter for QPM. However, some orientations are preferred due to larger d_{eff} (see Problem 5.19).

5.19 GaAs is a nonbirefringent crystal, which has a point group symmetry of $\bar{4}3m$. Assume the geometry of the interacting waves as shown in Figure 5.10. All the interacting waves have a k-vector in the y–z-plane at an angle of 45° to the y-axis and z-axis. The pump polarization is at an angle γ to the x-axis. Now let us calculate d_{eff} for a DFG process.

a. Write down the electric-field directions for the pump (λ_1), signal (λ_2), and idler (λ_3) where each is assumed to have an arbitrary angle with respect to the x-axis. Use the designations γ_1, γ_2, and γ_3 for the angles.

b. Using the pump and signal as the inputs, calculate the nonlinear polarization induced at the idler.

c. Show that the induced polarization is perpendicular to the k-vector of the waves.

d. Since the polarization is perpendicular to the k-vector, it automatically couples to a propagating mode. Hence, the polarization direction is the direction of the idler's field, γ_3. Use this result and the nonlinear polarization from (b) to find d_{eff}.

e. Plot d_{eff} as a function of γ_1 for the case where the pump and the signal are parallel, that is, $\gamma_1 = \gamma_2$. On the same graph include a plot of d_{eff} as a function of γ_1 for the case where the pump and the signal are perpendicular.

f. Finally show that if γ_1 is oriented at 90°, that for an arbitrary signal polarization, the output idler polarization is the same direction as the signal.

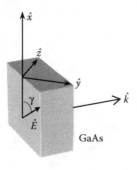

FIGURE 5.10   Orientation of GaAs for Problem 5.19.

The significance is that we may have a DFG interaction with an unpolarized input without throwing away 50% of the energy.

5.20 Derive an analytic expression for η_{QPM} for a QPM structure that has a periodicity of Λ, 50% duty cycle, and a length of L. In your calculation, consider only the first-order QPM. Does η_{QPM} depend on the specific crystal used?

5.21 Create a plot of η_{QPM} for a QPM structure that has a 25-μm periodicity, 50% duty cycle, a length of 1 cm, and a π phase change in the middle of the structure. You may solve this problem analytically or numerically. For the numerical approach, start by creating an array to represent d(z). A simple way to create a periodic square wave for d(z) is through a command like sign(cos($k_g z$)), where

$$\text{sign}(x) = \begin{cases} +1 & x > 0 \\ -1 & x < 0. \end{cases} \tag{5.37}$$

Hence, the overall structure is given by

$$d(z) = \begin{cases} \text{sign}(\cos(k_g z)) & 0 < x < \dfrac{L}{2} \\ \text{sign}(\cos(k_g z + \pi)) & \dfrac{L}{2} < x < L \end{cases}. \tag{5.38}$$

Next, take the FFT of the d(z) array to create $D(\Delta k)$. Finally, η_{QPM} is $|D(\Delta k)/L|^2$. Note that the resolution of the FFT depends on the size of the d(z) array, which can be increased by adding zeroes to the end of the array (called padding). In order to resolve all the pertinent features of $D(\Delta k)$, pad the d(z) array with zeroes equivalent to 10 times the number of points required to represent d(z) from 0 to L.

5.22 A chirped QPM structure has a periodicity ranging from 80 to 100 μm. An approximation of η_{QPM} for this structure is given by

$$\eta_{QPM} = \begin{cases} 0.01 & 11.5 \times 10^{-3} < \dfrac{\Delta k}{2\pi} < 12.5 \times 10^{-3} \\ 0 & \text{otherwise} \end{cases}. \tag{5.39}$$

The chirped QPM structure is fabricated using a GaAs crystal (see Appendix B for the Sellmeier equation). Plot the efficiency for a DFG interaction as a function of λ_2, given a fixed 2.3-μm pump laser. Plot over the range 2.8 μm $< \lambda_2 <$ 14 μm. Note that since GaAs is nonbirefringent the polarization combination does not affect the periodicity, but does determine the efficiency via d_{eff} (see Problem 5.19).

REFERENCE

Armstrong, J. A., N. Bloembergen, J. Ducuing, and P. S. Pershan. 1962. Interactions between light waves in a nonlinear dielectric. *Physical Review* 127: 1918–1939.

FURTHER READING

Eyres, L. A., P. J. Tourreau, T. J. Pinguet, C. B. Ebert, J. S. Harris, M. M. Fejer, L. Becouarn, B. Gerard, and E. Lallier. 2001. All-epitaxial fabrication of thick, orientation-patterned GaAs films for nonlinear optical frequency conversion. *Applied Physics Letters* 79: 904–906.

Fejer, M. M., G. A. Magel, D. H. Jundt, and R. L. Byer. 1992. Quasi-phase-matched second harmonic generation: Tuning and tolerances. *IEEE Journal of Quantum Electronics* 28: 2631–2654.

Gordon, L., G. L. Woods, R. C. Eckhardt, R. R. Route, R. S. Feigelson, M. M. Fejer, and R. L. Byer. 1993. Diffusion-bonded stacked GaAs for quasiphase-matched second-harmonic generation of a carbon dioxide laser. *Electronic Letters* 29: 1942–1944.

Miller, G. 1998. Periodically poled lithium niobate: Modeling, fabrication, and nonlinear-optical performance. PhD dissertation, Stanford University. http://www.stanford.edu/group/fejer/fejerpubs/Dissertations/Greg_Miller.pdf. Accessed May 30, 2010.

Myers, L. E., R. C. Eckardt, M. M. Fejer, R. L. Byer, W. R. Bosenberg, and J. W. Pierce. 1995. Quasi-phase-matched optical parametric oscillators in bulk periodically poled $LiNbO_3$. *Journal of the Optical Society of America B* 12: 2102–2116.

Powers, P. E., T. J. Kulp, and S. E. Bisson. 1998. Continuous tuning of a continuous-wave periodically poled lithium niobate optical parametric oscillator by use of a fan-out grating design. *Optics Letters* 23: 159–161.

Yamada, M., N. Nada, M. Saitoh, and K. Watanabe. 1993. First-order quasi-phase matched $LiNbO_3$ waveguide periodically poled by applying an external field for efficient blue second-harmonic generation. *Applied Physics Letters* 62: 435–436.

6

THREE-WAVE MIXING BEYOND THE SMALL-SIGNAL LIMIT

6.1 INTRODUCTION

Our treatment of the coupled wave equations in Chapter 4 introduces the general characteristics of phase matching and tolerances. We derive these properties by making several assumptions. The two principal assumptions are monochromatic plane-wave interactions and the small-signal approximation where the inputs are negligibly affected by the nonlinear interaction. We return again to the coupled amplitude equations but with less restrictive assumptions regarding the changes in field amplitudes. This treatment shows the important result that nonlinear interactions exhibit exponential gain. Hence, even weak nonlinearities may yield high conversion efficiency with a sufficiently long interaction length. Anything that reduces the exponential gain, such as absorption losses, has the potential to greatly reduce the conversion efficiency. A fundamental upper limit to the conversion efficiency is dictated by energy conservation, which we discuss in the context of the Manley–Rowe relations in Section 6.4. This chapter ends by covering spontaneous parametric scattering (SPS), which is a three-wave mixing process that requires a different formalism than that introduced in Chapter 4.

6.2 DFG WITH A SINGLE STRONG PUMP

We start with the coupled amplitude equations (Equations 4.25 through 4.27) and we work with the example of difference-frequency generation (DFG). Keep in mind that other three-wave processes such as sum-frequency generation (SFG) and second-harmonic generation (SHG) have nearly identical treatments. The three coupled equations are reduced to two when we assume that one of the input field amplitudes is constant. In the following treatment, we let the input corresponding to the highest frequency remain constant (the pump), while the other two fields are assumed to change appreciably. A common terminology for this situation is called the undepleted pump approximation.

6.2.1 DEFINING EQUATIONS FOR THE UNDEPLETED PUMP APPROXIMATION

The physical situation we consider is one where two fields with complex amplitudes A_1 and A_2 are present at the input, and through difference-frequency mixing, they generate a field with complex amplitude A_3. The evolution of A_2 and A_3 is governed by the coupled amplitude equations (Equations 4.26 and 4.27), which are repeated here:

$$\frac{dA_2}{dz} = i\frac{\omega_2}{n_2 c}d_{\text{eff}}A_1 A_3^* e^{i\Delta kz} = i\kappa_2 A_3^* e^{i\Delta kz},\qquad(6.1)$$

$$\frac{dA_3}{dz} = i\frac{\omega_3}{n_3 c}d_{\text{eff}}A_1 A_2^* e^{i\Delta kz} = i\kappa_3 A_2^* e^{i\Delta kz},\qquad(6.2)$$

where

$$\kappa_2 \equiv \frac{\omega_2}{n_2 c}d_{\text{eff}}A_1\qquad(6.3)$$

and

$$\kappa_3 \equiv \frac{\omega_3}{n_3 c}d_{\text{eff}}A_1.\qquad(6.4)$$

Since the pump field, A_1, is treated as a constant, then κ_2 and κ_3 are also constants; they are introduced to simplify the algebra. Equations 6.1 and 6.2 are two first-order coupled differential equations, and they are decoupled by trading the coupling for a second-order derivative. Decoupling the equations starts with Equation 6.2 where we solve for A_2^* or equivalently for A_2:

$$A_2 = \frac{i}{\kappa_3^*}\frac{dA_3^*}{dz}e^{i\Delta kz}.\qquad(6.5)$$

We take the derivative of Equation 6.5 and substitute it into Equation 6.1, leaving us with a single differential equation in terms of A_3:

$$\frac{dA_2}{dz} = \frac{i}{\kappa_3^*}\left(\frac{d^2 A_3^*}{dz^2}e^{i\Delta kz} + i\Delta k\frac{dA_3^*}{dz}e^{i\Delta kz}\right) = i\kappa_2 A_3^* e^{i\Delta kz}.$$ (6.6)

Simplifying Equation 6.6 gives

$$\frac{d^2 A_3}{dz^2} - i\Delta k\frac{dA_3}{dz} - \kappa_2^*\kappa_3 A_3 = 0.$$ (6.7)

So we have traded the two first-order coupled equations (Equations 6.1 and 6.2) for a single second-order equation, which allows for a straightforward solution.

6.2.2 SOLUTION FOR DIFFERENCE-FREQUENCY OUTPUT

Equation 6.7 is solved by assuming a solution of the form

$$A_3 \sim e^{mz}.$$ (6.8)

Inserting the trial solution (Equation 6.8) into Equation 6.7 gives the characteristic equation

$$m^2 - i\Delta km - \kappa_3\kappa_2^* = 0.$$ (6.9)

The two roots of this equation are

$$m_{1,2} = \frac{i\Delta k}{2} \pm \sqrt{\kappa_3\kappa_2^* - \left(\frac{\Delta k}{2}\right)^2}.$$ (6.10)

The general solution to Equation 6.7 includes both roots:

$$A_3 = Ce^{m_1 z} + De^{m_2 z},$$ (6.11)

where C and D are constants determined by initial conditions. To simplify the following discussion, we make variable substitutions in the expressions for m_1 and m_2. First, let

$$\Gamma^2 \equiv \kappa_3\kappa_2^* = \frac{\omega_2\omega_3 d_{eff}^2}{n_3 n_2 c^2}|A_1|^2 = \frac{2\omega_2\omega_3 d_{eff}^2}{n_1 n_2 n_3 \varepsilon_0 c^3}I_1.$$ (6.12)

Γ^2 is a purely real, positive quantity. $(\Gamma L)^2$ represents the parametric gain of the system. The second substitution is a "modified gain" $(gL)^2$ that shows a decrease in gain from the perfectly phase-matched situation. g is defined by

$$g \equiv \sqrt{\Gamma^2 - \left(\frac{\Delta k}{2}\right)^2}. \tag{6.13}$$

In terms of these variable substitutions, the roots to the characteristic equation (Equation 6.9) are

$$m_{1,2} = i\frac{\Delta k}{2} \pm g. \tag{6.14}$$

Hence, the solution to Equation 6.7 is

$$A_3(z) = e^{i(\Delta kz/2)}\left(Ce^{gz} + De^{-gz}\right). \tag{6.15}$$

This solution shows that if $\Delta k = 0$ then the field grows exponentially. The exponential nature of the field evolution is an important result. One of our key assumptions is that the nonlinear coupling is weak and hence the nonlinear gain should be small. However, with the exponential dependence on gain, a long interaction length can compensate to make the process highly efficient. Indeed, for a phase-matched interaction ($\Delta k = 0$) and for sufficiently long interaction lengths, the generated field becomes comparable to the undepleted pump. Of course, if the field grows to such a high magnitude, then our undepleted pump approximation breaks down. Note that if Δk is large enough, the modified gain becomes imaginary, and the solution to Equation 6.15 is oscillatory, just as in the non-phase-matched cases presented in Section 4.1.

6.2.3 SOLUTION WITH SPECIFIC BOUNDARY CONDITIONS

The solution presented in Equation 6.15 is not complete without values for the two constants C and D, which are determined from boundary conditions. For example, let us assume that only A_1 and A_2 are present at the input such that $A_1(0) = A_{10}$, $A_2(0) = A_{20}$, and $A_3(0) = 0$. At $z = 0$, Equation 6.15 becomes

$$A_3(0) = C + D = 0. \tag{6.16}$$

Equation 6.16 allows us to rewrite Equation 6.15 as

$$A_3(z) = 2Ce^{i\Delta kz/2}\sinh(gz), \tag{6.17}$$

where sinh(x) is the hyperbolic sine defined by

$$\sinh(x) = \frac{e^x - e^{-x}}{2}.$$

(6.18)

The constant, C, is obtained by equating the derivative of Equation 6.17 to Equation 6.2 at $z = 0$:

$$\left.\frac{dA_3}{dz}\right|_{z=0} = 2gC = i\kappa_3 A_2^*(0).$$

(6.19)

Hence,

$$C = \frac{i\kappa_3 A_{20}^*}{2g},$$

(6.20)

and

$$A_3(z) = i\frac{\kappa_3 A_{20}^* e^{i\Delta kz/2} \sinh(gz)}{g}.$$

(6.21)

A similar procedure to that used to obtain Equation 6.21 is applied to find A_2 (see Problem 6.1). Alternatively, Equation 6.21 is substituted into Equation 6.2 to give $A_2(z)$.

The solutions for A_2 and A_3 give even more insight if we look at the solutions from a standpoint of photon numbers and if we assume the process is phase matched with $\Delta k = 0$. We first find the intensity corresponding to A_3:

$$I_3(z) = n_3 \frac{A_3(z)A_3^*(z)}{2\mu_0 c} = \frac{n_3 \kappa_3 \kappa_3^*}{2\mu_0 c} \frac{A_{20}^* A_{20}}{\Gamma^2} \sinh^2(\Gamma z).$$

(6.22)

Equation 6.22 is in turn written in terms of the intensity of I_{20}:

$$I_3(z) = \frac{n_3 \kappa_3 \kappa_3^*}{n_2 \Gamma^2} I_{20} \sinh^2(\Gamma z).$$

(6.23)

Note that

$$\frac{\kappa_3 \kappa_3^*}{\Gamma^2} = \frac{\kappa_3^*}{\kappa_2^*} = \frac{n_2 \omega_3}{n_3 \omega_2}.$$

(6.24)

Hence, Equation 6.23 is written as

$$I_3(z) = \frac{\omega_3}{\omega_2} I_{20} \sinh^2(\Gamma z). \tag{6.25}$$

For a unit cross-sectional area and a unit time (Δt), the number of photons for a plane wave with the intensity given by Equation 6.25 is

$$N_3 = \frac{I_3}{\hbar \omega_3} \times \text{area} \times \Delta t. \tag{6.26}$$

Hence,

$$N_3(z) = \frac{I_{20}}{\hbar \omega_2} \sinh^2(\Gamma z) \times \text{area} \times \Delta t. \tag{6.27}$$

Note that Equation 6.27 contains N_{20} so that Equation 6.27 is rewritten as

$$N_3(z) = N_{20} \sinh^2(\Gamma z) \tag{6.28}$$

A similar procedure used for finding N_3 is used for finding N_2, resulting in

$$N_2(z) = N_{20} \cosh^2(\Gamma z). \tag{6.29}$$

If we have mixed boundary conditions (where both A_{20} and A_{30} are nonzero), then the solutions are[*]

$$N_3(z) = N_{20} \sinh^2(\Gamma z) + N_{30} \cosh^2(\Gamma z), \tag{6.30}$$

$$N_2(z) = N_{30} \sinh^2(\Gamma z) + N_{20} \cosh^2(\Gamma z). \tag{6.31}$$

A plot of the growth of N_2 and N_3 for the case $N_{20} \neq 0$ and $N_{30} = 0$ is shown in Figure 6.1. Figure 6.1 shows that both N_2 and N_3 grow exponentially. The exponential growth of both fields occurs because the difference-frequency process between the inputs A_1 and A_2 results in a photon at ω_1 splitting into two photons at ω_3 and ω_2. Hence, the difference frequency is generated at ω_3 while the second input at ω_2 is amplified. For this reason, DFG and OPA are equivalent.

---

[*] The solution assumes that the initial phases are all zero (see Problems 6.4 and 6.5).

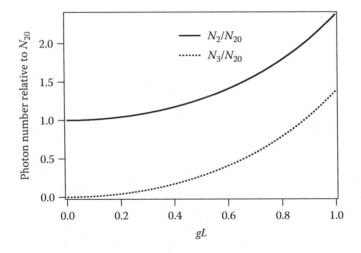

FIGURE 6.1   DFG exhibits exponential growth in the number of photons as a function of distance in the crystal. Coupled to this growth is the amplification of the second input to the crystal (the pump is assumed to be constant).

6.3   DFG WITH STRONG PUMP AND LOSS

The exponential gain of the above example is one of the reasons why the three-wave mixing processes can be highly efficient. In real-world cases, however, the exponential gain is degraded by factors such as absorption. A common scenario is a crystal that is completely transparent for two out of the three wavelengths but has some absorption for the third wavelength. Such cases occur when pushing deep into the ultraviolet or infrared. Terahertz (THz) generation is such an example where two closely spaced frequencies within the transparency range of a crystal mix to generate the difference frequency in the THz region (wavelengths beyond 100 μm), and the THz is absorbed to some extent. Finding a crystal that is completely transparent for such broadly spaced frequencies is a challenge indeed!

In most cases, absorption degrades the performance. Here, we are trying to estimate by how much. We ignore cases of resonant enhancement of the nonlinearity. To quantify the degradation, we rederive the coupled amplitude equations from Maxwell's equations while keeping loss terms. Loss is introduced through the free current density or the imaginary part of the susceptibility (or equivalently through the imaginary part of the permittivity). Loss via free carriers is typical in semiconductors due to electrons in the conduction band and holes in the valence band. However, most crystals used for nonlinear applications are insulators that have negligible free carriers and are typically transparent over a broad range. Outside of this range, absorption occurs due to the imaginary part of the susceptibility (see Chapter 2).

The effects of loss on a nonlinear process are understood by once again recalling Maxwell's equations. In particular,

$$\vec{\nabla} \times \underline{\vec{H}} = \underline{\vec{j}}_f + \frac{\partial}{\partial t}\left(\varepsilon_0 \underline{\vec{E}} + \underline{\vec{P}}\right). \tag{6.32}$$

The free current density is assumed to follow Ohm's law such that its complex amplitude obeys

$$\vec{j}_f = \sigma \vec{E}, \tag{6.33}$$

where σ is the complex conductivity of the medium. Our assumption here is that σ is independent of the field strength. Normally, σ is taken to be purely real, and in the cases of many nonlinear crystals, it is zero. The polarization's complex amplitude with losses is written as

$$\vec{P} = \varepsilon_0(\chi_R^{(1)} + i\chi_I^{(1)})\vec{E} + \vec{P}^{NL}, \tag{6.34}$$

where $\chi_R^{(1)}$ and $\chi_I^{(1)}$ are the real and the imaginary parts of the linear susceptibility, respectively.

Including the effects of $\vec{j}_f$ and/or a complex linear susceptibility results in the following set of coupled amplitude equations:

$$\frac{dA_1}{dz} + \frac{\alpha_1}{2}A_1 = i\frac{\omega_1}{n_1 c}d_{eff}A_2A_3e^{-i\Delta kz}, \tag{6.35}$$

$$\frac{dA_2}{dz} + \frac{\alpha_2}{2}A_2 = i\frac{\omega_2}{n_2 c}d_{eff}A_1A_3^*e^{i\Delta kz}, \tag{6.36}$$

$$\frac{dA_3}{dz} + \frac{\alpha_3}{2}A_3 = i\frac{\omega_3}{n_3 c}d_{eff}A_1A_2^*e^{i\Delta kz}. \tag{6.37}$$

Losses enter into the equations through α_1, α_2, and α_3 whose values are functions of σ and/or the imaginary part of $\tilde{\chi}^{(1)}$. Equations 6.35 through 6.37 represent the general case where all three fields are absorbed, but as mentioned above, more commonly only one of the fields experiences loss. Let us consider the case introduced in Section 6.2, difference frequency between an undepleted input at ω_1 and a weaker input at ω_2. We include absorption for the generated field at ω_3. Equations 6.35 through 6.37 reduce to

$$\frac{dA_2}{dz} = i\kappa_2 A_3^*e^{i\Delta kz} \tag{6.38}$$

$$\frac{dA_3}{dz} + \frac{\alpha_3}{2} A_3 = i\kappa_3 A_2^* e^{i\Delta kz},$$ (6.39)

where κ_2 and κ_3 are the same variable substitutions used in Equations 6.1 and 6.2. The solution follows the same procedure as in Section 6.2. First solve for A_2^* in Equation 6.39 and take its derivative. Next equate it to Equation 6.38, which after some algebra simplifies to

$$\frac{d^2 A_3}{dz^2} + \left(\frac{\alpha_3}{2} - i\Delta k\right)\frac{dA_3}{dz} - \left(\kappa_3\kappa_2^* + i\Delta k \frac{\alpha_3}{2}\right)A_3 = 0.$$ (6.40)

The solution to this equation is of the same form as the lossless case, namely,

$$A_3 = Ce^{m_1 z} + De^{m_2 z}.$$ (6.41)

m_1 and m_2 are the roots to the characteristic equation, which are found by substituting e^{mz} into Equation 6.40. The roots of the characteristic equation are

$$m = \frac{-(\alpha_3/2) + i\Delta k}{2} \pm \sqrt{\frac{((\alpha_3/2) + i\Delta k)^2}{4} + \kappa_3\kappa_2^*}.$$ (6.42)

All we need now for the solution given in Equation 6.41 are the boundary conditions. We use the same boundary conditions as in the lossless case presented in Section 6.2, where $A_1(0) = A_{10}$, $A_2(0) = A_{20}$, and $A_3(0) = 0$. To focus on the effects of loss on the nonlinear process, we eliminate the effects of phase mismatch; that is, we consider the solution to Equation 6.40 in the limit where $\Delta k = 0$. This approach lets us compare the performance of a perfectly phase-matched process with and without the absorption of the generated wave. With $\Delta k = 0$ and the stated boundary conditions, the solution to Equation 6.40 is

$$A_3 = \frac{i\kappa_{30} A_{20}^* e^{-(\alpha_3/4)z}}{\sqrt{(\alpha_3^2/16) + \Gamma^2}} \sinh\left(\left[\sqrt{(\alpha_3^2/16) + \Gamma^2}\right]z\right).$$ (6.43)

Γ has the same definition as in Equation 6.12.

A simple way to look at this expression is to first focus only on the exponential terms. For long interaction lengths, $\sinh(x) \to e^x$ so that

$$A_3(z) \sim \exp\left(\left[\sqrt{(\alpha_3/4)^2 + \Gamma^2} - (\alpha_3/4)\right]z\right).$$ (6.44)

The argument of the exponential term can be shown to be always less than Γz (Problem 6.9). Hence, the effect of loss is to reduce the amplitude and to slow the exponential growth. Anything that reduces the exponential gain greatly reduces the efficiency.

A simpler case involving loss is one where the two inputs are both assumed to be constant, which further reduces the coupled equations down to one:

$$\frac{dA_3}{dz} + \frac{\alpha_3}{2} A_3 = i\frac{\omega_3}{n_3 c} d_{\text{eff}} A_1 A_2^* e^{i\Delta kz}. \tag{6.45}$$

Note that we return to the case where Δk is not necessarily zero. The solution to the differential equation (Equation 6.45) is left as an exercise with the result

$$A_3(z) = \frac{i\omega_3 d_{\text{eff}} A_1 A_2^*}{n_3 c} \frac{e^{i\Delta kz}}{((\alpha_3/2) + i\Delta k)} (1 - e^{-((\alpha_3/2) + i\Delta k)z}). \tag{6.46}$$

The corresponding intensity for the difference-frequency amplitude given in Equation 6.46 is given by

$$I_3 = I_{\text{max}} \frac{(1 + e^{-\alpha_3 L} - 2e^{-(\alpha_3/2)L} \cos \Delta kL)}{\left(((\alpha_3/2)L)^2 + (\Delta kL)^2\right)}, \tag{6.47}$$

where

$$I_{\text{max}} = \frac{2\omega_3^2 d_{\text{eff}}^2 I_1 I_2}{n_1 n_2 n_3 \varepsilon_0 c^3} L^2 = \frac{8\pi^2 d_{\text{eff}}^2 I_1 I_2}{n_1 n_2 n_3 \varepsilon_0 c \lambda_3^2} L^2. \tag{6.48}$$

I_{max} is the intensity when the loss is zero and the process is perfectly phase matched, so that it represents the best performance. The expression given by (I_3/I_{max}) serves as a relative phase-matching efficiency function in the presence of loss. Figure 6.2 is

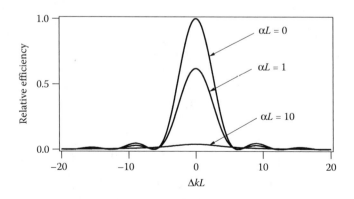

FIGURE 6.2   Plot of the relative efficiency as a function of ΔkL for different loss scenarios.

a plot of (I_3/I_{max}) for several different loss cases. As shown in Figure 6.2, as $\alpha_3 \rightarrow 0$, the phase-matching efficiency goes to the familiar $\text{sinc}^2(\Delta kL/2)$ shape.

6.4 SOLUTIONS FOR ALL THREE COUPLED AMPLITUDE EQUATIONS

Now we turn our attention to working with the coupled amplitude equations when all three fields may change appreciably. The result of this analysis is a behavior not observed in the cases presented so far. We also show that phase matching is still central to the overall efficiency of the process and that exponential gain allows weak inputs to grow substantially.

6.4.1 MANLEY–ROWE RELATIONS

A general property of three-wave interactions is one of energy conservation, which can be cast into the form of photon number relationships. In the following, we assume that the crystal is transparent at all wavelengths of interest. We start with the intensity of a given wave, written as

$$I = \frac{nAA^*}{2\mu_0 c}. \tag{6.49}$$

The rate of change of this intensity along the propagation direction is

$$\frac{dI}{dz} = \frac{n}{2\mu_0 c}\left(A\frac{dA^*}{dz} + A^*\frac{dA}{dz}\right). \tag{6.50}$$

The derivative terms are taken directly from the coupled amplitude equations (Equations 4.25 through 4.27). For example, the derivative of I_3 is

$$\frac{dI_3}{dz} = -i\omega_3\frac{A_1^* A_2 A_3 d_{eff}e^{-i\Delta kz}}{2\mu_0 c^2} + \text{c.c.} \tag{6.51}$$

Equation 6.51 is rearranged to give

$$\frac{1}{\omega_3}\frac{dI_3}{dz} = -i\frac{A_1^* A_2 A_3 d_{eff}e^{-i\Delta kz}}{2\mu_0 c^2} + \text{c.c.} \tag{6.52}$$

An analogous procedure for I_2 gives

$$\frac{1}{\omega_2}\frac{dI_2}{dz} = -i\frac{A_1^* A_2 A_3 d_{eff}e^{-i\Delta kz}}{2\mu_0 c^2} + \text{c.c.} \tag{6.53}$$

Similarly for I_1,

$$\frac{1}{\omega_1}\frac{dI_1}{dz} = i\frac{A_1^* A_2 A_3 d_{eff} e^{-i\Delta kz}}{2\mu_0 c^2} + \text{c.c.} \tag{6.54}$$

Note that it is possible to make the following equality from Equations 6.52 through 6.54:

$$\frac{1}{\omega_2}\frac{dI_2}{dz} = \frac{1}{\omega_3}\frac{dI_3}{dz} = -\frac{1}{\omega_1}\frac{dI_1}{dz}. \tag{6.55}$$

This set of equalities constitutes the Manley–Rowe relationship. Equation 6.55 is converted into a more convenient form by multiplying by a cross-sectional unit area and time interval Δt:

$$\frac{d}{dz}\left(\frac{I_2}{\hbar\omega_2}\times\text{area}\times\Delta t\right) = \frac{d}{dz}\left(\frac{I_3}{\hbar\omega_3}\times\text{area}\times\Delta t\right) = -\frac{d}{dz}\left(\frac{I_1}{\hbar\omega_1}\times\text{area}\times\Delta t\right) \tag{6.56}$$

The quantities in brackets in Equation 6.56 are photon numbers passing through the area in the time interval Δt, resulting in

$$\frac{dN_2}{dz} = \frac{dN_3}{dz} = -\frac{dN_1}{dz}. \tag{6.57}$$

Hence,

$$\Delta N_2 = \Delta N_3 = -\Delta N_1. \tag{6.58}$$

Equation 6.58 shows that for every pump photon at ω_1 that is destroyed, one is created at both ω_2 and ω_3, and vice versa. As an example, consider DFG with inputs at ω_1 and ω_2. In the difference-frequency process, the photon at ω_1 splits into two photons, one at ω_2 and one at ω_3 so that $\Delta N_2 = \Delta N_3 = +1$, while $\Delta N_1 = -1$. Another example is SFG, where two photons at frequencies ω_2 and ω_3 add to give one photon at frequency ω_1. For a single SFG interaction, $\Delta N_2 = \Delta N_3 = -1$, while $\Delta N_1 = +1$.

Another result that comes out of the Manley–Rowe derivation is that we recombine Equations 6.52 through 6.54 to come up with "constants of motion" for the coupled wave equations. That is,

$$\frac{d}{dz}\left(\frac{I_1}{\omega_1} + \frac{I_3}{\omega_3}\right) = 0 \quad \therefore N_1 + N_3 = \text{constant}, \tag{6.59}$$

$$\frac{d}{dz}\left(\frac{I_1}{\omega_1} + \frac{I_2}{\omega_2}\right) = 0 \quad \therefore N_1 + N_2 = \text{constant}, \tag{6.60}$$

$$\frac{d}{dz}\left(\frac{I_3}{\omega_3} - \frac{I_2}{\omega_2}\right) = 0 \quad \therefore N_3 - N_2 = \text{constant}, \tag{6.61}$$

where the constants are, in general, all different values. By combining any two out of the three equations (Equations 6.59 through 6.61), we can also show (see Problem 6.13)

$$I_1 + I_2 + I_3 = \text{constant}. \tag{6.62}$$

Equation 6.62 is equivalent to the statement that the total power remains constant throughout the crystal when no absorption losses are present.

6.4.2 ANALYTIC SOLUTION FOR THREE PLANE WAVES

Analytic solutions to the three coupled wave equations (Equations 4.25 through 4.27) are derived in Appendix D. Alternatively, the three equations may be numerically integrated (see Problem 6.15). The qualitative behavior of the analytic solutions is periodic with a periodicity dependent on the amplitudes of the input fields. Interestingly, the analytic solutions are similar to those of a physical pendulum. The periodic behavior for a three-wave interaction means that even when phase matched, energy flow from the inputs to a desired output changes direction periodically. In other words, the process *backconverts*.

To illustrate the general properties for the analytic solutions, we consider DFG between two lasers at 1.0 μm (pump) and 1.5 μm (signal) that generate a 3.0-μm (idler) output. The pump, signal, and idler terminology is commonly used for DFG and OPA applications, and we use it here to simplify our discussion. For all of the following examples, we start with the same number of photons at the input so that the sum of $N_1(0)$, $N_2(0)$, and $N_3(0)$ equals a constant and we normalize this constant to 1. We also assume perfect phase matching, $\Delta k = 0$.

The first example, shown in Figure 6.3, is the typical situation where two beams are present at the input. For this specific example, we consider $N_1 = N_2 = 1/2$ in normalized units. Figure 6.3 shows that the 1.5-μm beam is amplified and that the difference-frequency beam at 3.0 μm grows with it. The growth of these two beams is exponential at first but slows down when the pump begins to deplete significantly. Eventually, the 1.0-μm beam is completely depleted, and after this point the 1.5- and 3.0-μm beams then *backconvert* to the 1.0-μm beam through SFG. The distance for the pump to completely deplete, L_D, depends on the input intensities. The specific length of L_D is not important here, rather the key point is the general behavior of the periodic energy flow.

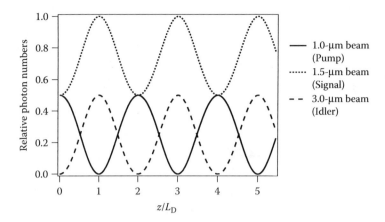

FIGURE 6.3 Three-wave mixing process where the two inputs have equal photon numbers. L_D is the "depletion length."

Figure 6.3 shows that three-wave mixing processes such as DFG and SFG are linked. Even though the process in Figure 6.3 starts off as a difference-frequency process, it turns into a sum-frequency process after the point where the pump is zero. At some later location, when the sum-frequency process runs to completion, the process turns back into DFG.

Ideally, we would stop the difference-frequency process at the point where the pump is completely depleted, but this is rarely practical. In real systems, the input beams typically have Gaussian cross sections as opposed to the plane waves assumed here. Hence, with real beams, back conversion may occur inhomogeneously across the beam front. That is, back conversion occurs first in the center of the beam where the intensity is highest. Back conversion makes it difficult to achieve high conversion efficiency everywhere across a beam profile. The variation in efficiency leads to beam profile degradation; consider the case of the pump where the center of the beam is depleted while the wings are not. It is possible that in a given crystal the center of the beam experiences several back conversion to reconversion cycles. The process as presented in Figure 6.3 is not the final word on conversion efficiency since other factors such as beam diffraction and time dependence need to be considered. We consider these effects in Chapter 10.

The next example, shown in Figure 6.4, considers the case where all three frequencies are present at the input. In this example, the idler input is much smaller than the pump or signal, and the input phases are all set to zero. The same pattern as in Figure 6.3 occurs but note that the maximum conversion efficiency is reduced since the pump never completely depletes.

The phases of the inputs for the pump, signal, and idler in Figure 6.4 were all set to zero. Now let us consider the same scenario except where the phase of the idler field is shifted 45° with respect to the other two fields. Interestingly, the result (Figure 6.5) shows that the process starts out as a sum frequency instead of a difference frequency. So the phases of the inputs matter. In fact, a scenario with three fields

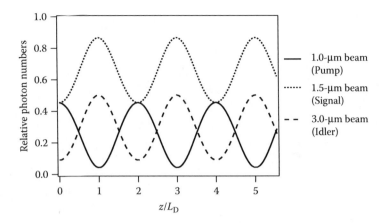

FIGURE 6.4 Three-wave mixing process when inputs at all three frequencies are present. The phases of the waves at the input are all the same.

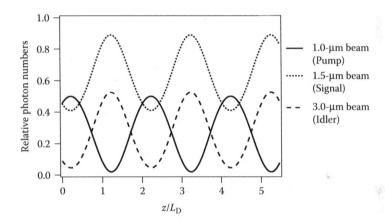

FIGURE 6.5 Three-wave mixing process when inputs at all three frequencies are present. The idler phase is shifted 45° with respect to the other inputs.

present at the input, and with different phases, is not that uncommon, since in real systems it is possible that the third input comes from a double reflection from the crystal itself. Such reflections degrade the performance or lead to instabilities and so most crystals are antireflection (AR)-coated and/or wedged.

In the next example, we look at SHG. SHG is interesting because it is possible to obtain an asymptotic 100% conversion efficiency, shown in Figure 6.6. The normalization in Figure 6.6 is chosen such that $2N_F + N_{SHG} = 1$, where N_F is the number of photons at the fundamental frequency and N_{SHG} is the number of photons at the second harmonic.

If a second harmonic is present at the input, it changes the asymptotic behavior to one that is periodic, as shown in Figure 6.7. A second harmonic present at the input has a significant effect on the process such that AR coatings and wedged crystals are

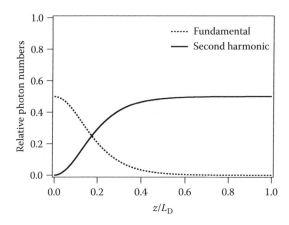

FIGURE 6.6   SHG when only the fundamental is present at the input.

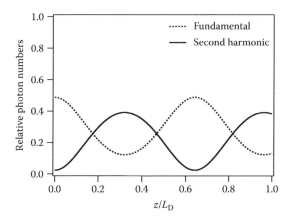

FIGURE 6.7   SHG when a small SHG signal is present with the fundamental at the input. The fundamental and SHG both start with a zero phase. Note that the depletion length, L_D, is the same distance as in Figure 6.6.

advisable. Similar to the case of DFG, the SHG behavior when both the fundamental and SHG are present at the input is strongly dependent on the phases of the inputs.

6.5   SPONTANEOUS PARAMETRIC SCATTERING (OPTICAL PARAMETRIC GENERATION)

Until this point, we consider only difference frequency and optical parametric interactions that have two inputs present at the input. Understanding the process classically is clear in that these two inputs generate a nonlinear polarization that then drives the output frequency. However, an additional noise signal is present with the DFG/OPA process where the pump spontaneously generates a signal and idler.

This process is called spontaneous parametric scattering (SPS), spontaneous down-conversion, or optical parametric generation (OPG). Describing the SPS process classically runs into problems since we have no way to obtain a nonlinear polarization without a second input. The spontaneous process is described quantum mechanically as a pump photon incident on a crystal that spontaneously splits into a signal and idler photon (Giallorenzi and Tang, 1968). The specific signal–idler pairing in the spontaneous process is determined by phase matching, and consequently the OPG bandwidth is essentially the same as the phase-matching bandwidth. Hence, an excess bandwidth is added to DFG/OPA mixing processes. The noise introduced into a DFG or OPA signal by the OPG process is not noticeable for low pump intensities, but it can dominate for high intensities. In many cases, the OPG process is an intentional means of frequency converting a pump laser. With nanosecond or shorter pulsed lasers coupled with high nonlinearity crystals, the conversion efficiency of the pump to signal and idler can be highly efficient, greater than 20%.

Explaining the OPG process requires a deviation from our previous classical calculations. So far, we assume two fields present at the input to generate a nonlinear polarization. However, as mentioned above, a classical description of DFG fails when only the pump input is present. We skirt this problem by introducing a second input that has a quantum origin due to coupling with the vacuum electromagnetic field (zero-point energy). The problem is set up in the following manner. A pump field is incident on a crystal, and we assume it is mixing with several idler fields, the origin of which we at first ignore. The idler inputs consist of k-vectors pointing in arbitrary directions. One particular idler k-vector is shown in Figure 6.8.

Figure 6.8 shows a specific signal–idler pair, but in our analysis, we consider a range of k-vectors with magnitudes between k_i, and $k_i + dk_i$ and polar angles ranging between ψ and $\psi + d\psi$ shown in Figure 6.9. The volume of the annulus region shown in Figure 6.9 is

$$dV_i = 2\pi k_i^2 \sin\psi \, d\psi \, dk. \tag{6.63}$$

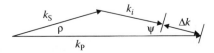

FIGURE 6.8 k-Vector diagram for a specific idler energy and direction.

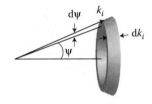

FIGURE 6.9   Diagram showing geometry of idler k-vectors with magnitudes in the range between k_i and $k_i + dk_i$ and polar angle in the range between ψ and $\psi + d\psi$. The azimuthal angle (not shown) is integrated around the central axis forming the volume in the annulus.

In order for the idler fields to exist in the range of k-vectors shown in Figure 6.9, they must be allowed modes of the EM field in the volume. The possible number of modes is the volume given in Equation 6.63 divided by the volume a single mode takes up in k-space.

We calculate the volume of a single mode by considering that an electric field can be expanded in terms of plane waves:

$$\underline{\vec{E}}(\vec{r},t) = \frac{1}{2} \sum_k i \varepsilon_{\vec{k}} \vec{A}_{\vec{k}} e^{i(\vec{k}\cdot\vec{r}-\omega t)} + \text{c.c.,} \tag{6.64}$$

where the summation is over all plane-wave modes, $\varepsilon_{\vec{k}}$ is an expansion coefficient with the units of the electric field, and $\vec{A}_{\vec{k}}$ expresses the mode's field strength and polarization. The modes are restricted by the condition that they satisfy periodic boundary conditions, for example,

$$\underline{\vec{E}}(x + L_x, y, z, t) = \underline{\vec{E}}(x, y, z, t), \tag{6.65}$$

where L_x is the x-dimension of the volume being considered. This condition and similar ones for y and z are inserted into Equation 6.64 giving the following relationships:

$$k_x = \frac{2\pi}{L_x} m_x, \tag{6.66}$$

$$k_y = \frac{2\pi}{L_y} m_y, \tag{6.67}$$

$$k_z = \frac{2\pi}{L_z} m_z, \tag{6.68}$$

where m_x, m_y, and m_z are integers. Therefore, a mode in k-space takes up a volume

$$\frac{8\pi^3}{V}, \tag{6.69}$$

where $V = L_x L_y L_z$. A more detailed derivation of Equation 6.69 may be found in texts that derive blackbody radiation equations (Loudon, 1983).

The number of allowed modes for the k-vector annulus volume shown in Figure 6.9 is

$$\text{Number of modes} = \frac{dV_i}{(8\pi^3/V)} = \frac{k_i^2 \sin\psi \, d\psi \, dk_i}{4\pi^2} V. \tag{6.70}$$

Note that Equation 6.70 has no factor of two for polarization dependence; we enumerate the number of modes for a single polarization. In the derivations for blackbody radiation, another factor of two is needed to account for two allowed polarizations. However, for nonlinear interactions, phase-matching couples to only one of the polarization modes. Equation 6.70 gives the number of allowed modes, but we need to know how many of these modes are populated with photons, and we need to address where the photons come from. A first guess is that the photons come from the environment through blackbody radiation. However, calculations show that a blackbody input is far too weak to account for measured outputs when the SPS signal is in the visible and near-IR.

The source of SPS in the visible region is described quantum mechanically. A quantum mechanical treatment gives the result that the zero-point energy of an allowed mode is $(1/2)\hbar\omega$, equivalent to 1/2 photon per mode in the ground state. These zero-point photons may not be extracted from the system, but they stimulate the SPS process. A fully quantum mechanical calculation is examined in Chapter 11. We can obtain the same result as the quantum calculation by using a semiclassical approach. We assume that every allowed idler mode is occupied with one photon and that the corresponding signal mode has zero. With this input, we then calculate the output using the classical equations for DFG derived in the previous chapters. The result is precisely the same as those obtained using a fully quantum calculation.

The classical equations are written in terms of field amplitudes and so we need to convert the quantum input into a classical field amplitude. For a plane wave traveling in the crystal, the intensity is given by

$$I = \frac{N}{V}\frac{\hbar\omega}{2}\frac{c}{n}, \tag{6.71}$$

where N is the number of photons and V is the same volume as in Equation 6.69. The intensity comes from a photon density multiplied by the photon speed (c/n). Equation 6.70 gives the number of photons, N, and so the idler input to the classical equations is

$$dI_i = \frac{ck_i^2\hbar\omega_i \sin\psi \, d\psi \, dk_i}{8\pi^2 n_i}. \tag{6.72}$$

To avoid the issue of accounting for the zero-point energy, we calculate the signal that is generated due to the idler input given in Equation 6.72, since we put all the zero-point energy into the idler mode. At this point, the calculation of the signal output is carried out the same way as for the difference-frequency case presented in Section 6.2 with a solution similar to Equation 6.22. The difference is that here we work with the signal and not with the idler, and we do not set Δk equal to zero. The solution is given by

$$dI_S(z) = \frac{2\omega_S^2 d_{\text{eff}}^2}{n_P n_S n_i \varepsilon_0 c^3}\frac{\sinh^2(gz)}{g^2}(dI_i)I_P, \tag{6.73}$$

where I_P is the incident pump intensity, dI_i is given by Equation 6.72, and g is the same as in Equation 6.13. Substituting Equation 6.72 into Equation 6.73 gives

$$dI_S(z) = \frac{\hbar n_i \omega_S^2 \omega_i^3 d_{\text{eff}}^2}{4\pi^2 c^5 n_P n_S \varepsilon_0} \frac{\sinh^2(gz)}{g^2} I_P \sin\psi \, d\psi \, d\omega_i. \tag{6.74}$$

Since we are measuring the signal, we would like the bandwidth and angles to be measured with respect to the signal. Assuming that the input pump is mono-chromatic, then the bandwidth of the signal is related to that of the idler through $|d\omega_2| = |d\omega_3|$. For small angles $\psi = (k_S/k_i)\rho$ (using the law of sines and Figure 6.8), and hence $\sin\psi \, d\psi \sim (k_S/k_i)^2 \rho d\rho$. Furthermore, an experiment measures the collected power, which we obtain by multiplying both sides of Equation 6.74 by an area. With these substitutions, Equation 6.74 becomes

$$dP_S(z) = \frac{\hbar n_S \omega_S^2 \omega_i d_{\text{eff}}^2}{4\pi^2 \varepsilon_0 c^5 n_P n_i} \frac{\sinh^2(gz)}{g^2} P_P \, \rho \, d\rho \, d\omega_S. \tag{6.75}$$

Remarkably, a full quantum calculation gives the same result as Equation 6.75. The power measured with a detector is the integral of Equation 6.75 over frequency and collection angle. The bandwidth and collection angle are not arbitrary and are determined primarily by phase matching through g's dependence on Δk, since frequencies and angles that result in a large Δk will not contribute substantially to the integrated power.

One interesting consequence of our parametric down-conversion result is that spontaneously emitted light in Type I phase matching of both ordinary signal and idler waves will be emitted in concentric annuli at different angles, which are determined by the phase-matching angles. By placing narrow band filters in the emission, the ring can be readily observed. The quantum correlated nature of the emitted signal and idler paired photons have deep consequences for our understanding of nature—a topic we will return to in Chapter 11.

Equation 6.75 is a small-signal calculation. When the OPG output power becomes appreciable to the incident pump, the undepleted pump approximation built into Equation 6.73 breaks down. Calculating the OPG power for the case of high conversion efficiency relies on numerical integrations of the coupled amplitude equations for all the fields involved in the OPG process.

PROBLEMS

6.1 Consider a difference-frequency interaction where the field corresponding to ω_1 is treated as a constant throughout the crystal (undepleted pump approximation). For the initial conditions, $A_1(0) = A_{10}$, $A_2(0) = A_{20}$, and $A_3(0) = 0$, find $A_2(z)$ assuming $\Delta k = 0$.

6.2 Consider a low-efficiency difference-frequency interaction where the field corresponding to ω_2 is treated as a constant throughout the crystal. The initial conditions are $A_1(0) = A_{10}$, $A_2(0) = A_{20}$, and $A_3(0) = 0$. Find $A_3(z)$ assuming $\Delta k = 0$. How does the solution differ qualitatively to the result given in Equation 6.21?

6.3 Repeat Problem 6.2 for the case where $\Delta k \neq 0$.

6.4 Equations 6.30 and 6.31 result when we encounter mixed boundary conditions in the undepleted pump approximation (A_1 approximately constant). Derive Equations 6.30 and 6.31 by using the boundary conditions, $A_2(0) = A_{20}$ and $A_3(0) = A_{30}$, which are both nonzero. Furthermore, assume that A_{10}, A_{20}, and A_{30} have the same phase, which we set to zero.

6.5 Repeat Problem 6.4 for the case when A_{10}, A_{20}, and A_{30} have different phases.

6.6 In the undepleted pump approximation (i.e., $A_1 \sim$ constant), Equations 6.28 and 6.29 govern the growth of $N_2(z)$ and $N_3(z)$ with the initial conditions $N_1(0) = N_{10}$, $N_2(0) = N_{20}$, and $N_3(0) = 0$. The undepleted pump approximation breaks down in the limit of high conversion efficiency, which occurs when $N_3(z)$ becomes comparable to N_{10}. Calculate the crystal length required to obtain $N_3(L) = N_{10}/4$. Give your answer in terms of Γ (see Equation 6.12), N_{10}, and N_{20}. Show that the length required to deplete the pump decreases logarithmically with N_{20}. That is, the length to obtain pump depletion is a weak function of the starting number of photons, N_{20}.

6.7 Consider a low-efficiency sum-frequency interaction such that the fields corresponding to the two inputs at ω_2 and ω_3 do not change appreciably during the interaction. Calculate the output intensity for the sum frequency at ω_1 when the crystal has absorption (α_1) at ω_1. Assume $A_1(0) = 0$.

6.8 Show that the solution to Equation 6.40 is Equation 6.43, given $A_1(0) = A_{10}$, $A_2(0) = A_{20}$, $A_3(0) = 0$, and $\Delta k = 0$.

6.9 Equation 6.43 is the expression for the difference-frequency output when the DFG frequency is absorbed by the crystal. This expression has a sinh dependence, which seems to show a larger signal for higher absorption. Verify that the overall exponential growth of $|A|$ is less than $e^{\Gamma z}$.

6.10 Consider DFG when the two inputs remain essentially constant and where the difference-frequency output is absorbed. This situation is described by Equation 6.45. Verify that the solution to this equation is given by Equation 6.46, and that the intensity for the DFG output is given by Equation 6.47.

6.11 Plot I_3/I_{max} (see Equation 6.47) as a function of ΔkL for the cases when $\alpha L = 0$ and $\alpha L = 10$. Based on the plots, which case has the larger phase-matching bandwidth?

6.12 Derive an expression for the absorption coefficient, α, when (a) σ is nonzero and $\tilde{\chi}^{(1)}$ is purely real, and (b) when $\sigma = 0$ and $\tilde{\chi}^{(1)}$ is complex. Assume a scalar, monochromatic plane wave in a homogeneous, isotropic medium and as usual take the SVEA. (*Hint*: Start with Maxwell's equations, $\vec{\nabla} \times (\vec{\nabla} \times \vec{E})$ and the appropriate constitutive relationships.)

6.13 Use the results of the Manley–Rowe relationships as given in Equations 6.59 through 6.61 to show that $I_1 + I_2 + I_3 = $ constant for a three-wave mixing interaction.

6.14 Consider difference-frequency mixing between two pulsed lasers at ω_1 and ω_2. The pulses have Gaussian temporal profiles with an FWHM of ~10 ns. Make a qualitative sketch of the temporal profile of the pump (corresponding to ω_1) at the end of the crystal in the limit when the pump experiences significant depletion. Compare the output profile to the input Gaussian profile. (*Hint*: Break up the nanosecond pulse duration into several temporal slices and consider the conversion efficiency of each slice. This process works because the crystal length is much shorter than the spatial extent of a 10-ns pulse (~3 m).)

6.15 As an alternative to using the exact solutions to the three coupled amplitude equations for three collinear monochromatic plane waves (Equations 4.25 through 4.27), it is possible to numerically integrate the three coupled amplitude equations. Consider a phase-matched ($\Delta k = 0$) DFG interaction between two plane waves with wavelengths 1.0 and 3.0 μm. The coupled amplitude equations are written as

$$\frac{dA_1}{dz} = i\kappa_1 A_2 A_3, \tag{6.76}$$

$$\frac{dA_2}{dz} = i\kappa_2 A_1 A_3^*, \tag{6.77}$$

$$\frac{dA_3}{dz} = i\kappa_3 A_1 A_2^*. \tag{6.78}$$

For the wavelengths involved and for a typical nonlinearity on the order of 2 pm/V, $\kappa_1 = 1.3 \times 10^{-5}$, $\kappa_2 = 8.4 \times 10^{-6}$, and $\kappa_3 = 4.2 \times 10^{-6}$ (SI units). The starting field amplitudes are $|A_1| = 2.0 \times 10^8$ V/m, $A_2 = 0$, and $|A_3| = 6.0 \times 10^7$ V/m, which are easily attainable for loosely focused nanosecond-pulsed lasers. Integrate the equations using a Runge–Kutta (RK) algorithm (second-order is good enough) to propagate the three fields through a 50-mm crystal. Most mathematical packages have the RK routine built in, but if such a routine is not available, then programming the RK algorithm is straightforward (see Appendix C). To perform the integral, break up the crystal into 1000 steps for Δz. After each step of the integration routine, save the value of the intensities (i.e., calculate it based on the field amplitude) and arrange them into three arrays. For calculating the intensities, assume $n = 2.0$. These arrays are a record of how the field intensities change through the crystal. Plot the three intensities on one graph as a function of distance in the crystal to study how they change relative to each other. In a second plot, graph the three intensities multiplied by the wavelength associated with each one. The second plot is related to the flow of photons at each wavelength as a function of propagation distance.

6.16 Numerically integrate Equations 6.25 through 6.27 using the parameter values given in Problem 6.15. Add the capability to include Δk in the equations and consider several values $\Delta k = \pi$, 10π, and 40π (SI units).

REFERENCES

Giallorenzi, T. G. and C. L. Tang. 1968. Quantum theory of spontaneous parametric scattering of intense light. *Physical Review* 166: 225–233.

Loudon, R. 1983. *The Quantum Theory of Light*. New York, NY: Oxford University Press.

FURTHER READING

Armstrong, J. A., N. Bloembergen, J. Ducuing, and P. S. Pershan. 1962. Interactions between light waves in a nonlinear dielectric. *Physical Review* 127: 1918–1939.

Boyd, R. W. 2008. *Nonlinear Optics*. Boston, MA: Academic Press.

Boyd, G. D. and D. A. Kleinman. 1968. Parametric interaction of focused Gaussian light beams. *Journal of Applied Physics* 39: 3597.

Byer, R. L. and S. E. Harris. 1968. Power and bandwidth of spontaneous parametric emissions. *Physical Review* 168: 1064–1068.

Manley, J. M. and H. E. Rowe. 1956. Some general properties of nonlinear elements— Part I. General energy relationships. *Proceedings of the Institute of Radio Engineers* 44: 904–913.

Shen, Y. R. 1984. *The Principles of Nonlinear Optics*. New York, NY: John Wiley & Sons.

Sutherland, R. L. 2003. *Handbook of Nonlinear Optics* (2nd edition). New York, NY: Marcel Dekker, Inc.

7

$\chi^{(2)}$ DEVICES

7.1 INTRODUCTION

A principal application of nonlinear optics is frequency-converting laser outputs. As discussed in the previous chapters, the addition of a single nonlinear crystal to a laser system has the potential to convert a fixed frequency laser into a broadly tunable, coherent light source. Adding a frequency-converting crystal to an existing laser system is practical because lasers are widely available and, for the most part, reliable. The concepts from earlier chapters are critical to the successful design of frequency converters, but the idealized plane-wave treatment needs some modifications to account for real-world laser sources. To address practical issues, the first part of this chapter looks at modifying the plane-wave equations so that they are applicable to Gaussian laser beams. The second part introduces optical cavities to resonate one or more of the frequencies involved. In addition to looking at resonator cavities with the aim of increasing the efficiency, we introduce a type of device in common use called an optical parametric oscillator (OPO).

7.2 OPTIMIZING DEVICE PERFORMANCE: FOCUSING

An important consideration with any nonlinear device is to choose a crystal that both phase matches and is transparent for the relevant wavelengths. In many cases, several crystals satisfy phase matching; so factors such as angular acceptance, phase-matching bandwidth, and temperature tolerance can help narrow down the choice of crystals. For example, short-pulse applications require a large phase-matching bandwidth. Once we decide on a particular crystal, the next consideration is how to achieve the highest conversion efficiency. A first approach is to look at the plane-wave results from previous chapters and see how the nonlinear gain can be increased.

Specifically, the plane-wave results show that in the small-signal limit, the output intensity is proportional to a product of intensities. In the case of difference-frequency generation (DFG),

$$I_{DFG} \sim I_1 I_2. \tag{7.1}$$

For second-harmonic generation (SHG),

$$I_{SHG} \sim I_F^2 \tag{7.2}$$

and for sum-frequency generation (SFG),

$$I_{SFG} \sim I_2 I_3. \tag{7.3}$$

Since the desired output is proportional to a product of intensities, a simple idea is to focus the beams tightly. In Chapter 4, we see that the phase-matching angular acceptance places a limit on such focusing. Moreover, the intensity of focused beams is high only in a region near the beam waist, roughly plus or minus one Rayleigh range, z_R. Outside the Rayleigh range, the intensity falls off rapidly so that for tight focusing, when $2z_R$ is less than the crystal length, the effective nonlinear interaction length is approximately $2z_R$. To clarify the following discussion, we call the condition "confocal focusing" when the crystal length is equal to $2z_R$. In our initial discussion, we assume that the beams are Gaussian and satisfy the confocal focusing condition. We assume the process is phase matched, and then in Section 7.2.3, we relax this assumption to include phase-matching issues that arise with tight focusing.

7.2.1 OVERLAP OF GAUSSIAN BEAMS WITH NONLINEAR POLARIZATION

Cylindrically symmetric Gaussian beams have field envelopes of the form

$$A(r,z) = A_0(z)e^{-r^2/w^2(z)}, \tag{7.4}$$

where A_0 is the complex field amplitude on axis, $w(z)$ is the field $1/e$ radius, and r is the transverse dimension. In the confocal region, the beam size changes slowly, and so we approximate $w(z)$ as constant and equal to w_0. Another way of stating this approximation is that we ignore diffraction. In the confocal region, the coupled wave equations (Equations 4.25 through 4.27) with the field envelope expanded as in Equation 7.4 become

$$\frac{dA_{10}}{dz}e^{-r^2/w_1^2} = i\frac{\omega_1}{n_1 c}d_{eff}A_{20}A_{30}e^{-r^2\left((1/w_2^2)+(1/w_3^2)\right)}e^{-i\Delta kz}, \tag{7.5}$$

$$\frac{\mathrm{d}A_{20}}{\mathrm{d}z}e^{-r^2/w_2^2} = i\frac{\omega_2}{n_2 c}d_{\text{eff}}A_{10}A_{30}^*e^{-r^2\left((1/w_1^2)+(1/w_3^2)\right)}e^{i\Delta kz}, \tag{7.6}$$

$$\frac{\mathrm{d}A_{30}}{\mathrm{d}z}e^{-r^2/w_3^2} = i\frac{\omega_3}{n_3 c}d_{\text{eff}}A_{10}A_{20}^*e^{-r^2\left((1/w_1^2)+(1/w_2^2)\right)}e^{i\Delta kz}. \tag{7.7}$$

The right-hand side of Equations 7.5 through 7.7 has a field radius determined by nonlinear polarization. The left-hand side shows the radial dependence of the driven field. The field radii of the nonlinear polarization terms are more simply written as

$$\frac{1}{\overline{w}_3^2} \equiv \frac{1}{w_1^2} + \frac{1}{w_2^2}, \tag{7.8}$$

$$\frac{1}{\overline{w}_2^2} \equiv \frac{1}{w_1^2} + \frac{1}{w_3^2}, \tag{7.9}$$

$$\frac{1}{\overline{w}_1^2} \equiv \frac{1}{w_2^2} + \frac{1}{w_3^2}. \tag{7.10}$$

Note that the nonlinear polarization radius is always smaller than either of the two field radii that generate it.

According to Poynting's theorem, the rate of energy exchange between a field and nonlinear polarization is proportional to

$$\int \vec{E} \cdot \frac{\mathrm{d}\vec{P}}{\mathrm{d}t}\mathrm{d}V. \tag{7.11}$$

Equation 7.11 reinforces the idea that the spatial overlap of the field and nonlinear polarization feed directly into the efficiency of the process. We use Equation 7.11 to justify a radial overlap integral where we multiply Equations 7.5 through 7.7 by e^{-r^2/w_1^2}, e^{-r^2/w_2^2}, and e^{-r^2/w_3^2}, respectively, and then integrate both sides of the equations over the radial dimension. Note that if the nonlinear polarization radius and the field radius are the same, then the radial integrations divide out to unity. Performing the integrals leads to the following set of equations:

$$\frac{\mathrm{d}A_{10}}{\mathrm{d}z} = ig_1\frac{\omega_1}{n_1 c}d_{\text{eff}}A_{20}A_{30}e^{-i\Delta kz}, \tag{7.12}$$

$$\frac{\mathrm{d}A_{20}}{\mathrm{d}z} = ig_2\frac{\omega_2}{n_2 c}d_{\text{eff}}A_{10}A_{30}^*e^{i\Delta kz}, \tag{7.13}$$

$$\frac{dA_{30}}{dz} = ig_3 \frac{\omega_3}{n_3 c} d_{\text{eff}} A_{10} A_{20}^* e^{i\Delta kz},$$ (7.14)

where

$$g_1 = \frac{2\bar{w}_1^2}{\bar{w}_1^2 + w_1^2},$$ (7.15)

$$g_2 = \frac{2\bar{w}_2^2}{\bar{w}_2^2 + w_2^2},$$ (7.16)

$$g_3 = \frac{2\bar{w}_3^2}{\bar{w}_3^2 + w_3^2}.$$ (7.17)

The g parameters are reduction factors that account for the imperfect overlap of the nonlinear polarization and the field that it drives. The reduction depends on the particular experimental setup. For example, in DFG, the two inputs' radial sizes are w_1 and w_2, determined by the input focusing optics, whereas the output difference-frequency beam size, w_3, is unconstrained and has the same size as the nonlinear polarization. Hence, $\bar{w}_3 = w_3$ and using Equations 7.15 through 7.17 along with Equations 7.8 through 7.10,

$$g_1 = \frac{1}{1 + (w_1^2/w_2^2)},$$ (7.18)

$$g_2 = \frac{1}{1 + (w_2^2/w_1^2)},$$ (7.19)

$$g_3 = 1.$$ (7.20)

The coupled amplitude equations for the Gaussian beams are of the same form as the plane-wave equations in Chapter 4 (Equations 4.25 through 4.27). The equations differ in the additional multiplicative constants g_1, g_2, and g_3, and the equations are written in terms of the on-axis amplitudes. Hence, we draw the solutions from previous chapters and include the reduction factors and the on-axis intensities. The solutions to the coupled amplitude equations found previously relate the intensity of one beam to a product of the other two. In the present case of Gaussian beams, we go one step further to write the equations in terms of the power in the beams, which is more useful for experiments.

7.2.2 PARAMETRIC INTERACTIONS WITH FOCUSED GAUSSIAN BEAMS

For a Gaussian laser beam, the peak intensity at the focus is proportional to the total power in the beam. The intensity at the beam waist is given by

$$I = I_0 e^{-2r^2/w_0^2}, \tag{7.21}$$

where I_0 is the on-axis intensity, r is the radial distance, and w_0 is the field $1/e$ radius. The total power is obtained by integrating over the intensity distribution, giving

$$P = \int_0^\infty 2\pi I_0 e^{-2r^2/w_0^2} r \, \mathrm{d}r = \frac{\pi w_0^2}{2} I_0. \tag{7.22}$$

Hence, the peak intensity based on the laser power is

$$I_0 = \frac{2P}{\pi w_0^2}. \tag{7.23}$$

Furthermore, confocal focusing with Gaussian beams provides an analytic expression that relates the beam waist, w_0, to the crystal length,

$$z_R = \frac{n\pi w_0^2}{\lambda} = \frac{L}{2}. \tag{7.24}$$

Hence,

$$w_{0,\text{confocal}}^2 = \frac{\lambda L}{2n\pi}. \tag{7.25}$$

Substituting Equation 7.25 for w_0^2 into Equation 7.23 gives the on-axis intensity in terms of power and crystal length,

$$I_0 = \frac{4n}{\lambda L} P = \frac{2n\omega}{\pi c L} P. \tag{7.26}$$

As mentioned above, the solutions to coupled amplitude equations (Equations 7.12 through 7.14) are the same as in the previous chapters, provided we use the Gaussian beam's on-axis intensity and that we include the reduction factors. Therefore, we may go straight to solutions already derived in the previous chapters. As an example, let us consider a DFG interaction in the small-signal limit. The plane-wave result from Chapter 4 (Equation 4.32) is repeated here:

$$I_3 = \frac{2\mu_0\omega_3^2 L^2}{n_1 n_2 n_3 c} d_{\text{eff}}^2 I_1 I_2 \operatorname{sinc}^2\left(\frac{\Delta k L}{2}\right). \tag{7.27}$$

Equation 7.26 is substituted in for the intensities I_1 and I_2 in Equation 7.27. However, the reduction factor, g_3, is unity since the DFG output is unconstrained and matches the nonlinear driving polarization. Hence, $w_3 = \bar{w}_3$ and is not confocal. The equation for the Gaussian DFG interaction is

$$P_3 = \frac{4\omega_1\omega_2\omega_3^2}{\pi\varepsilon_0 n_3 c^4 (n_1\omega_1 + n_2\omega_2)} d_{\text{eff}}^2 P_1 P_2 L \operatorname{sinc}^2\left(\frac{\Delta k L}{2}\right). \tag{7.28}$$

In a similar fashion, confocal focusing for SHG leads to the following equation for a Gaussian beam (starting with Equation 4.42):

$$P_{\text{SHG}} = \frac{2 d_{\text{eff}}^2 \omega_F^3}{\pi\varepsilon_0 c^4 n_{\text{SHG}} n_F} P_F^2 L \operatorname{sinc}^2\left(\frac{\Delta k L}{2}\right). \tag{7.29}$$

The reduction factor is unity since the SHG output is unconstrained, and it overlaps with the nonlinear polarization. Later in this chapter, starting with Section 7.3.3, we encounter cases where the reduction factor is not unity.

A significant distinction between the plane-wave result and that for Gaussian beams is that their length dependencies differ. Equations 7.28 and 7.29 show a linear dependence of the efficiency on crystal length, whereas the plane-wave result is quadratic with length. Tight focusing does increase the intensities and the efficiency, but under the condition of confocal focusing, the crystal length and beam waist size are not independent, leading to the linear dependence of efficiency on crystal length.

An important exception to the linear dependence rule for tight focusing is frequency conversion in a waveguide. The waveguide confines the beam over large distances. Since the beam size is constant, the intensity is also constant, and we regain the quadratic length dependence for efficiency. An extra complication with waveguides is that phase matching needs to account for mode dispersion in addition to material dispersion. A more difficult problem to manage is mode overlap between the differing frequency components. Waveguides are typically designed for a relatively narrow frequency range. With three-wave mixing, the large frequency differences require special designs to ensure that all frequencies excite the same waveguide mode.

7.2.3 OPTIMIZING GAUSSIAN BEAM INTERACTIONS

The confocal condition outlined above serves as a guide to estimating the output power when tightly focusing the input lasers. Its range of validity is limited to low conversion efficiency, confocally focused Gaussian beams, and walk-off free interactions. However, confocal focusing is not the optimum focusing condition; so we now look at extending the discussion to arbitrary focusing.

The critical parameters in determining the optimum focusing are crystal length, focused spot size, w_0, for each of the frequencies, and the amount of phase mismatch, Δk. Surprisingly, with focused beams, $\Delta k = 0$ is not the optimum. Rather a negative nonzero value results in the highest efficiency.[*] The reason why $\Delta k = 0$ is not the optimum is that a focused Gaussian beam has a range of k-vector directions and only one of these components is collinear with the beam direction. When Δk is negative, the collinear component is detuned; however, many of the off-axis components satisfy the noncollinear phase-matching condition, as illustrated in Figure 7.1. In practice, Δk is tuned to the optimum by varying the crystal angle, temperature, and so forth to achieve the highest output efficiency.

Optimum focusing for certain interactions with Gaussian beams is calculated by Boyd and Kleinman (1968). Their technique takes into account focusing, birefringent walk-off, and optimizing Δk. We look at their optimization routine for Type I SHG in negative uniaxial crystals (o + o → e). They introduce the dimensionless parameters,

$$\xi \equiv \frac{L}{b}, \tag{7.30}$$

$$\sigma \equiv z_R \Delta k. \tag{7.31}$$

ξ and σ contain the focusing and phase mismatch parameters, respectively, and they are parameters under our control in an experiment. b is the confocal parameter defined by

$$b = 2z_R \tag{7.32}$$

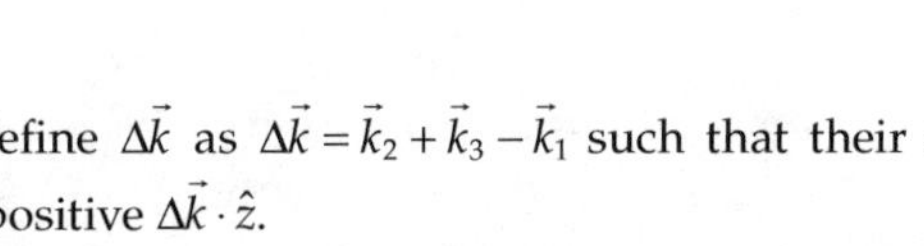

FIGURE 7.1 Negative $\vec{\Delta k} \cdot \hat{z}$ for collinear phase matching leads to $\vec{\Delta k} = \vec{0}$ for noncollinear interactions.

---

[*] Note that other authors define $\vec{\Delta k}$ as $\vec{\Delta k} = \vec{k}_2 + \vec{k}_3 - \vec{k}_1$ such that their sign convention leads to an optimum focusing with a positive $\vec{\Delta k} \cdot \hat{z}$.

Birefringent walk-off comes into the formalism through the parameter

$$B = \frac{\rho}{2} \sqrt{\frac{2\pi L n_F}{\lambda_F}}, \qquad (7.33)$$

where λ_F is the fundamental wavelength, n_F is the index, and ρ is the walk-off angle.

Boyd and Kleinman show that the SHG power for Type I interactions is proportional to a function, $h_m(\sigma, B, \xi)$, that depends on parameters we control experimentally. As a given crystal typically has only one angle that satisfies phase matching, the only way to change B (walk-off) for a given SHG frequency pairing is to use a different crystal. Hence, the optimization routine fixes B at a particular value; σ, on the other hand, is a free variable since it corresponds to the phase mismatch, which is affected by experimental parameters such as crystal temperature or orientation. The numerical optimization routine performs the same task in that σ is varied for fixed B and ξ until the maximum value of h_m is obtained. This procedure is repeated for a wide range of focusing parameters, ξ. The end result is a plot of h_m as a function of ξ for a fixed B (see Figure 7.2). When walk-off is not present, the optimum ξ is 2.84; however, we note that $\xi = 1$ (confocal focusing) is fairly close to the maximum. In many cases, the $\xi = 1$ is easier to implement than the tighter focusing; so a slight decrease in performance is traded off for operational convenience.

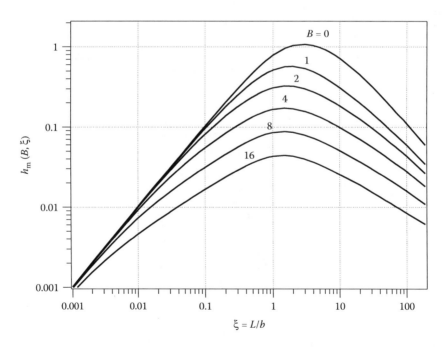

FIGURE 7.2 Dependence of Type I SHG on focusing and walk-off. (Adapted and reprinted with permission from Boyd, G. D. and D. A. Kleinman. 1968. *Journal of Applied Physics* 39: 3597–3639. Copyright 1968, American Institute of Physics.)

Type I SHG is a special case because of the relationship between the fundamental and second-harmonic wavelengths and polarizations. DFG and SFG have optimum focusing that depends on the specific three frequencies as well as the three polarizations involved. The optimization procedure used for Type I SHG does not carry over to the general DFG and SFG cases, or to the Type II SHG case. As a general rule in devices, high efficiency (but perhaps not the absolute highest) is obtained by matching the confocal parameter, b, of the three beams and using a value of $\xi = 1$. When obtaining the absolute optimum performance is necessary, numerical techniques give the most reliable approach.

Optimum focusing is not always practical. Consider the case of a nanosecond-pulsed laser with 100 mJ of energy per laser pulse. The peak power of the pulse is roughly 10^8 W. For common crystal lengths, on the order of 1 cm, confocal focusing gives $w_0 \sim 30$ µm so that the fluence is ~3 kJ/cm² or, in terms of the peak intensity, ~7 GW/cm²! Considering that most materials damage at levels less than 10 J/cm², tightly focused beams and high-energy pulsed lasers do not mix. In such cases, the beam cross section is increased to drop the intensity below the crystal's damage threshold. Another example where optimum focusing may not work is with ultrashort pulses in the picosecond and femtosecond regions. With ultrashort pulses, the peak power turns on other nonlinear effects that adversely affect the intended process. As a final cautionary remark, we remind the reader that all optimization calculations presented here assume low conversion efficiency. As the efficiency increases to the point at which back conversion occurs, numerical approaches are more appropriate for optimizing the process.

7.3   RESONATOR DEVICES

As shown in Equations 7.1 through 7.3, the efficiency of the nonlinear interactions is proportional to the product of intensities. With single-pass devices, the only way to increase the intensity is to turn up the power or to focus more tightly in the crystal. Another option is to resonate one or more of the interacting beams in a cavity since nonlinear conversion efficiency increases with increasing intracavity intensity.

7.3.1 RESONANT SHG

To guide our initial discussion, consider the case of placing a second-harmonic crystal in a cavity where only the fundamental wavelength is resonated (see Figure 7.3). The mirrors are designed to reflect the fundamental while transmitting the second harmonic. In this way, the second harmonic is coupled out of the cavity after being generated in the crystal. The ring configuration ensures that the pump is unidirectional in the crystal. We assume that the input mirror has a reflectivity R at the fundamental, while the other two mirrors reflect 100% of the fundamental and 0% of the SHG. Figure 7.3 shows flat mirrors, but in most systems, curved mirrors are

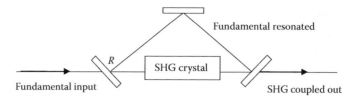

FIGURE 7.3 Resonantly enhanced SHG. The fundamental is resonated in a cavity, thereby increasing its intensity and the conversion efficiency to the SHG.

used to form a stable resonator mode. More details on cavity design may be found in Section 7.3.5.

By tuning the fundamental frequency to a cavity resonance (i.e., the round-trip phase is a multiple of 2π), the field amplitude inside the cavity is increased by a factor of

$$\frac{\sqrt{1-R}}{1-\sqrt{R}}. \tag{7.34}$$

The numerator contains the transmission amplitude at the input mirror is $t = \sqrt{1-R}$. For each round trip, the fundamental wave in the cavity mixes with the input wave by an amount $\sqrt{R}$. Since the SHG output has a quadratic dependence on the fundamental intensity, the resonant enhancement magnifies the efficiency. The trade-off with a resonant cavity is that the phase of the fundamental wave must be locked to the cavity length or vice versa. That is, the cavity length must remain an integral number of half-wavelengths of the fundamental wavelength. If the cavity length drifts off of the resonance condition, then the fundamental wave is rejected from the cavity and the SHG wave amplitude falls. A way around the phase-locking condition is to place the doubling crystal inside the laser cavity itself. As the cavity resonance drifts, then the fundamental laser wavelength shifts and the SHG frequency shifts as well. The downside to this approach is that the SHG process may drive the laser to an unstable output condition characterized by SHG frequency drifts.

The same strategy of resonating one or more of the inputs applies to SFG and DFG. For DFG and SFG, the output is proportional to the product of the intensities of the inputs (see Equations 7.1 and 7.3). Hence, if the crystal is placed in a cavity where one of the inputs is matched to a resonant cavity mode, then the improvement is linear in the intensity enhancement in the cavity. Both inputs may resonate with longitudinal cavity modes, but the double-locking requirement complicates the system considerably.

7.3.2 OPTICAL PARAMETRIC OSCILLATOR

An interesting effect occurs with OPA or DFG in a resonant cavity. In discussing these effects, we use the common terminology of pump (λ_1), signal (λ_2), and idler (λ_3). When the losses are low enough, resonating the signal leads to an oscillation in which the signal is self-sustaining and only the pump input is required to keep the process going. In fact, an oscillation builds up from noise when only the pump is incident on the cavity. Noise photons generated by the SPS mechanism (see Section 6.5) start the process. Such a device is called an OPO.

Consider a resonant Fabry–Perot-type cavity that reflects the signal and idler beams and transmits the pump (shown schematically in Figure 7.4). When the signal and idler beams pass through the nonlinear crystal, they experience amplification at the expense of the pump. Losses to the resonant beams are incurred at the mirrors since the mirrors are not 100% reflectors. The OPO output comes from the partial transmission of the end mirrors. Other losses include absorption in the crystal, scattering, and clipping due to the finite size of the crystal aperture and cavity mirrors. If the amplification of the resonated field(s) is greater than or equal to the cavity loss, then an oscillation occurs and the process is self-sustaining.

To calculate an OPO threshold, we start with a simplified model that treats the fields in the cavity as plane waves. We assume that the resonant cavity condition is met for both the signal and the idler. Further simplifying assumptions are the OPA process is perfectly phase matched ($\Delta k = 0$), the crystal is AR-coated, and the pump is undepleted. With these assumptions, the OPA process in the crystal is described as two coupled amplitude equations, the same as in Section 6.2. Solving the two coupled equations follows the same procedure as in Section 6.2, except that we now impose boundary conditions where both the signal and the idler are present at the input along with the pump. Solving for the signal and idler fields with the more general boundary conditions gives

$$A_I(z) = A_{I0} \cosh(\Gamma z) + i \frac{\kappa_I A_{S0}^* \sinh(\Gamma z)}{\Gamma}, \tag{7.35}$$

$$A_S(z) = A_{S0} \cosh(\Gamma z) + i \frac{\kappa_S A_{I0}^* \sinh(\Gamma z)}{\Gamma}, \tag{7.36}$$

where A_{S0} and A_{I0} are the signal and idler field amplitudes at the front surface of the crystal, $z = 0$. The parameter Γ is defined in Equation 6.12, repeated here in terms of the pump, signal, and idler notation:

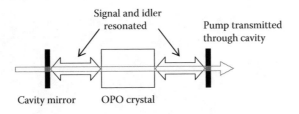

FIGURE 7.4 Schematic of a doubly resonant (DR)-OPO. The signal and idler are resonated while the pump is transmitted. The OPO output comes from the two end mirrors, which partially transmit the signal and idler.

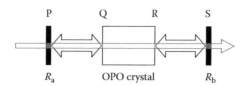

FIGURE 7.5 Simple two-mirror OPO. R_a and R_b are the reflectivities of the two mirrors. For each mirror, the reflectivity may be different for the signal and idler. P, Q, R, and S mark different locations in the cavity and are used for the discussion in the text.

$$\Gamma^2 = \frac{2\omega_S\omega_I}{n_P n_S n_I \varepsilon_0 c^3} d_{\text{eff}}^2 I_P.$$

(7.37)

An equivalent but perhaps more convenient form of Equation 7.37 for OPO calculations is in terms of vacuum wavelengths instead of angular frequencies:

$$\Gamma^2 = \frac{8\pi^2}{n_P n_S n_I \varepsilon_0 c \lambda_S \lambda_I} d_{\text{eff}}^2 I_P.$$

(7.38)

Equations 7.35 and 7.36 show that the fields grow exponentially in the crystal; hence, the fields are amplified.

After one round trip in the simplified cavity with arbitrary mirror reflectivities as shown in Figure 7.5, the fields experience reflection losses and linear absorption losses in the crystal. Figure 7.6 shows both the gain and loss processes by tracing the signal amplitude for one round trip through the OPO cavity, starting at position Q in Figure 7.5. Figure 7.6 shows that the field is amplified in the crystal in going from Q to R. Since the crystal is AR-coated, the amplitude remains unchanged until position S where a fraction of the field is coupled out of the cavity through the mirror, which is the reason for the drop in amplitude. The field then propagates back to the crystal

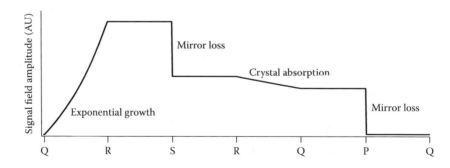

FIGURE 7.6 Signal experiences gain and loss as it traverses the OPO cavity. This plot shows the change in the signal field amplitude in one traversal of the OPO cavity. P, Q, R, and S refer to different locations in the cavity (see Figure 7.5).

(position R). The signal is attenuated on the second pass through the crystal and it does not see gain, since the interaction is not phase matched (the pump is going in the opposite direction). The signal beam propagates to the second cavity mirror at P where a fraction is coupled out of the cavity. Finally, the field propagates back to the crystal at point Q and the entire process starts over.

When the OPO reaches threshold, the gain and losses per round trip are equal and we may apply a self-consistent condition for the resonated field. Hence, the field at the input to the crystal ($z = 0$) is equal to the field at the crystal output ($z = L$) multiplied by reflection losses and linear loss in the crystal (assumed to be two passes through the crystal):

$$A_{I0} = \sqrt{R_{aI}R_{bI}}\,e^{-2\alpha_I L}A_I(L) = \rho_I A_I(L),$$ (7.39)

$$A_{S0} = \sqrt{R_{aS}R_{bS}}\,e^{-2\alpha_S L}A_S(L) = \rho_S A_S(L).$$ (7.40)

The loss terms are lumped together into ρ_S and ρ_I. Losses due to clipping on the crystal or the cavity optics are not considered in Equations 7.39 and 7.40, but such losses could be included in the factors ρ_I and ρ_S.

Equations 7.39 and 7.40 together with Equations 7.35 and 7.36 allow us to relate the gain and loss processes. Substituting Equation 7.39 into Equation 7.35 gives

$$A_I(L) = \rho_I A_I(L)\cosh(\Gamma L) + i\frac{\kappa_I A_{S0}^* \sinh(\Gamma L)}{\Gamma}.$$ (7.41)

Solving for $A_I(L)$ gives

$$A_I(L) = i\frac{\kappa_I A_{S0}^* \sinh(\Gamma L)}{\Gamma(1 - \rho_I \cosh(\Gamma L))}.$$ (7.42)

Next, A_{S0} is written in terms of $A_S(L)$ with the help of Equation 7.40 so that Equation 7.42 becomes

$$A_I(L) = i\frac{\kappa_I \rho_S A_S^*(L)\sinh(\Gamma L)}{\Gamma(1 - \rho_I \cosh(\Gamma L))}.$$ (7.43)

Similarly,

$$A_S(L) = i\frac{\kappa_S \rho_I A_I^*(L)\sinh(\Gamma L)}{\Gamma(1 - \rho_S \cosh(\Gamma L))}.$$ (7.44)

Rearranging Equation 7.44 gives

$$A_I(L) = iA_S^*(L) \frac{\Gamma(1 - \rho_S \cosh(\Gamma L))}{\kappa_S^* \rho_I \sinh(\Gamma L)}.$$

(7.45)

Equating Equations 7.45 and 7.43 gives the resonator condition:

$$\Gamma^2 (1 - \rho_I \cosh(\Gamma L))(1 - \rho_S \cosh(\Gamma L)) = \kappa_I \rho_S \kappa_S^* \rho_I \sinh^2(\Gamma L).$$

(7.46)

Note that the field amplitudes have dropped out of the expression. Equation 7.46 is rewritten more clearly as

$$\cosh(\Gamma L) \geq \frac{1 + \rho_S \rho_I}{\rho_S + \rho_I},$$

(7.47)

where $\Gamma^2 = \kappa_S^* \kappa_I$ and the trigonometric identity $\cosh^2(x) - \sinh^2(x) = 1$ are used in the simplification. Equation 7.47 separates the gain term on the left-hand side from the loss terms on the right-hand side, and the inequality is introduced to indicate when the oscillation condition is met.

Equation 7.47 is simplified further by assuming that the gain in the undepleted pump limit is low and $\Gamma L \ll 1$. Hence, the cosh term in Equation 7.47 is expanded resulting in

$$1 + \frac{(\Gamma L)^2}{2} \geq \frac{1 + \rho_S \rho_I}{\rho_S + \rho_I}.$$

(7.48)

Simplifying leads to the condition

$$\frac{(\Gamma L)^2}{2} \geq \frac{(1 - \rho_S)(1 - \rho_I)}{\rho_S + \rho_I}.$$

(7.49)

A more convenient threshold condition is in terms of the pump power required. We treat the OPO beams as segments of infinite plane waves where each beam has a cross-sectional area A. To this end, Γ is expressed using Equation 7.38, with the pump intensity written as P_P/A. Although plane waves are an idealization, many OPO devices operate with cavity lengths that are much smaller than the beam's Rayleigh range. In this limit, applying the plane-wave approximation to treat the OPO is reasonable. The threshold condition for a plane-wave OPO is given by the equality

$$P_{TH} = A \frac{n_P n_S n_I \varepsilon_0 c \lambda_S \lambda_I}{4\pi^2 d_{eff}^2 L^2} \frac{(1 - \rho_S)(1 - \rho_I)}{\rho_S + \rho_I}.$$

(7.50)

Note that λ_S and λ_I are not independent since they are connected to each other by energy conservation. A way to show the connection more clearly is to write the wavelengths with respect to the degenerate wavelength, $\lambda_d = 2\lambda_P$. Energy conservation dictates that the signal and the idler are symmetric about the degenerate frequency as we observe in the following expressions:

$$\omega_S = \frac{\omega_P}{2}(1+\delta), \tag{7.51}$$

$$\omega_I = \frac{\omega_P}{2}(1-\delta), \tag{7.52}$$

where δ is the degeneracy factor ranging from 0 to 1. Equations 7.51 and 7.52 are written in terms of vacuum wavelengths as

$$\lambda_S = \frac{2\lambda_P}{(1+\delta)}, \tag{7.53}$$

$$\lambda_I = \frac{2\lambda_P}{(1-\delta)}. \tag{7.54}$$

The threshold condition written in terms of the degeneracy factor is

$$P_{TH} = A\frac{n_P n_S n_I \varepsilon_0 c\lambda_P^2}{\pi^2 d_{eff}^2 L^2(1-\delta^2)}\frac{(1-\rho_S)(1-\rho_I)}{\rho_S+\rho_I}. \tag{7.55}$$

Equation 7.55 shows that the lowest threshold for the OPO occurs for a signal idler pairing at degeneracy ($\delta = 0$). Equations 7.50 and 7.55 show the dependencies of threshold on mirror reflectivities, crystal length, nonlinearity, and oscillation wavelength. Achieving the lowest possible threshold is not always the driving motivation. Usually, we trade-off a higher threshold to allow for higher-output coupling (lower mirror reflectivity).

7.3.3 OPO WITH GAUSSIAN BEAMS

The plane-wave threshold calculation is modified for Gaussian beams in a similar fashion to that introduced in Section 7.2. Moreover, the calculations from Chapter 6 carry over to the Gaussian case as well. Following the same derivation as in Section 6.2, but now including the reduction factors introduced in Section 7.2, gives a parametric gain of

$$(\Gamma_{Gaussian}L)^2 = g_S g_I(\Gamma_{plane\,wave}L)^2. \tag{7.56}$$

g_S and g_I are relabeled versions of the reduction factors g_2 and g_3, respectively, introduced in Section 7.2. Expanding out $\Gamma_{\text{plane wave}}$ as in Equation 7.38 gives

$$\Gamma^2_{\text{Gaussian}} = g_S g_I \frac{8\pi^2}{n_P n_S n_I \varepsilon_0 c \lambda_S \lambda_I} d^2_{\text{eff}} I_{P0}. \tag{7.57}$$

Note that in switching from the plane-wave equations to the Gaussian case, Γ is written in terms of the on-axis intensity. We assume that the pump beam is focused to the center of the crystal (not necessarily confocal) with the on-axis intensity at that point given by I_{P0}. We relate the on-axis intensity to the total power and the focused spot size, w_P, via Equation 7.23, so that

$$\Gamma^2 = g_S g_I \frac{16\pi^2}{n_P n_S n_I \varepsilon_0 c \lambda_S \lambda_I} d^2_{\text{eff}} \frac{P_P}{\pi w_P^2}. \tag{7.58}$$

The product, $g_S g_I$, is expanded using Equations 7.8 through 7.10 and Equations 7.15 through 7.17 resulting in

$$g_S g_I = 4 w_P^2 \left[\frac{w_P w_S w_I}{w_P^2 w_S^2 + w_P^2 w_I^2 + w_S^2 w_I^2} \right]^2 = 4 \frac{w_P^2}{W^2}, \tag{7.59}$$

where[*]

$$\frac{1}{W^2} = \left[\frac{w_P w_S w_I}{w_P^2 w_S^2 + w_P^2 w_I^2 + w_S^2 w_I^2} \right]^2. \tag{7.60}$$

Now Equation 7.58 becomes

$$\Gamma^2 = \frac{64\pi}{n_P n_S n_I \varepsilon_0 c \lambda_S \lambda_I W^2} d^2_{\text{eff}} P_P. \tag{7.61}$$

The OPO threshold condition (P_{TH}) for an OPO with Gaussian beams is given by the equality below. For oscillation, the pump power (P_P) exceeds this condition:

$$P_P \geq P_{\text{TH}} = \frac{n_P n_S n_I \varepsilon_0 c \lambda_S \lambda_I W^2}{32\pi L^2 d^2_{\text{eff}}} \frac{(1 - \rho_S)(1 - \rho_I)}{\rho_S + \rho_I}. \tag{7.62}$$

---

[*] Other authors use the terminology M^2 instead of $1/W^2$.

In terms of the degeneracy factor introduced in Equations 7.51 and 7.52, Equation 7.62 becomes

$$P_P \geq P_{TH} = \frac{n_P n_S n_I \varepsilon_0 c \lambda_P^2 W^2}{8\pi L^2 d_{eff}^2 (1-\delta^2)} \frac{(1-\rho_S)(1-\rho_I)}{\rho_S + \rho_I}. \tag{7.63}$$

The threshold power depends on the overlap of the pump, signal, and idler through the quantity W^2. In an OPO, the cavity determines one or more of the beam sizes, w_S and w_I, so that w_P is adjusted to optimize $1/W^2$. For cases in which both the signal and the idler are resonant in the cavity (known as a doubly resonant OPO [DR-OPO]), w_S and w_I are determined by the resonator modes and $1/W^2$ is optimized with respect to w_p with the result

$$\frac{1}{w_P^2} = \frac{1}{w_S^2} + \frac{1}{w_I^2}. \tag{7.64}$$

The corresponding calculation for the value of W^2 is (see Problem 7.13)

$$\frac{1}{W^2} = \frac{1}{4(w_S^2 + w_I^2)}. \tag{7.65}$$

A common scenario is an OPO where the signal and idler cavity modes are confocally focused; that is, $2z_R$ is equal to the crystal length. For this scenario, the optimum beam size for the pump is also confocal. Note that substituting in the confocal beam sizes, $w^2 = (\lambda L / 2\pi n)$, $1/W^2$ is proportional to $1/L$. Hence, the threshold for a confocal DR-OPO is inversely proportional to L and not L^2.

7.3.4 DOUBLY RESONANT OPOs

The simplest DR-OPO is one where the mirror reflectivity is the same for both the signal and idler wavelengths, and where the crystal has no absorption losses. Therefore, $\rho_1 = \rho_2 = R$, where R is the mirror reflectivity. The threshold condition for the plane-wave DR-OPO follows from Equation 7.50:

$$P_{TH,plane\ wave} = A \frac{n_p n_S n_I \varepsilon_0 c \lambda_S \lambda_I}{8\pi^2 d_{eff}^2 L^2} \frac{(1-R)^2}{R}. \tag{7.66}$$

The threshold for the Gaussian beam case given in Equation 7.62 becomes

$$P_{TH,Gaussian} = \frac{n_p n_S n_I \varepsilon_0 c \lambda_S \lambda_I W^2}{64\pi d_{eff}^2 L^2} \frac{(1-R)^2}{R}. \tag{7.67}$$

Assume that the Gaussian beam DR-OPO is configured such that the signal and idler modes are confocal, then $w_S^2 = \lambda_S L/2\pi n_S$ and $w_I^2 = \lambda_I L/2\pi n_I$. Approximating $n_S \approx n_I = n$ and substituting w_S and w_I into Equation 7.65 gives

$$\frac{1}{W^2} = \frac{\pi n}{2L(\lambda_S + \lambda_I)}.$$ (7.68)

The OPO threshold under the confocal focusing condition is

$$P_{\text{TH,confocal}} = \frac{n^2 \varepsilon_0 c \lambda_S \lambda_I (\lambda_S + \lambda_I)}{32 \pi^2 d_{\text{eff}}^2 L} \frac{(1-R)^2}{R}.$$ (7.69)

Finally, we write an equivalent expression in terms of the degeneracy factor (see Equations 7.51 and 7.52):

$$P_{\text{TH,confocal}} = \frac{n^2 \varepsilon_0 c \lambda_P^3}{2\pi^2 d_{\text{eff}}^2 L(1-\delta^2)^2} \frac{(1-R)^2}{R}.$$ (7.70)

As an example, consider a confocal DR-OPO with a 5-cm periodically poled lithium niobate crystal, $R \sim 0.98$, $\lambda_I = 3.66$ μm, $\lambda_S = 1.5$ μm, $\lambda_P = 1.064$ μm, $n = 2.2$, $L = 5$ cm, and $d_{\text{eff}} = 16$ pm/V. Substituting numerical values into Equation 7.69 gives $P_{\text{TH}} = 37$ mW, which is attainable with low-power lasers.

The cost of going to a DR-OPO is that we have to simultaneously resonate the signal and idler at the same time as satisfying energy conservation among the pump, signal, and idler. The constraints are such that energy conservation dictates that when the signal is resonant, there has to be a resonator mode at the corresponding idler wavelength. Such an opportunistic overlap is uncommon and unstable. Typically, when one wavelength is resonant, the other one is not and the OPO does not turn on at all. A characteristic of a free-running DR-OPO is that the output flickers on and off as the signal and idler come into and out of resonance. One example of a scheme that mitigates this problem using a coupled cavity approach is shown in Figure 7.7. Length

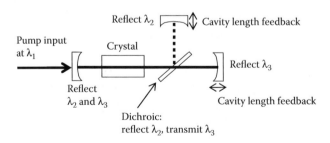

FIGURE 7.7 DR-OPO where the two resonated frequencies may be cavity locked independently. If the phase matching is Type II, then the dichroic mirror may be replaced with a polarizing beam splitter.

controls of the two cavities shown are decoupled making the problem of locking both frequencies simpler.

Optimization of the OPO performance requires appropriate focusing. As discussed above, when the signal and idler beams are confocally focused, the pump beam should be confocally focused. However, this analysis is a special case for noncritical phase matching (no walk-off). When walk-off is included, the optimum focusing condition changes slightly. As an example, consider the same focusing condition where the pump, signal, and idler all have the same confocal parameter (not necessarily matched to the crystal length). The OPO crystal is positive uniaxial where the signal and the idler are e-polarized and the pump is o-polarized. Similar to the second-harmonic optimization introduced in Section 7.2.3, Boyd and Kleinman (1968) derive an expression, $\bar{h}(B,\xi)$, proportional to the inverse threshold for this specific example. Figure 7.8 shows $\bar{h}$ as a function of the focusing parameter ξ for several different walk-off conditions, B.

When walk-off is not present ($B = 0$), the analysis applies to other polarization combinations. Hence, noncritically phase-matched DR-OPOs have an optimum focusing parameter of $\xi = 2.94$. Figure 7.8 shows that when walk-off is present, the threshold becomes somewhat insensitive to focusing for a large range of ξ. Similar to SHG, Figure 7.8 should only be used as a guide to determine a reasonable focusing condition when applied to polarization combinations other than the Type I configuration assumed here.

A variation of the DR-OPO, where the pump and the signal are resonantly coupled to cavity modes by mirrors and the idler is coupled out of the cavity, has certain advantages. Resonating the pump as well as the signal has the advantage of increasing the

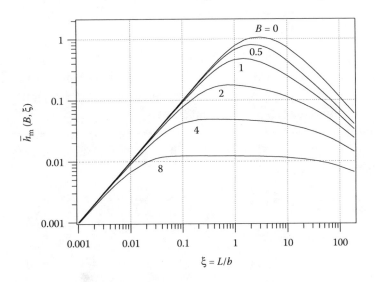

FIGURE 7.8 Dependence of the inverse threshold of an OPO on focusing and walk-off. (Adapted and reprinted with permission from Boyd, G. D. and D. A. Kleinman. 1968. *Journal of Applied Physics* 39: 3597–3639. Copyright 1968, American Institute of Physics.)

effective power of the pump laser. The disadvantage of this DR-OPO design is that the resonator has to be locked to the pump laser, which adds complexity to the system. However, this scheme is simpler to implement than the other DR-OPOs discussed earlier in this chapter. The reason is that when the cavity is locked to the pump laser, a cluster of resonator modes lies within the signal bandwidth. One or more of these modes resonates and the idler has a wavelength determined by energy conservation. The idler is free to change wavelength, since it is not tied to a cavity mode. Hence, in this scheme, only one locking control loop is required instead of two as is the case for the previous DR-OPO.

7.3.5 SINGLY RESONANT OPOs

When only the signal is resonant with the cavity, then $\rho_S = R$ and $\rho_I = 0$ in Equations 7.50 and 7.62. A technicality of OPO language is that in a singly resonant OPO (SR-OPO) the convention is to label the resonated wave as the signal. However, either the lower-energy or higher-energy photon in the parametric process may be resonated. The results of Sections 7.3.2 and 7.3.3 carry over to the singly resonant case. The plane-wave threshold of the SR-OPO is

$$P_{\text{TH,plane wave}} = A \frac{n_P n_S n_I \varepsilon_0 c \lambda_S \lambda_I}{4\pi^2 d_{\text{eff}}^2 L^2} \frac{(1-R)}{R}. \tag{7.71}$$

The Gaussian beam threshold is

$$P_{\text{TH,Gaussian}} = \frac{n_P n_S n_I \varepsilon_0 c \lambda_S \lambda_I W^2}{32\pi d_{\text{eff}}^2 L^2} \frac{(1-R)}{R}. \tag{7.72}$$

W^2 depends on the particular focusing parameters. In the case of the SR-OPO, only the signal beam size is constrained by the cavity. The pump beam size is adjusted with external focusing optics and the idler beam size is then determined by the nonlinear polarization that drives it. That is, $\bar{w}_I = w_I$ as would be the case for difference-frequency mixing. When the resonator gives a confocally focused signal beam and the pump beam is confocally focused with external optics, the idler beam size is, according to Equation 7.8,

$$\frac{1}{w_I^2} = \frac{1}{w_S^2} + \frac{1}{w_P^2}. \tag{7.73}$$

The corresponding value of W^2 is

$$\frac{1}{W^2} = \frac{1}{4(w_P^2 + w_S^2)}. \tag{7.74}$$

For the case when both the pump and the signal are confocally focused, the threshold is

$$P_{\text{TH,confocal}} = \frac{n^2 \varepsilon_0 c \lambda_S \lambda_I (\lambda_P + \lambda_S)}{16 \pi^2 d_{\text{eff}}^2 L} \frac{(1-R)}{R}.$$

(7.75)

By inspection of Equations 7.71 and 7.72, the threshold of the SR-OPO is found higher than the DR-OPO by a factor of $2/(1-R)$. However, the SR-OPO is a far more popular design because it oscillates more stably. The SR-OPO freely runs without length stabilization. If the cavity length changes, then the oscillating frequency shifts, and the OPO frequency shifts with it, or the OPO mode hops. Since the idler wave is not resonant in the cavity, its frequency drifts to maintain energy conservation ($\omega_I = \omega_P - \omega_S$). Similarly, if the pump frequency changes, then the signal frequency is free to oscillate unchanged, while the idler frequency shifts. The lack of stringent resonant conditions leads to a much more stable performance when compared to the DR-OPO. In fact, SR-OPOs easily tune by changing the crystal phase-matching angle or its temperature. The tuning is typically limited by the phase matching of the crystal or the cavity mirror reflectivity range.

7.3.6 CAVITY DESIGN

Up to this point, we assumed specific pump and signal beam sizes in the OPO without looking at how to obtain these beam sizes. When the pump wave has no cavity feedback, its spot size is determined by the pump laser's beam profile and formatting optics outside of the OPO cavity. Typically, we directly measure the pump laser's beam profile to obtain its Gaussian beam parameter, q_1, at the output of the laser. A lens sequence transforms the pump beam from the q_1 plane to the crystal plane, which is calculated using the ABCD Gaussian beam propagation method introduced in Chapter 2. In many OPOs, the pump is injected into the cavity through one of the OPO mirrors. If the OPO mirror has a radius of curvature, then it acts as a negative lens for the pump beam. Part of the pump focusing calculation should include the transmission through the OPO optic as well as the air–crystal interface.

The signal beam size is determined by the resonant cavity mode. As an example, consider the cavity shown in Figure 7.9. The cavity mode size is calculated with a

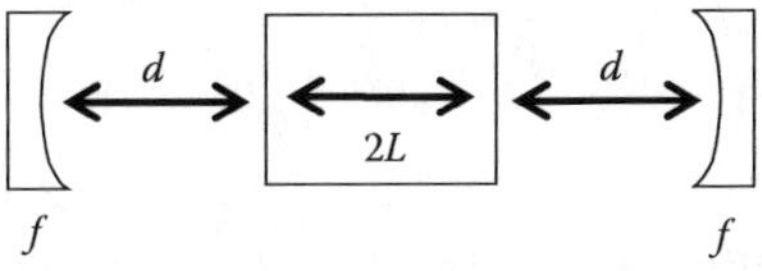

FIGURE 7.9   Simple OPO cavity consisting of a crystal and two curved mirrors. The radius of curvature of the mirrors is related to the focal length by $R = 2f$.

self-consistent ABCD approach. The mode size calculation starts with the Gaussian beam parameter found in Chapter 2 (Equation 2.130), which is repeated here:

$$\frac{1}{q} = \frac{1}{R} + \frac{i\lambda}{\pi n w^2} .$$

(7.76)

Note that R in this context is the radius of curvature and not mirror reflectivity and w is the $1/e$ field radius. For the resonator shown in Figure 7.9, symmetry dictates that the beam waist is located at the center of the cavity. Hence, the radius of curvature at this location is infinite and the Gaussian beam parameter is purely imaginary:

$$q_0 = -i\frac{\pi n w_0^2}{\lambda} .$$

(7.77)

Note that a stable, symmetric resonator requires that if we start with q_0 at the center and then propagate to one of the cavity mirrors and back, then we end up with the same q_0. The ABCD matrix for starting at the center of the cavity, going to one of the mirrors and back again, is given by

$$
\begin{bmatrix} A & B \\ C & D \end{bmatrix} = \begin{bmatrix} 1 & L \\ 0 & 1 \end{bmatrix} \begin{bmatrix} 1 & 0 \\ 0 & \frac{1}{n} \end{bmatrix} \begin{bmatrix} 1 & d \\ 0 & 1 \end{bmatrix} \begin{bmatrix} 1 & 0 \\ -\frac{1}{f} & 1 \end{bmatrix} \begin{bmatrix} 1 & d \\ 0 & 1 \end{bmatrix} \begin{bmatrix} 1 & 0 \\ 0 & n \end{bmatrix} \begin{bmatrix} 1 & L \\ 0 & 1 \end{bmatrix}
$$

$$
= \begin{bmatrix} 1 - \frac{1}{f}L_{\text{eff}} & 2nL_{\text{eff}} - \frac{n}{f}L_{\text{eff}}^2 \\ \frac{-1}{nf} & 1 - \frac{1}{f}L_{\text{eff}} \end{bmatrix},
$$

(7.78)

where

$$L_{\text{eff}} = d + \frac{L}{n} .$$

(7.79)

The Gaussian beam transforms according to (see Section 2.6.2)

$$q' = \frac{Aq_0 + B}{Cq_0 + D} .$$

(7.80)

However, the self-consistent condition imposed by the resonator is

$$q' = q_0 .$$

(7.81)

Combining Equations 7.80 and 7.81 with the ABCD values found in Equation 7.78 results in

$$q_0 = -i \left[n \sqrt{f L_{\text{eff}} \left(2 - \frac{L_{\text{eff}}}{f} \right)} \right]. \tag{7.82}$$

The beam waist found by equating Equation 7.82 to Equation 7.77 is

$$w_0^2 = \frac{\lambda}{\pi} \sqrt{f L_{\text{eff}} \left(2 - \frac{L_{\text{eff}}}{f} \right)}. \tag{7.83}$$

Now we are in a position to match the resonator mode size to the desired focusing parameter. As a specific example, let us consider an SR-OPO operating at 1.5 μm with $\xi = 2.8$ and a crystal that is 5 cm long. For the sake of this example, we assume that the index of refraction is 2.0. The focusing parameter definition in Equation 7.30 is used to calculate the spot size,

$$w_0^2 = \frac{L_{\text{crystal}} \lambda}{2 \pi n \xi}, \tag{7.84}$$

where L_{crystal} is the crystal length. Equation 7.84 gives the desired beam waist size for the resonator, and based on the numbers for our example, the spot size is $w_0 = 46$ μm. Next we need to come up with resonator parameters that result in a beam waist size of 46 μm. A number of possible cavity mirror focal lengths and cavity lengths give the desired spot size. Typically, we choose a convenient focal length for the mirrors, where the deciding factor may be availability or cost. For this example, assume that the focal length of the mirrors is 5 cm (10 cm radius of curvature). Substituting numerical values into Equation 7.83 gives two values of L_{eff}, but only one of the values is experimentally realizable, $L_{\text{eff}} = 9.98$ cm. Equation 7.79 lets us calculate the distance between the crystal and the cavity mirrors (d in Figure 7.9), giving $d = 8.73$ cm.

Further consideration of the cavity design should include a stability analysis. As can be found in most laser textbooks, a stable cavity requires that $-1 < (A + D)/2 < 1$, where A and D are the elements of the ABCD matrix corresponding to a complete round trip in the cavity. Note that the ABCD matrix given in Equation 7.78 is only one half of a round trip, and the matrix should be squared before calculating the stability condition. For our example, $(A_{\text{round trip}} + D_{\text{round trip}})/2 = 0.98$, which is on the edge of stability. A slight increase in the cavity length takes this cavity out of the stable operation region, making for tricky alignment. Moving the cavity mirrors closer together increases the focused spot size and leads to a more stable cavity. Increasing the spot size, or decreasing ξ, is a reasonable trade-off since the OPO threshold is not a strong function of ξ (see Figure 7.8).

7.3.7 PULSED OPOs

Pulsed OPOs operating in the nanosecond regime do not have stringent resonator mode restrictions. In fact, many OPOs, using a nanosecond pulse width pump, operate with unstable resonators, such as a simple plane-parallel mirror configuration. Consider an Nd:YAG laser with a 10-ns pulse duration and 10 mJ of energy per pulse. The peak power of this laser is roughly 1 MW, easily exceeding the OPO threshold even when the cavity mirrors are simple plane-parallel mirrors. The difficulty with high-pulse-energy-pumped OPOs is avoiding crystal damage, which forces large beam sizes when compared to stable resonators. With a large pump beam and an unstable resonator, a challenge in OPO operation is controlling the OPO's output divergence and bandwidth. Injection seeding an OPO with a narrow-bandwidth laser that has a wavelength within the signal's phase-matching bandwidth is one strategy to overcome these challenges.

As the pump pulse duration decreases, we need to pay attention to the number of round trips that the resonated signal actually overlaps with the pump. A 1-ns pulse corresponds to a 30-cm spatial extent in air, so if the OPO cavity round-trip optical path length is greater than 30 cm, then on the second pass through the resonator, the signal experiences no overlap with the pump and hence sees no gain. With even shorter pulses in the picosecond and femtosecond regimes, OPOs are synchronously pumped. In a synch-pumped OPO, the pump pulse repetition rate is matched to the round-trip time of the signal in the OPO cavity. Hence, on every round trip that the signal pulse travels through the crystal, it overlaps a pump pulse. The limitation on synch-pumping is the cavity length of the OPO resonator. For example, when the pump repetition rate is 100 MHz, the OPO cavity must have a round-trip optical path length of 3 m. Lower repetition rates mean longer cavity lengths, which at some point become impractical. Because of the relatively long cavity lengths, stable cavity designs are employed, and the previous discussion on mode matching the pump to the cavity mode applies.

Synchronously pumped ultrashort-pulse OPOs require additional intracavity optics to compensate for material dispersion. Without the compensation, short pulses (100 fs and shorter) do not form. A common compensating scheme is a two-prism sequence, as depicted in Figure 7.10. The prisms are cut at Brewster's angle to

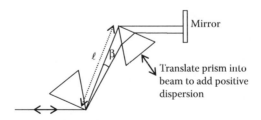

FIGURE 7.10   Two-prism sequence pulse compressor. An optical pulse incident from the left has its spectrum dispersed over an angular width β by the first prism and collimated by the second prism. The same dispersion is added to the pulse on the reverse pass through the prism pair. Overall dispersion is adjustable by the prism separation, ℓ, and by the amount of prism glass added to the sequence.

minimize the insertion loss, and the prism sequence by itself is a pulse compressor (or stretcher). Light entering the first prism is deflected according to the refractive index of each wavelength. The angular spread of the beam is denoted by β in Figure 7.10. Longer wavelengths are deflected less than shorter wavelengths, and in the space between the prisms, the longer wavelengths travel a longer optical path. Hence, longer wavelengths are delayed more than shorter wavelengths, which constitutes anomalous dispersion. The effect of the prism pair is determined by its effect on group-velocity dispersion (GVD). The GVD of an optical system is related to the second-order dispersion, defined as the second derivative with respect to the angular frequency of the cavity phase, $\phi = (2\pi/\lambda)P$

$$\frac{d^2\phi}{d\omega^2} = \frac{\lambda^3}{2\pi c^2} \frac{d^2 P}{d\lambda^2}$$

(7.85)

where P is the optical path length. In a slab of optical material of length L and index n, the optical path length is given by $P = nL$. However, for the prism sequence shown in Figure 7.10, the optical path length is related to the angular spread of the beam; the dispersion is calculated as (Fork et al., 1984)

$$\frac{d^2 P}{d\lambda^2} \approx 8 \left[\frac{d^2 n}{d\lambda^2} + \left(2n - \frac{1}{n^3} \right) \left(\frac{dn}{d\lambda} \right)^2 \right] h - 8\ell \left(\frac{dn}{d\lambda} \right)^2 ,$$

(7.86)

where h is the beam diameter at the second prism and ℓ is the tip-to-tip distance between the prisms. Equation 7.86 shows that the overall dispersion is negative for sufficiently large prism separations. The calculation assumes tip-to-tip propagation, so positive dispersion is added to the sequence when the optical path encounters more prism glass. By translating one or more of the prisms further into the beam, more positive dispersion is added to the sequence. With a long enough distance between the prisms, the geometric and material dispersions combine to give an overall net positive or negative dispersion, depending on the amount of prism glass added to the sequence.

When inserted into an OPO cavity, the prism sequence adds either positive or negative dispersion. Normally, the dispersion from the nonlinear crystal and the cavity mirrors is net positive, so the prism sequence is adjusted to compensate with negative dispersion. As an example, consider an OPO operating at 1.3 μm with a 1-mm thickness KTP crystal and an intracavity SF10 prism pair. The Sellmeier equations for both KTP and SF10 are found in Appendix B. The Sellmeier equations for KTP give

$$\frac{d^2 n_{KTP}}{d\lambda^2} \approx 0.02 \, \mu m^{-2}.$$

(7.87)

Hence, for a 1-mm crystal alone,

$$\frac{d^2 P_{KTP}}{d\lambda^2} \approx 20\ \mu m^{-1}. \tag{7.88}$$

Before calculating the prism spacing, we estimate the amount of extra glass for the prism sequence. Typically, the OPO beam is not exactly at the tips of the prism and a reasonable estimate is a total of 6 mm of extra glass. The Sellmeier equations for SF10 give

$$\frac{d^2 n_{SF10}}{d\lambda^2} \approx 0.016\ \mu m^{-2}. \tag{7.89}$$

For the 6-mm path length,

$$\frac{d^2 P_{SF10}}{d\lambda^2} \approx 96\ \mu m^{-1}. \tag{7.90}$$

Therefore, the total positive dispersion in the cavity is $\lambda^3/2\pi c^2$ multiplied by

$$\frac{d^2 P_{total}}{d\lambda^2} \approx +116\ \mu m^{-1}. \tag{7.91}$$

The prism sequence is designed so that the net dispersion of the cavity is zero, which is the case when

$$\frac{d^2 P_{prism\,sequence}}{d\lambda^2} \approx -116\ \mu m^{-1}. \tag{7.92}$$

The prism separation is calculated by substituting Equation 7.92 into Equation 7.86, and by estimating the beam size h. For a typical synch-pumped OPO, as shown in Figure 7.11, the beam size at the second prism is ~1 mm. Solving Equation 7.86 for the prism separation gives $\ell \approx 11$ cm. In practice, the prism separation may be increased

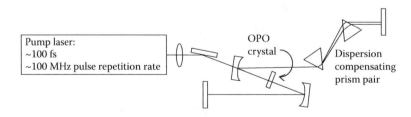

FIGURE 7.11 Synchronously pumped OPO with an intracavity prism pair to provide dispersion compensation. The cavity length of the OPO is matched to the pulse repetition rate of the pump laser.

to allow for uncertainties in the dispersion values and the beam size. Once the prism spacing is set, the duration of the OPO pulses is adjusted by adding more or less glass, as shown in Figure 7.10. By monitoring the OPO output with a pulse-width-measuring device such as an autocorrelator or a frequency-resolved optical gate (FROG), the amount of glass is adjusted for minimum pulse duration. If a minimum is not reached, then the prism spacing needs to be adjusted. An important consideration for achieving the shortest possible pulses is that higher-order dispersion is ignored in the calculations shown here.

7.3.8 BACKWARD OPOS

A different type of OPO is one where the signal and the idler are antiparallel, called a backward OPO (BOPO). The backward-wave configuration leads to an oscillation even when cavity mirrors are not present. The BOPO was first predicted in 1966 (Harris, 1966) and first demonstrated in 2007 (Canalias and Pasiskevicius, 2007). The BOPO phase-matching picture is shown in Figure 7.12. Satisfying $\Delta k = 0$ requires an exceptionally large birefringence for most signal–idler pairings. However, when the idler wavelength becomes large, such that its k-vector is small, then birefringent phase matching is theoretically possible. An alternative is quasi-phase matching, which requires a short poling periodicity. The first BOPO, based on periodically poled KTP, used a QPM period of 800 nm.

The backward-wave analysis is a straightforward solution to the coupled amplitude equations with appropriate boundary conditions. The derivations for the backward-wave case are similar to the calculations for OPA found in Section 6.2. Consider an undepleted pump scenario in a crystal with counterpropagating signal and idler beams and where the signal and the pump propagate in the same direction. The boundary conditions are the signal field incident from the left, $A_S(0)$, and since the idler is propagating in the backward direction, the other boundary condition is the idler input from the right, $A_I(L)$. Assume that the process is phase matched, then the coupled amplitude equations are written as

$$\frac{dA_S}{dz} = i\frac{\omega_S}{n_S c} d_{\text{eff}} A_P A_I^* = i\kappa_S A_I^*, \tag{7.93}$$

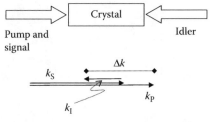

FIGURE 7.12   Phase matching for a BOPO.

$$\frac{dA_I}{dz} = -i\frac{\omega_I}{n_I c} d_{\text{eff}} A_P A_S^* = i\kappa_I A_S^*. \tag{7.94}$$

The minus sign in the Equation 7.94 comes about because the idler k-vector is traveling in the $-z$ direction (in Equation 2.120 $k \to -k$). Taking the derivative of the first equation leads to

$$\frac{d^2 A_S}{dz^2} = -\kappa_S \kappa_I^* A_S. \tag{7.95}$$

The solution to Equation 7.95 is of the form

$$A_S \sim e^{mz}. \tag{7.96}$$

Substituting Equation 7.96 into Equation 7.95 gives the characteristic equation:

$$m = \pm i\sqrt{\kappa_S \kappa_I^*}. \tag{7.97}$$

The characteristic equation has imaginary roots; hence, the solutions to the coupled amplitude equations are

$$A_S(z) = Ce^{i\Gamma z} + De^{-i\Gamma z}, \tag{7.98}$$

$$A_I(z) = Ee^{i\Gamma z} + Fe^{-i\Gamma z}, \tag{7.99}$$

where $\Gamma^2 = \kappa_S \kappa_I^*$ is the same as in Equation 7.37. Applying the boundary conditions, let us solve C, D, E, and F. The boundary conditions at $z = 0$ are

$$C + D = A_S(0), \tag{7.100}$$

$$\left.\frac{dA_I}{dz}\right|_{z=0} = i\kappa_I A_S^*(0) = i\Gamma(E - F). \tag{7.101}$$

The boundary conditions at $z = L$ are

$$Ee^{i\Gamma L} + Fe^{-i\Gamma L} = A_I(L), \tag{7.102}$$

$$\left.\frac{dA_S}{dz}\right|_{z=L} = i\kappa_S A_I^*(L) = i\Gamma(Ce^{i\Gamma L} - De^{-i\Gamma L}). \tag{7.103}$$

Solving for the coefficients gives the solution

$$A_S(z) = \frac{1}{\cos(\Gamma L)}\left[A_S(0)\cos(\Gamma(L-z)) + \frac{\kappa_S A_I^*(L)}{\Gamma}\sin(\Gamma z)\right].$$ (7.104)

The solution is somewhat different from earlier ones in that we do not have an exponential gain for the field. Rather, the field amplitude oscillates with distance. Interestingly, the amplitude of the oscillation can be large, depending on the magnitude of $\cos(\Gamma L)$. Note that as ΓL approaches $\pi/2$, the signal field, $A_S(z)$, diverges, which corresponds to an oscillation threshold much like the threshold with a traditional OPO. Calculating a BOPO threshold condition in terms of intensity or power follows from setting $\Gamma L = \pi/2$. Although the linearized solution blows up, the field amplitude remains bounded because the pump depletion has to be taken into account. The fundamental difference between a BOPO and an OPO is that the oscillation occurs in a BOPO without mirrors to provide feedback.

PROBLEMS

7.1 The reduction factors g_1, g_2, and g_3 given in Equations 7.15 through 7.17 are the result of an overlap integral. For example,

$$g_1 = \frac{\displaystyle\int_0^\infty 2\pi r e^{-r^2\left((1/w_1^2)+(1/w_2^2)+(1/w_3^2)\right)}dr}{\displaystyle\int_0^\infty 2\pi r e^{-(2r^2/w_1^2)}dr}.$$ (7.105)

Perform the integrations to verify that the gain reduction factors are as given in Equations 7.15 through 7.17.

7.2 Calculate the gain reduction factors g_1, g_2, and g_3 for (a) a DFG process where the input beam sizes are w_1 and w_2 (corresponding to the beams with frequencies ω_1 and ω_2, respectively) and (b) a sum-frequency process where the input sizes are w_2 and w_3 (corresponding to the beams with frequencies ω_2 and ω_3, respectively).

7.3 (a) Use the coupled wave equation for Gaussian beams given in Equations 7.12 through 7.14 to derive expressions for the on-axis intensities $I_{20}(z)$ and $I_{30}(z)$ given $\Delta k = 0$ (see parallel derivation in Section 6.2). Assume that the on-axis intensity of the pump $I_{10}(z)$ is a constant and that $I_{30}(0) = 0$. (b) Consider the case when the two inputs are confocally focused into the crystal with beam sizes of w_1 and w_2, respectively. Derive an expression for $P_2(z) - P_2(0)$, where $P_2(z)$ is the output power of the amplifier at ω_2. Assume that the gain of the amplifier is low so that we may approximate $\cosh(x) \sim 1 + x^2$.

7.4 Consider a second-harmonic process with a fundamental frequency of ω_F and with a Gaussian beam shape with a field radius of w_F. The fundamental input is confocally focused with respect to the SHG crystal. Verify that the output SHG power is given by Equation 7.29.

7.5 A 1-cm crystal is cut such that a fundamental wavelength of 1.064 μm is frequency doubled when it is normally incident on the crystal. The crystal cut accommodates a critically phase-matched Type I interaction (o + o → e). The SHG has a walk-off angle of 1° with respect to the fundamental input and the index of refraction is 1.7 for the fundamental and second harmonic. What is the optimum focusing parameter and hence the optimum spot size, w_F, for the interaction? (*Hint*: Use Figure 7.2.)

7.6 Consider a pulsed laser with a Gaussian beam profile that has a repetition rate of f_{rep} and an average output power of 1 W. The laser is confocally focused ($\xi = 1$) in a 5-cm crystal. If the crystal has a damage threshold of 2 J/cm², what is the lowest repetition rate possible without damaging the crystal? The laser wavelength is 1.064 μm, the pulse duration is 10 ns, and the index of refraction of the crystal is 1.5. Assume that the laser's temporal profile is rectangular, that is, when it is on, it has a constant power.

7.7 Section 7.3 shows that the conversion efficiency of an SHG process is enhanced by resonating the fundamental and coupling out the SHG. What would happen if the resonator mirrors did not completely couple out the SHG such that some of the SHG is fed back to the input of the crystal?

7.8 Prove that the lowest threshold for plane-wave OPOs occurs for a signal idler pairing at degeneracy.

7.9 Prove that, for a fixed signal wavelength, the plane-wave OPO threshold decreases as the pump wavelength decreases.

7.10 What is the signal–idler wavelength pair that gives the lowest threshold for a confocally focused SR-OPO (see Equation 7.75). How does this compare with the plane-wave case (Problem 7.8)?

7.11 Calculate the threshold for a DR-OPO (Gaussian beams) for a 2-cm crystal, $\lambda_P = 0.9$ μm, $\lambda_S = 1.5$ μm, $d_{eff} = 1.2$ pm/V, and a mirror reflectivity of 0.9. Assume that the signal and idler are confocally focused and that $n = 1.7$ for all wavelengths.

7.12 Calculate the threshold for a confocally focused DR-OPO for a 5-cm crystal with $\lambda_P = 1.064$ μm, $\lambda_S = 1.5$ μm, $n = 2.2$, $L = 5$ cm, and $d_{eff} = 16$ pm/V. The mirror reflectivity for the signal is 0.95 and the reflectivity for the idler is 0.6.

7.13 In a DR-OPO using a stable resonator, the beam sizes of the signal and the idler (w_S and w_I) are determined by the resonator cavity. Prove that the optimum pump beam size is given by Equation 7.64. Given this optimized pump beam size, show that the optimum W^2 is given by Equation 7.65.

7.14 For a Gaussian beam SR-OPO, the signal beam and pump beam sizes are fixed. The signal size is determined by the resonator and the pump size is determined by focusing optics. Show that the idler size given by Equation 7.73 minimizes the threshold.

7.15 Verify that when the pump and signal beams are confocally focused in an SR-OPO, the threshold is given by Equation 7.75.

7.16 When the cavity length of a two-mirror DR-OPO increases, the resonant modes shift to longer wavelengths. Use energy conservation (assuming the pump wavelength is fixed) to show that it is impossible for both the signal and idler frequencies to both tune with the cavity modes as the cavity length changes.

7.17 What mirror reflectivity is required for an SR confocally focused OPO given a 5-cm crystal with $\lambda_P = 1.064$ μm, $\lambda_S = 1.5$ μm, $n = 2.2$, $L = 5$ cm, $d_{eff} = 16$ pm/V, and a pump power of 10 W.

7.18 A laser with $w_0 = 500$ μm located 5 cm outside of its exit port needs to be focused to a 90-μm spot size at the center of the OPO crystal shown in Figure 7.13. A simple way to determine an appropriate focusing lens and lens position is to trace the beam from the center of the cavity back toward the laser. At the location where the backward-propagated beam has $w = 500$ μm, we place a lens of the appropriate focal length to collimate the beam (form a beam waist). This procedure turned around then takes the collimated output of the laser and focuses it to the correct beam size in the cavity. Working from the center of the cavity outward, and using Gaussian ABCD beam propagation (see Section 7.3.6), calculate the beam radius, $w(x)$. The crystal length is 5 cm and it has an index of refraction of 1.5. The cavity mirrors are plano-concave with a concave radius of curvature of $R = 10$ cm. Treat the mirror as a thin lens, using $1/f = (1 - n)/|R|$, where n is the index of refraction, which is equal to 1.5. The pump laser wavelength is 1.064 μm. What value of x (see Figure 7.13) gives $w = 500$ μm? What lens focal length should be placed at this location to attain proper pump focusing? Note that to collimate a beam, the focal length matches the beam's radius of curvature (see Problem 7.19).

7.19 Show that a diverging beam with a radius of curvature, R, is collimated by a lens of focal length $f = R$.

7.20 Consider an SR-OPO as shown in Figure 7.9 with a crystal length of 2 cm and a crystal index of 1.7. The desired focusing condition for the resonated signal (at 1.5 μm) is $\xi = 2.0$. What w_0 is required for this focusing condition? What mirror-to-crystal distance gives an OPO cavity mode with a beam waist of w_0 at the center of the crystal? The mirror radius of curvature is 10 cm.

7.21 For the cavity shown in Figure 7.14, the beam waist must lie at the center of the crystal ($n = 2.0$). Use the self-consistent analysis of Section 7.3.6 to find w_0, given that $\lambda = 1.064$ μm.

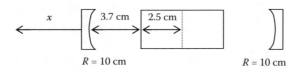

FIGURE 7.13   OPO cavity for Problem 7.18.

FIGURE 7.14   OPO cavity for Problem 7.21.

7.22 Consider an OPO cavity designed for ultrashort-pulse operation, as shown in Figure 7.11. The OPO crystal is a 1-mm KTP crystal. The resonated signal beam is o-polarized such that its index of refraction in the KTP crystal is $n_y(\lambda_S)$. Given a signal wavelength of 1.4 µm and a beam size of $w \sim 500$ µm, what is the minimum prism spacing (SF10 prisms) that will obtain a net zero GVD for the cavity? Assume that the total path length in the prism sequence is 6 mm.

7.23 Verify the GVD formula given in Equation 7.85. Using the Sellmeier equation for the refractive indices of fused silica and of SF10 in Appendix B, calculate $d^2n/d\lambda^2$ over a range of wavelengths from 1 to 1.6 µm. The dispersion in fused silica becomes anomalous; note the anomalous regime from the wavelength where the dispersion vanishes.

7.24 Calculate the BOPO threshold intensity for a 1-cm crystal with $n \sim 2.2$, and $d_{eff} = 16$ pm/V. The BOPO pump wavelength is 0.780 µm and the signal and idler wavelengths are at the degenerate wavelength, 1.560 µm.

7.25 Calculate the QPM grating periodicity required in lithium niobate to design a BOPO interaction for a pump wavelength of 0.780 µm and a signal wavelength of 1.5 µm. Assume an e → e + e interaction so that $n = n_z(\lambda)$.

REFERENCES

Boyd, G. D. and D. A. Kleinman. 1968. Parametric interaction of focused Gaussian light beams. *Journal of Applied Physics* 39: 3597–3639.

Canalias, C. and V. Pasiskevicius. 2007. Mirror-less optical parametric oscillator. *Nature Photonics* 1: 459–462.

Fork, R. L., O. E. Martinez, and J. P. Gordon. 1984. Negative dispersion using pairs of prisms. *Optics Letters* 9: 150–152.

Harris, S. E. 1966. Proposed backward wave oscillation in the infrared. *Applied Physics Letters* 9: 114–116.

FURTHER READING

Bloembergen, N. 1965. *Nonlinear Optics*. New York, NY: W. A. Benjamin, Inc.

Giordmaine, J. A. and R. C. Miller. 1965. Tunable coherent parametric oscillation in LiNbO₃ at optical frequencies. *Physical Review Letters* 14: 973–976.

Harris, S. E. 1969. Tunable optical parametric oscillators. *Proceedings of the IEEE* 12: 2096–2113.

Rabin, H. and C. L. Tang (Eds). 1975. *Quantum Electronics: A Treatise*. New York, NY: Academic Press.

Yariv, A. 1975. *Quantum Electronics* (2nd edition). New York, NY: John Wiley & Sons.

8

$\chi^{(3)}$ PROCESSES

8.1 INTRODUCTION

This chapter extends our discussions of optical effects to those due to third-order susceptibilities $\chi^{(3)}$. The occurrence of third-order nonlinear phenomena is far less restrictive than for $\chi^{(2)}$ processes discussed in the previous chapters; $\chi^{(3)}$ processes occur to some degree in all materials. Historically, the effect was first reported in 1875 by the Scottish physicist, John Kerr, who discovered materials with induced birefringence that exhibits a quadratic dependence on the applied field. The mechanism for the $\chi^{(3)}$ nonlinearity is attributed to distortions in the electron cloud surrounding an atom or molecule. A simple, classical model for the effect is presented in Chapter 3 using the anharmonic oscillator.

Throughout the rest of this chapter, we investigate $\chi^{(3)}$ processes that are electronic in origin. The analysis methods we develop here also apply to processes where $\chi^{(3)}$ has a different origin, some of which are treated in Chapter 9. $\chi^{(3)}$ parametric interactions give rise to a host of frequency conversion processes analogous to $\chi^{(2)}$ interactions. However, a particularly important distinction is that $\chi^{(3)}$ interactions also include self-induced effects, which are interactions where a nonlinear polarization is generated at the same frequency as the input. The corresponding nonlinear phase shifts lead to phenomena such as self-focusing, modulation instability, and solitons.

Section 8.2 introduces the nonlinear polarization for $\chi^{(3)}$ processes, which are proportional to the cube of the input field. Following this formalism, in Section 8.3, we derive the envelope wave equations for $\chi^{(3)}$ interactions, which is similar to that derived for $\chi^{(2)}$. The wave equation forms the foundation to describe several $\chi^{(3)}$ effects. The first process we discuss in Section 8.4 is the nonlinear index of refraction and associated phenomena of self-focusing and nonlinear absorption. We also show how a $\chi^{(3)}$ material can be used in a resonator geometry to make devices that can have two (or more) stable output intensities for the same input intensity; a phenomenon called optical bistability.

Section 8.5 describes third-order parametric amplifiers, which describes the amplification of an input signal as well as the generation of new frequencies. The parametric amplifier can lead to mirrorless resonators as we discuss in Section 8.7 on degenerate four-wave mixing (DFWM). Another unusual and interesting effect we study is called optical-phase conjugation, where the reflected field has a phase that can be described as a time-reversed version of the input wave's phase. Optical-phase conjugation is demonstrated using a special DFWM arrangement. Section 8.8 discusses a technique called z-scan, which is a method to measure the real and imaginary coefficients of the $\chi^{(3)}$ nonlinearity of a medium.

8.2 NONLINEAR POLARIZATION FOR $\chi^{(3)}$ PROCESSES

8.2.1 DEFINING RELATIONSHIPS

The extension from $\chi^{(2)}$ processes to higher-order parametric processes is a straightforward one. We consider the case where the nonlinear polarization is separable:

$$P = P^{(1)} + P^{\text{NL}}. \tag{8.1}$$

By splitting the polarization into linear and nonlinear parts, we derive a linear wave equation and add the nonlinear contribution to it. The linear part of the wave equation is the same as in Chapter 4 where the plane-wave SVEA was expressed with a general nonlinear polarization. The SVEA gives an equation in terms of complex amplitude; Equation 4.18 is repeated here:

$$2ik \frac{dA}{dz} e^{ikz} = -\omega^2 \mu_0 P^{\text{NL}}. \tag{8.2}$$

Now, instead of the nonlinearity originating with $\chi^{(2)}$, we consider $\chi^{(3)}$ effects. In this chapter, we assume that the origin of $\chi^{(3)}$ is parametric (no energy exchange with the medium), but the techniques we develop also apply to processes where $\chi^{(3)}$ is nonparametric. The nonlinear polarization for $\chi^{(3)}$ processes is proportional to the total input field cubed (see Equation 3.3), which gives rise to many possible output frequencies. Table 8.1 lists the frequencies present after cubing a field that has three frequency components at the input: ω_2, ω_3, and ω_4. Fewer distinct frequencies may be present depending on the input frequencies. For example, if the three inputs are 3ω, 2ω, and ω, then only 10 frequencies are present in the nonlinear polarization.

Each of the entries listed in Table 8.1 potentially has a different nonlinear susceptibility. Before writing a general expression for the nonlinear polarization,

TABLE 8.1   Frequencies Present in the Nonlinear Polarization for an $\chi^{(3)}$ Process When Three Frequencies, ω_2, ω_3, and ω_4, are Present at the Input

Multiplicity	New Frequencies		
1	$3\omega_2$	$3\omega_3$	$3\omega_4$
3	$2\omega_2 + \omega_3$	$2\omega_2 + \omega_4$	$\omega_2 + 2\omega_4$
	$2\omega_2 - \omega_3$	$2\omega_2 - \omega_4$	$\omega_2 - 2\omega_4$
	$2\omega_4 + \omega_3$	$2\omega_3 + \omega_2$	$2\omega_3 + \omega_4$
	$2\omega_4 - \omega_3$	$2\omega_3 - \omega_2$	$2\omega_3 - \omega_4$
	$\omega_2 + \omega_2 - \omega_2$	$\omega_3 + \omega_3 - \omega_3$	$\omega_4 + \omega_4 - \omega_4$
6	$\omega_2 + \omega_3 + \omega_4$	$\omega_2 + \omega_3 - \omega_4$	$\omega_2 + \omega_4 - \omega_3$
	$\omega_3 + \omega_4 - \omega_2$		
	$\omega_2 + \omega_3 - \omega_3$	$\omega_2 + \omega_4 - \omega_4$	$\omega_3 + \omega_2 - \omega_2$
	$\omega_3 + \omega_4 - \omega_4$	$\omega_4 + \omega_2 - \omega_2$	$\omega_4 + \omega_3 - \omega_3$

Note: The multiplicity is the number of times the particular term appears in E^3. Pairings such as $\omega_2 + \omega_3 - \omega_3$ are not simplified; they are written to show their origin.

let us consider a few specific examples from Table 8.1 to see how degeneracy is included in the polarization expression. To simplify our discussion, we consider a situation where the fields and nonlinear polarization point in the same direction so that we treat them as scalars. As a first example, consider a nonlinear polarization at $\omega_1 = \omega_2 + \omega_3 + \omega_4$, which has a complex amplitude proportional to the component of the field cubed at ω_1,

$$P^{(3)}(\omega_1 = \omega_2 + \omega_3 + \omega_4) \sim 6\chi^{(3)}(\omega_1; \omega_2, \omega_3, \omega_4) A(\omega_2) A(\omega_3) A(\omega_4) e^{i(k_2 + k_3 + k_4)z}. \quad (8.3)$$

The factor of 6 indicates that this particular frequency, $\omega_2 + \omega_3 + \omega_4$, occurs 6 times when we cube the field (see multiplicity factor in Table 8.1). We use the same notation for $\chi^{(3)}$ as in Chapter 3 where the first frequency in $\chi^{(3)}$ ($\omega_1; \omega_2, \omega_3, \omega_4$) corresponds to the nonlinear polarization and the other three correspond to the inputs such that the first frequency within brackets is the sum of the other three.

Similarly, consider a second example with a frequency combination of $2\omega_2 + \omega_3$, which occurs 3 times (has a multiplicity of 3) so that its nonlinear polarization complex amplitude is proportional to

$$P^{(3)}(\omega_{new} = 2\omega_2 + \omega_3) \sim 3\chi^{(3)}(\omega_{new}; \omega_2, \omega_2, \omega_3) A^2(\omega_2) A(\omega_3) e^{i(2k_2 + k_3)z}. \quad (8.4)$$

Note that two of the inputs for this interaction are indistinguishable.

A third example of a possible nonlinear polarization is the case for $3\omega_2$, which occurs only once when we cube the field resulting in

$$P^{(3)}(\omega_{\text{new}} = 3\omega_2) \sim \chi^{(3)}(\omega_{\text{new}}; \omega_2, \omega_2, \omega_2) A^3(\omega_2) e^{i3k_2 z}. \quad (8.5)$$

Note that the three fields considered here are indistinguishable.

A pattern emerges from Equations 8.3 through 8.5; the multiplicity is given by the number of distinct permutations of the three input fields. In Equation 8.3, all three fields are different giving six distinct permutations. In Equation 8.4, two fields are indistinguishable so that the number of distinct permutations is 3. Equation 8.5, where all three fields are indistinguishable, gives only one unique combination of fields.

The multiplicity tells us the number of occurrences for a specific frequency combination. In some cases, different frequency combinations lead to the same output frequency. When two or more of the frequencies in Table 8.1 are the same (degenerate), then the nonlinear polarization for that frequency includes all the different degenerate combinations and their multiplicity factors. Consider the case where the frequency of interest is ω_2; then, the nonlinear polarization is proportional to

$$P^{(3)}(\omega_2) \sim \begin{bmatrix} 3\chi^{(3)}(\omega_2; \omega_2, \omega_2, -\omega_2)|A(\omega_2)|^2 A(\omega_2) e^{ik_2 z} \\ + 6\chi^{(3)}(\omega_2; \omega_2, \omega_3, -\omega_3)|A(\omega_3)|^2 A(\omega_2) e^{ik_2 z} \\ + 6\chi^{(3)}(\omega_2; \omega_2, \omega_4, -\omega_4)|A(\omega_4)|^2 A(\omega_2) e^{ik_2 z} \end{bmatrix}. \quad (8.6)$$

We see that the nonlinear polarization has three separate frequency combinations that lead to ω_2; for each combination, we include the multiplicity factor. Other terms may need to be added if other sum- or difference-frequency combinations yield a nonlinear polarization at ω_2.

In addition to the different frequency combinations, if the fields and nonlinear polarization point in different directions, then we also need to consider couplings that involve different combinations of field directions. We account for the different directional couplings by writing the susceptibility as a Rank 4 tensor so that the third-order nonlinear polarization (real) is written as

$$P_i^{(3)}(t) = \frac{\varepsilon_0}{8} \sum_{n,m,p} \sum_{j,k,l} \chi_{ijkl}^{(3)}(\omega_m + \omega_n + \omega_p; \omega_n, \omega_m, \omega_p) A_j(\omega_m) A_k(\omega_n) A_l(\omega_p) e^{i(k_m + k_n + k_p)z}$$

$$\times e^{-i(\omega_m + \omega_n + \omega_p)t}, \quad (8.7)$$

where, for example, n, m, and p range over $-4, -3, -2, 2, 3, 4$ when the input frequencies are ω_2, ω_3, and ω_4. i, j, k, and l are the Cartesian coordinates. Similar to the approach in Chapter 3, we simplify the nonlinear polarization by choosing a specific frequency of interest, and we take advantage of permutation symmetries. We justify the approach of picking one frequency component of the nonlinear polarization by noting that the interactions are weak, and therefore we treat the different frequencies as independent of each other. However, we need to make one exception to this rule. As shown in Table 8.1, self-induced effects are possible with $\chi^{(3)}$ interactions, where the nonlinear polarization is at the same frequency as the input frequency. In some cases, we need to include the self-induced effects in addition to a frequency conversion process, as we discuss on a case-by-case basis in the coming sections.

8.2.2 PERMUTATION SYMMETRIES FOR $\chi^{(3)}$

The permutation symmetries for $\chi_{ijkl}^{(3)}$ are the same as in Chapter 3, and we briefly outline them here. The last three Cartesian indices of $\chi_{ijkl}^{(3)}$ may be freely permuted as long as we permute the frequencies with the indices. This statement is equivalent to saying that the order of the fields in Equation 8.7 does not matter. If we assume that the material is lossless at all frequencies involved, then we may freely permute all four Cartesian indices as long as we permute the corresponding frequencies. In this case, the permutation connects different processes. For example, the following sum- and difference-frequency processes have the same nonlinear susceptibility:

$$\chi_{ijkl}^{(3)}(\omega_1;\omega_2,\omega_3,\omega_4)=\chi_{jikl}^{(3)}(\omega_2;\omega_1,-\omega_3,-\omega_4)=\chi_{klij}^{(3)}(\omega_3;-\omega_4,\omega_1,-\omega_2). \tag{8.8}$$

The Kleinman symmetry may be invoked if the dispersion in $\chi^{(3)}$ is negligible. This symmetry allows us to permute the Cartesian indices without permuting the frequencies with them. If one or more of the frequencies is near an absorption feature of the material, then the Kleinman symmetry should be used with caution.

Following the pattern introduced above (starting at Equation 8.3), and by assuming a lossless medium, the complex amplitude of the nonlinear polarization is given by

$$P_i^{(3)}(\omega_1)=\frac{\varepsilon_0}{4}\sum_{j,k,l}D\chi_{ijkl}^{(3)}(\omega_2+\omega_3+\omega_4;\omega_2,\omega_3,\omega_4)A_j(\omega_2)A_k(\omega_3)A_l(\omega_4)e^{i(k_2+k_3+k_4)z}, \tag{8.9}$$

where D is a multiplicity factor given by

$$\begin{cases}1:\text{all fields in the sum are indistinguishable}\\3:\text{two fields in the sum are indistinguishable}\\6:\text{all fields in the sum are distinguishable}\end{cases} \tag{8.10}$$

Note that we treat $+\omega$ and $-\omega$ as distinct from each other so that, for example, $\omega_2+\omega_2-\omega_2$ has $D=3$ and $\omega_2+\omega_3-\omega_3$ has $D=6$. Also note that we consider $A_x(\omega)$ as distinct from $A_y(\omega)$ so that, for example, the field combination $A_x(\omega)A_y(\omega)A_z(\omega)$ has a degeneracy factor of 6.

8.2.3 SYMMETRY CONSIDERATIONS FOR CENTROSYMMETRIC MEDIA

Material symmetry considerations also apply to $\chi^{(3)}$. The fourth-order tensor is tedious to express its 81 components. However, like the $\chi^{(2)}$ tensor components, the number of independent components is reduced for different crystal classes. The list of independent components for crystals of different frequencies is given in Table 8.2. A significant difference between $\chi^{(2)}$ and $\chi^{(3)}$ processes is that $\chi^{(3)}$ effects occur in centrosymmetric media (symmetric under inversion) such as gases, liquids, amorphous solids (glass), and so on. We show that $\chi^{(3)}$ effects are allowed in centrosymmetric media by following the formalism introduced in Chapter 3, and we look at their symmetry properties.

Consider a coordinate transformation for an inversion:

$$R = \begin{bmatrix} -1 & 0 & 0 \\ 0 & -1 & 0 \\ 0 & 0 & -1 \end{bmatrix}. \tag{8.11}$$

In component form, Equation 8.11 is

$$R_{ij} = -\delta_{ij}, \tag{8.12}$$

where δ_{ij} is the Kronecker delta (see Equation 2.53). Since $\chi^{(3)}_{ijkl}$ is a tensor, it transforms according to the rules given in Section 2.2:

$$\chi^{(3)'}_{ijkl}(\omega_1;\omega_2,\omega_3,\omega_4) = R_{i\alpha}R_{j\beta}R_{k\gamma}R_{\ell\nu}\chi^{(3)}_{\alpha\beta\gamma\nu}(\omega_1;\omega_2,\omega_3,\omega_4). \tag{8.13}$$

Note that this expression is written in the Einstein notation so that we sum over repeated indices. Also note that, unlike permutation symmetries, the frequency dependence of the susceptibility remains unchanged under a coordinate transformation since we are looking at the same interaction from a different vantage point. To simplify the discussion, we drop the frequency dependence for the rest of this section.

Substituting Equation 8.12 into Equation 8.13 gives

$$\chi^{(3)'}_{ijkl} = (-\delta_{i\alpha})(-\delta_{j\beta})(-\delta_{k\gamma})(-\delta_{l\nu})\chi^{(3)}_{\alpha\beta\gamma\nu}. \tag{8.14}$$

Hence,

$$\chi^{(3)'}_{ijkl} = \chi^{(3)}_{ijkl}. \tag{8.15}$$

TABLE 8.2 Third-Order Susceptibility Tensor Components

Triclinic 1 $\bar{1}$	81 independent elements	
Monoclinic 2 m 2/m	41 independent elements	3: *xxxx yyyy zzzz* 18: *xxyy xyxy xyyx*, etc. 12: *yyxz yxyz xyxz*, etc. 4: *xxxz xxzx xzxx zxxx* 4: *xzzz zxzz zzxz zzzx*
Orthorhombic 222 mm2 mmm	21 independent elements	3: *xxxx yyyy zzzz* 18: *xxyy xyxy xyyx*, etc.
Tetragonal 4 $\bar{4}$ 4/m	41 nonzero elements 21 independent elements	2: *xxxx = yyyy zzzz* 3: $xxyy = yyxx$ $xyxy = yxyx$ $xyyx = yxxy$ 6: $zzxx = zzyy$ $xxzz = yyzz$ $zxzx = zyzy$ $xzxz = yzyz$ $zxxz = zyyz$ $xzzx = yzzy$ 6: $xyzz = yxzz$ $zzxy = \overline{zzyx}$ $xzyz = \overline{yzxz}$ $zxzy = \overline{zyzx}$ $zxyz = \overline{zyxz}$ $xzzy = \overline{yzzx}$ 4: $xxxy = \overline{yyyx}$ $xxyx = \overline{yyxy}$ $xyxx = \overline{yxyy}$ $yxxx = \overline{xyyy}$
Tetragonal 422 4mm 4/mmm	21 nonzero elements 11 independent elements	2: *xxxx = yyyy zzzz* 9: $zzxx = zzyy$ $xxzz = yyzz$ $zxzx = zyzy$ $xzxz = yzyz$ $zxxz = zyyz$ $xzzx = yzzy$ $xxyy = yyxx$ $xyxy = yxyx$ $xyyx = yxxy$
Cubic 23 m3	21 nonzero elements 7 independent elements	1: *xxxx = yyyy = zzzz* 1: *xxyy = yyzz = zzxx* 1: *xxzz = yyxx = zzyy* 1: *xyxy = yzyz = zxzx* 1: *yxyx = zyzy = xzxz* 1: *xxyx = yzzy = zxxz* 1: *yxxy = zyyz = xzzx*
Cubic 432 $\bar{4}$3m m3m	21 nonzero elements 4 independent elements	1: *xxxx=yyyy=zzzz* 1: *xxyy = yyxx = xxzz*, etc. 1: *xyxy = yxyx = xzxz*, etc. 1: *xyyx = yxxy = zxxz*, etc.
Trigonal 3 $\bar{3}$	73 nonzero elements 27 independent elements	1: *zzzz* *xxxx = yyyy = xxyy + xyyx + xyxy* 3: *xxyy = yyxx* *xyxy = yxyx* *xyyx = yxxy* 12: $yyzz = xxzz$ $xyzz = \overline{yxzz}$ $zzyy = zzxx$ $zzxy = \overline{zzyx}$ $zyyz = zxxz$ $zxyz = \overline{zyxz}$ $yzzy = xzzx$ $xzzy = \overline{yzzx}$ $yzyz = xzxz$ $xzyz = \overline{yzxz}$ $zyzy = zxzx$ $zxzy = \overline{zyzx}$

(Continued)

TABLE 8.2 (*Continued*)   Third-Order Susceptibility Tensor Components

$$xxxy = \overline{yyyx} = yyxy + yxyy + xyyy$$

$$3: \quad yyxy = \overline{xxyx}$$

$$yxyy = \overline{xyxx}$$

$$xyyy = \overline{yxxx}$$

$$8: \quad yyyz = \overline{yxxz} = \overline{xyxz} = \overline{xxyz}$$

$$yyzy = \overline{yxzx} = \overline{xyzx} = \overline{xxzy}$$

$$yzyy = \overline{yzxx} = \overline{xzyx} = \overline{xzxy}$$

$$zyyy = \overline{zyxx} = \overline{zxyx} = \overline{zxxy}$$

$$xxxz = \overline{xyyz} = \overline{yxyz} = \overline{yyxz}$$

$$xxzx = \overline{xyzy} = \overline{yxzy} = \overline{yyzx}$$

$$xzxx = \overline{yzxy} = \overline{yzyx} = \overline{xzyy}$$

$$zxxx = \overline{zxyy} = \overline{zyxy} = \overline{zyyx}$$

Trigonal	37 nonzero elements	$1: zzzz$
$3m\ \bar{3}m\ 32$	14 independent elements	$xxxx = yyyy = xxyy + xyyx + xyxy$

$$3: \quad xxyy = \overline{yyxx} \quad xyxy = \overline{yxyx} \quad xyyx = \overline{yxxy}$$

$$6: \quad yyzz = \overline{xxzz} \quad zzyy = \overline{zzxx} \quad zyyz = \overline{zxxz}$$

$$yzzy = \overline{xzzx} \quad yzyz = \overline{xzxz} \quad zyzy = \overline{zxzx}$$

$$4: \quad xxxz = \overline{xyyz} = \overline{yxyz} = \overline{yyxz}$$

$$xxzx = \overline{xyzy} = \overline{yxzy} = \overline{yyzx}$$

$$xzxx = \overline{yzxy} = \overline{yzyx} = \overline{xzyy}$$

$$zxxx = \overline{zxyy} = \overline{zyxy} = \overline{zyyx}$$

Hexagonal	41 nonzero elements	$1: zzzz$
$6\ \bar{6}\ 6/m$	19 independent elements	$xxxx = yyyy = xxyy + xyyx + xyxy$

$$3: xxyy = \overline{yyxx} \quad xyxy = \overline{yxyx} \quad xyyx = \overline{yxxy}$$

$$12: yyzz = \overline{xxzz} \quad xyzz = \overline{yxzz}$$

$$zzyy = \overline{zzxx} \quad zzxy = \overline{zzyx}$$

$$zyyz = \overline{zxxz} \quad zxyz = \overline{zyxz}$$

$$yzzy = \overline{xzzx} \quad xzzy = \overline{yzzx}$$

$$yzyz = \overline{xzxz} \quad xzyz = \overline{yzxz}$$

$$zyzy = \overline{zxzx} \quad zxzy = \overline{zyzx}$$

$$xxxy = \overline{yyyx} = yyxy + yxyy + xyyy$$

$$3: \quad yyxy = \overline{xxyx}$$

$$yxyy = \overline{xyxx}$$

$$xyyy = \overline{yxxx}$$

Hexagonal	21 nonzero elements	$1: zzzz$
$622\ 6mm\ 6/mmm\ \bar{6}m21$	10 independent elements	$xxxx = yyyy = xxyy + xyyx + xyxy$

$$3: xxyy = \overline{yyxx} \quad xyxy = \overline{yxyx} \quad xyyx = \overline{yxxy}$$

$$6: yyzz = \overline{xxzz} \quad zzyy = \overline{zzxx} \quad zyyz = \overline{zxxz}$$

$$yzzy = \overline{xzzx} \quad yzyz = \overline{xzxz} \quad zyzy = \overline{zxzx}$$

(*Continued*)

TABLE 8.2 (*Continued*)   Third-Order Susceptibility Tensor Components

Isotropic	21 nonzero elements	1: $xxxx = yyyy = zzzz$
	3 independent elements	2: $xxxx = xxyy + xyyx + xyxy$
		$yyzz = zzyy = zzxx = xxzz = xxyy = yyxx$
		$yzyz = zyzy = zxzx = xzxz = xyxy = yxyx$
		$yzzy = zyyz = zxxz = xzzx = xyyx = yxxy$

Equation 8.15 shows that the third-order susceptibility is unchanged by an inversion of the coordinate system for any material. Hence, centrosymmetric media, which require that $\chi^{(3)}$ remain unchanged under an inversion operation, automatically allow $\chi^{(3)}$ interactions.

Not all elements of the $\chi^{(3)}_{ijkl}$ tensor are independent, and many of the elements are zero. In a centrosymmetric crystal, other symmetries are satisfied, such as mirror reflections in the yz-, xz-, and xy-planes. Consider a reflection in the yz-plane,

$$R = \begin{bmatrix} -1 & 0 & 0 \\ 0 & 1 & 0 \\ 0 & 0 & 1 \end{bmatrix}. \tag{8.16}$$

Under this transformation, any combination of $i, j, k,$ and l, where there are an odd number of indices with the Cartesian coordinate, x, results in

$$\chi^{(3)'}_{ijkl} = -\chi^{(3)}_{ijkl}. \tag{8.17}$$

But according to the symmetry condition, the susceptibility looks the same before and after the transformation; therefore, Equation 8.17 is satisfied only if the susceptibility is zero. A similar argument for reflection symmetries in the xz and xy-planes results in zero susceptibility for any combination of indices $i, j, k,$ and l with an odd number of the Cartesian coordinates $x, y,$ or z.

Still other simplifications for the centrosymmetric case arise from 90° rotations such as

$$R = \begin{bmatrix} 0 & 1 & 0 \\ -1 & 0 & 0 \\ 0 & 0 & 1 \end{bmatrix}. \tag{8.18}$$

We note a pattern for this operation by considering the transformation of the specific tensor element:

$$\chi^{(3)'}_{xxzz} = (R_{yx})(R_{yx})(R_{zz})(R_{zz})\chi^{(3)}_{yyzz} = \chi^{(3)}_{yyzz}, \tag{8.19}$$

where only the nonzero term in the sum 8.13 is shown. Since Equation 8.18 is also a symmetry operation for the material, $\chi_{xxzz}^{(3)} = \chi_{xxzz}^{(3)'} = \chi_{yyzz}^{(3)}$. By considering 90° rotations about the x-, y-, and z-axes, we obtain the following equalities:

$$\chi_{xxzz}^{(3)} = \chi_{zzxx}^{(3)} = \chi_{zzyy}^{(3)} = \chi_{yyzz}^{(3)} = \chi_{yyxx}^{(3)} = \chi_{xxyy}^{(3)}, \tag{8.20}$$

$$\chi_{xzzx}^{(3)} = \chi_{zxxz}^{(3)} = \chi_{zyyz}^{(3)} = \chi_{yzzy}^{(3)} = \chi_{yxxy}^{(3)} = \chi_{xyyx}^{(3)}, \tag{8.21}$$

$$\chi_{xzxz}^{(3)} = \chi_{zxzx}^{(3)} = \chi_{zyzy}^{(3)} = \chi_{yzyz}^{(3)} = \chi_{yxyx}^{(3)} = \chi_{xyxy}^{(3)}, \tag{8.22}$$

$$\chi_{xxxx}^{(3)} = \chi_{yyyy}^{(3)} = \chi_{zzzz}^{(3)}. \tag{8.23}$$

The $\chi_{ijkl}^{(3)}$ elements listed in Equations 8.20 through 8.23 show four independent susceptibilities. However, a final symmetry links the four to each other in an isotropic material. Consider an arbitrary rotation about the z-axis. This operation gives (see Problem 8.2)

$$\chi_{xxxx}^{(3)} = \chi_{xxyy}^{(3)} + \chi_{xyyx}^{(3)} + \chi_{xyxy}^{(3)}. \tag{8.24}$$

Therefore, isotropic materials have three independent susceptibility tensor elements. A similar exercise is carried out for materials with other symmetry properties; however, in this chapter, we generally treat $\chi^{(3)}$ effects in centrosymmetric media.

Note that the results based on inversion and other material symmetries are different from the permutation symmetries given in Section 8.2.2. For example, contrast the following two statements:

$$\chi_{xxyy}^{(3)}(\omega_1; \omega_2, \omega_3, \omega_4) = \chi_{yyxx}^{(3)}(\omega_1; \omega_2, \omega_3, \omega_4) \quad \text{Inversion symmetry,} \tag{8.25}$$

$$\chi_{xxyy}^{(3)}(\omega_1; \omega_2, \omega_3, \omega_4) = \chi_{yyxx}^{(3)}(\omega_3; -\omega_4, \omega_1, -\omega_2) \quad \text{Permutation symmetry.} \tag{8.26}$$

Note that with permutation symmetry, the frequencies are permuted with the Cartesian indices so that the equated susceptibilities refer to different processes.

8.3   WAVE EQUATION FOR $\chi^{(3)}$ INTERACTIONS

8.3.1   FOUR DISTINCT FREQUENCIES

The nonlinear polarization given in Equation 8.9 is inserted into Equation 8.2 to give a wave equation for a particular frequency. The nonlinear polarization couples multiple fields so that we obtain a set of coupled amplitude equations. To simplify the discussion and to see the general behavior, we assume that the fields and the nonlinear polarization are scalars so that we drop the Cartesian indices from $\chi^{(3)}$ and the fields.

As a first example, consider the nonlinear polarization at $\omega_1 = \omega_2 + \omega_3 + \omega_4$ that results from three linearly polarized superposed fields given by

$$\underline{E} = \frac{1}{2}\left(A_1 e^{i(k_1 z - \omega_1 t)} + A_2 e^{i(k_2 z - \omega_2 t)} + A_3 e^{i(k_3 z - \omega_3 t)} + A_4^{i(k_4 z - \omega_4 t)} + \text{c.c.}\right), \quad (8.27)$$

where we introduce the notation: $A_2 = A(\omega_2)$, and so on. When the field in Equation 8.27 is cubed, we must pay attention to all field combinations that yield a particular frequency. In the case of a nonlinear polarization at ω_1, we need to consider a total of five combinations:

$$\omega_1 = \omega_2 + \omega_3 + \omega_4,$$
$$\omega_1 = \omega_1 + \omega_1 - \omega_1,$$
$$\omega_1 = \omega_1 + \omega_2 - \omega_2, \quad (8.28)$$
$$\omega_1 = \omega_1 + \omega_3 - \omega_3,$$
$$\omega_1 = \omega_1 + \omega_4 - \omega_4.$$

The nonlinear polarization includes contributions from each combination. Assuming no dispersion in $\chi^{(3)}$, the nonlinear polarization is given by Equation 8.9 with degeneracy factors taken from Table 8.1:

$$P^{(3)}(\omega_1) = \frac{3\varepsilon_0 \chi^{(3)}}{4}\left[\begin{array}{l}(|A_1|^2 + 2|A_2|^2 + 2|A_3|^2 + 2|A_4|^2)A_1 e^{ik_1 z} \\ + 2A_2 A_3 A_4 e^{i(k_2 + k_3 + k_4)z}\end{array}\right]. \quad (8.29)$$

Equation 8.29 shows two types of contributions to the nonlinear polarization. The contributions proportional to $A_1 e^{ik_1 z}$ have the same spatial wavelength as a freely propagating field at ω_1. Hence, these terms are *automatically phase matched*. When $\chi^{(3)}$ is real valued, the contribution to the nonlinear polarization proportional to $A_1 e^{ik_1 z}$ does not lead to energy exchange between the fields, as can be confirmed using Poynting's theorem (see Problem 8.4) and as discussed in Section 8.3.2. The automatically phase-matched terms give rise to self-phase shifts and cross-phase shifts (see Section 8.4). With nonparametric interactions, these same types of combinations can lead to energy exchange between different optical fields (see Chapter 9).

The second type of contribution to the nonlinear polarization in Equation 8.29, given by

$$\frac{3\varepsilon_0 \chi^{(3)}}{2} A_2 A_3 A_4 e^{i(k_2 + k_3 + k_4)z}, \quad (8.30)$$

involves energy exchange between the different fields. This contribution to the nonlinear polarization has a spatial wavelength, determined by $k_2 + k_3 + k_4$, which may be different from the field at m_1, and this term requires phase matching in order to have an appreciable effect. The same techniques for phase matching applied to $\chi^{(2)}$ processes apply here, but four-wave processes typically are easier to phase match as discussed later in Section 8.6.

Equation 8.29 shows that all four fields are coupled to each other. Nonlinear polarizations at frequencies of ω_2, ω_3, and ω_4 have similar coupled expressions. Four coupled amplitude equations are found by substituting the nonlinear polarization for each frequency into Equation 8.2 resulting in

$$\frac{dA_1}{dz} = i\frac{3\omega_1\chi^{(3)}}{8n_1c}\left[\begin{array}{l}(|A_1|^2 + 2|A_2|^2 + 2|A_3|^2 + 2|A_4|^2)A_1 \\ + 2A_2A_3A_4e^{-i\Delta kz}\end{array}\right], \tag{8.31}$$

$$\frac{dA_2}{dz} = i\frac{3\omega_2\chi^{(3)}}{8n_2c}\left[\begin{array}{l}(|A_2|^2 + 2|A_1|^2 + 2|A_3|^2 + 2|A_4|^2)A_2 \\ + 2A_1A_3^*A_4^*e^{i\Delta kz}\end{array}\right], \tag{8.32}$$

where

$$\frac{dA_3}{dz} = i\frac{3\omega_3\chi^{(3)}}{8n_3c}\left[\begin{array}{l}(|A_3|^2 + 2|A_1|^2 + 2|A_2|^2 + 2|A_4|^2)A_3 \\ + 2A_1A_2^*A_4^*e^{i\Delta kz}\end{array}\right], \tag{8.33}$$

$$\frac{dA_4}{dz} = i\frac{3\omega_4\chi^{(3)}}{8n_4c}\left[\begin{array}{l}(|A_4|^2 + 2|A_1|^2 + 2|A_2|^2 + 2|A_3|^2)A_4 \\ + 2A_1A_2^*A_3^*e^{i\Delta kz}\end{array}\right], \tag{8.34}$$

where

$$\Delta k = k_1 - k_2 - k_3 - k_4. \tag{8.35}$$

Note that the four coupled equations above are specific to the process described by $\omega_1 = \omega_2 + \omega_3 + \omega_4$ and rearrangements of this relationship. A set of coupled amplitude equations are found using the same procedure for other frequency combinations. The coupled amplitude equations are solved in various limits. For example, in the small-signal limit where the inputs have a negligible change in amplitude, we are left with one equation that we can integrate. Solutions to the coupled amplitude equations given above parallel the same techniques we use for $\chi^{(2)}$ effects.

8.3.2 MANLEY–ROWE RELATIONS

As with three-wave interactions, we can visualize four-wave interactions in terms of photons. The process given by $\omega_1 = \omega_2 + \omega_3 + \omega_4$ indicates that, at the most basic level, three photons at ω_2, ω_3, and ω_4 are destroyed and one photon at ω_1 is created.

We validate this idea by following the same procedure as deriving the Manley–Rowe relationships for a three-wave process. The rate of change in the intensity is given by

$$\frac{dI}{dz} = \frac{n}{2\mu_0 c}\left(A\frac{dA^*}{dz} + A^*\frac{dA}{dz}\right). \tag{8.36}$$

Hence, we see that, for example,

$$\frac{1}{\omega_1}\frac{dI_1}{dz} = i\frac{3\chi^{(3)}}{16\mu_0 c^2}\left[\begin{array}{l}(|A_1|^2 + 2|A_2|^2 + 2|A_3|^2 + 2|A_4|^2)|A_1|^2 \\ + 2A_1^* A_2 A_3 A_4 e^{-i\Delta kz}\end{array}\right] + \text{c.c.} \tag{8.37}$$

For a lossless media, $\chi^{(3)}$ is a real quantity. Because the automatically phase-matched terms in Equation 8.37 are also real, they cancel with the complex conjugate term in Equation 8.37 resulting in

$$\frac{1}{\omega_1}\frac{dI_1}{dz} = i\frac{3\chi^{(3)}}{8\mu_0 c^2}A_1^* A_2 A_3 A_4 e^{-i\Delta kz} + \text{c.c.} \tag{8.38}$$

Similar expressions for I_2, I_3, and I_4 lead to the result

$$\frac{1}{\omega_1}\frac{dI_1}{dz} = -\frac{1}{\omega_2}\frac{dI_2}{dz} = -\frac{1}{\omega_3}\frac{dI_3}{dz} = -\frac{1}{\omega_4}\frac{dI_4}{dz}. \tag{8.39}$$

Equation 8.39 has an equivalent description in terms of photon numbers (see similar discussion for $\chi^{(2)}$ interactions in Section 6.4.1),

$$\Delta N_2 = \Delta N_3 = \Delta N_4 = -\Delta N_1. \tag{8.40}$$

where ΔN_1, ΔN_2, ΔN_3, and ΔN_4 are the changes in the number of photons for each frequency, with ΔN_1 corresponding to the highest-energy photons. Figure 8.1 shows

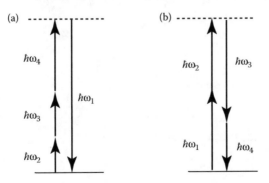

FIGURE 8.1 Energy diagrams for two parametric interactions: (a) SFG, $\Delta N_2 = \Delta N_3 = \Delta N_4 = -\Delta N_1$, and (b) two-pump parametric amplifier, $\Delta N_3 = \Delta N_4 = -\Delta N_1 = -\Delta N_2$.

the same process diagrammatically. Note that Equations 8.39 and 8.40 are specific to the $\omega_1 = \omega_2 + \omega_3 + \omega_4$ process. Similar expressions for other processes yield expressions that we expect based on the energy conservation equation. For example, the process $\omega_1 + \omega_2 = \omega_3 + \omega_4$ results in

$$\Delta N_3 = \Delta N_4 = -\Delta N_1 = -\Delta N_2. \tag{8.41}$$

This equation shows that photons are annihilated at ω_1 and ω_2, while new photons at ω_3 and ω_4 are created.

In the case of self-induced processes where the nonlinear polarization is the same as the inputs, such as for $\omega = \omega + \omega - \omega$, we obtain the result that the number of photons remains unchanged:

$$\Delta N = 0. \tag{8.42}$$

8.4 SELF-INDUCED EFFECTS

8.4.1 NONLINEAR INDEX OF REFRACTION

A distinct difference between $\chi^{(3)}$ and $\chi^{(2)}$ interactions is that $\chi^{(3)}$ processes have self-induced effects. Consider a $\chi^{(3)}$ process in a lossless medium involving a single-input frequency at m for the case where the nonlinear polarization's frequency is also at ω. For this example, we assume propagation along the z-direction with the field pointing in the x-direction. The nonlinear polarization is given by Equation 8.9 with $D = 3$:

$$P_x^{(3)}(\omega) = \frac{3\varepsilon_o}{4} \chi_{xxxx}^{(3)} \, |A_x|^2 \, A_x e^{i(k-k+k)z}. \tag{8.43}$$

To simplify the following discussion we relabel $\chi_{xxxx}^{(3)} = \chi^{(3)}$ and drop the Cartesian subscripts on $P^{(3)}$ and A. Substituting the nonlinear polarization, Equation 8.43, into the wave equation, Equation 8.2, gives

$$\frac{dA}{dz} = i\frac{3\omega}{8nc} \chi^{(3)} \, |A|^2 \, A. \tag{8.44}$$

Note that the exponential phase term, e^{ikz}, cancels out of this equation, which indicates that the process automatically phase matches. Equation 8.44 is solved by writing the amplitude, $A(z)$, as

$$A(z) = u(z)e^{i\phi(z)}, \tag{8.45}$$

where $u(z)$ and $\phi(z)$ are real-valued functions. Substituting Equation 8.45 into 8.44 gives

$$\frac{du}{dz} + iu\frac{d\phi}{dz} = i\frac{3\omega}{8nc}\chi^{(3)}u^3.$$ (8.46)

Assuming that the material is lossless, then $\chi^{(3)}$ is a real quantity so that Equation 8.46 is split into two equations, one for the real part and the other for the imaginary part:

$$\frac{du}{dz} = 0,$$ (8.47)

$$\frac{d\phi}{dz} = \frac{3\omega}{8nc}\chi^{(3)}u^2.$$ (8.48)

Not surprisingly, Equation 8.47 shows that $u(z)$ is constant, which is consistent with the lossless assumption. Equation 8.48 is integrated to give the nonlinear phase shift for a thin sample:

$$\Delta\phi = \frac{3\omega}{8nc}\chi^{(3)}u^2\Delta z.$$ (8.49)

In a linear medium, the phase shift of a given wave that traverses a distance Δz is

$$\phi^{(L)} = \frac{n_0\omega}{c}\Delta z,$$ (8.50)

where n_0 is the linear index of refraction. The total phase shift in a nonlinear medium is therefore

$$\phi = \frac{\omega}{c}\left(n_0 + \frac{3}{8n}\chi^{(3)}u^2\right)\Delta z.$$ (8.51)

As written in Equation 8.51, the nonlinear effect is a perturbation to the index of refraction:

$$n = n_0 + \frac{3}{8n_0}\chi^{(3)}u^2.$$ (8.52)

Note that u^2 is directly related to intensity; hence, Equation 8.52 may be rewritten as

$$n = n_0 + \frac{3}{4\varepsilon_0 n_0^2 c} \chi^{(3)} I. \tag{8.53}$$

The nonlinear change in index is commonly written as

$$n = n_0 + n_2^I I, \tag{8.54}$$

n_2^I is called the nonlinear index:

$$n_2^I = \frac{3}{4\varepsilon_0 n_0^2 c} \chi^{(3)}. \tag{8.55}$$

The intensity-dependent change in the index of refraction is called the Kerr effect or the quadratic EO effect. The quadratic EO effect differs from the linear EO effect discussed in Chapter 3 in that the field causing the change in the index is the incident laser itself instead of an externally applied field.

In some cases, the nonlinear change in index is written in terms of the field amplitude:

$$n = n_0 + \frac{n_2}{2} |A|^2. \tag{8.56}$$

This second expression for the nonlinear index requires special attention since the field amplitude depends on how we define the complex amplitude (see Section 3.1.2). We distinguish between the two notations by including the superscript n_2^I, denoting the intensity relationship in Equation 8.54.

8.4.2 NONLINEAR ABSORPTION

If $\chi^{(3)}$ is complex, as occurs when the frequency is close to an absorption band, then $\chi^{(3)} = \chi_R^{(3)} + i\chi_{Im}^{(3)}$ where $\chi_R^{(3)}$ and $\chi_{Im}^{(3)}$ are the real and imaginary parts of $\chi^{(3)}$, respectively. Equation 8.46 becomes

$$\frac{du}{dz} + iu\frac{d\phi}{dz} = -\frac{3\omega}{8nc}\chi_{Im}^{(3)}u^3 + i\frac{3\omega}{8nc}\chi_R^{(3)}u^3. \tag{8.57}$$

This equation is rewritten by separating its real and imaginary parts:

$$\frac{du}{dz} = -\frac{3\omega}{8nc}\chi_{Im}^{(3)}u^3, \tag{8.58}$$

$$\frac{d\phi}{dz} = -\frac{3\omega}{8nc}\chi_R^{(3)}u^2. \tag{8.59}$$

Equation 8.58 shows that the amplitude of the field changes, that is, a nonlinear absorption is present. Absorption is more typically associated with intensity, which when written in terms of u is

$$I = \frac{\varepsilon_0 nc}{2} u^2. \tag{8.60}$$

The rate of change of the intensity is given by

$$\frac{dI}{dz} = \varepsilon_0 nc\, u \frac{du}{dz}. \tag{8.61}$$

Substituting Equation 8.58 into Equation 8.61 gives

$$\frac{dI}{dz} = -\beta I^2, \tag{8.62}$$

where

$$\beta = \frac{3\omega}{2\varepsilon_0 n^2 c^2} \chi_{\text{Im}}^{(3)}. \tag{8.63}$$

When both linear and nonlinear absorptions are present in the medium, the rate of change in intensity is

$$\frac{dI}{dz} = -\alpha I - \beta I^2, \tag{8.64}$$

where α is the linear absorption coefficient. We note that a nonlinear absorption may increase or decrease the absorption over the linear case, depending on the sign of β.

8.4.3 CROSS-PHASE SHIFTS

When two beams at ω_1 and ω_2 are present at the input instead of one, another nonlinear phase shift occurs due to a cross-phase term. Consider again a lossless medium ($\chi^{(3)}$ real) with the fields propagating in the z-direction and polarized in the x-direction. The nonlinear polarization at ω_1 is

$$P^{(3)}(\omega_1) = \frac{3\varepsilon_0}{4} \chi^{(3)} \left(|A_1|^2 + 2|A_2|^2 \right) A_1 e^{ik_1 z}. \tag{8.65}$$

Here, we assume the Kleinman symmetry and neglect dispersion in $\chi^{(3)}$. A similar expression results for the nonlinear polarization at ω_2, and the two fields are coupled to each other through the nonlinear polarization. We define the fields in magnitude and phase form, as in Equation 8.45, and substitute Equation 8.65 (and a similar expression for ω_2) into Equation 8.2 to give the coupled equations

$$\frac{du_1}{dz} + iu_1\frac{d\phi_1}{dz} = i\frac{3\omega_1\chi^{(3)}}{8nc}\left(u_1^2 + 2u_2^2\right)u_1, \tag{8.66}$$

$$\frac{du_2}{dz} + iu_2\frac{d\phi_2}{dz} = i\frac{3\omega_2\chi^{(3)}}{8nc}\left(u_2^2 + 2u_1^2\right)u_2. \tag{8.67}$$

In these equations and in the following equations, we ignore dispersion in the index of refraction. These two equations become uncoupled if we assume a lossless medium so that u_1 and u_2 are constant (see discussion around Equation 8.47). In this case, the nonlinear phase at ω_1 is

$$\Delta\phi(\omega_1) = \frac{\omega_1}{c}\left[\frac{3}{4\varepsilon_0 n^2 c}\chi^{(3)}(I_1 + 2I_2)\right]\Delta z. \tag{8.68}$$

Equation 8.68 shows both a self-phase and cross-phase term. The index of refraction including the cross-phase term is

$$n = n_0 + n_2^I I(\omega_1) + n_{2,\text{cross}}^I I(\omega_2). \tag{8.69}$$

Note that according to Equation 8.68, the cross-phase nonlinear index is twice that of the self-induced one.

In our analysis, the two beams have the same polarization but different frequencies. Another form of cross-phase modulation (not covered here) occurs when two orthogonal polarizations are present in a medium at the same frequency. The intensity modulation in one polarization component can then affect the phase of the other orthogonal component.

8.4.4 SELF-FOCUSING

The intensity-dependent index of refraction given in Equation 8.54 is derived assuming plane-wave interactions. We extend the discussion to Gaussian beams by assuming that diffraction is negligible over the length of the medium (thin-sample approximation). In this approximation, we treat each point on the beam's cross section independent of the others. The effect on a beam traversing a nonlinear material is an intensity-dependent phase delay. For a sample with a thickness of L, the phase delay is given by Equation 8.51, which we rewrite as

$$\phi = \left(n_0 + n_2^I I\right)\frac{\omega}{c}L. \tag{8.70}$$

Consider a Gaussian beam with a beam waist, w_{inc}, incident on the sample with a field given by

$$A(r, z = 0) = A_0 \exp\left(-\frac{r^2}{w_{inc}^2}\right). \tag{8.71}$$

The radius of curvature of the Gaussian beam at the waist is infinite (flat phase front). The intensity corresponding to Equation 8.71 is

$$I(r) = I_0 \exp\left(-2\frac{r^2}{w_{inc}^2}\right). \tag{8.72}$$

where I_0 is the on-axis intensity at the beam waist. The output beam is the same as the input beam multiplied by an exponential phase term due to the nonlinear phase shift. We can think of this process as equivalent to superposing a phase mask on the original Gaussian beam:

$$A(r, z) = A_0 \exp\left(-\frac{r^2}{w_{inc}^2}\right) \exp\left(i\left(n_0 + n_2^I I_0 e^{-2\left(r^2/w_{inc}^2\right)}\right)\frac{\omega}{c} z\right). \tag{8.73}$$

Equation 8.73 shows that the nonlinear index leads to an optical path length that varies from the center of the beam outward (see Figure 8.2). The optical path length

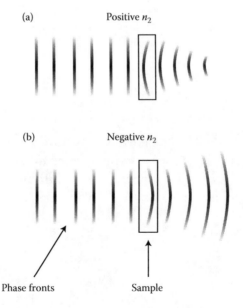

(a)   Positive n_2

(b)   Negative n_2

Phase fronts   Sample

FIGURE 8.2   A Gaussian beam incident on a nonlinear medium with (a) a positive n_2 and (b) a negative n_2. Because the center of the beam has higher intensity, it obtains the largest nonlinear phase shift.

across the beam profile is similar to that of a thin lens, which varies quadratically across a beam. Hence, the additional nonlinear index term leads to focusing or defocusing depending on the sign of the nonlinearity.

The focal length of the lens is estimated by using the paraxial approximation, which is valid for beam rays close to the axis of propagation. For these rays, the nonlinear index's Gaussian distribution is approximated as

$$\exp\left(-2\frac{r^2}{w_{inc}^2}\right) \approx 1 - 2\frac{r^2}{w_{inc}^2}. \tag{8.74}$$

In this limit, Equation 8.73 becomes

$$A(r, z) = A_0 \exp\left(-\frac{r^2}{w_{inc}^2}\right)\exp\left(i\left(n_0 + n_2^I I_0\right)\frac{\omega}{c}L\right)\exp\left(-i\frac{2n_2^I \omega I_0 L}{w_{inc}^2 c}r^2\right). \tag{8.75}$$

The first phase exponential is a constant phase delay for the entire beam, and this term has no effect on focusing. Equation 8.75 is equivalent to a Gaussian beam of beam size w_{inc}, but now with a finite radius of curvature. The radius of curvature is found by comparing Equation 8.75 to the canonical Gaussian beam equation (see Section 2.6),

$$A(r, z) = A_0 \frac{w_0}{w(z)}\exp\left(\frac{-r^2}{w^2(z)}\right)\exp\left(ikz + ik\frac{r^2}{2R(z)} - i\zeta(z)\right). \tag{8.76}$$

Comparing Equation 8.75 with Equation 8.76 gives the radius of curvature induced by the nonlinearity:

$$R_{NL} = -\frac{n_0 w_{inc}^2}{4n_2^I I_0 L}. \tag{8.77}$$

For small intensities, we relate the radius of curvature to a focal length by noting that if we start with a Gaussian beam waist and propagate for a distance much larger than the Rayleigh range, then $R_{NL} \sim z$. Turning this process around, knowing R gives the distance from the lens location to the beam waist. Therefore, the focal length of the lens is

$$|f| \approx |R_{NL}| = \left|\frac{n_0 w_{inc}^2}{4n_2^I I_0 L}\right|. \tag{8.78}$$

The sign of the focal length depends on the sign of n_2^I; the focal length is positive for $n_2^I > 0$.

In a thick sample with a positive n_2^I, as the beam focuses, the intensity increases and the self-focusing becomes stronger. Eventually, the beam comes to a focus in the material. A proper calculation of self-focusing in thick media requires that we include the effects of diffraction as the beam propagates through the sample. The radius of curvature due to diffraction is given in Section 2.6; for strong intensity, the propagation distances are much less than the Rayleigh range, so we approximate the diffraction effect as

$$R_D = \frac{z_R^2}{L} = \frac{\pi^2 n_0^2 w_{\text{inc}}^4}{\lambda^2 L}. \tag{8.79}$$

At a *critical power*, the self-focusing balances diffraction ($R_D \approx -R_{NL}$) so that the beam size remains constant in a sample

$$P_{cr} = \frac{1}{2} I_0 \pi w_{\text{inc}}^2 = 0.146 \frac{\lambda^2}{n_0 n_2^I}. \tag{8.80}$$

The numerical prefactor is adjusted based on numerical estimates. Interestingly, the balancing of the nonlinearity with diffraction is determined by the total beam power in the beam and not intensity (see Problem 8.16). The stability of the beam at the critical power depends on the number of transverse dimensions. For dependence on two dimensions, the wave is unstable and the nonlinearity leads to a collapse of the beam into a filament. We defer further discussion of self-focusing until Chapter 10, where diffraction effects are included in the wave equation.

The self-focusing description below the critical power for thin samples is a powerful technique for characterizing the complex $\chi^{(3)}$ nonlinear coefficient using a technique, called z-scan, described in Section 8.8. The z-scan technique also gives the sign of the real part of $\chi^{(3)}$.

8.4.5 OPTICAL BISTABILITY

The self-induced nonlinear phase shift leads to the phenomena of optical bistability where the output of a device depends on its past history. In other words, the device exhibits hysteresis. Optical bistability occurs, for example, when a nonlinear medium is placed inside a Fabry–Perot etalon. The transmission of the etalon is calculated by adding the contributions from successive reflections in the cavity. We consider an etalon consisting of the nonlinear medium with reflection coatings on its input and output faces (see Figure 8.3). The field's reflection and transmission coefficient from the coating are r and t, respectively. At normal incidence, the power reflectivity and transmission are $R = |r|^2$ and $T = |t|^2$, respectively, and they are related by $R + T = 1$.

The complex amplitude of the output field is related to that of the input field via

$$A_{\text{Trans}} = t^2 A_0 e^{ikL} [1 + r^2 e^{i2kL} + r^4 e^{i4kL} + \cdots] = \frac{t^2 A_0 e^{ikL}}{1 - r^2 e^{i2kL}}. \tag{8.81}$$

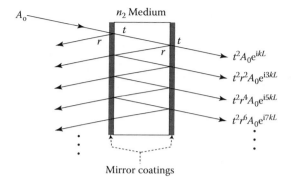

FIGURE 8.3 A nonlinear medium has two mirrored surfaces such that the output field is the superposition of the fields from each successive reflection. The field's reflection and transmission coefficient are r and t, respectively.

For a nonlinear medium, the k-vector magnitude is given by

$$k = \frac{2\pi(n_0 + n_2^I I_{\text{int}})}{\lambda}. \qquad (8.82)$$

I_{int} is the intensity inside the cavity. The round trip phase shift for the cavity is hence intensity dependent. We calculate the output intensity using Equation 8.81 to give

$$I_{\text{Trans}} = \frac{T^2}{T^2 + 4R \sin^2{(kL)}} I_0. \qquad (8.83)$$

The output and the internal intensities are related by $I_{\text{Trans}} = T I_{\text{int}}$. This relationship combined with Equations 8.82 and 8.83 gives

$$I_0 = \left[1 + \frac{4R}{T^2} \sin^2\left(\frac{2\pi\left(n_0 + n_2^I(I_{\text{Trans}}/T)\right)L}{\lambda} \right) \right] I_{\text{Trans}}. \qquad (8.84)$$

This equation is written with the transmitted intensity as the independent variable. In this form, we can plot I_0 as a function of I_{Trans} and then rotate the plot to show I_{Trans} as a function of I_0. Another approach is to look at Equation 8.84 in terms of normalized variables as follows:

$$\eta = \frac{1}{\left[1 + \left(4R/(1-R)^2\right)\sin^2\left((2\pi n_0/\lambda)\left(1 + \eta f/(1-R)\right)L\right)\right]}, \qquad (8.85)$$

where

$$\eta \equiv \frac{I_{\text{Trans}}}{I_0} \tag{8.86}$$

and

$$f \equiv \frac{n_2^I I_0}{n_0}. \tag{8.87}$$

η is a measure of the output intensity scaled by the input intensity and f is a rough measure of the fractional change in index, but the actual change in index is different since the intracavity intensity is not the same as I_0. Equation 8.85 is transcendental; however, we may plot its solution by the method of contours. For a given reflectivity, n_0, wavelength, and sample length, we define a new function in terms of normalized variables:

$$g(f, \eta) \equiv \eta - \frac{1}{\left[1 + (4R/(1-R)^2)\sin^2\left((2\pi n_0/\lambda)(1 + f\eta/(1-R))L\right)\right]}. \tag{8.88}$$

The solution to Equation 8.85 occurs when $g(f, \eta) = 0$, which defines a contour in $f - \eta$ space. The contour is found by first scaling a two-dimensional array such that the x-coordinate corresponds to f and the y-coordinate corresponds to η. Next, we assign to each element of the array the value of $g(f, \eta)$. The final step is to create a contour plot based on the array and only plot the zero-contour (most mathematical packages have built-in capabilities to find contours). The zero-contour is shown in Figure 8.4 for the case where $R = 0.4$, $n_0 = 1.5$, $L = 5.5$ mm, and $\lambda = 1.55$ μm.

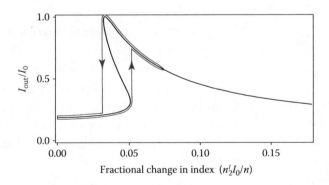

FIGURE 8.4 Hysteresis curve for the transmission from an etalon containing a nonlinear medium.

Figure 8.4 shows a classic nonlinear signature where the output has two possible values for a given input. The output state depends on the past history of the system, which is shown in Figure 8.4 as a superposed hysteresis curve. The path starts with I_0 at a small value. As we follow the curve for increasing input intensity, the output increases slightly faster until $f \sim 0.06$, where the output state changes discontinuously. As the input increases further, the transmitted output now decreases. If the input intensity is then decreased, the output increases until it reaches a peak where $I_{\text{Trans}} = I_0$. As the input decreases further, the output jumps to a lower value. The bistable transmission curve also depends on cavity length and wavelength, both of which are held fixed in Figure 8.4 (see Problems 8.22 and 8.23).

Figure 8.4 shows that the discontinuity occurs for a fractional change in the index of about 5%. To put this number in context, consider a birefringent crystal with a fractional change in index defined by

$$f = \frac{n_e - n_0}{n_0}. \tag{8.89}$$

A moderately birefringent crystal (e.g., lithium niobate) has $f \sim 0.02$; so, the required fractional change in index shown in Figure 8.4 is greater than the birefringence of many crystals. Obtaining such a large change in index requires materials exhibiting a large nonlinear index, and an n_2^I with an electronic origin (nonresonant) is not large enough to achieve optical bistability. Other mechanisms, such as ones involving a resonant excitation of an electronic or molecular energy level, can have effective nonlinearities orders of magnitude larger than those with a nonresonant electronic origin making optical bistability possible.

8.5 PARAMETRIC AMPLIFIERS

8.5.1 INTRODUCTION

The diversity of frequency combinations of four-wave interactions gives rise to an equally diverse number of four-wave devices. One of these devices is a parametric amplifier, which performs much like its three-wave counterpart. This device has two inputs at ω_P and ω_S called the pump and the signal, respectively. In some cases, the signal is called the anti-Stokes. The output is an amplified signal and a newly generated frequency at ω_I called the idler, which is sometimes called the Stokes frequency. Because a new frequency is generated, the parametric amplifier is also a means of frequency conversion. At first glance, the parametric interaction looks like a three-wave $\chi^{(2)}$ process since three frequencies at the pump, signal, and idler are involved. In the four-wave parametric amplifier, the difference comes in its statement of energy conservation:

$$2\hbar\omega_P = \hbar\omega_S + \hbar\omega_I. \tag{8.90}$$

Equation 8.88 describes a process where two pump photons are converted into a signal and an idler, shown diagrammatically in Figure 8.5. A key difference between a four-wave mixing parametric amplifier and a three-wave mixing version is that photon energy of the signal is greater than the input pump photon energy. Because the four-wave mixing parametric amplifier is a $\chi^{(3)}$ process, it occurs in centrosymmetric media. Therefore, long interaction lengths are possible by confining the process in optical fibers.

Our approach to calculating amplification of the signal input is the same as in the general four-wave mixing problem in Section 8.3. We use the nonlinear polarization for the three fields to derive the coupled amplitude equations. In most parametric amplifiers, the pump, signal, and idler are close in frequency, and so we ignore dispersion in the index of refraction. The nonlinear polarization's complex amplitude at the signal frequency is, from Equation 8.9,

$$P^{(3)}(\omega_S) = \frac{3\varepsilon_0\chi^{(3)}}{4}\left[\left(|A_S|^2 + 2|A_I|^2 + 2|A_P|^2\right)A_S e^{ik_S z} + A_P^2 A_I^* e^{i(2k_P - k_I)z}\right]. \tag{8.91}$$

Similar expressions result for the nonlinear polarizations at ω_P and ω_I. Substituting the nonlinear polarizations into Equation 8.2 gives the coupled amplitude equations for the pump, signal, and idler:

$$\frac{dA_P}{dz} = i\frac{3\omega_P\chi^{(3)}}{8nc}\left[\left(|A_P|^2 + 2|A_S|^2 + 2|A_I|^2\right)A_P + 2A_S A_I A_P^* e^{-i\Delta kz}\right], \tag{8.92}$$

$$\frac{dA_S}{dz} = i\frac{3\omega_S\chi^{(3)}}{8nc}\left[\left(|A_S|^2 + 2|A_I|^2 + 2|A_P|^2\right)A_S + A_P^2 A_I^* e^{i\Delta kz}\right], \tag{8.93}$$

$$\frac{dA_I}{dz} = i\frac{3\omega_I\chi^{(3)}}{8nc}\left[\left(|A_I|^2 + 2|A_S|^2 + 2|A_P|^2\right)A_I + A_P^2 A_S^* e^{i\Delta kz}\right], \tag{8.94}$$

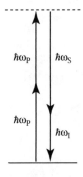

FIGURE 8.5   Energy diagram for a parametric interaction.

where

$$\Delta k = 2k_P - k_S - k_I. \tag{8.95}$$

The next sections look at solutions to these coupled amplitude equations.

8.5.2 TWO UNDEPLETED INPUTS

A straightforward way to solve these coupled equations is numerically integrating them; however, we defer the numerical results until after looking at a small-signal example where a pump and a signal are present at the input with $|A_P| \sim$ constant and $|A_S| \sim$ constant, respectively, and where $|A_P| \gg |A_S|$. Even though the inputs in this example have constant amplitudes, they experience a nonlinear phase shift via the $\chi^{(3)}$ interaction. Therefore, instead of setting 8.92 and 8.93 to 0, we keep the dominant nonlinear phase shift term. Under the constant amplitude approximation, the coupled amplitude equations, Equations 8.92 through 8.94, become decoupled, making it possible to solve each independent of the others:

$$\frac{dA_I}{dz} = i\frac{3\omega_I\chi^{(3)}}{8nc}\left[2|A_P|^2 A_I + A_P^2 A_S^* e^{i\Delta kz}\right], \tag{8.96}$$

$$\frac{dA_S}{dz} = i\frac{3\omega_S\chi^{(3)}}{4nc}|A_P|^2 A_S, \tag{8.97}$$

$$\frac{dA_P}{dz} = i\frac{3\omega_P\chi^{(3)}}{8nc}|A_P|^2 A_P. \tag{8.98}$$

Equations 8.97 and 8.98 are in the same form as Equation 8.44 and have the same type of solution, namely a phase-shifted field:

$$A_S(z) = A_S(0)e^{2i\gamma\omega_S z}, \tag{8.99}$$

$$A_P(z) = A_P(0)e^{i\gamma\omega_P z}, \tag{8.100}$$

where

$$\gamma \equiv \frac{3\chi^{(3)}}{8nc}|A_P|^2. \tag{8.101}$$

The idler field experiences a nonlinear phase shift and parametric gain. To simplify the algebra for the idler, Equation 8.96, we factor out of the idler field the nonlinear phase that would occur if no parametric gain were present:

$$A_I = A_I' \exp(2i\gamma\omega_I z). \tag{8.102}$$

This variable substitution and Equations 8.99 and 8.100 are substituted into the equation for the idler field, Equation 8.96, giving

$$\frac{dA_I'}{dz} = i\gamma\omega_I A_S^*(0)e^{i\kappa z}, \tag{8.103}$$

where

$$\kappa \equiv \Delta k + 2\gamma(\omega_P - \omega_S - \omega_I) = \Delta k - 2\gamma\omega_P. \tag{8.104}$$

Integrating Equation 8.103 results in

$$A_I = i\gamma\omega_I A_S^*(0)L \,\text{sinc}\left(\frac{\kappa L}{2}\right)\exp\left\{i\left(\frac{\kappa L}{2} + 2\gamma\omega_I L\right)\right\}. \tag{8.105}$$

The corresponding intensity is, after writing out the variable substitution in Equation 8.101,

$$I_I(z) = \frac{9\left(\chi^{(3)}\right)^2 \omega_I^2}{8n^5 c^5 \varepsilon_0^3} I_P^2 I_S L^2 \text{sinc}^2\left(\frac{\kappa L}{2}\right). \tag{8.106}$$

In the small-signal limit, the output idler has a familiar sinc^2 dependence, but the $\kappa L/2$ term includes contributions from both material dispersion and a nonlinear phase shift. Maximum efficiency occurs for $\kappa = 0$, and for a range $|\kappa L/2| < \pi/2$, the efficiency of the process remains within 60% of the peak (see Chapter 4 for similar tolerance calculations). Hence, κ has a range given by

$$|\kappa| < \frac{\pi}{L}. \tag{8.107}$$

We use these criteria to estimate the importance of the nonlinear phase shift ($2\gamma\omega_P L$, see Equation 8.104) to the overall conversion efficiency. The contribution of $2\gamma\omega_P L$ is estimated using $\lambda_P = 1.5\,\mu\text{m}$, $\chi^{(3)} \sim 10^{-22}\,\text{m}^2/\text{V}^2$, and $n \sim 1.5$:

$$2\gamma\omega_P = \frac{3\pi\chi^{(3)}}{\varepsilon_0 n^2 c\lambda_P} I_P \approx I_P \times 10^{-13}. \tag{8.108}$$

For the case where $\Delta k = 0$, Equation 8.107 combined with Equation 8.108 shows that the nonlinear phase shift starts to influence the performance in this example when

$$I_P \approx \frac{\pi}{L} \times 10^{13} \text{ W/m}^2.$$ (8.109)

Nonlinear phase shifts may be appreciable for pulsed lasers focused to tight spots sizes. However, another way to obtain appreciable effects is by utilizing long interaction lengths. In fiber-optic-based devices, interaction lengths of 100–1000 m are not uncommon. The intensity required to satisfy Equation 8.109 (using a 100 m fiber) is ~30 MW/cm², which is easily attainable with a pulsed laser in a fiber. For example, many fiber lasers operate with pulse energies on the order of 1 μJ and 10 ns pulse durations. Fiber sizes vary, but a typical mode-field diameter is 10 μm. Under these conditions, the intensity is roughly 130 MW/cm² so that nonlinear phase shifts are important for fiber lengths >100 m. For situations where the fiber length or I_P are less than in this example, the nonlinear contribution to κ is negligible and device performance is dominated by Δk. For fiber devices, Δk must include the fiber's waveguide dispersion.

A key consideration with fiber devices operating with multiple wavelengths is to ensure that the mode associated with each field has significant overlap. A particularly attractive choice is a photonic crystal fiber engineered for endlessly single-mode operation. These fibers are single mode for a broad wavelength range and hence ensure beam overlap for a parametric process involving widely separated frequencies.

8.5.3 ONE UNDEPLETED INPUT

The next step in analyzing $\chi^{(3)}$ parametric amplifiers is to consider the case when only the pump field is undepleted. By making variable substitutions as in Equation 8.102, the coupled amplitude equations, Equations 8.93 and 8.94, for the signal and idler fields become

$$\frac{dA_S'}{dz} = i\gamma\omega_S A_I'^* e^{i\kappa z},$$ (8.110)

$$\frac{dA_I'}{dz} = i\gamma\omega_I A_S'^* e^{i\kappa z},$$ (8.111)

where Equation 8.100 has been substituted in for the pump field and γ is given by Equation 8.101. Equations 8.110 and 8.111 are solved in the same way as the parametric equations that arise with $\chi^{(2)}$ processes (see Chapter 6). The decoupled equation for A_S' is

$$\frac{d^2 A_S'}{dz^2} - i\kappa \frac{dA_S'}{dz} - \gamma^2 \omega_I \omega_S A_S' = 0.$$ (8.112)

The solution to this differential equation depends on the boundary conditions. As a specific example, we choose $A_I(0) = 0$ and $A_S(0) \neq 0$. The solution is

$$A_S(z) = \left[\cosh(gz) - i \frac{\kappa}{2g} \sinh(gz) \right] A_S(0) \exp\left(i\left(\frac{\kappa}{2} + 2\gamma\omega_S \right) z \right), \qquad (8.113)$$

where

$$g^2 = \gamma^2 \omega_S \omega_I - \left(\frac{\kappa}{2} \right)^2. \qquad (8.114)$$

The signal's corresponding intensity is

$$I_S = \left[1 + \frac{\gamma^2 \omega_S \omega_I}{g^2} \sinh^2(gz) \right] I_{S0}. \qquad (8.115)$$

A similar derivation for the idler intensity gives

$$I_I = \frac{\gamma^2 \omega_I^2}{g^2} I_{S0} \sinh^2(gz). \qquad (8.116)$$

These expressions show that both the signal and idler experience exponential gain, provided that $\kappa < 2\gamma\sqrt{\omega_S \omega_I}$. Even though $\chi^{(3)}$ effects are small, the exponential gain allows us to compensate with long interaction lengths such as are possible in optical fibers. In fact, optical fiber-based parametric amplifiers can have high conversion efficiencies to the point where the pump depletes.

8.5.4 PUMP DEPLETION

For cases where the pump field depletes, solutions to the coupled amplitude equations, Equations 8.92 through 8.94, are integrated numerically using, for example, a Runge–Kutta algorithm (see Appendix C). As an example to show general trends, we consider a parametric interaction in a fiber with a mode-field diameter of 10 μm. The input pump intensity is calculated assuming 10 μJ pulse energy and 100 ns pulse duration. The input idler intensity assumes a 1 μJ pulse energy and the same pulse duration as the pump. The pump, signal, and idler wavelengths are 1.5, 1.45, and 1.55 μm, respectively. Figure 8.6 shows the energy that would be measured for different fiber lengths with $\Delta k = 0$ (including waveguide dispersion). Both the signal and the idler build up until the pump is depleted. The growth of the signal and the idler is slightly asymmetric due to their different photon energies. In this interaction, the pump depletes followed by a back conversion. Note that the total energy remains constant.

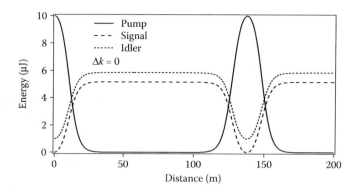

FIGURE 8.6   Parametric amplifier showing the growth of the signal and the idler at the expense of the pump followed by back conversion.

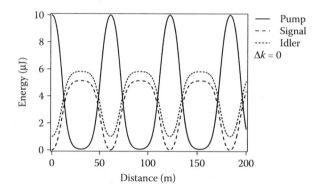

FIGURE 8.7   Parametric amplifier when three fields are present at the input.

If a portion of the signal field is present at the input, the dynamic of the interaction changes. Figure 8.7 shows that back conversion occurs sooner when three fields are present at the input, even for a small input at the signal wavelength.

8.6   NONCOLLINEAR PROCESSES

So far, we assumed collinear plane-wave interactions, and for processes that are not automatically phase matched, material dispersion dictates the phase mismatch. Noncollinear schemes offer more flexibility in phase matching. Consider phase matching for a parametric amplifier with the energy conservation statement, $2\hbar\omega_p = \hbar\omega_S + \hbar\omega_{AS}$. The vector phase-matching condition is

$$\Delta\vec{k} = 2\vec{k}_P - \vec{k}_S - \vec{k}_I.\tag{8.117}$$

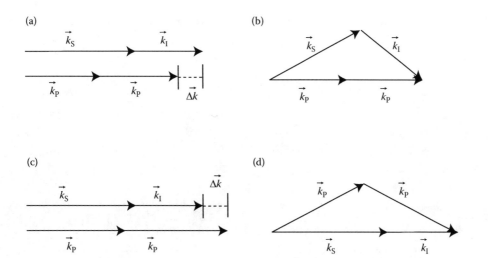

FIGURE 8.8 k-Vector diagrams for a parametric amplifier when $\Delta k < 0$: (a) collinear and (b) noncollinear and when $\Delta k > 0$: (c) collinear and (d) noncollinear.

For a material exhibiting normal dispersion, it is possible to show (see Problem 8.9)

$$\Delta k_{\text{collinear}} = \frac{1}{v_g^2} \frac{\partial v_g}{\partial \omega} \Delta \omega^2, \tag{8.118}$$

where v_g is the group velocity and $\Delta\omega$ is the frequency separation between the pump and the signal. Hence, the GVD determines the sign of Δk. When $\Delta k_{\text{collinear}} < 0$, tilting the signal and the idler with respect to the pump makes it possible to obtain $\Delta k = 0$ (see Figure 8.8).

Alternatively, if $\Delta k > 0$, then the pump beam can be split up and sent in to the sample as shown in Figure 8.8d. In all the noncollinear interactions, the three input waves create a nonlinear polarization that propagates in the direction of the generated field.

Another useful noncollinear interaction involves counter-propagating signal and idler beams, similar to the BOPO presented in Chapter 7. Unlike the BOPO, the four-wave counterpart phase matches much more easily (see Figure 8.9). The phase-matching condition is

$$\vec{k}_{P1} + \vec{k}_{P2} = \vec{k}_S + \vec{k}_I. \tag{8.119}$$

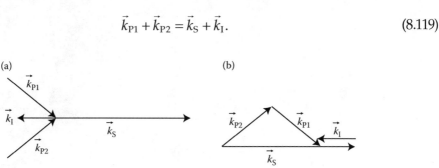

FIGURE 8.9 k-Vector diagram for a backward-wave scheme where the signal and the idler counter-propagate. Two equivalent representations of the k-vector diagram are shown in (a) and (b).

Although two pumps are present in the interaction, it is possible to use a single pump laser with a beam splitter to provide both beams. The sum of the two pump k-vectors can be made to be parallel to, and less in magnitude than, the signal k-vector. Hence, for the proper pump angle, a counter-propagating idler makes $\Delta k = 0$ (see Problem 8.15).

The counter-propagating signal and the idler provide feedback, and like the BOPO in Chapter 7, an oscillation may occur without cavity mirrors. In fact, assuming that the two pump fields are undepleted and $\Delta k = 0$, the analysis is the same as the BOPO calculation (see Problem 8.14).

8.7 DEGENERATE FOUR-WAVE MIXING

8.7.1 INTRODUCTION

Phase matching a backward interaction is particularly simple for the case when all frequencies are the same, called DFWM. In terms of energy conservation, the process is described as $\omega = \omega - \omega + \omega$, and the phase-matching condition is $\Delta \vec{k} = \vec{k}_{P1} + \vec{k}_{P2} - \vec{k}_{Probe} - \vec{k}_{Signal}$, where we use the terminology of P1 (Pump1), P2 (Pump2), probe, and signal to distinguish the different fields. The k-vector diagram and a possible experimental setup are shown in Figure 8.10. Since the frequency of each beam is the same, the k-vector magnitudes are also the same (in an isotropic medium), and phase matching, $\Delta k = 0$, is automatically satisfied for the counter-propagating geometry shown in Figure 8.10.

In an experimental arrangement, three fields are incident on the medium: two counter-propagating pump beams and the probe beam. Figure 8.10b shows one possible arrangement; however, in most experiments, the angle between the probe beam and Pump1 is made small to achieve longer interaction lengths. The three

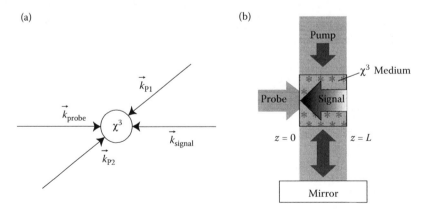

FIGURE 8.10 (a) k-Vector diagram for DFWM, where the frequency of each beam is the same. (b) A possible experimental arrangement with a pump and probe input. The pump is retro-reflected to provide two counter-propagating beams.

fields generate a nonlinear polarization at ω that counter-propagates with respect to the probe. The two pumps are treated as undepleted and are much greater in amplitude than the probe beam. Another way to look at the interaction is in terms of a nonlinear grating set up by the interacting beams. Through the nonlinear index, a diffraction grating is formed from the interference of Pump1 and the probe. A small portion of Pump2 diffracts off the grating into the signal direction, which counter-propagates with respect to the probe beam. Similarly, Pump2 and the probe set up a grating, which causes a portion of Pump1 to diffract into the signal direction.

8.7.2 PUMP PHASE SHIFTS

Although the two pump beams are undepleted, we consider their nonlinear phase shifts due to self-phase and cross-phase modulations. For example, the complex amplitude of the nonlinear polarization of the pump field, P1, is

$$P_{P1}^{(3)}(\omega) = \frac{3\varepsilon_0 \chi^{(3)}}{4} \left(|A_{P1}|^2 + 2|A_{P2}|^2 \right) A_{P1} e^{ik_{P1}z}, \tag{8.120}$$

where terms involving the probe or signal are neglected since they are assumed to be much lower in amplitude. Combinations of P1 and P2 that require phase matching are also not included. The nonlinear polarization is substituted into Equation 8.2 giving the equations for the two pump fields:

$$\frac{dA_{P1}}{dz} = i \frac{3\omega \chi^{(3)}}{8nc} \left[\left(|A_{P1}|^2 + 2|A_{P2}|^2 \right) \right) A_{P1} \right], \tag{8.121}$$

$$\frac{dA_{P2}}{dz} = -i \frac{3\omega \chi^{(3)}}{8nc} \left[\left(|A_{P2}|^2 + 2|A_{P1}|^2 \right) \right) A_{P2} \right]. \tag{8.122}$$

In the undepleted pump approximation, the two pumps experience nonlinear phase shifts, but no change in intensity. Assuming that the two pump magnitudes are equal, $|A_P|^2 \equiv |A_{P1}|^2 = |A_{P2}|^2$, then Equations 8.121 and 8.122 simplify to

$$\frac{dA_{P1}}{dz} = i \frac{9\omega \chi^{(3)} |A_P|^2}{8nc} A_{P1}, \tag{8.123}$$

$$\frac{dA_{P2}}{dz} = -i \frac{9\omega \chi^{(3)} |A_P|^2}{8nc} A_{P2}. \tag{8.124}$$

The solutions to these two equations are

$$A_{P1}(z) = A_{P1}(0)e^{i\kappa' z}, \tag{8.125}$$

$$A_{P2}(z) = A_{P2}(L)e^{i\kappa'(z-L)}, \tag{8.126}$$

where

$$\kappa' = \frac{9\omega\chi^{(3)}|A_P|^2}{8nc}. \tag{8.127}$$

Note that since the second pump propagates in the backward direction, Equation 8.126 is defined in terms of the boundary at $z = L$ (see Figure 8.10b).

8.7.3 PROBE AND SIGNAL FIELDS

The nonlinear polarizations for the probe and signal fields contain terms for a nonlinear phase shift as well as a parametric interaction. We use Equation 8.9 to obtain the nonlinear polarization for the signal ($k_s > 0$, $k_{Probe} = k_s$)

$$\begin{aligned}
P_S^{(3)}(\omega) &= \frac{\varepsilon_0\chi^{(3)}}{4}\left(6\left(|A_{P1}|^2 + |A_{P2}|^2\right)A_S e^{-ik_S z} + 6A_{P1}A_{P2}A_{Probe}^* e^{-ik_{Probe} z}\right), \\
&= \frac{3\varepsilon_0\chi^{(3)}}{2}\left(2|A_P|^2 A_S e^{-ik_S z} + A_{P1}A_{P2}A_{Probe}^* e^{-ik_S z}\right),
\end{aligned} \tag{8.128}$$

where $|A_S|^2 \ll |A_P|^2$ and $\vec{k}_{P1} + \vec{k}_{P2} = 0$, $\vec{k}_{Probe} = -\vec{k}_{Signal}$, and the degeneracy factor in Equation 8.9 is determined from Table 8.1. Substituting the nonlinear polarization for the signal and a similar expression for the probe into Equation 8.2 gives

$$\frac{dA_{Probe}}{dz} = i\frac{3\omega\chi^{(3)}}{4nc}\left[2|A_P|^2 A_{Probe} + A_{P1}A_{P2}A_{Signal}^*\right], \tag{8.129}$$

$$\frac{dA_{Signal}}{dz} = -i\frac{3\omega\chi^{(3)}}{4nc}\left[2|A_P|^2 A_{Signal} + A_{P1}A_{P2}A_{Probe}^*\right]. \tag{8.130}$$

The minus sign in Equation 8.130 comes out of deriving Equation 8.2 for a counter-propagating field with an $e^{i(-kz-\omega t)}$ dependence instead of an $e^{i(kz-\omega t)}$ dependence.

As in the case of the parametric amplifier in Section 8.5, we simplify the probe and signal equations by changing variables to factor out the nonlinear phase that occurs in the absence of the parametric term

$$A_{\text{Probe}} = A'_{\text{Probe}}e^{i\kappa''z}, \tag{8.131}$$

$$A_{\text{Signal}} = A'_{\text{Signal}}e^{i\kappa''(L-z)}, \tag{8.132}$$

where

$$\kappa'' = \frac{3\omega\chi^{(3)}|A_P|^2}{2nc}. \tag{8.133}$$

With these variable substitutions and the equations for the two pump fields, Equations 8.125 and 8.126, the coupled amplitude equations for the probe and signal fields, Equations 8.129 and 8.130, become

$$\frac{dA'_{\text{Probe}}}{dz} = i\frac{6\omega\chi^{(3)}}{8nc}|A_P|^2 A'^*_{\text{Signal}} = i\eta A'^*_{\text{Signal}}, \tag{8.134}$$

$$\frac{dA'_{\text{Signal}}}{dz} = -i\frac{6\omega\chi^{(3)}}{8nc}|A_P|^2 A'^*_{\text{Probe}} = -i\eta A'^*_{\text{Probe}}. \tag{8.135}$$

The solution to these equations follows the same approach as for the backward-propagating OPO discussed in Chapter 7. The boundary conditions for the probe and signal fields are assumed to be $A_{\text{Probe}}(0) = A_{\text{Probe,in}}$ and $A_{\text{Signal}}(L) = 0$.

We combine Equations 8.134 and 8.135 by taking the derivative of Equation 8.134 and using Equation 8.135 in the resulting expression to give

$$\frac{d^2 A'_{\text{Probe}}}{dz^2} = -\eta^2 A'_{\text{Probe}}. \tag{8.136}$$

Similarly,

$$\frac{d^2 A'_{\text{Signal}}}{dz^2} = -\eta^2 A'_{\text{Signal}}. \tag{8.137}$$

The solutions to Equations 8.136 and 8.137 are

$$A'_{\text{Signal}} = Ce^{i\eta z} + De^{-i\eta z}, \tag{8.138}$$

$$A'_{\text{Probe}} = Ee^{i\eta z} + Fe^{-i\eta z}. \tag{8.139}$$

C, D, E, and F are constants determined from the boundary conditions. We first solve for C and D,

$$A'_{\text{Signal}}(L) = Ce^{i\eta L} + De^{-i\eta L} = 0.$$ (8.140)

Hence,

$$C = -De^{-2i\eta L}.$$ (8.141)

The second boundary condition relates to the derivative at $z = 0$, which from Equation 8.136 is

$$\left.\frac{dA'_{\text{Signal}}}{dz}\right|_0 = -i\eta A'^*_{\text{Probe,in}} = i\eta(C - D).$$ (8.142)

Combining Equations 8.141 and 8.142 allows us to solve for C and D:

$$C = -\frac{A'^*_{\text{Probe,in}} e^{-i\eta L}}{2\cos(\eta L)},$$ (8.143)

$$D = \frac{A^*_{30} e^{i\eta L}}{2\cos(\eta L)}.$$ (8.144)

With these constants, Equation 8.138 yields

$$A'_{\text{Signal}}(z) = \frac{i\sin(\eta(L-z))}{\cos(\eta L)} A'^*_{\text{Probe,in}}.$$ (8.145)

In terms of the original signal and probe field variables, we have

$$A_{\text{Signal}}(z) = \frac{i\sin(\eta(L-z))}{\cos(\eta L)} A^*_{\text{Probe0}} e^{i\kappa''(L-z)}.$$ (8.146)

A particularly interesting aspect of this solution is obtained by looking at the fields at $z = 0$:

$$A_{\text{Signal}}(0) = -ie^{-i\kappa''L} \tan(\eta L) A^*_{\text{Probe},in}.$$ (8.147)

Written in this form, we see that the signal is related to the phase conjugate of the probe; hence, the DFWM acts as a phase-conjugate mirror. Moreover, the signal beam is an amplified version of the probe, depending on the value of ηL. In the limit that $\eta L \to \pi/2$, our assumption of undepleted pumps breaks down.

8.7.4 OPTICAL-PHASE CONJUGATION

The phase conjugation of the probe field given in Equation 8.147 is based on $\chi^{(3)}$ with an electronic origin. Although not covered here, other mechanisms, such as the Brillouin scattering, can also lead to a phase-conjugate–reflected beam. The action of a phase-conjugate mirror compared to a normal mirror is illustrated in Figure 8.11. To understand the behavior of the phase-conjugate mirrors, we consider the probe field just before, and the signal field just after the phase-conjugate mirror. The probe beam has a beam profile with a complex amplitude given by $A_{\text{Probe}}(x, y, z)$, where x and y are the transverse coordinates. The probe beam accumulates an additional phase, $\Delta\phi(x, y)$, as it travels to the phase-conjugate mirror where the beam's complex amplitude is given by

$$A_{\text{Probe}}(x,y,z+\Delta z) = \left|A_{\text{Probe}}(x,y,z)\right| e^{i\phi_{\text{Probe}}(x,y)} e^{i\Delta\phi(x,y)}, \tag{8.148}$$

where ϕ_{Probe} is the phase of the probe field at z. The phase-conjugate–reflected signal beam is therefore

$$A_{\text{Signal}}(x,y,z+\Delta z) = \gamma \left|A_{\text{Probe}}(x,y,z)\right| e^{-i\phi_{\text{Probe}}(x,y)} e^{-i\Delta\phi(x,y)}, \tag{8.149}$$

where γ is an amplification factor (see Equation 8.147). The key difference between the signal and the probe is the minus sign in the exponent. As the signal propagates in the backward direction, it accumulates an additional $e^{+i\Delta\phi(x,y)}$ phase such that the extra accumulated phase term cancels:

$$A_{\text{Signal}}(x,y,z) = \gamma \left|A_{\text{Probe}}(x,y,z)\right| e^{-i\phi_{\text{Probe}}(x,y)} = \gamma A_{\text{Probe}}^*(x,y,z). \tag{8.150}$$

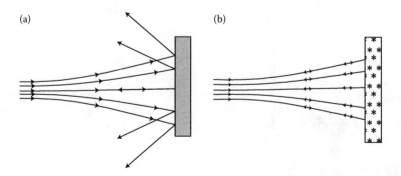

FIGURE 8.11 A diverging waveform is incident on (a) a regular mirror and (b) a phase-conjugate mirror. The pump beams for the phase-conjugate mirror are not shown.

This example shows how the backward-propagating signal beam matches the incoming probe beam.

Another way to show the same effect is to consider the signal's real field:

$$E_{\text{Signal}} = \frac{1}{2}\gamma A_{\text{Probe}}^*(x,y,z)\,e^{i(-kz-\omega t)} + \text{c.c.} \qquad (8.151)$$

We can choose to rewrite this equation as

$$E_{\text{Signal}} = \frac{1}{2}\gamma^* A_{\text{Probe}}(x,y,z)e^{i(kz+\omega t)} + \text{c.c.} \qquad (8.152)$$

Now note that this is the same field as the probe field with the substitution of $t \rightarrow -t$. That is, the signal is a scaled and "time-reversed" version of the probe field, and it therefore reproduces the input probe beam as shown in Figure 8.11.

An application for phase-conjugate mirrors is coherent image restoration. For example, a laser passing through an aberrating medium ends up with a distorted phase front. However, if the aberrated beam is reflected with a phase conjugation, the phase due to the aberration cancels out when the beam propagates back through the medium. One example is shown in Figure 8.12. Here, a seed laser with a high-quality beam profile is amplified in a gain medium where the beam profile is significantly degraded. The phase-conjugate mirror allows the amplified beam to propagate backward and cancel out the aberrations introduced by the gain medium. The backward-propagating beam is separated from the input beam with a polarizing beam splitter. This type of scheme has been applied to pulsed lasers using phase conjugation based on the Brillouin effect (which is introduced in Chapter 9). The same idea also applies to image restoration where, instead of the laser propagating through an amplifier, it passes through a passive distorting medium.

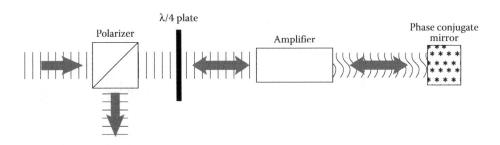

FIGURE 8.12   Phase-conjugate mirror setup for amplifying an optical beam with good beam quality. The QWP and polarizer allow the input and output beams to be separated.

8.8 z-SCAN

8.8.1 INTRODUCTION

The self-focusing effect discussed in Section 8.4 is the basis for a technique to measure the nonlinear index of refraction. The idea, introduced by M. Sheik-Bahae and coworkers, is based on the change in divergence of a beam after passing through a nonlinear medium (Sheik-Bahae et al., 1990). Because the nonlinear index is proportional to intensity, the beam's radius of curvature changes as its intensity is increased or decreased. By measuring the change in divergence, it is possible to back out the nonlinear index. Changing the beam intensity distribution on a sample by changing the power introduces two variables, a change in overall power and a change in divergence due to nonlinearity. To get away from such a coupled measurement, the z-scan approach keeps the overall power constant and obtains a variable intensity by focusing the beam. A sample is placed in the beam and translated through the focal region where the beam's intensity distribution changes because the beam size changes. This technique is called a z-scan because the sample is translated through the focus in the z-direction. A concept of the experiment is shown in Figure 8.13.

The change in divergence is measured by using a pinhole to mask off all but the center of the beam. As the beam divergence changes, the power that passes through the pinhole also changes. For a material exhibiting no nonlinearity, the translating sample has no effect on the beam's divergence and the power passing through the pinhole is constant.

Consider a material with a positive nonlinear index, $n_2 > 0$ (see Figure 8.13a) incident on the sample. The Gaussian beam has an intensity distribution that changes across its beam cross section, and the intensity-dependent index forms a effective index profile that is equivalent to adding a thin lens with a positive focal length at the sample location (see Section 8.4). The effective focal length of the nonlinear lens is given by Equation 8.78. As shown in Figure 8.13a, on the side of the focus where the beam is converging (sample position labeled by 1), the nonlinear lens shortens the beam's waist position to a negative z value. As the beam passes through the shifted

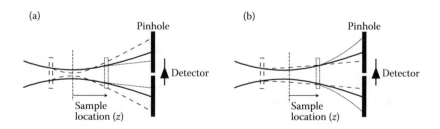

FIGURE 8.13 Beam divergence for various sample locations for (a) positive n_2^I and (b) negative n_2^I, z is measured with respect to the original beam waist location.

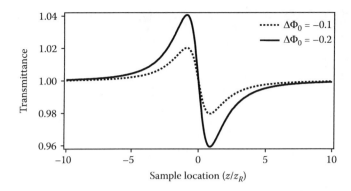

FIGURE 8.14 Transmittance as a function of sample position for a material with a negative nonlinearity. $\Delta\Phi_0$ is the on-axis nonlinear phase shift at the focus, defined in Equation 8.161.

focus, it diverges at a greater diffraction angle and the beam's power is spread over a wider area in the far field so that the power passing through the pinhole decreases. Placing the sample of the positive z side (Sample Position 2) where the beam is diverging, the nonlinear lens tends to decrease the radius of curvature and the beam divergence is smaller so that in the far field, the power passing through the pinhole increases. The effective focal length generated by the nonlinear lens depends on the peak intensity so that when the sample is placed far from the focus, the intensity is low and nonlinear focusing effects are negligible. Figure 8.14 shows the characteristic signature of a z-scan for a material with a negative nonlinearity.

8.8.2 MEASURING THE NONLINEAR INDEX OF REFRACTION

Z-scan analysis centers on modifications to the focused Gaussian beam through the action of the nonlinearity. In practice, obtaining appreciable nonlinear effects requires short-pulsed lasers; hence, a detailed calculation includes the temporal profile of the incident laser. However, we consider only the continuous-wave case to show the basic principles; for an analysis that includes time dependence, see Sheik-Bahae et al. (1990). The complex amplitude of a focused Gaussian beam is given in Chapter 2 and is repeated here:

$$A(r,z) = A_0 \frac{q_0}{q} \exp\left(i \frac{kr^2}{2q} \right),$$ (8.153)

where A_0 is the on-axis field amplitude at the focus, and we use the Gaussian beam parameter defined for beams of the form $e^{i(kz-\omega t)}$:

$$\frac{1}{q} = \frac{1}{R(z)} + \frac{2i}{kw^2(z)}.$$ (8.154)

$z = 0$ is the location of the beam waist where R is infinite (flat phase front) and $q_0 = -iz_R$ (see Section 2.6). Note that $q(z) = q_0 + z$, which lets us write

$$\frac{q_0}{q} = \frac{-iz_R}{z - iz_R}.$$ (8.155)

We substitute this relationship into Equation 8.153 and calculate the corresponding intensity:

$$I(z, r) = \frac{I_0}{1 + (z^2 / z_R^2)} \exp\left(\frac{-2r^2}{w^2(z)}\right).$$ (8.156)

When a sample is placed at a location, z, then the Gaussian beam is modified by the absorption of the sample and by its nonlinear phase shift. We assume a linear absorption in the sample that follows Beer's law so that

$$I(z + z', r) = I(z, r) e^{-\alpha_0 z'},$$ (8.157)

where z' is the distance in the sample.

As the beam passes through the sample, it accumulates a nonlinear phase shift given by Equation 8.48, which we rewrite in terms of the nonlinear index and intensity:

$$\frac{d\phi^{NL}}{dz'} = \frac{\omega}{c} n_2^I I(z + z', r) = \frac{\omega}{c} n_2^I I(z, r) e^{-\alpha_0 z'}.$$ (8.158)

Integrating this equation gives the nonlinear phase shift for the sample:

$$\Delta\phi^{NL} = \frac{\omega}{c} n_2^I I(z, r) \frac{1 - e^{-\alpha_0 L}}{\alpha_0}.$$ (8.159)

The linear absorption results in an effective interaction length defined by

$$L_{eff} = \frac{1 - e^{-\alpha_0 L}}{\alpha_0}.$$ (8.160)

This equation shows that the effective interaction length is reduced by the presence of absorption. In the limit, $\alpha L \ll 1$, note that $L_{eff} \to L$. The nonlinear phase shift in terms of L_{eff} is

$$\Delta\phi^{\mathrm{NL}} = \frac{\omega}{c} n_2^I I(z, r) L_{\mathrm{eff}}. \tag{8.161}$$

The intensity is rewritten in terms of its values at the focus ($z = 0$ and $r = 0$) using Equation 8.156:

$$\Delta\phi^{\mathrm{NL}} = \frac{1}{1+(z^2/z_R^2)} \frac{\omega}{c} n_2^I L_{\mathrm{eff}} I_0 \exp\left(-2r^2/w^2\right) = \Delta\phi_0 \exp\left(-2r^2/w^2\right), \tag{8.162}$$

where

$$\Delta\phi_0 = \frac{\omega n_2^I I_0 L_{\mathrm{eff}}}{c\left(1+(z^2/z_R^2)\right)}. \tag{8.163}$$

$\Delta\phi_0$ at $z = 0$ is the on-axis nonlinear phase shift at the focus, which we define as

$$\Delta\Phi_0 = \frac{\omega}{c} n_2^I I_0 L_{\mathrm{eff}}. \tag{8.164}$$

In terms of $\Delta\Phi_0$, Equation 8.163 is

$$\Delta\phi_0 = \frac{\Delta\Phi_0}{1+(z^2/z_R^2)}. \tag{8.165}$$

Written in this form, note that the nonlinear phase shift is directly proportional to n_2^I through $\Delta\Phi_0$ and that it is a function of z, the position of the sample in the focused beam. We now have a connection between the nonlinear index and the experimental variable z. The following calculations show how the nonlinear phase shift leads to a change in the signal on a detector and how we can extract n_2^I from that signal.

In a thin-sample approximation, the Gaussian beam that emerges from the sample is a phase-shifted version of the beam that enters it, with the phase shift varying across the beam cross section:

$$A(z, r) = \frac{A_0}{1+i(z/z_R)} \exp\left(-\frac{r^2}{w^2(z)}\right) \exp\left(i\frac{kr^2}{2R}\right) \exp\left((i\Delta\phi^{\mathrm{NL}})\right). \tag{8.166}$$

In Section 8.4, we approximated the nonlinear phase shift as a thin lens. Another approach is to rewrite the nonlinear phase exponential as a Gaussian beam decomposition. In this way, the modified beam can be thought of as a superposition of the Gaussian beams and each of the Gaussian beams is propagated to the detector plane using the standard Gaussian beam propagation equations given in Section 2.6. The Gaussian beam decomposition starts with expanding the nonlinear phase exponential

$$\exp\left(i\Delta\phi^{NL}\right) = 1 + i\Delta\phi^{NL} + \frac{\left(i\Delta\phi^{NL}\right)^2}{2!} + \cdots = \sum \frac{\left(i\Delta\phi^{NL}\right)^m}{m!}. \tag{8.167}$$

We substitute in Equation 8.167 for $\Delta\phi^{NL}$ to obtain

$$e^{i\Delta\phi^{NL}} = \sum \frac{\left(i\Delta\phi_0\right)^m}{m!} \exp\left(-2mr^2/w^2\right). \tag{8.168}$$

Equation 8.168 is substituted into Equation 8.166 to give the Gaussian beam decomposition:

$$A(z,r) = \frac{A_o}{1 + i(z/z_R)} \sum \frac{\left(i\Delta\phi_0\right)^m}{m!} \exp\left\{-\frac{r^2}{w^2}(1+2m) + i\frac{kr^2}{2R}\right\}. \tag{8.169}$$

Each term, m, in the summation is in the form of a Gaussian field, and we can extract the beam size and radius of curvature. The beam size for a given m is

$$w_{m0}^2 = \frac{w^2}{1+2m}. \tag{8.170}$$

The radius of curvature for each m is given by R. From w_{m0} and R, we obtain a Gaussian beam parameter:

$$\frac{1}{q_{m0}} = \frac{1}{R} + \frac{2i}{kw_{m0}^2}. \tag{8.171}$$

In terms of the Gaussian beam parameter, q_{m0}, Equation 8.169 becomes

$$A(z,r) = \frac{A_0}{1 + i(z/z_R)} \sum \frac{\left(i\Delta\phi_0\right)^m}{m!} \exp\left\{i\frac{kr^2}{2q_{m0}}\right\}. \tag{8.172}$$

Equation 8.172 gives the field immediately after the thin sample. Now, we are in a position to propagate each Gaussian beam component (each m) to the aperture plane a distance d away using the Gaussian beam propagation equation (see Section 2.6):

$$q'_m = q_{m0} + d. \tag{8.173}$$

The total field at the detector plane is a superposition of the Gaussian beam components propagated a distance d:

$$A(z+d,r) = \frac{A_0}{1+\mathrm{i}(z/z_R)} \sum \frac{(\mathrm{i}\Delta\phi_0)^m}{m!} \frac{q_{m0}}{q_{m0}+d} \exp\left\{-\mathrm{i}\frac{kr^2}{2(q_{m0}+d)}\right\}, \tag{8.174}$$

where $A(z + d, r)$ is the field amplitude at the detector plane distance d away from the sample and where the sample is located at z. The ratio

$$\frac{q_{m0}}{q_{m0}+d} \tag{8.175}$$

in Equation 8.174 accounts for the change in the on-axis amplitude as each Gaussian beam component propagates. Substituting Equation 8.171 into Equation 8.175 gives

$$\frac{q_{m0}}{q_{m0}+d} = \frac{1}{1+(d/R)+(2d\mathrm{i}/kw_{m0}^2)} = \frac{1}{g+\mathrm{i}(d/d_m)}, \tag{8.176}$$

where

$$g \equiv 1 + d/R \tag{8.177}$$

and

$$d_m \equiv kw_{m0}^2/2. \tag{8.178}$$

We now calculate the transmission of the field through a pinhole placed in the far field, $d \gg z$. The transmission through a pinhole is approximated by considering only the on-axis component of the field:

$$T(z,\Delta\Phi_0) = \frac{|A(z+d,r=0,\Delta\phi_0)|^2}{|A(z+d,r=0,\Delta\phi_0=0)|^2}. \tag{8.179}$$

A further approximation is to assume small nonlinearities ($|\Delta\Phi_0| \ll 1$) such that the contribution from higher-order terms in expansion 8.174 may be neglected. Keeping two terms in the expansion gives

$$T(z, \Delta\Phi_0) \approx \frac{|(g+i(d/d_0))^{-1} + i\Delta\phi_0(g+i(d/d_1))^{-1}|^2}{|(g+i(d/d_0))^{-1}|^2}. \tag{8.180}$$

This expression simplifies in the far field where $d \gg z_R$. In this limit,

$$g \approx d/R. \tag{8.181}$$

Hence, Equation 8.180 is approximated as

$$T(z, \Delta\Phi_0) \approx \left|1 + i\Delta\phi_0\left(\frac{d_0 + iR}{d_0 + 3iR}\right)\right|^2, \tag{8.182}$$

where we use $d_1 = d_0/3$ from Equations 8.178 and 8.170. Furthermore, it is possible to show

$$\frac{d_0}{R} = \frac{z}{z_R}. \tag{8.183}$$

With this substitution and for small $\Delta\Phi_0$, Equation 8.182 becomes

$$T(z, \Delta\Phi_0) \approx 1 + \frac{4\Delta\Phi_0 x}{(x^2+9)(x^2+1)}, \tag{8.184}$$

where $x = z/z_R$. The location of the peak and valley of $T(z, \Delta\Phi_0)$ are found by solving for the maximum and minimum of Equation 8.184. The transmission difference between these two points can be shown to be

$$\Delta T_{p-v} = 0.406\Delta\Phi_0. \tag{8.185}$$

By measuring the characteristic curve in Figure 8.14, we may extract the nonlinear phase shift, $\Delta\Phi_0$, which with Equation 8.163 gives the nonlinear index coefficient.

8.8.3 NONLINEAR ABSORPTION

In addition to measuring the nonlinear index, the z-scan technique can also measure nonlinear absorption. By removing the pinhole and measuring the total power as a

---

function of sample position z, the z-scan technique provides a measure of changes in the integrated power hitting the detector.

For an absorption in the form of $\alpha = \alpha_o + \beta I$, we integrate Equation 8.64 (see Problem 8.18),

$$I(z+L, r) = \frac{I(z,r)e^{-\alpha_0 L}}{1 + \beta I(z,r)L_{\text{eff}}},$$

$$(8.186)$$

where L_{eff} is given by Equation 8.160 and $L \ll z_R$ is the sample thickness and

$$I(z, r) = I(z)\, e^{-2r^2/w(z)^2}.$$

$$(8.187)$$

$I(z)$ is the on-axis intensity and is given by

$$I(z) = \frac{I_0}{1 + (z^2/z_R^2)},$$

$$(8.188)$$

where I_0 is the on-axis intensity at the focus. The total power transmitted through the sample is found by integrating over the radial dimension:

$$P_{\text{out}}(z) = \int_0^\infty \frac{I(z,r)\, e^{-\alpha_0 L}}{1 + \beta I(z,r)L_{\text{eff}}}\, 2\pi r\, dr$$

$$= \frac{\pi w^2(z)e^{-\alpha_0 L}}{2\beta L_{\text{eff}}} \ln\left(1 + \beta I(z)\, L_{\text{eff}}\right).$$

$$(8.189)$$

We note that the total power in the incident beam is related to the on-axis intensity $I(z)$ by Equation 7.23, repeated here:

$$P = \frac{\pi w^2(z)}{2} I(z).$$

$$(8.190)$$

We substitute this equation into Equation 8.189:

$$\frac{P_{\text{out}}(z)}{P} = \frac{e^{-\alpha_0 L}}{I(z)\,\beta L_{\text{eff}}} \ln\left(1 + \beta I(z)\, L_{\text{eff}}\right).$$

$$(8.191)$$

This equation is the transmittance, T, which we rewrite in terms of z by using in Equation 8.188:

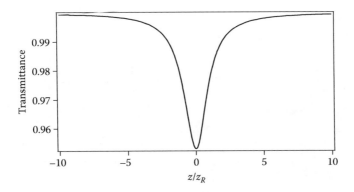

FIGURE 8.15   Plot of Equation 8.190 corresponding to an open-aperture z-scan showing nonlinear absorption.

$$T = \frac{e^{-\alpha_0 L}}{\beta I_0 L_{\text{eff}}} \left(1 + \frac{z^2}{z_R^2} \right) \ln \left(1 + \frac{\beta I_0 L_{\text{eff}}}{1 + (z^2/z_R^2)} \right). \tag{8.192}$$

Equation 8.192 is plotted in Figure 8.15 as a function of sample location, z, for the conditions where $\alpha_0 = 0$ and $\beta I_0 L = 0.1$. The transmission function given by Equation 8.192 may be used to fit experimental data, and by doing so, determine the nonlinear absorption coefficient, β.

In fact, both the open aperture (β) and the closed aperture (n_2^I) may be measured simultaneously by using a camera to image the entire beam. Instead of using a pinhole for the closed aperture, we use software to single out a few pixels at the center of the beam. The open aperture is obtained by integrating the signal contribution from the entire image.

PROBLEMS

8.1   Use the relationship for the complex amplitude of the third-order nonlinearity (see below) to prove that $\chi_{ijkl}^{(3)}$ transforms according to the rules for a Rank 4 tensor (see Section 2.2):

$$P_i^{(3)} = \frac{\varepsilon_0}{4} D \chi_{ijkl}^{(3)} A_j A_k A_l. \tag{8.193}$$

8.2   Show that for an isotropic material

$$\chi_{xxxx}^{(3)} = \chi_{xxyy}^{(3)} + \chi_{xyyx}^{(3)} + \chi_{xyxy}^{(3)}. \tag{8.194}$$

(Hint: Consider the transformation operation corresponding to a rotation of an arbitrary angle θ about the z-axis.)

8.3 A 3 m crystal has a reflection symmetry given by Equation 8.16 (reflection in the yz-plane). Similar to the centrosymmetric case presented in Section 8.2.3, this symmetry requires that $\chi_{ijkl}^{(3)} = 0$ for the case when i, j, k, and l are restricted to x and y and when there is either an odd number of x's or an odd number of y's, that is, $\chi_{xxxy}^{(3)} = 0; \chi_{yxxx}^{(3)} = 0; \ldots$ Another symmetry for the 3 m crystal is a rotation of $120°$ about the z-axis. The transformation matrix for this rotation is

$$
\begin{bmatrix}
-\dfrac{1}{2} & \dfrac{\sqrt{3}}{2} & 0 \\[2mm]
-\dfrac{\sqrt{3}}{2} & -\dfrac{1}{2} & 0 \\[2mm]
0 & 0 & 1
\end{bmatrix}.
\tag{8.195}
$$

   a. Use the R-matrix in Equation 8.195 to calculate a relationship between $\chi_{xxxx}^{(3)}$ and other tensor elements. Note: $\chi_{xxxy}^{(3)} = 0; \chi_{yyyx}^{(3)} = 0; \ldots$ and R is a symmetry operation for the crystal $\chi_{xxxx}^{(3)}{}' = \chi_{xxxx}^{(3)}$.
   b. Repeat part (a) for $\chi_{yyyy}^{(3)}$.
   c. Subtract the results for (a) and (b) to show that $\chi_{xxxx}^{(3)} = \chi_{yyyy}^{(3)}$ for 3 m crystals.
   d. Optional. Use the same procedure to show that $\chi_{xxyy}^{(3)} = \chi_{yyxx}^{(3)}$.

8.4 Consider a four-wave mixing problem defined by $\omega_1 = \omega_2 + \omega_3 + \omega_4$ in a medium that has a purely real $\chi^{(3)}$. Use Poynting's theorem to show that the self-phase-matched contribution to the nonlinear polarization (see Equation 8.196) at ω_1 does not involve energy exchange with the field at ω_1.

$$
P_{\text{self-phase matched}}^{(3)}(\omega_1) = \frac{3\varepsilon_0 \chi^{(3)}}{4} \left[(|A_1|^2 + 2|A_2|^2 + 2|A_3|^2 + 2||A_4|^2) A_1 e^{ik_1 z} \right].
\tag{8.196}
$$

8.5 A parametric four-wave mixing process is defined by $\omega_1 + \omega_2 = \omega_3 + \omega_4$. Follow a similar procedure as that used to derive Equation 8.40 to show that when the medium is lossless:

$$
\Delta N_3 = \Delta N_4 = -\Delta N_1 = -\Delta N_2.
\tag{8.197}
$$

8.6 In the case of self-induced processes where the nonlinear polarization frequency is the same as the input, $\omega = \omega + \omega - \omega$, follow a similar procedure as that used to derive Equation 8.40 to show that $\Delta N = 0$.

8.7 Prove that in a material exhibiting normal dispersion, the process defined by $\omega_1 = \omega_2 + \omega_3 + \omega_4$ results in $k_1 > k_2 + k_3 + k_4$ when all fields have the same polarization.

8.8 Prove that the signal and idler frequencies in a four-wave mixing parametric amplifier process are symmetrical about the pump frequency.

8.9 Consider a parametric interaction in an isotropic medium defined by $2\omega_P = \omega_S + \omega_I$. In Problem 8.8, we show that ω_S and ω_I are symmetrically located about ω_P. In many cases, the frequency separation between the pump, signal, and idler is small enough to allow us to write k_S and k_I as the Taylor series expansions about k_P (e.g., $k_S = k_P + (\partial k / \partial \omega)\big|_{\omega_P} \Delta\omega + \cdots$). Show that in this situation and when the pump frequency is fixed, Δk is given by

$$\Delta k = \frac{1}{v_g^2} \frac{\partial v_g}{\partial \omega} \Delta\omega^2, \tag{8.198}$$

where $\Delta\omega$ is the frequency separation between the pump and the signal (and between the pump and the idler) and v_g is the group velocity.

8.10 A parametric amplifier defined by the relationship $2\omega_P = \omega_S + \omega_I$ occurs in fused silica (see the Sellmeier equation in Appendix B). In this interaction, there is a single pump laser incident on the glass. Determine the wavelength range where noncollinear phase matching is possible. To find the range, use Equation 8.196 to plot Δk as a function of λ_P over $0.3~\mu m < \lambda_P < 2.0~\mu m$. The region where $\Delta k < 0$ is the wavelength range where noncollinear phase matching is possible for the collinear pump.

8.11 Find the noncollinear angle (angle between the signal and the pump) that makes $\Delta k = 0$ in fused silica (use the Sellmeier equation in Appendix B) for a parametric amplifier defined by the relationship $2\omega_P = \omega_S + \omega_I$. The pump wavelength is $0.5~\mu m$, and the signal wavelength is $0.45~\mu m$.

8.12 Show that the solution to the parametric amplifier equation with an undepleted pump, Equation 8.112, with the initial conditions $A_I(0) = 0$ and $A_S(0) = A_{S0}$ is given by Equation 8.113. Show that the corresponding signal intensity is as given in Equation 8.115.

8.13 Numerically integrate the coupled amplitude equations for the parametric amplifier, Equations 8.92 through 8.94 given the initial conditions: $A_P(0) = 2.5 \times 10^7$ V/m, $A_S(0) = 8 \times 10^6$ V/m, and $A_I(0) = 0$. Use a Runge–Kutta or similar algorithm (see Appendix C) to perform the numerical integration. Assume that the interaction occurs in a fiber so that we can have a long interaction length. Plot the resultant pump, signal, and idler intensities as a function of distance in the fiber from 0 to 200 m. Start with a step size of 0.5 m and reduce it until the solution stops changing.

8.14 Consider a parametric amplifier defined by the relationship $2\omega_P = \omega_S + \omega_I$ where the pump is undepleted. Use the two coupled amplitude equations, Equations 8.110 and 8.111, to solve for the signal field, A_S', for a backward geometry as shown in Figure 8.9. Note that Equation 8.111 should have a sign change to account for the backward-propagating wave. Assume that the process is phase matched, the interaction length is L, and the boundary conditions are $A_S(0) = A_{S0}$ and $A_I(L) = 0$.

Does this backward parametric amplifier have an oscillation condition similar to the BOPO discussed in Chapter 7?

8.15 For the backward parametric interaction shown in Figures 8.9a and b derive an expression for the angle between the pump and the signal that satisfies the noncollinear phase-matching condition. Use this formula to calculate the noncollinear angle in fused silica (see the Sellmeier equations in Appendix B) between the pump and the signal for a backward parametric interaction when the pump wavelength is 1.5 μm and the signal wavelength is 1.4 μm.

8.16 Consider a Gaussian beam with the front face of a sample located at the beam waist. The beam's radius is w_0, and the sample has a positive n_2^I. In the first segment of the sample, the nonlinearity induces a negative radius of curvature as given in Equation 8.77. The focusing effect of the nonlinearity can be balanced by diffraction for a critical beam power.

a. Calculate the radius of curvature due to diffraction by neglecting the sample's nonlinearity. This calculation can be accomplished by directly calculating the radius of curvature from the Gaussian beam parameter, q, after propagating a distance Δz. Note that $z_R \gg \Delta z$, where z_R is the Rayleigh range.

b. When the radius of curvature of the diffraction is equal but opposite to that due to the nonlinearity, the two effects cancel and the beam waist is unchanged in the sample. Calculate the beam power where this balancing occurs and show that the balance depends only on the beam power and not the beam radius.

8.17 Estimate the change in the index of refraction ($n_2^I I$) in a medium with $n_2^I = 10^{-16}\,\text{cm}^2/\text{W}$ for

a. A continuous-wave laser beam

b. A 10 ns width pulsed from a laser operating at a 30 Hz pulse repetition rate

c. A 100 fs pulse width from a laser operating at a 1 kHz pulse repetition rate

For all three lasers the average power is 1 W and the beam is focused to an area of $10^{-4}\,\text{cm}^2$. Assume a flat-top spatial beam profile and a rectangular pulse shape in time to simplify your intensity calculations.

8.18 Solve for intensity as a function of distance in a nonlinear medium using Equation 8.64.

8.19 Two laser beams, one at angular frequency ω_1 and the other at ω_2, propagate in a long fiber-optic cable.

a. If the beam at ω_2 is phase modulated, what is the effect on the beam at ω_1?

b. If the beam at ω_2 is amplitude modulated, what is the effect on the beam at ω_1?

Note that a phase-modulated laser beam's field is in form, $E = E_0 \cos(kz - \omega t + \delta \sin \Omega t)$, where δ is called the modulation index and Ω is the phase modulation frequency.

8.20 Consider the equation for the optical bistable etalon, Equation 8.84, for the case when $R = 0.7$, $n_2^I = 10^{-10}\,\text{cm}^2/\text{W}$, $n_0 = 1.5$, $L = 2$ cm, and $\lambda = 1.5$ μm. Plot I_0 as a function of I_{Trans} over the range $0 < I_{Trans} < 5 \times 10^9$ W/m^2. Include on the plot the line $I_0 = I_{Trans}$. Interpret the resulting plot.

8.21 Reproduce the optical bistability curve given in Figure 8.4 by plotting the zero-contour for the function given in Equation 8.88 for $R = 0.4$, $n_0 = 1.5$, $L = 5.5$ mm, and $\lambda = 1.55$ µm. Note that several branches of solutions appear that are not shown in Figure 8.4. Explain why these branches are inaccessible.

8.22 Determine the sensitivity of the optical bistability curve given in Figure 8.4 to a change in cavity length by plotting the zero-contour of Equation 8.88 for $R = 0.4$, $n_0 = 1.5$, and $\lambda = 1.55$ µm, but with cavity lengths that differ from the starting value of 5.5 mm. We define the sensitivity as the change in cavity length that leads to a clear change in the bistability curve. Plot the bistability curve for $L = 5.5$ mm and for $L = 5.5 + \Delta L$, where ΔL is the value you determine that leads to an appreciable change in the bistability curve. Is the requisite change in cavity length likely to occur due to thermal expansion? For glass, the thermal expansion is $\Delta L/L \sim 10^{-4}$ for a 10°C temperature change.

8.23 Determine the sensitivity of the optical bistability curve given in Figure 8.4 to a change in wavelength by plotting the zero-contour for $R = 0.4$, $n_0 = 1.5$, and $L = 5.5$ mm, but with wavelengths that differ from 1.55 µm. We define the sensitivity as the change in wavelength that leads to a clear change in the bistability curve. Plot the bistability curve for $\lambda = 1.55$ µm and for $\lambda = 1.55$ µm $+ \Delta\lambda$, where $\Delta\lambda$ is the value you determine that leads to an appreciable change in the bistability curve.

8.24 Verify that the nonlinear polarization for signal field in a DFWM problem is as given in Equation 8.128. Use this result in combination with the wave equation in the form of Equation 8.2 to derive Equations 8.129 and 8.130.

8.25 In a z-scan experiment, a key constraint is translating the sample. If we have a translation stage that allows for a 20 cm travel, how tight a focus is required such that we could create the signature z-scan transmission curve (see Figure 8.14)? Estimate the maximum sample thickness such that the thin-sample approximation, $L \ll z_R$, is valid. Assume a wavelength of 0.800 µm.

8.26 In a z-scan experiment, the modified beam is written as the sum of modified Gaussian beams. Show that for each Gaussian beam component, m, in the beam size at the detector plane is $w_m^2 = w_{m0}^2 \left(g^2 + (d^2/d_m^2) \right)$. See Section 8.8 for variable definitions.

8.27 Verify that the expression for the on-axis transmission is given by Equation 8.182 by first showing that $d_0/R = z/z_R$ and then use Equation 8.180.

8.28 Use Equation 8.184 to find the displacement between the maximum and minimum of the z-scan's transmission curve. Scale the displacement as a factor times z_R.

8.29 Estimate the signal-to-noise ratio required to make an n_2^I measurement when $\Delta\Phi_0 = 0.1$. (Hint: see Equation 8.185.)

REFERENCE

Sheik-Bahae, M., A. A. Said, T.-H. Wei, D. J. Hagan, and E. W. VanStryland. 1990. Sensitive measurement of optical nonlinearities using a single beam. *IEEE Journal of Quantum Electronics* 26: 760–769.

FURTHER READING

Agrawal, G. P. 2013. *Nonlinear Fiber Optics (5th edition)*. New York: Academic Press.

Boyd, R. W. 2008. *Nonlinear Optics (3rd edition)*. Burlington, MA: Academic Press.

Butcher, P. N. and D. Cotter. 1990. *The Elements of Nonlinear Optics*. Cambridge: Cambridge University Press.

Gibbs, H. M., S. L. McCall, and T. N. C. Venkatesan. 1976. Differential gain and bistability using a sodium-filled Fabry–Perot interferometer. *Physical Review Letters* 36: 1135–1138.

He, G. S. and S. H. Lui. 1999. *Physics of Nonlinear Optics*. Singapore: World Scientific Publishing Co. Pte. Ltd.

Manley, J. M. and H. E. Rowe. 1956. Some general properties of nonlinear elements—Part I. General energy relationships. *Proceedings of the Institute of Radio Engineers* 44: 904–913.

New, G. 2014. *Introduction to Nonlinear Optics*. Cambridge: Cambridge University Press.

Shen, Y. R. 1984. *The Principles of Nonlinear Optics*. New York: John Wiley & Sons.

Stegeman, G. I. and Stegeman, R. A. 2012. *Nonlinear Optics: Phenomena, Materials and Devices*. Hoboken, NJ: John Wiley & Sons.

Stolen, R. and J. Bjorkholm. 1982. Parametric amplification and frequency conversion in optical fibers. *JEEE Journal of Quantum Electronics* 18: 1062–1072.

Sutherland, R. L. 2003. *Handbook of Nonlinear Optics (2nd edition)*. New York: Marcel Dekker, Inc.

9

RAMAN AND BRILLOUIN SCATTERING

9.1 INTRODUCTION

We turn our attention now to study optical processes where the interaction of light with the medium is described as an inelastic scattering process. The wavelength of the scattered light is changed in a process that does not conserve the optical energy so that the net exchange of energy is either given to or taken from the material. The scattered energy loss characteristics are attributed to the internal structure of the medium. All materials, whether gas, liquid, or solid, exhibit specific frequency shifts, which are signatures or fingerprints revealing their internal quantum structure.

The optical interactions considered in the previous chapters exploited phenomena where total optical energy is conserved by splitting or combining photons such that no net energy goes to or leaves the host medium. More specifically, the phenomena considered so far involve only nonresonant transitions. That is, the photon energies of the various waves do not match any quantized energy-level transitions and hence are not absorbed. The nonlinear absorption introduced in Chapter 8 is an exception to this rule, but we did not consider where the energy went in this case.

In this chapter, we examine processes where energy exchange with the medium produces a specific shift of the scattered photon energy. The scattering effects may involve atomic or molecular properties or an excited state of the system. We first consider the Raman effect where an optical field interacts with molecular vibrations and rotations. We follow this with a treatment of the related Brillouin effect, which couples optical fields to acoustic modes of the material.

9.2 SPONTANEOUS RAMAN SCATTERING

The Raman effect, named after the codiscoverer Chandrasekhara V. Raman (Raman and Krishnan, 1928), describes the process where photons scattered from a medium have an energy different from the incident beam. The photons in the incident beam, called the pump, lose energy such that the photons in the scattered output, called the Stokes photons, have a lower frequency. Moreover, the scattered photon energy difference is specific to the material so that by measuring a spectrum of the scattered light, it is possible to determine what materials are present in a given sample as well as crystal orientation, purity, stresses, and strains. The Stokes wave spectrum is distinct from molecular fluorescence since the frequency shift is a constant characteristic of the material and independent of the pump frequency and the process is quenched when the pump is turned off. Also, the scattered pump photons can also gain energy such that some of the scattered outputs, called the anti-Stokes photons, have a higher frequency than the pump photon.

The energy exchange originates with an excitation of the medium. In a gas-phase sample, the interactions with molecular vibrations and rotations lead to lower energy when the input beam excites a vibration and higher energy when a molecule gives up some of its vibrational or rotational energy. Because different gases have different rotational and vibrational spectra, their Raman frequency shifts are specific to that gas. In solid- and liquid-phase samples, the energy exchange comes from individual molecular vibrations and from phonon modes. A phonon is a quantized lattice vibration in a material. The bandwidth in the solid and liquid phases tends to be much broader than that in the gas phase, but distinct spectral signatures, especially for crystalline samples, are present. A generalization of the Raman Stokes and anti-Stokes processes is shown in the energy diagram in Figure 9.1 for an interaction with a molecular vibration. Note that, in general, many vibrational states are present; hence, a typical Raman spectrum contains many Stokes-shifted lines. Figure 9.1 shows that in the Stokes process, an incident photon excites a virtual state, and subsequently a

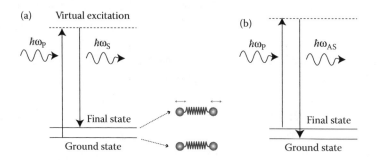

FIGURE 9.1   Energy diagram for a Raman process: (a) Stokes and (b) anti-Stokes. Also shown is an example molecule, represented as two masses connected with springs, in its ground state and excited vibrational state.

TABLE 9.1 **Frequency Shifts and Differential Cross Sections Relative to N₂ for the Strongest (*Q*-Branch) Vibrational Raman Transitions in Several Gas-Phase Molecules**

Molecule	Frequency Shift (cm⁻¹)	$\dfrac{d\sigma}{d\Omega} \Big/ \dfrac{d\sigma}{d\Omega}\Big|_{N_2}$
N_2	2330	1
H_2	4156	3.2
O_2	1556	1.23
CO_2	1388	1.51
CO	2143	0.98
NO	1876	0.49
H_2O	3657	2.5
SO_2	1151	4.9
O_3	1103	3.0
CH_4	2917	7.3

Source: With kind permission from Danichkin, S. A. et al. 1981. *Journal of Applied Spectroscopy* 35: 1057–1066, Table 3.

Note: Measurements of the differential cross section were made at $\lambda_p = 488$ nm:

$$\frac{d\sigma}{d\Omega}\bigg|_{N_2} = 5.5 \times 10^{-31} \, cm^2/Sr/molecule$$

photon with lower energy is emitted, and the medium is left in an excited vibrational state. Similarly in the anti-Stokes process, an incident photon transfers energy from the vibrational state by exciting a virtual state in the molecule, which then relaxes back to the ground state by giving off a photon at the anti-Stokes frequency. Note that the frequency shifts are most commonly reported in cm⁻¹ units, which is

$$\frac{1}{\lambda_S} = \frac{1}{\lambda_P} - \Delta\left(\frac{1}{\lambda}\right)_R , \tag{9.1}$$

where λ_S are the vacuum wavelengths measured in centimeters. The last term represents the Raman frequency shifts; selected values of Raman frequency shifts are listed in Table 9.1.

9.2.1 CLASSICAL MODEL OF SPONTANEOUS RAMAN SCATTERING

A complete description of spontaneous Raman scattering is calculated on a microscopic level using quantum mechanics. This calculation goes beyond the scope of this text, and so we look at a classical model that gives some of the general features of spontaneous Raman scattering. We consider the simplest case where a monochromatic field is incident on an isolated molecule inducing a dipole moment. Furthermore, we assume that the molecule is in an excited vibrational or rotational state. The field and induced polarization are treated as scalars with the field given as

$$\underline{E} = E_0 \cos(\omega_P t). \tag{9.2}$$

As a reminder, the underscore indicates the real field as opposed to the field's complex amplitude. The induced dipole moment and the field are related via the microscopic polarizability, α,

$$\underline{p} = \varepsilon_0 \alpha \underline{E}. \tag{9.3}$$

The polarizability is assumed to be a weak function of the vibrational or rotational state of the molecule. A way to generalize the vibration, rotation, phonon mode, and so on is to describe the mode in terms of a generalized coordinate, Q, which represents the independent variable for a given mode. For example, a linear molecule vibrating about its center of mass (see Figure 9.1) is described by a single variable x (the separation between the two atoms) instead of two variables representing the location of each atom. For the two-mass example, the generalized coordinate, Q, represents a scalar mass displacement. In the rest of this chapter, we refer to all of these modes as generalized "vibrations."

The polarizability of the molecule, α, is assumed to be a weak function of Q so that we can expand it in a Taylor series to first-order in Q,

$$\alpha = \alpha_0 + \frac{d\alpha}{dQ}\bigg|_0 Q + \cdots. \tag{9.4}$$

Classically, when the molecule is excited, it vibrates or rotates at the angular frequency Ω, and hence Q is of the form

$$\underline{Q} = Q_0 \cos(\Omega t). \tag{9.5}$$

The dipole moment is found by substituting the field (Equation 9.2) and the polarizability (Equation 9.4) into Equation 9.3:

$$\underline{p} = \varepsilon_0 \alpha_0 E_0 \cos(\omega_P t) + \varepsilon_0 E_0 \frac{d\alpha}{dQ}\bigg|_0 Q_0 \cos(\Omega t)\cos(\omega_P t). \tag{9.6}$$

The pump wave oscillates in time with an angular frequency ω_P. Expanding the product of the two cosines gives

$$p = \varepsilon_0 \alpha_0 E_0 \cos(\omega_P t) + \varepsilon_0 \frac{E_0 Q_0}{2} \left.\frac{d\alpha}{dQ}\right|_0 (\cos(\omega_P - \Omega)t + \cos(\omega_P + \Omega)t). \tag{9.7}$$

This equation shows that the dipole moment has three frequencies present. The first term, at the same frequency as the input, is the linear result. The other two terms show that the incident field couples with Q leading to energy exchange between the field and the molecule. In the case that the angular frequency is downshifted from the pump wave $\omega_P - \Omega$, it is called the Stokes frequency. The incident field loses energy, and the molecule gains energy. The anti-Stokes angular frequency at $\omega_P + \Omega$ corresponds to the case where the field gains energy at the expense of energy stored in the molecular excited state.

9.2.2 RAMAN SCATTERING CROSS SECTION

The above analysis shows a key point of the Raman scattering: new frequencies are generated at the Stokes and anti-Stokes. However, the classical model is too simplistic to extend much further. Recall that the model assumes the molecule to be in an excited state. If the molecule is in its ground state, the classical model has no mechanism for generating the new frequencies. Although the full derivation is beyond the scope of our first treatment here, the quantum mechanical result gives further insight into the properties of the spontaneous Raman scattering. The quantum calculation results in a scattering cross section for the Raman process. The cross section relates the total scattered power to the incident intensity,

$$P_{\text{Scattered}} = \sigma_R I_0, \tag{9.8}$$

where σ_R is the Raman scattering cross section, which has units of area. The quantum result for the cross section of a single molecule is (Demtroder, 1996)

$$\sigma_R = \frac{8\pi\omega_S^4}{9\hbar c^4} \left| \sum_j \left[\frac{M_{\text{initial},j} M_{j,\text{final}}}{\omega_{\text{initial},j} - \omega_P - i\Gamma_j} + \frac{M_{j,\text{initial}} M_{j,\text{final}}}{\omega_{j,\text{final}} - \omega_P - i\Gamma_j} \right] \right|^2, \tag{9.9}$$

where a summation over all higher-energy excited states, j, is carried out. In Figure 9.2, an energy diagram with four high-energy excited states illustrates the nonresonant nature of the Raman scattering process; the excited state near the ground state is the final state of the electron. In Equation 9.9, Γ_j is the linewidth for the given state (inversely related to the lifetime of that state), and $\omega_{\text{initial},j}$ corresponds to the frequency

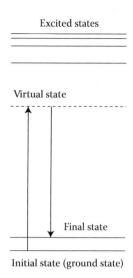

Excited states

Virtual state

Final state

Initial state (ground state)

FIGURE 9.2 Energy diagram showing excited states. According to Equation 9.9, the Raman cross section depends on the coupling of the excited states with the ground state and with the final state.

difference between the initial state and a given excited state, j. $M_{\text{initial},j}$ refers to a "matrix element," which is a coupling coefficient between the initial state and state j. The matrix element's magnitude depends on the specific molecule and the polarization of the incident and scattered light. The symmetries of a given molecule strongly influence the magnitude of these elements.

The scattering cross section given in Equation 9.9 shows the general trend that the Raman scattering is proportional to ω_S^4, just like the Rayleigh scattering. So, it is advantageous to use higher-frequency (shorter-wavelength) laser excitations to detect a stronger signal. Another important point is that when the incident laser frequency at ω_p approaches a transition corresponding to $\omega_{\text{initial},j}$, then the scattering cross section is resonantly enhanced.

Equation 9.9 shows how to calculate a given cross section if the resonant frequencies and the matrix elements are known, but more typically a direct measurement is made in the form of a differential scattering cross section. The total scattering cross section is the integral of the differential scattering cross section over all scattering angles:

$$\sigma = \int \frac{d\sigma}{d\Omega} d\Omega. \tag{9.10}$$

The differential scattering cross section is a useful measurement because it shows us how much scattering takes place and its angular distribution. The differential cross section is typically reported in units of $cm^2/Sr/molecule$; see Table 9.1 for a listing of the differential scattering cross section of several gases. In many experimental configurations, the collection solid angle, $\Delta\Omega_C$, is small such that the differential cross section is approximately constant, and so the power scattered into a solid angle $\Delta\Omega_C$ is

$$P_R = I_0 N \int \frac{d\sigma}{d\Omega} d\Omega \approx I_0 N \frac{d\sigma}{d\Omega} \Delta\Omega_C, \qquad (9.11)$$

where N is the number of scatterers in the observation volume.

As an example, we consider measuring the Raman signal generated from nitrogen using the experimental arrangement shown in Figure 9.3. In this particular experimental arrangement, a lens collects scattered light and images it into the entrance slits of a monochromator (spectrometer), and we assume that the angular acceptance of the monochromator is matched to collection solid angle imaged by the lens. The way to visualize which region of the sample contributes to the Raman signal is to project an image of the entrance slits onto the sample since light originating from outside this region is not imaged into the monochromator (see also Figure 9.15 in the Problems section). The number of molecules that contribute to the Raman signal is ηAL, where η is the number density of the molecules, A is the effective area of the pump laser beam, and L is the width of the slits. Note that we assume a 1:1 image from the sample to the monochromator. With these conditions, we rewrite Equation 9.11 by using $P_0 = I_o A$ to give

$$\frac{P_R}{P_0} = \eta L \frac{d\sigma}{d\Omega} \Delta\Omega_C. \qquad (9.12)$$

Consider a laser beam focused in air. The atmospheric number density of nitrogen molecules is ~2.4×10^{19} mol/cm³. The differential scattering cross section for nitrogen measured with a laser operating at 488 nm in the configuration is shown in Figure 9.3 (i.e., at 90°) is 5.5×10^{-31} cm²/Sr/molecule. The collection solid angle is matched to

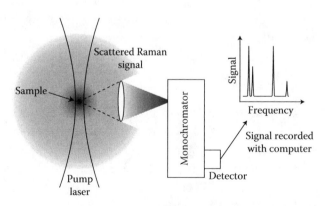

FIGURE 9.3   A setup to measure the spontaneous Raman-scattered signal. Other more efficient collection geometries are possible (see Figure 9.4).

that of the monochromator, and for this example, we assume that this solid angle is $\pi/16$. Assuming an entrance slit width of 100 μm, Equation 9.12 gives $P_R/P_0 = 3 \times 10^{-14}$. Detecting such a small fraction of the incident power requires high-sensitivity detectors, for example, photomultipliers. Raman-scattered signals from solids and liquids are much larger because these samples have a much higher number density.

Measuring a spontaneous Raman signal poses several challenges. Already mentioned is the low signal level of the scattered light, which requires the use of sensitive photodetectors. However, detector technology has improved to the point where sensitive detectors are commonly available. Another challenge is blocking the pump light from reaching the detector. In the configuration shown in Figure 9.3, some of the pump laser is Rayleigh scattered into the same solid angle as the Raman system. Inside the monochromator, a diffraction grating disperses the input beam into a spectrum on the output port. In this way, the pump beam and Raman signal are spatially separated by sufficient distance that allows the pump photons to be blocked from hitting the detector at the position that the Raman photons are collected. Hence, the monochromator helps to reject the pump laser, and at the same time, it measures the Raman spectrum.

Even when the pump laser beam is effectively filtered out of the detection path, other signals may interfere with measuring the Raman signal. When pumping with, for example, a 488 nm laser, the Raman signal is often accompanied by a fluorescence signal. In many cases, the fluorescence is much larger than the Raman signal. This is particularly true with biological samples. For this reason, many systems use a near-IR laser at ~785 nm as the excitation source. The trade-off for these systems is that detectors responsive at the near-IR wavelengths are not as sensitive as ones for visible wavelengths. Moreover, because of the ω^4 scaling of the Raman signal (see Equation 9.9), a near-IR laser wavelength has a much smaller scattering cross section. Another strategy is to use deep-UV wavelengths such that the Raman signal lies outside of the autofluorescence band. The deep-UV excitation benefits from removing the fluorescence signal and increasing the cross section through the ω^4 frequency dependence for the scattered signal. The downside to UV excitation is that it may introduce undesired and uncontrollable effects since the photon energy is large enough to break chemical bonds.

9.2.3 RAMAN MICROSCOPE

The ability to probe materials and to determine chemical composition using the Raman effect is a powerful technique applied to essentially all scientific and engineering disciplines. A particularly useful Raman system is one built around a microscope (see Figure 9.4 for a schematic representation). The microscope architecture has several advantages. The first is the ability to collect a large solid angle of the scattered Raman signal. Microscope objectives with numerical apertures, for example, NA = 0.95, collect large solid angles. The microscope can probe small areas of a sample, which allows for measuring high-resolution Raman images by scanning the sample. Additionally,

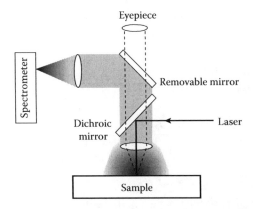

FIGURE 9.4 Schematic representation of a Raman microscope. The dichroic mirror reflects the laser but transmits the Raman signal. The scattered Raman signal passes through the dichroic mirror and is directed to the spectrometer (monochromator). The removable mirror allows for visual inspection of the surface using the microscope eyepiece.

the microscope approach allows for measuring the Raman signals from irregularly shaped or opaque samples. A second lens with long focal length is used to focus the collected Raman signal within the acceptance cone of the spectrometer.

9.3 STIMULATED RAMAN SCATTERING

9.3.1 INTRODUCTION

It is possible to enhance the weak spontaneous Raman effect by introducing two laser beams: one at the pump and the other at the Stokes frequency. This type of arrangement is called the stimulated Raman scattering. The pump and Stokes together stimulate, for example, a molecular vibration in a medium, which in turn provides a means to couple the pump and Stokes so that they can exchange energy with each other. Furthermore, it is possible to describe the process classically in terms of a nonlinear polarization and nonlinear susceptibility. The origin of the nonlinearity lies in the coupling of the optical fields and vibration, rotation, and so on, which we will hereafter refer in terms of the generalized coordinate, Q.

9.3.2 CLASSICAL CALCULATION FOR INDUCING A MOLECULAR VIBRATION

We start by considering an optical field incident on a dilute molecular gas. The field is given by

$$E_{\text{inc}} = \frac{1}{2}\left(A_P e^{i(k_P z - \omega_P t)} + A_S e^{i(k_S z - \omega_S t)} + \text{c.c.}\right), \tag{9.13}$$

where A_P and A_S are the complex amplitudes for the pump and Stokes fields, respectively. A dipole moment is induced in the medium when a field is incident, given by Equation 9.3.

We expand the polarizability in a Taylor series expansion as in Equation 9.4 so that the macroscopic polarization is given by

$$\underline{P} = N\varepsilon_0 \left(\alpha_0 + \frac{d\alpha}{dQ}\bigg|_0 \underline{Q} \right) \underline{E}_{inc}, \tag{9.14}$$

where N is the number of molecules per unit volume. We observe that the polarization in Equation 9.14 has a nonlinear part proportional to QE_{inc}. The incident field is known, and in this stimulated Raman case, we now show that $\underline{Q}$ is induced by the incident fields.

A vibrational mode is modeled classically by a driven harmonic oscillator where the oscillator's displacement is given by $\underline{Q}$. The general equation of motion for a harmonic oscillator is

$$\frac{d^2\underline{Q}}{dt^2} + \Gamma \frac{d\underline{Q}}{dt} + \omega_R^2 \underline{Q} = \frac{F}{m}, \tag{9.15}$$

where Γ is a damping term, ω_R is the resonant frequency of the oscillator, F is a driving force, and m is the mass of the oscillator. The driving force in this case is a result of the incident field. The force on a microscopic dipole is derived from the potential energy stored in a dielectric,

$$U = -\frac{1}{2}\varepsilon \underline{E}_{inc}^2, \tag{9.16}$$

where ε is the material's permittivity. The macroscopic (vector) force is defined as

$$\vec{\underline{F}}_M = -\vec{\nabla} U. \tag{9.17}$$

For our one-dimensional system using the generalized coordinate, $\underline{Q}$, Equation 9.17 gives

$$\underline{F}_M = \frac{\partial}{\partial Q}\left(\frac{1}{2}\varepsilon \underline{E}_{inc}^2 \right). \tag{9.18}$$

We relate the macroscopic permittivity, ε, to the microscopic polarizability, α, by relating $\underline{D}$ to $\underline{P}$. For an isotropic medium,

$$D = \varepsilon_0 E + P = \varepsilon_0(1 + N\alpha)E = \varepsilon E. \tag{9.19}$$

Therefore,

$$\varepsilon = \varepsilon_0(1 + N\alpha). \tag{9.20}$$

Substituting Equation 9.20 into Equation 9.18 and expanding α as in Equation 9.4 yield an expression for the force on a single dipole:

$$F = \frac{F_M}{N} = \frac{1}{2}\varepsilon_0 \left.\frac{d\alpha}{dQ}\right|_0 E_{inc}^2. \tag{9.21}$$

For the driven harmonic oscillator model, Equation 9.15, the force in Equation 9.21 is a source term that drives the oscillator. Because the force is proportional to E_{inc}^2, the driving term has many frequencies present. We use the same arguments as with $\chi^{(2)}$ and $\chi^{(3)}$ effects (see, e.g., Section 4.1.1) to treat each of these frequencies separately. The vibrational coordinate, Q, is expected to be a relatively low-frequency oscillation when compared with the high-frequency optical inputs; hence, we focus on the difference-frequency term in E_{inc}^2, $\omega_\Delta = \omega_P - \omega_S$. Keeping only this difference frequency, the driving force on a single oscillator is

$$F = \frac{1}{4}\varepsilon_0 \left.\frac{d\alpha}{dQ}\right|_0 \left(A_P A_S^* e^{i(k_P - k_S)z} e^{-i\omega_\Delta t} + c.c.\right). \tag{9.22}$$

We assume that the harmonic oscillator is in steady state so that the generalized vibrational coordinate Q is given by

$$Q = \frac{1}{2} X e^{-i\omega_\Delta t} + c.c., \tag{9.23}$$

where X is the complex amplitude of Q. Inserting Equations 9.23 and 9.22 into the harmonic oscillator equation, Equation 9.15, yields

$$X = \frac{\varepsilon_0}{2mL(\omega_\Delta)} \left.\frac{d\alpha}{dQ}\right|_0 A_P A_S^* e^{i(k_P - k_S)z}, \tag{9.24}$$

where

$$L(\omega_\Delta) \equiv \left[-\omega_\Delta^2 - i\Gamma\omega_\Delta + \omega_R^2\right]. \tag{9.25}$$

9.3.3 NONLINEAR POLARIZATION FOR A STIMULATED RAMAN PROCESS

Equation 9.24 shows that the two incident fields combine to drive the vibrational coordinate, Q, in the material. Now, this material excitation interacts with the optical fields to allow energy exchange between the pump and Stokes inputs. The mechanism for energy exchange is the nonlinear polarization. Expanding Equation 9.14 gives

$$\underline{P}^{NL} = N\varepsilon_0 \frac{d\alpha}{dQ}\bigg|_0 \left(\frac{1}{2}Xe^{-i\omega_\Delta t} + c.c.\right)\left(\frac{1}{2}A_Pe^{i(k_Pz - \omega_Pt)} + \frac{1}{2}A_Se^{i(k_Sz - \omega_St)} + c.c.\right). \quad (9.26)$$

Because of the product QE_{inc}, the nonlinear polarization has angular frequencies shifted from the input wave angular frequency by an amount plus or minus the vibrational angular frequency ω_Δ. Figure 9.5 gives a visualization of the classical calculation showing that an incident field drives a molecular vibration, Q, which in turn mixes with the incident field to create a nonlinear polarization.

The nonlinear polarization at the Stokes frequency is

$$\underline{P}^{NL}(\omega_S) = N\varepsilon_0 \frac{1}{4}\frac{d\alpha}{dQ}\bigg|_0 A_PX^*e^{i(k_Pz)}e^{-i(\omega_P - \omega_\Delta)t} + c.c. \quad (9.27)$$

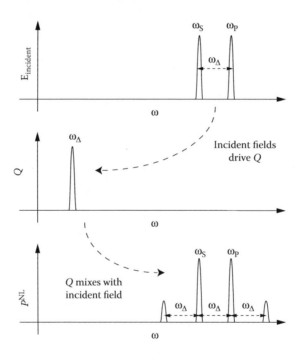

FIGURE 9.5   A visualization of the classical stimulated Raman process. The incident fields drive a molecular vibration, Q. Once Q is set up, it mixes with the incident field to generate a nonlinear polarization. The nonlinear polarization shows additional frequency components at a second Stokes frequency and at an anti-Stokes frequency.

We substitute the expression for X found in Equation 9.24 and write out the nonlinear polarization's complex amplitude:

$$P^{\mathrm{NL}}(\omega_S) = N\varepsilon_0^2 \frac{1}{4mL^*(\omega_\Delta)}\left(\frac{d\alpha}{dQ}\bigg|_0\right)^2 A_P A_P^* A_S e^{ik_S z}. \tag{9.28}$$

Note that the nonlinear polarization is the same spatial frequency as the Stokes field, and hence, this process is automatically phase matched, even if the pump field counter-propagates with respect to the Stokes field.

The nonlinear polarization is proportional to three fields, and hence, it looks like a $\chi^{(3)}$ process. That is, we can rewrite Equation 9.28 as

$$P^{\mathrm{NL}}(\omega_S) = \frac{3}{2}\varepsilon_0 \chi_Q^{(3)}(\omega_S; \omega_P, -\omega_P, \omega_S) A_P A_P^* A_S e^{ik_S z}, \tag{9.29}$$

where

$$\chi_Q^{(3)}(\omega_S; \omega_P, -\omega_P, \omega_S) = N\varepsilon_0 \frac{1}{6mL^*(\omega_\Delta)}\left(\frac{d\alpha}{dQ}\bigg|_0\right)^2. \tag{9.30}$$

The factor of 3/2 in Equation 9.29 is to maintain consistency with four-wave mixing notation where the degeneracy factor indicates the number of distinct permutations of the fields (see Equation 8.9). Similarly, the pump wave also has a nonlinear susceptibility, whose derivation is left as a homework problem.

The nonlinear susceptibility has a frequency dependence determined by L^*:

$$\frac{1}{L^*(\omega_\Delta)} = \frac{\left(\omega_R^2 - \omega_\Delta^2\right) - i\Gamma\omega_\Delta}{\left(\omega_R^2 - \omega_\Delta^2\right)^2 + \Gamma^2\omega_\Delta^2}. \tag{9.31}$$

From this expression, we see that the nonlinear susceptibility has a positive real component and a negative imaginary component:

$$\chi_Q^{(3)}(\omega_S; \omega_P, -\omega_P, \omega_S) = \frac{N\varepsilon_0}{2m} \frac{\left(\omega_R^2 - \omega_\Delta^2\right)}{\left(\omega_R^2 - \omega_\Delta^2\right) + \Gamma^2\omega_\Delta^2}\left(\frac{d\alpha}{dQ}\bigg|_0\right)^2$$

$$- i\frac{N\varepsilon_0}{2m} \frac{\Gamma\omega_\Delta}{\left(\omega_R^2 - \omega_\Delta^2\right)^2 + \Gamma^2\omega_\Delta^2}\left(\frac{d\alpha}{dQ}\bigg|_0\right)^2. \tag{9.32}$$

When $\hbar\omega_\Delta$ matches the energy needed to excite a vibrational mode, $\hbar\omega_R$, $\chi_Q^{(3)}$ is a maximum. Moreover, at resonance, $\chi_Q^{(3)}$ has a pure imaginary value. As we discover in the following section, the fact that the susceptibility has a negative imaginary component leads to amplification of the Stokes field.

9.3.4 WAVE EQUATION FOR THE STOKES FIELD

We assume plane-wave interactions where the SVEA is valid. The wave equation for this scenario reduces to (see Chapter 4, Equation 4.22)

$$2ik\frac{dA}{dz}e^{ikz} = -\omega^2\mu_0 P^{NL}. \tag{9.33}$$

Substituting in the nonlinear polarization from Equation 9.29 results in a copropagating wave equation for the Stokes frequency:

$$\frac{dA_S}{dz} = i\frac{3\omega_S}{4n_Sc}\chi_{QS}^{(3)}A_P A_P^* A_S, \tag{9.34}$$

where we simplify the notation for the nonlinear susceptibility by the substitution,

$$\chi_{QS}^{(3)} = \chi_Q^{(3)}(\omega_S; \omega_P, -\omega_P, \omega_S). \tag{9.35}$$

Note that the $e^{ik_S z}$ term divides out of Equation 9.34 so that the Raman process for the Stokes field is automatically phase matched. When the pump and Stokes inputs to the system have an energy spacing that matches a vibrational excitation, then as shown in Equation 9.32, $\chi_Q^{(3)}$ is a negative imaginary number:

$$\chi_{QS}^{(3)} = -i\left|\chi_{QS}^{(3)}\right|. \tag{9.36}$$

We substitute this equation into the wave equation, Equation 9.34, yielding

$$\frac{dA_S}{dz} = +\frac{3\omega_S}{4n_Sc}\left|\chi_{QS}^{(3)}\right|\left|A_P\right|^2 A_S. \tag{9.37}$$

In the undepleted pump approximation, the solution to this equation is

$$A_S(z) = A_S(0)e^{g I_P z/2}, \tag{9.38}$$

where the coefficient is

$$g = \frac{3\omega_S}{\varepsilon_0 n_S n_P c^2} \left| \chi_{QS}^{(3)} \right|. \tag{9.39}$$

On resonance, the Stokes field experiences exponential growth. Moreover, even if the angular frequency spacing between the pump and Stokes, $\omega_\Lambda = \omega_P - \omega_S$, does not match a vibrational frequency in the medium, then we show in the next section that the Stokes field still yields gain.

9.3.5 AMPLIFICATION OF THE STOKES FIELD OFF RESONANCE

To calculate the stimulated Raman effect when ω_Λ is off resonance, we work with the intensity of the Stokes field:

$$I_S = \frac{n_S \varepsilon_0 c A_S A_S^*}{2}. \tag{9.40}$$

The rate at which the Stokes intensity changes is

$$\frac{dI_S}{dz} = \frac{n_S \varepsilon_0 c}{2} \left(A_S^* \frac{dA_S}{dz} + A_S \frac{dA_S^*}{dz} \right). \tag{9.41}$$

Substituting Equation 9.34 into Equation 9.41 and simplifying give

$$\frac{dI_S}{dz} = i \frac{3\omega_S}{2 n_S n_P \varepsilon_0 c^2} I_S I_P \left(\chi_{QS}^{(3)} - \chi_{QS}^{(3)*} \right). \tag{9.42}$$

Note that $\chi - \chi^* = 2i \, \text{Im}(\chi)$ so that Equation 9.42 becomes

$$\frac{dI_S}{dz} = -\frac{3\omega_S}{n_S n_P \varepsilon_0 c^2} I_S I_P \text{Im}\left(\chi_{QS}^{(3)} \right). \tag{9.43}$$

Now, recall that the imaginary part of $\chi_Q^{(3)}$ is always negative (see Equation 9.32), and hence,

$$\frac{dI_S}{dz} = +\left(\frac{3\omega_S}{n_S n_P \varepsilon_0 c^2} I_P \left| \text{Im}\left(\chi_{QS}^{(2)} \right) \right| \right) I_S. \tag{9.44}$$

This equation shows that the rate of change of the Stokes field is positive, and hence, it yields gain even when ω_Λ is off resonance. In the undepleted pump approximation, the solution to Equation 9.44 shows an exponential gain:

$$I_S(z) = I_S(0) \exp g' I_P z, \tag{9.45}$$

where

$$g' = \left(\frac{3\omega_S}{n_S n_P \varepsilon_0 c^2} \left| \mathrm{Im}\left(\chi_{QS}^{(3)}\right) \right| \right). \tag{9.46}$$

The combination of exponential gain and automatic phase-matching makes the Raman process a popular choice for optical amplifiers.

9.3.6 STOKES AMPLIFICATION WITH A DEPLETED PUMP

Sections 9.3.4 and 9.3.5 demonstrate that the Stokes field has exponential gain. In the limit of high conversion efficiency, clearly the undepleted pump approximation breaks down. To understand the behavior of the Stokes field when the pump depletes, we include a wave equation for the pump field. The nonlinear polarization for the pump is

$$P^{\mathrm{NL}}(\omega_P) = \frac{3}{2} \varepsilon_0 \chi_{QP}^{(3)} A_S A_S^* A_P e^{ik_P z}, \tag{9.47}$$

where $\chi_{QP}^{(3)} \equiv \chi_Q^{(3)}(\omega_P; \omega_S, -\omega_S, \omega_P)$. Following the same procedure as in Section 9.3.2 to derive $\chi_{QP}^{(3)}$, it is possible to show

$$\chi_{QP}^{(3)} = \chi_{QS}^{(3)*}. \tag{9.48}$$

Substituting the nonlinear polarization into the wave equation, Equation 9.33, leads to the following differential equation:

$$\frac{dI_P}{dz} = -\left(\frac{3\omega_P}{4 n_S n_P \varepsilon_0 c^2} \left| \mathrm{Im}\left(\chi_{QS}^{(3)}\right) \right| \right) I_S I_P. \tag{9.49}$$

Equations 9.49 and 9.44 form a set of coupled differential equations in I_P and I_S, repeated here in simplified form,

$$\omega_P \frac{dI_S}{dz} = +\gamma I_S I_P, \tag{9.50}$$

$$\omega_S \frac{dI_P}{dz} = -\gamma I_S I_P, \tag{9.51}$$

where

$$\gamma \equiv \frac{3\omega_S \omega_P}{4n_S n_P \varepsilon_0 c^2} \left| \mathrm{Im}\left(\chi_{QS}^{(3)}\right) \right|. \tag{9.52}$$

We note that in Equation 9.50 the Stokes intensity remains zero unless a seed field initiates the gain process. The seed is provided by the spontaneous Stokes emission of a single photon as discussed in Section 6.5. The treatment of spontaneous emission will be treated in more detail in Chapter 11.

Equations 9.50 and 9.51 are equated to give

$$\frac{1}{\omega_P} \frac{dI_P}{dz} = -\frac{1}{\omega_S} \frac{dI_S}{dz}. \tag{9.53}$$

This result is similar to the Manley–Rowe relations derived in the previous chapters. Here, the result is transformed into photon numbers as

$$\Delta N_S = -\Delta N_P. \tag{9.54}$$

This equation makes sense for the Raman process since for every pump photon destroyed ($\Delta N_P = -1$), one is created at the Stokes.

We also use Equation 9.53 to find the Stokes and pump intensities as a function of z. Rearranging Equation 9.53,

$$\frac{dI_P}{dI_S} = -\frac{\omega_P}{\omega_S}. \tag{9.55}$$

Integrating this equation yields, after rearrangement,

$$\frac{I_P(z)}{\omega_P} + \frac{I_S(z)}{\omega_S} = \frac{I_P(0)}{\omega_P} + \frac{I_S(0)}{\omega_S}. \tag{9.56}$$

If we multiply this equation by a cross-sectional area and an integration time, it represents the total number of photons passing through the area in the integration time. Equation 9.56 shows that the total number of pump and Stokes photons is a constant throughout the interaction.

We solve for the intensity as a function of distance by first identifying the constant

$$C = \frac{I_P(0)}{\omega_P} + \frac{I_S(0)}{\omega_S}. \tag{9.57}$$

C is substituted back into the original differential equation, Equation 9.50, to obtain

$$\frac{dI_S}{dz} = \gamma I_S \left(C - \frac{I_S}{\omega_S} \right). \tag{9.58}$$

Separating variables and integrating give

$$\int_{I_S(0)}^{I_S} \frac{dI_S}{I_S \left(1 - (I_S/C\omega_S) \right)} = \gamma C \int_0^L dz \tag{9.59}$$

$$\ln \left[\frac{I_S(I_{S0} - C\omega_S)}{I_{S0}(I_S - C\omega_S)} \right] = C\gamma L. \tag{9.60}$$

Equation 9.60 is rearranged to provide $I_S(L)$,

$$I_S(L) = \frac{\left((I_{S0}/\omega_S) + (I_{P0}/\omega_P) \right) I_{S0} e^{C\gamma L}}{(I_{S0}/\omega_S) e^{C\gamma L} + (I_{P0}/\omega_P)}. \tag{9.61}$$

A similar calculation for the pump intensity yields

$$I_P(L) = \frac{\left((I_{P0}/\omega_P) + (I_{S0}/\omega_S) \right) I_{P0}}{(I_{P0}/\omega_P) + (I_{S0}/\omega_S) e^{C\gamma L}}. \tag{9.62}$$

Consider the limit of a long interaction length, high incident intensity, or any combination that yields a large value of $C\gamma L$. In this limit, Equations 9.61 and 9.62 show that the pump is totally converted to Stokes. Another way to look at Equations 9.61 and 9.62 is in terms of photon numbers. The equations can be written in terms of the ratio of the number of Stokes photons to the total number of photons,

$$\eta \equiv \frac{N_S}{N_{Tot}}. \tag{9.63}$$

Converting Equations 9.61 and 9.62 into photon numbers using the substitution in Equation 9.63 gives

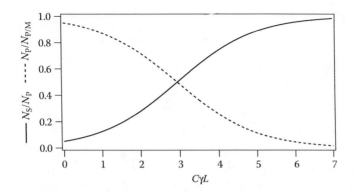

FIGURE 9.6   Stokes amplification when the pump depletes plotted as a function of the exponential factor $C\gamma L$ (see Equations 9.64 and 9.65).

$$\frac{N_S(L)}{N_{\text{Tot}}} = \frac{\eta_o e^{C\gamma L}}{1 + \eta_o(e^{C\gamma L} - 1)},\tag{9.64}$$

$$\frac{N_P(L)}{N_{\text{Tot}}} = \frac{1 - \eta_o}{1 + \eta_o(e^{C\gamma L} - 1)},\tag{9.65}$$

where

$$\eta_o = \frac{N_S(0)}{N_{\text{Tot}}}.\tag{9.66}$$

Figure 9.6 shows the Stokes and pump in terms of relative photon numbers for the initial condition, $\eta_o = 0.05$. The figure shows that eventually the pump completely depletes, provided that the interaction length is long enough. Long interaction lengths in fibers make such scenarios possible.

In cases where the pump depletes, it becomes likely that an anti-Stokes component is present (see Section 9.4). Furthermore, as the Stokes field grows in amplitude, it acts as a pump to drive more Stokes-shifted frequencies. In this way, a cascade of Stokes-shifted frequencies may appear in the output (see Figure 9.7).

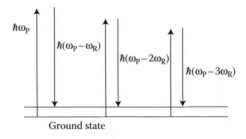

FIGURE 9.7   When the Stokes field becomes appreciable, it acts as a pump to drive other Stokes-shifted outputs. Each output is evenly spaced by ω_R.

9.4 ANTI-STOKES GENERATION

9.4.1 CLASSICAL DERIVATION OF THE ANTI-STOKES NONLINEAR POLARIZATION

The spontaneous Raman scattering contains both a Stokes and an anti-Stokes component. As the previous sections show, when a pump and Stokes input are present, the Stokes input is amplified at the expense of the pump. We may ask the same question regarding a situation where a pump and anti-Stokes are present. In fact, this situation is precisely the same as the stimulated Raman process discussed above, except that since the anti-Stokes has higher photon energy, it acts as the pump. Therefore, the anti-Stokes input decays, and the pump is amplified. A more interesting stimulated case is one where a pump, Stokes, and anti-Stokes are present. In fact, this situation is the most common because the pump and Stokes create a population of molecules in the excited state so that it becomes more probable that an anti-Stokes photon is generated (see Figure 9.1).

We consider the case when the difference frequencies between the pump and the Stokes and between the anti-Stokes and the pump correspond to a Raman transition. That is,

$$\omega_P - \omega_S = \omega_R, \tag{9.67}$$

$$\omega_{AS} - \omega_P = \omega_R. \tag{9.68}$$

In the present analysis, the pump is treated as undepleted. Note that Equations 9.67 and 9.68 indicate that we now have two means of exciting the Raman transition. To understand how the two mechanisms influence the Raman process, we use the classical model introduced in Section 9.3.2. Recall that we treat the molecular vibration, rotation, and so on in terms of a generalized coordinate, Q. The generalized vibration is modeled as a simple harmonic oscillator with a driving term (see Equation 9.15). The driving force for the classical oscillator is given by Equation 9.21, repeated here,

$$\underline{F} = \frac{F_M}{N} = \frac{1}{2}\varepsilon_0 \left.\frac{d\alpha}{dQ}\right|_0 \underline{E}_{inc}^2. \tag{9.69}$$

In the present scenario, the incident field contains three fields:

$$\underline{E}_{inc}^2 = \frac{1}{2}A_{AS}e^{i(k_{AS}z-\omega_{AS}t)} + \frac{1}{2}A_Pe^{i(k_Pz-\omega_Pt)} + \frac{1}{2}A_Se^{i(k_Sz-\omega_St)} + \text{c.c.} \tag{9.70}$$

Squaring the field gives two distinct field combinations at the difference frequency,

$$\underline{E}_{inc}^2 = \frac{1}{4}\left(2A_{AS}A_P^*e^{i((k_{AS}-k_P)z-\omega_Rt)} + 2A_PA_S^*e^{i((k_P-k_S)z-\omega_Rt)} + \cdots + \text{c.c.}\right), \tag{9.71}$$

where only terms at ω_R are displayed. We again look only at the component of Q at ω_R such that Q has the same form as in Equation 9.23. Hence, the solution to the harmonic oscillator equation (with $\omega_\Delta = \omega_R$), Equation 9.15, is

$$X = i \frac{\varepsilon_0}{2m\Gamma\omega_R} \left.\frac{d\alpha}{dQ}\right|_0 \left(A_P A_S^* e^{i(k_P - k_S)z} + A_{AS} A_P^* e^{i(k_{AS} - k_P)} \right). \tag{9.72}$$

Note that the difference between this result and the one obtained in Equation 9.24 is an additional term with the anti-Stokes field. The nonlinear polarization with three fields is given by

$$\underline{P}^{NL} = N\varepsilon_0 \left.\frac{d\alpha}{dQ}\right|_0 \left(\frac{1}{2} X e^{-i\omega_\Delta t} + \text{c.c.} \right) \left(\frac{1}{2} A_{AS} e^{i(k_{AS}z - \omega_{AS}t)} + \frac{1}{2} A_P e^{i(k_P z - \omega_P t)} + \frac{1}{2} A_S e^{i(k_S z - \omega_S t)} + \text{c.c.} \right). \tag{9.73}$$

The nonlinear polarization's complex amplitudes at the Stokes and anti-Stokes frequencies are separated out from Equation 9.73:

$$P^{NL}(\omega_S) = -i \frac{N\varepsilon_0^2}{4m\Gamma\omega_R} \left(\left.\frac{d\alpha}{dQ}\right|_0 \right)^2 \left(|A_P|^2 A_S e^{ik_S z} + A_P^2 A_{AS}^* e^{i(2k_P - k_{AS})z} \right), \tag{9.74}$$

$$P^{NL}(\omega_{AS}) = +i \frac{N\varepsilon_0^2}{4m\Gamma\omega_R} \left(\left.\frac{d\alpha}{dQ}\right|_0 \right)^2 \left(|A_P|^2 A_{AS} e^{ik_{AS} z} + A_P^2 A_S^* e^{i(2k_P - k_S)z} \right). \tag{9.75}$$

These expressions can be rewritten in terms of third-order nonlinear susceptibilities, which to be consistent with four-wave mixing terminology include degeneracy factors to account for the number of distinct permutations of the fields (see Equation 8.9). The third-order susceptibilities are given by

$$\left| \chi^{(3)}(\omega_S; \omega_P, -\omega_P, \omega_S) \right| = \left| \chi^{(3)}(\omega_{AS}; \omega_P, -\omega_P, \omega_{AS}) \right| = \frac{N\varepsilon_0}{6m\Gamma\omega_R} \left(\left.\frac{d\alpha}{dQ}\right|_0 \right)^2, \tag{9.76}$$

$$\left| \chi^{(3)}(\omega_S; \omega_P, \omega_P - \omega_{AS}) \right| = \left| \chi^{(3)}(\omega_{AS}; \omega_P, \omega_P, -\omega_S) \right| = \frac{N\varepsilon_0}{3m\Gamma\omega_R} \left(\left.\frac{d\alpha}{dQ}\right|_0 \right)^2. \tag{9.77}$$

To simplify Equations 9.74 and 9.75, they are expressed as

$$\chi_R^{(3)} \equiv \left| \chi^{(3)}(\omega_S; \omega_P, -\omega_P, \omega_S) \right|. \tag{9.78}$$

This substitution yields

$$P^{NL}(\omega_S) = -i\frac{3\varepsilon_0 \left|\chi_R^{(3)}\right|}{2}\left(|A_P|^2 A_S e^{ik_S z} + A_P^2 A_{AS}^* e^{i(2k_P - k_{AS})z}\right),\tag{9.79}$$

$$P^{NL}(\omega_{AS}) = +i\frac{3\varepsilon_0 \left|\chi_R^{(3)}\right|}{2}\left(|A_P|^2 A_{AS} e^{ik_{AS} z} + A_P^2 A_S^* e^{i(2k_P - k_S)z}\right).\tag{9.80}$$

Equations 9.79 and 9.80 give the Raman contribution to the nonlinear polarization. Note that a nonresonant four-wave mixing contribution occurs as well with the same frequency combinations. For example, the four-wave mixing interaction defined by $\omega_S = \omega_P - \omega_P + \omega_S$ has a nonlinear polarization given by

$$P^{NL}(\omega_S) = \frac{3\varepsilon_0 \chi_{NR}^{(3)}}{2}|A_P|^2 A_S e^{ik_S z}.\tag{9.81}$$

This expression looks the same as in Equation 9.80, but its physical interpretation is different. In the nonresonant case, the system starts and ends in the ground state, and results in a cross-phase shift between the field at ω_P and the field at ω_S. In this interaction, the number of pump and Stokes photons remains constant. In Raman Stokes generation, the system starts in the ground state but ends in an excited vibrational state. In the Raman case, a pump photon is annihilated and a Stokes photon is created. Because $\chi_{NR}^{(3)}$ relates to a nonresonant interaction, it is weaker than the Raman susceptibility, and hence, we consider only the Raman contributions in the following section.

9.4.2 WAVE EQUATION FOR STOKES AND ANTI-STOKES IN THE UNDEPLETED PUMP APPROXIMATION

The nonlinear polarizations in Equations 9.79 and 9.80 are substituted into the wave equation, Equation 9.33. In the undepleted pump approximation this yields

$$\frac{dA_S}{dz} = \kappa\left(A_S + A_{AS}^* e^{i(\Delta k z + 2\phi_P)}\right),\tag{9.82}$$

$$\frac{dA_{AS}}{dz} = -\frac{n_S \omega_{AS}}{n_{AS} \omega_S}\kappa\left(A_{AS} + A_S^* e^{i(\Delta k z + 2\phi_P)}\right),\tag{9.83}$$

where several variable substitutions are used. The first is

$$\kappa \equiv \frac{3\omega_S \left|\chi_R^{(3)}\right|}{4n_S c}|A_P|^2.\tag{9.84}$$

The k-vector mismatch is given by

$$\Delta k = 2k_\mathrm{P} - k_\mathrm{AS} - k_\mathrm{S} \tag{9.85}$$

ϕ_P is the pump phase obtained from

$$A_\mathrm{P} = |A_\mathrm{P}| e^{i\phi_\mathrm{P}}. \tag{9.86}$$

We leave the solution to Equations 9.82 and 9.83 as a problem at the end of this chapter (see Problem 9.11). However, we examine two limiting cases followed by a numerical solution. We first consider the case where Δk is large such that Equations 9.82 and 9.83 become effectively decoupled. As shown in the previous chapters, nonphase-matched interactions do not grow appreciably. The same is true here, and hence, we drop the second term from the right-hand side of Equations 9.82 and 9.83 that contains Δk. The resulting equations are the same as in Section 9.3.4 for the case of the stimulated Stokes generation. In this limit where Δk is large, the Stokes field grows exponentially, and the anti-Stokes decays exponentially.

The second limiting case we consider is $\Delta k = 0$. In an isotropic medium with normal dispersion, $\Delta k = 0$ occurs at a specific noncollinear angle (see Problem 9.13). Making $\Delta k = 0$ for a collinear process is possible using the techniques discussed in the previous chapters. To simplify the following discussion, we make the approximation that

$$\frac{n_\mathrm{S}\omega_\mathrm{AS}}{n_\mathrm{AS}\omega_\mathrm{S}} \approx 1. \tag{9.87}$$

We also simplify Equations 9.82 and 9.83 by using the variable substitutions

$$A_\mathrm{S} = A_\mathrm{S}' e^{\kappa z}, \tag{9.88}$$

$$A_\mathrm{AS} = A_\mathrm{AS}' e^{-\kappa z} e^{-2i\phi_\mathrm{P}}. \tag{9.89}$$

With these substitutions, Equations 9.82 and 9.83 become

$$\frac{\mathrm{d}A_\mathrm{S}'}{\mathrm{d}z} = \kappa A_\mathrm{AS}'^{*} e^{-2\kappa z}, \tag{9.90}$$

$$\frac{\mathrm{d}A_\mathrm{AS}'}{\mathrm{d}z} = -\kappa A_\mathrm{S}'^{*} e^{2\kappa z}. \tag{9.91}$$

Equations 9.90 and 9.91 are decoupled by using Equation 9.90 to solve for A'^*_{AS}. Next, we differentiate this expression for A'^*_{AS} and equate it to the complex conjugate of Equation 9.91. After completing these steps and similar ones for A'^*_{AS}, we arrive at the following two second-order differential equations:

$$\frac{d^2 A'_S}{dz^2} + 2\kappa \frac{dA'_S}{dz} + \kappa^2 A'_S = 0, \tag{9.92}$$

$$\frac{d^2 A'_{AS}}{dz^2} - 2\kappa \frac{dA'_{AS}}{dz} + \kappa^2 A'_{AS} = 0. \tag{9.93}$$

The solution to Equation 9.92 is

$$A'_S = (C + Dz)e^{-\kappa z}, \tag{9.94}$$

where C and D are determined from the boundary conditions. Changing back to the original variable, A_S, using Equation 9.88 results in

$$A_S = C + Dz. \tag{9.95}$$

Similarly, the solution to Equation 9.93 gives

$$A_{AS} = E + Fz, \tag{9.96}$$

where E and F are determined from the boundary conditions. Surprisingly, when $\Delta k = 0$, we observe that the Stokes and anti-Stokes have a linear growth as opposed to the previous cases where the growth is exponential. The Stokes and anti-Stokes are strongly coupled in this limit with the result that neither grows exponentially.

9.4.3 STOKES AND ANTI-STOKES GENERATION WITH PUMP DEPLETION

Another approach to solving the above system of equations is by numerical integration. The numerical approach is no more difficult if we add in the pump wave equation so that we do not need to keep the undepleted pump approximation. To a good approximation, $n_S = n_P = n_{AS}$, so that the three coupled amplitude equations become (see Problems 9.12 and 9.14)

$$\frac{dA_S}{dz} = (1 - \delta)\kappa' |A_P|^2 \left(A_S + A^*_{AS} e^{i(\Delta kz + 2\phi_P)} \right), \tag{9.97}$$

$$\frac{dA_P}{dz} = \kappa' \left(|A_{AS}|^2 - |A_S|^2 \right) A_P,$$ (9.98)

$$\frac{dA_{AS}}{dz} = -(1+\delta)\kappa' |A_P|^2 \left(A_{AS} + A_S^* e^{i(\Delta kz + 2\phi_P)} \right),$$ (9.99)

where

$$\delta = 1 - \frac{\omega_S}{\omega_P}$$ (9.100)

and

$$\kappa' = \frac{3\omega_P \left| \chi_R^{(3)} \right|}{4 n_S c}.$$ (9.101)

We numerically integrate these equations using the Runge–Kutta algorithm (see Appendix C). As an example, we consider an optical fiber with a path length of 20 m. Figure 9.8 compares the numerical $\Delta k = 0$ result to the analytic undepleted pump case given by Equation 9.95. Given a long-enough interaction length, both the Stokes and the anti-Stokes become large enough such that pump depletion effects must be included.

In the $\Delta k = 0$ case, the Stokes and anti-Stokes fields are strongly coupled and eventually both grow to an appreciable magnitude. For a large Δk, the Stokes and anti-Stokes are decoupled such that the Stokes field grows at the expense of the pump and the anti-Stokes field decays. Figure 9.9 shows the growth of the Stokes field for the $\Delta k = 0$ case and the large Δk case. For further comparison, see Problems 9.11 through 9.14.

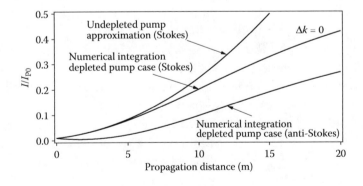

FIGURE 9.8 Plot of the ratio of the Stokes and anti-Stokes intensity to the incident pump intensity for the $\Delta k = 0$ case. The undepleted pump approximation for the Stokes is also shown (Equation 9.95). The numerical integration includes the effects of all three coupled amplitude equations, Equations 9.97 through 9.99.

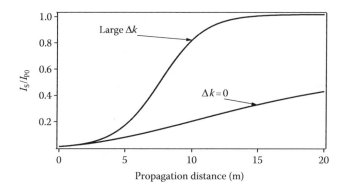

FIGURE 9.9 A comparison of the $\Delta k = 0$ to the large Δk case shows that the Stokes grows much more quickly when it is effectively decoupled from the anti-Stokes field.

As mentioned in Section 9.3.6, when the pump field depletes, it becomes likely that a cascade of Stokes fields are generated. This situation greatly complicates the calculation of the fields. When several Stokes fields are present, the number of ways to generate ω_A increases. These pathways to creating ω_A couple the fields to each other through the nonlinear polarization. Hence, if we consider N Stokes, pump, and anti-Stokes fields, we will have N coupled differential equations, which are solved numerically.

9.5 RAMAN AMPLIFIERS

An important application for the stimulated Raman generation is a fiber amplifier. Long interaction lengths combined with the Stokes exponential gain leads to appreciable amplification. In telecommunication systems where fibers are already in place, signals can be amplified using the transmit fiber itself by coupling in a pump laser. The Raman gain, gI_P, is given in Equation 9.46 where g is related to $\chi_R^{(3)}$, but in the literature, it is more common to find values for g alone. Figure 9.10 shows the Raman gain spectrum for quartz, soda–lime–silicate, and pyrex.

To take advantage of the Raman gain for an amplifier, we must ensure that we stimulate the Raman transition. In the case of a glass optical fiber, Figure 9.10 shows that the optimum frequency separation between the pump and the Stokes should be approximately 400 cm^{-1} resulting in $g\sim4.8 \times 10^{-13}$ m/W. If we would like to amplify a laser at, for example, 1.5 μm, then we need to choose a pump laser with a frequency that is 400 cm^{-1} higher at 1.42 μm. The Raman gain, gI_P, depends on the pump intensity in the fiber. Consider a single-mode glass fiber with a mode-field diameter of 10 μm and a 100 mW pump laser:

$$gI_P = 4.8\times10^{-13}\,(W/m)\left(\frac{0.1\,W}{\pi(5\times10^{-6}\,m)^2}\right) = 6\times10^{-4}\,m^{-1}. \tag{9.102}$$

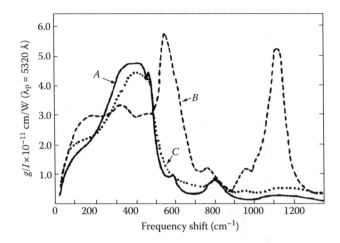

FIGURE 9.10　Raman gain curves for three representative glasses: (A) fused quartz, (B) soda–lime–silicate (20:10:70), and (C) pyrex. (Reprinted with permission from Stolen, R. H., E. P. Ippen, and A. R. Tynes. 1972. Raman oscillation in glass optical waveguide. *Applied Physics Letters* 20: 62–64. Copyright 1972, American Institute of Physics.)

A fiber length of 1 km gives an amplification of

$$\text{amplification} = e^{g I_P L} = e^{0.6} = 1.82. \tag{9.103}$$

Although a 1 km fiber length is rather long, as mentioned earlier, these same fibers are already in place for many telecommunication connections.

The Raman gain grows considerably when pulsed lasers are used. For example, the first Raman laser in a fiber was demonstrated by Stolen et al. (1972) using just a 190 cm fiber length (core diameter of 4 μm) by pumping the fiber with a pulsed laser that had a 500 W peak power. Other approaches to increasing the Raman gain are to use different materials that have a larger gain as well as using smaller diameter fiber cores.

9.6　PHOTOACOUSTIC EFFECTS: RAMAN-NATH DIFFRACTION

Energy exchange between optical fields and a material is not limited to single molecules. Indeed, the previous sections on the Raman effect include couplings between a field and phonon modes. On a much larger scale, coupling occurs between acoustic waves and optical fields. Because an acoustic wave gives rise to a periodic change in the density of a material, it produces a time and space periodic change in the index of refraction. An optical beam incident on the medium then diffracts from the index of refraction grating. The following analysis shows that the diffracted beams are shifted in frequency by the acoustic frequency. Hence, photoacoustic devices are a fairly simple means of generating frequency-shifted outputs and find applications with generating frequency shifts for heterodyne systems.

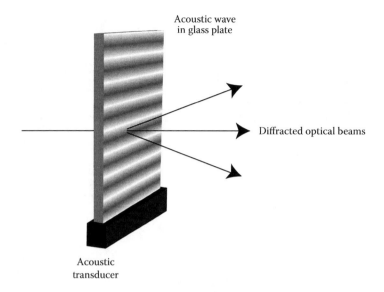

Acoustic wave
in glass plate

Diffracted optical beams

Acoustic
transducer

FIGURE 9.11 Raman-Nath diffraction. An incident beam is diffracted from an index grating generated from an acoustic wave.

As an example, we consider a thin sheet of material with a transducer that generates an acoustic wave as shown in Figure 9.11. This type of configuration is called the Raman-Nath acousto-optic effect.

The acoustic wave gives rise to the following index of refraction in the material:

$$n = n_0 + \Delta n_0 \sin\left(\vec{K}_{ac} \cdot \vec{r} - \Omega_{ac} t\right), \tag{9.104}$$

where n_0 is the index of refraction in the material when no acoustic wave is present, $\vec{K}_{ac}$ is the acoustic k-vector, and Ω_{ac} is the acoustic angular frequency. Δn_0 is the amplitude of the change in the refractive index, and its magnitude depends on the material, the transducer geometry, and the acoustic power supplied to the material. The dispersion relation for the acoustic wave is

$$\left|\vec{K}_{ac}\right| = \frac{\Omega_{ac}}{v_{ac}}, \tag{9.105}$$

where v_{ac} is the acoustic wave velocity (speed of sound in the medium).

A plane wave is incident on the material at an angle, θ, to the normal. The incident plane-wave field is

$$\underline{\vec{E}}(\vec{r},t) = \frac{\vec{A}_0}{2} e^{i(\vec{k}_{opt} \cdot \vec{r} - \omega_{opt} t)} + c.c. \tag{9.106}$$

Assuming that the material is a thin sheet, the optical k-vector magnitude in the material is

$$\left|\vec{k}_{opt}\right| = (n_0 + \Delta n)\frac{\omega}{c}, \tag{9.107}$$

where we use the substitution

$$\Delta n = \Delta n_0 \sin\left(\vec{K}_{ac} \cdot \vec{r} - \Omega_{ac}t\right). \tag{9.108}$$

The path length of the field through the material is

$$d = \frac{L}{\cos\theta_{int}}, \tag{9.109}$$

where θ_{int} is the optical k-vector's angle with respect to the normal inside the medium. θ_{int} is related to θ by Snell's law. After the optical field passes through the thin sheet, the optical field accumulates a phase given by

$$\phi = \frac{(n_0 + \Delta n)L}{\cos\theta_{int}}\frac{\omega}{c}. \tag{9.110}$$

Hence, the field after passing through the thin sheet is given by

$$\begin{aligned}\underline{\vec{E}}(\vec{r},t) &= \frac{\vec{A}_0}{2}\exp i\left(\vec{k}_{opt} \cdot \vec{r} - \omega_{opt}t\right)e^{i\phi} + c.c. \\ &= \frac{\vec{A}_0}{2}\exp i\left(\vec{k}_{opt} \cdot \vec{r} + \phi_0 + \delta\sin\left(\vec{K}_{ac} \cdot \vec{r} - \Omega_{ac}t\right) - \omega_{opt}t\right) + c.c.,\end{aligned} \tag{9.111}$$

where ϕ_0 is a constant phase offset given by

$$\phi_0 = \frac{n_0 L}{\cos\theta_{int}}\frac{\omega}{c}, \tag{9.112}$$

and δ is a variable substitution, called the modulation depth,

$$\delta = \frac{\Delta n_0 L}{\cos\theta_{int}}\frac{\omega}{c}. \tag{9.113}$$

The acoustic wave in the material induces a modulated phase in the field given in Equation 9.111. The standard approach to working with phase-modulated fields of the form given in Equation 9.111 is to expand the exponential using the following identity:

$$e^{i\delta\sin\theta} = \sum_{m=-\infty}^{\infty} J_m(\delta)\, e^{im\theta}, \tag{9.114}$$

where $J_m(\delta)$ is a Bessel function. Bessel functions are usually built-in functions included with most mathematical packages. Figure 9.12 shows a plot of a few of the lowest-order Bessel functions. In many cases, only a few terms in the Bessel expansion are required to give a good approximation to the phase-modulated waveform. Bessel functions find significant use in describing modes of optical fibers.

Using the Bessel expansion, Equation 9.111 is expressed as a sum over standard exponential terms:

$$\vec{E}(\vec{r},t) = \frac{\vec{A}_0 e^{i\phi_0}}{2} \sum_m J_m(\delta)\exp\left\{i\left(\vec{k}_{\text{opt}} + m\vec{K}_{\text{ac}}\right)\cdot\vec{r} - (\omega_{\text{opt}} + m\Omega_{\text{ac}})t\right\} + \text{c.c.} \tag{9.115}$$

This expansion shows that the output of the material is a set of diffracted plane waves propagating with a k-vector given by

$$\vec{k}_{\text{opt}} + m\vec{K}_{\text{ac}} \tag{9.116}$$

and a frequency

$$\omega_{\text{opt}} + m\Omega_{\text{ac}}. \tag{9.117}$$

This effect, resulting from a thin medium, is called the Raman-Nath diffraction. Similar to the Raman process, the diffracted outputs are shifted up or down in

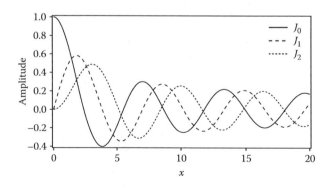

FIGURE 9.12 Plot of the first three Bessel functions: $J_0(\delta)$, $J_1(\delta)$, and $J_2(\delta)$.

frequency. However, the shifts are much smaller than for the Raman case. The Raman-Nath frequency shifts can be thought of in terms of a Doppler shift. The diffracted orders that propagate with a component of their direction parallel to the acoustic wave are frequency upshifted, while those propagating with a component antiparallel to the acoustic wave are frequency downshifted.

We note that the acoustic frequency shift is limited to frequency ranges of <1 GHz, whereas typical Raman shifts are much greater, on the order of THz and higher. The relative strength of each diffracted component in a Raman-Nath diffraction pattern is given by $J_m(\delta)$. From Figure 9.12, we see that for small δ the J_0 curve is largest and hence most of the energy is in the zeroth order. As the modulation depth increases, more energy goes to the higher-order diffracted orders, and for $\delta = 2.405$, no zero-order term is present in the output.

9.7 BRILLOUIN SCATTERING

9.7.1 SPONTANEOUS BRILLOUIN SCATTERING

The acousto-optic effects described in Section 9.6 assume that an acoustic wave is present in the medium, driven by an external transducer. The inelastic, acoustic effect was first theoretically described by Leon Brillouin (Brillouin 1922), who predicted both Stokes and anti-Stokes acoustic frequency shifts. The so-called Brillouin scattering effect was subsequently measured about a decade later. In the absence of an external acoustic source, density fluctuations in the medium itself lead to a variation in the index of refraction. An incident beam then scatters from the index profile and can gain or lose energy to the acoustic fluctuations and form a doublet of lines around the central Rayleigh-scattered line. This process is called the spontaneous Brillouin scattering. Consider a density fluctuation in the medium given by $\Delta\underline{\rho}$. The density fluctuation leads to a change in the permittivity via

$$\Delta\underline{\varepsilon} = \frac{\partial\varepsilon}{\partial\rho}\Delta\underline{\rho}. \tag{9.118}$$

We note that the change in susceptibility is related to change in permittivity:

$$\Delta\underline{\chi} = \frac{\Delta\varepsilon}{\varepsilon_0}. \tag{9.119}$$

The change in susceptibility in turn results in a change in polarization:

$$\Delta\underline{P} = \Delta\underline{\varepsilon}\underline{E}. \tag{9.120}$$

Substituting Equation 9.118 into Equation 9.120 gives

$$\Delta \underline{P} = \frac{\partial \varepsilon}{\partial \rho} \Delta \rho \underline{E}. \tag{9.121}$$

Now consider that a general density fluctuation propagating with an acoustic wave vector of $\vec{K}$ and a frequency, Ω_B,

$$\Delta \underline{\rho} = \Delta \rho_0 \cos\left(\vec{K} \cdot \vec{r} - \Omega_B t\right). \tag{9.122}$$

The polarization therefore contains terms proportional to

$$\exp\left\{i\left((\vec{k} \pm \vec{K}) \cdot \vec{r} - (\omega \pm \Omega_B)t\right)\right\}. \tag{9.123}$$

The Brillouin (acoustic) angular frequency shift, Ω_B, is small compared to the optical frequency. Hence, the scattered frequency is essentially the same as the incident beam, and therefore the scattered k-vector also has essentially the same length as the incident one (see Figure 9.13). In order for the Brillouin process to phase match, the relationship that $\left|\vec{k}_{scattered}\right| \approx \left|\vec{k}_{inc}\right|$ dictates the Brillouin acoustic k-vector length and hence the Brillouin frequency shift.

Referring to Figure 9.13, we use the law of cosines to obtain

$$|\vec{K}| = 2k \sin\frac{\theta}{2}, \tag{9.124}$$

where $\left|\vec{k}_{scattered}\right| \approx \left|\vec{k}_{inc}\right| = k$. The dispersion relationship for an acoustic wave is given by

$$|\vec{K}| = \frac{\Omega_B}{v_{ac}}, \tag{9.125}$$

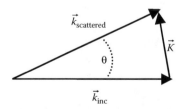

FIGURE 9.13 k-Vector diagram for the Brillouin scattering. Note that the magnitude of the incident and scattered optical k-vectors are nearly equal.

where v_{ac} is the acoustic velocity. Equating Equations 9.124 and 9.125 gives the spontaneous Brillouin frequency shift:

$$\Omega_B = 2 \frac{n \omega v_{ac}}{c} \sin \frac{\theta}{2}. \tag{9.126}$$

Note that the maximum frequency shift occurs at $\theta = \pi$ and also that no scattering occurs in the $\theta = 0$ direction. In an optical fiber, this means that the Brillouin scattering occurs in the backward direction with respect to optical mode traveling in the fiber. In this case, the forward and backward waves form a standing wave, which drives the acoustic wave. Although not indicated in Equation 9.126, a small Brillouin signal is forward scattered in optical fibers, but we neglect this weak effect.

9.7.2 CLASSICAL MODEL FOR THE STIMULATED BRILLOUIN SCATTERING

Perhaps the most important form of the Brillouin scattering is in the backward direction. This case is important because of the large frequency shift (see Equation 9.126) and because the backward wave is collinear with the incident beam resulting in long interaction lengths. For other angles, the scattered Brillouin wave has less overlap with the incident field. As mentioned in Section 9.7.1, no scattering in the forward direction occurs. We consider a backward wave that has a frequency of ω_B initially formed by the spontaneous process. Once this backward-propagating field forms, it mixes with the incident field to reinforce the acoustic wave.

The stimulated Brillouin scattering is similar to the stimulated Raman case where a pump and Stokes field induce a molecular vibration, which in turn mixes with the optical fields (see Figure 9.5). In the case of the stimulated Brillouin scattering, the incident field (pump) and the backward-propagating Brillouin field mix to generate an acoustic wave. This acoustic wave then mixes with the pump and Brillouin fields.

The mechanism for the pump and Brillouin field to generate an acoustic wave is via the electrostriction effect, which describes the process where a material compresses in the presence of a nonuniform electric field. The simple harmonic oscillator model that we use for describing Raman does not work here; instead, we describe the medium using the acoustic wave equation

$$\frac{\partial^2 \Delta \rho}{\partial t^2} - \Gamma_{ac} \nabla^2 \frac{\partial^2 \Delta \rho}{\partial t^2} - v_{ac}^2 \nabla^2 \Delta \rho = -\vec{\nabla} \cdot \vec{f}. \tag{9.127}$$

$\Delta \rho$ is the change in density of the medium away from its unperturbed value, Γ_{ac} accounts for damping of the acoustic wave, v_{ac} is the acoustic velocity, and $\vec{f}$ is an external force per unit volume on the medium. The force per unit volume at acoustic frequencies is related to the stored energy in the medium and is given by

$$\vec{f} = \frac{1}{2} \gamma_e \vec{\nabla} (\vec{E} \cdot \vec{E}). \tag{9.128}$$

γ_e is called the electrostrictive coefficient given by

$$\gamma_e = \rho_0 \left. \frac{\partial \varepsilon}{\partial \rho} \right|_{\rho_0}. \tag{9.129}$$

In an isotropic medium (Sutherland, 2003),

$$\gamma_e = \frac{1}{3} \varepsilon_0 (n^2 - 1)(n^2 + 2). \tag{9.130}$$

Substituting Equation 9.128 into Equation 9.127 shows a connection between an acoustic wave ($\Delta\underline{\rho}$) and the incident field:

$$\frac{\partial^2 \Delta\underline{\rho}}{\partial t^2} - \Gamma_{ac} \nabla^2 \frac{\partial \Delta\underline{\rho}}{\partial t} - v_{ac}^2 \nabla^2 \Delta\underline{\rho} = -\frac{1}{2} \gamma_e \nabla^2 \vec{\underline{E}} \cdot \vec{\underline{E}}. \tag{9.131}$$

The electric field includes the forward-propagating pump and the backward Brillouin wave, both of which we assume propagate along the z-axis and both are linearly polarized in the x-direction:

$$\vec{\underline{E}} = \left(\frac{1}{2} A_P e^{i(k_P z - \omega_P t)} + \frac{1}{2} A_B e^{i(-k_B z - \omega_B t)} + \text{c.c.} \right) \hat{x}. \tag{9.132}$$

When we take the dot product of this field with itself, we see that it has a difference-frequency component at the acoustic frequency, $\Omega = \omega_P - \omega_B$, given by

$$\vec{\underline{E}} \cdot \vec{\underline{E}} \Big|_{ac} = \frac{1}{2} A_P A_B^* e^{i((k_P + k_B)z - \Omega t)} + \text{c.c.} \tag{9.133}$$

In steady state, the acoustic wave takes the form

$$\Delta\underline{\rho} = \left(\frac{1}{2} \Delta\rho_0 e^{i(K_z - \Omega t)} + \text{c.c.} \right), \tag{9.134}$$

where K is the acoustic wave vector. Here, we assume that the acoustic wave's amplitude, $\Delta\rho_0$, does not grow substantially on the scale of the source term (Equation 9.133). We also note that the optical fields driving the acoustic wave determine the acoustic k-vector, $K = k_P + k_B \sim 2k_P$. Substituting Equations 9.133 and 9.134 into the acoustic wave equation, Equation 9.131, yields

$$\Delta\rho_0 = \frac{\gamma_e K^2 A_P A_B^*}{2\left(v_{ac}^2 K^2 - \Omega^2 - i\Gamma\Omega K^2\right)}. \tag{9.135}$$

Note that it is common to rewrite this equation as

$$\Delta\rho_0 = \frac{\gamma_e K^2 A_P A_B^*}{2\left(\Omega_B^2 - \Omega^2 - i\Gamma_B\Omega\right)}, \tag{9.136}$$

where Equation 9.125 is used, and

$$\Gamma_B = \Gamma K^2. \tag{9.137}$$

We expect that the frequency, Ω, will be close to the Brillouin frequency, Ω_B, especially for the case when the backward Brillouin wave is initiated by a spontaneous process. In this limit, we rewrite Equation 9.136 as

$$\Delta\rho_0 = \frac{\gamma_e K^2 A_P A_B^*}{2((\Omega_B - \Omega)(\Omega_B + \Omega) - i\Gamma_B\Omega)} \approx \frac{\gamma_e K^2 A_P A_B^*}{4\Omega((\Omega_B - \Omega) - i(\Gamma_B/2))}. \tag{9.138}$$

Furthermore, we take

$$K^2 \approx 2k_P K = 2\frac{n\omega_P}{c}\frac{\Omega}{v_{ac}}. \tag{9.139}$$

Hence, the acoustic wave's amplitude is

$$\Delta\rho_0 = \frac{\gamma_e n_P \omega_P}{2v_{ac} c ((\Omega_B - \Omega) - i(\Gamma_B/2))} A_P A_B^*. \tag{9.140}$$

This expression shows that the input optical fields induce an acoustic wave with an amplitude of $\Delta\rho_0$.

9.7.3 NONLINEAR POLARIZATION FOR THE STIMULATED BRILLOUIN SCATTERING

With the acoustic wave present in the medium, it then interacts with the incident field via the polarization expression given in Equation 9.121:

$$\Delta P = \frac{\partial\varepsilon}{\partial\rho}\bigg|_{\rho_0} \left(\frac{\Delta\rho_0}{2}e^{i(Kz-\Omega t)} + c.c.\right)\left(\frac{1}{2}A_P e^{i(k_P z - \omega_P t)} + \frac{1}{2}A_B e^{i(-k_B z - \omega_B t)} + c.c.\right). \tag{9.141}$$

This polarization gives frequency up- and down-shifted fields just as with the Raman case. Our primary interest is in the downshifted pump because, similar to the Raman case, we show that this field is amplified at the expense of the pump. Although the total field includes both a pump wave and a Brillouin wave, only the pump wave is incident on the medium and the Brillouin wave is initially generated spontaneously. The nonlinear polarization at $\omega_P - \Omega$ is

$$\Delta \underline{P} = \frac{1}{2} \Delta P(\omega_B) e^{-i(\omega_P - \Omega)t} + \text{c.c.} \tag{9.142}$$

Hence, the nonlinear polarization's complex amplitude of the downshifted pump field is, from Equations 9.142 and 9.141,

$$\Delta P = \frac{1}{2} \frac{\partial \varepsilon}{\partial \rho}\bigg|_{\rho_0} \Delta \rho_0^* A_P e^{i(k_P - K)z}. \tag{9.143}$$

Note that $k_P - K \approx -k_P$. Substituting Equation 9.136 into Equation 9.143 allows us to rewrite Equation 9.143 in terms of nonlinear susceptibility (following the form of Equation 8.9 with a multiplicity factor of 6):

$$\Delta P(\omega_B) = \frac{\partial \varepsilon}{\partial \rho}\bigg|_{\rho_0} \frac{\gamma_e n_P \omega_P |A_P|^2 A_B^*}{4 v_{ac} c((\Omega_B - \Omega) + i(\Gamma_B/2))} e^{i(k_P - K)z}$$

$$= \frac{3}{2} \varepsilon_0 \chi_B^{(3)}(\omega_B; \omega_P, -\omega_P, \omega_B) |A_P|^2 A_B e^{-ik_P z}. \tag{9.144}$$

Hence, the Brillouin nonlinear susceptibility is

$$\chi_B^{(3)}(\omega_B; \omega_P, -\omega_P, \omega_B) = \frac{\gamma_e^2 n_P \omega_P}{6 \varepsilon_0 \rho_0 v_{ac} c((\Omega_B - \Omega) + i(\Gamma_B/2))}, \tag{9.145}$$

where we use Equation 9.129 for $(\partial \varepsilon / \partial \rho)|_{\rho_0}$. In the following, we define

$$\chi_B^{(3)} \equiv \chi_B^{(3)}(\omega_B; \omega_P, -\omega_P, \omega_B). \tag{9.146}$$

A similar calculation gives the nonlinear susceptibility at the pump frequency:

$$\chi_{BP}^{(3)}(\omega_P; \omega_B, -\omega_B, \omega_P) = \frac{\gamma_e^2 n_P \omega_P}{6 \varepsilon_0 \rho_0 v_{ac} c((\Omega_B - \Omega) - i(\Gamma_B/2))} = \chi_B^{(3)*}. \tag{9.147}$$

The nonlinear polarization at the pump frequency is

$$\Delta P(\omega_P) = \frac{3}{2}\varepsilon_0 \chi_{BP}^{(3)}(\omega_P; \omega_B, -\omega_B, \omega_P)|A_B|^2 A_P e^{i(K-k_B)z}$$

$$= \frac{3}{2}\varepsilon_0 \chi_{BP}^{(3)*}|A_B|^2 A_P e^{i(K-k_B)z}. \tag{9.148}$$

Note that nonlinear polarization at the Brillouin frequency has a k-vector given by $-k_P \approx -k_B$ and that nonlinear polarization at the pump frequency has a k-vector given by $K - k_B \approx k_P$. Hence, both nonlinear polarizations are phase matched with the fields they are driving.

9.7.4 COUPLED INTENSITY EQUATIONS AND SOLUTIONS FOR THE STIMULATED BRILLOUIN SCATTERING

The pump and Brillouin wave evolution are determined by the wave equation, which in the SVEA is given by Equation 9.33. For the backward wave, Equation 9.33 is

$$-2ik\frac{dA}{dz}e^{-ikz} = -\omega^2\mu_0 P^{NL}. \tag{9.149}$$

The nonlinear polarization at the Brillouin frequency is dominated by the expression given in Equation 9.144, and contributions from nonresonant four-wave mixing are assumed to be negligible. Substituting the nonlinear polarization for the Brillouin field into Equation 9.149 and for the pump field into Equation 9.33 results in the following two coupled amplitude equations:

$$\frac{dA_B}{dz} = -i\frac{3\omega_B}{4nc}\chi_B^{(3)}|A_P|^2 A_B = -i\frac{3\omega_B}{2\varepsilon_0 n^2 c^2}\chi_B^{(3)}I_P A_B, \tag{9.150}$$

$$\frac{dA_P}{dz} = i\frac{3\omega_P}{2\varepsilon_0 n^2 c^2}\chi_B^{(3)*}I_B A_P. \tag{9.151}$$

Next, we transform these two equations into differential equations in the intensities. We follow the same procedure as the Raman case, where we use

$$\frac{dI}{dz} = \frac{n\varepsilon_0 c}{2}\left(A*\frac{dA}{dz} + c.c.\right). \tag{9.152}$$

The two coupled equations for the intensities are therefore

$$\frac{dI_B}{dz} = -\frac{3\omega_B}{\varepsilon_0 n^2 c^2}\left(\mathrm{Im}\,\chi_B^{(3)}\right)I_P I_B = -g_B I_P I_B, \tag{9.153}$$

$$\frac{dI_P}{dz} = -g_B I_P I_B. \tag{9.154}$$

g_B is obtained from Equation 9.153 and the expression for susceptibility in Equation 9.145,

$$g_B = \frac{\gamma_e^2 \omega^2}{\varepsilon_0^2 n \rho_0 v_{ac} c^3 \Gamma_B} \left[\frac{(\Gamma_B/2)^2}{(\Omega_B - \Omega)^2 + (\Gamma_B/2)^2} \right]. \tag{9.155}$$

The line shape function of g_B in square brackets is called a Lorentzian function; its FWHM is Γ_B, which is typically of the order of 10 MHz.

The simplest solution to Equation 9.153 is for the undepleted pump case ($I_P =$ constant). In this case, Equation 9.153 integrates to

$$I_B = I_B(L) e^{g_B I_P (L-z)}. \tag{9.156}$$

Note that the Brillouin field grows exponentially as it travels in the *negative* z-direction.

In the limit of high conversion efficiency to the Brillouin wave, we must consider pump depletion. We obtain solutions in the depleted pump regime by following an approach similar to the Raman case (see Section 9.3.6). First, note that by equating Equations 9.153 and 9.154, we obtain

$$dI_B = dI_P. \tag{9.157}$$

Therefore,

$$I_B(z) = I_P(z) + C, \tag{9.158}$$

where C is determined from boundary conditions. The boundary at $z = 0$ is applied to find C:

$$C = I_B(0) - I_P(0) = I_{B0} - I_{P0}. \tag{9.159}$$

We substitute Equation 9.158 into Equation 9.153, separate variables, and integrate:

$$\int_{I_{B0}}^{I_B} \frac{dI_B'}{I_B'(I_B' - C)} = -g_B \int_0^z dz'. \tag{9.160}$$

Evaluating the integral results in

$$I_B(z) = \frac{(I_{B0} - I_{P0})I_{B0}}{I_{B0} - I_{P0}\exp(-g_B(I_{B0} - I_{P0})z)}. \tag{9.161}$$

This equation is written in terms of the boundary condition at $z = 0$. We know the pump intensity at this location; however, the Brillouin intensity boundary condition is known at $z = L$. We can solve for I_{B0} in terms of $I_B(L)$ by substituting $z = L$ into Equation 9.161, but the resultant equation is transcendental. We get around this problem by rearranging Equation 9.161 as

$$g_B I_{P0} L = \frac{\ln(y - (y(y-1)/f))}{1-y}, \tag{9.162}$$

where

$$y \equiv \frac{I_{B0}}{I_{P0}} \tag{9.163}$$

and

$$f \equiv \frac{I_B(L)}{I_{P0}}. \tag{9.164}$$

Ideally, we would invert Equation 9.162 and write the ratio $y \equiv (I_{B0}/I_{P0})$ as a function of $g_B I_{P0} L$; however, the equation is transcendental and does not lend itself to inversion. Instead, we plot $g_B I_{P0} L$ as a function of y and then swap the axes (see Figure 9.14). The plot in Figure 9.14 shows that when the gain product, $g_B I_{P0} L$, becomes larger than 20, significant pump depletion occurs even for tiny Brillouin intensity inputs.

9.7.5 BRILLOUIN WITH LINEAR ABSORPTION

In most cases, the pump and Brillouin waves experience loss, and in this case, Equations 9.153 and 9.154 become

$$\frac{dI_B}{dz} = -g_B I_P I_B + \alpha I_B, \tag{9.165}$$

$$\frac{dI_P}{dz} = -g_B I_P I_B - \alpha I_P. \tag{9.166}$$

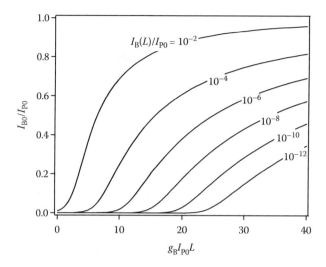

FIGURE 9.14 Brillouin output at $z = 0$ as a function of the gain product, $g_B P_0 L$.

The Brillouin equation has $+\alpha I_B$ because it propagates in the backward direction. We assume that the losses are the same at the pump and Brillouin frequencies. Equations 9.165 and 9.166 do not have a closed-form solution and are solved numerically (Chen and Bao, 1998). However, a solution exists in the limit that the pump intensity is much greater than the Brillouin intensity. In this case, we approximate Equation 9.166 as

$$\frac{dI_P}{dz} \approx -\alpha I_P. \tag{9.167}$$

The pump intensity integrates to $I_{P0}e^{-\alpha z}$, which we substitute into Equation 9.165 yielding

$$\frac{dI_B}{I_B} = \left(-g_B I_{P0} e^{-\alpha z} + \alpha\right) dz. \tag{9.168}$$

Integrating this equation gives the Brillouin intensity:

$$I_B(0) = I_B(L) \exp(g_B I_{P0} L_{eff} - \alpha L), \tag{9.169}$$

where

$$L_{eff} = \frac{1 - e^{-\alpha L}}{\alpha}. \tag{9.170}$$

The effect of introducing loss reduces the interaction length, and it also reduces the exponential gain.

9.7.6 MITIGATING BRILLOUIN EFFECTS

The Brillouin scattering is a potential problem for fiber-based systems. The backward-propagating Brillouin wave is amplified at the expense of the forward-propagating pump, introducing a loss for the desired signal. The problem is even more extreme for fiber lasers and fiber amplifiers where the fiber is doped to provide active gain. The exponential gain of the backward-propagating Brillouin wave is further enhanced by the laser/amplifier gain. In addition to reducing the useful output of a fiber-based system, the Brillouin wave may even damage components in an amplifier system. For example, consider amplifying a 10 mW laser using an erbium-doped fiber amplifier with a gain of 20 dB. Ideally, the output of the amplifier is 1 W; however, when Brillouin effects are not mitigated, it is possible that 1 W is reflected back into the original laser source. Such a situation results in damaging the laser source, a potentially expensive proposition.

A common approach to eliminating the Brillouin effects is to frequency modulate the pump beam. Let us assume that the acoustic wave at frequency Ω is generated from the stimulated Brillouin scattering between a pump and a Brillouin wave. For the Brillouin wave to build up coherently throughout the medium, its frequency and k-vector must remain constant. As the pump laser frequency changes, so does the Brillouin frequency (see Equation 9.126). If the pump frequency is modulated rapidly enough, then the Brillouin wave does not have a chance to build up throughout the medium. A typical frequency modulation bandwidth is 1 GHz. The disadvantage of frequency modulating the source is that it adds bandwidth to the output. Applications such as high-resolution spectroscopy and laser radar require narrower linewidths than 1 GHz.

A similar strategy is to engineer an optical fiber that will disrupt the Brillouin conversion process. If the properties of the fiber change along its length such that the optical or acoustic k-vector changes, then a specific Brillouin frequency cannot build up coherently throughout the fiber length. For example, by doping the fiber differently along its length, the acoustic velocity also varies, which according to Equation 9.126 changes the Brillouin frequency along the length of the fiber. Temperature gradients along the length of the fiber accomplish the same effect.

Another way to engineer the fiber to reduce the Brillouin effects is to increase the mode diameter. By doing so, the intensity drops and the gain factor, gI_pL, also drops. However, increasing the mode diameter may introduce higher-order fiber modes. Photonic crystal fibers get around this problem by using a periodic structure to obtain a single mode with a large mode-field diameter. In the case of fiber amplifiers, a final strategy is to reduce the fiber length by doping the fiber with higher concentrations of active ions.

PROBLEMS

9.1 A laser is incident on a hydrogen cell. What is the Stokes-shifted output wavelength if the laser wavelength is (a) 488 nm and (b) 600 nm?

9.2 A Raman spectrum is generated with a 488 nm laser, and two strong peaks are detected at 528 and 594 nm. Use Table 9.1 to determine what gases are present in this sample.

9.3 Monochromators and spectrometers are typically specified in terms of $f/\#$. $f/\#$ is defined as the ratio of focal length, f, to the mirror or lens diameter, d (see Figure 9.15). A 1 cm diameter lens is used to collect a Raman signal. What should the focal length and placement of the lens be such that all the light collected by the lens is matched to the monochromator's $f/\# = 8$?

9.4 Consider the setup shown in Figure 9.3. A lens of 2.5 cm diameter and 2.5 cm focal length is used to collect the scattered Raman signal and images it (1:1) into the monochromator. Assume that all of the light collected by the lens falls within the collection aperture of the monochromator. The monochromator slits are opened to 0.5 mm. The incident laser has a wavelength of 488 nm, has a power of 1 W, and is focused into hydrogen gas with a number density of 1×10^{19} molecules/cm^3. Estimate the number of photons collected in 1 s by a detector at the output of the monochromator. Assume that the monochromator throughput is 100%.

9.5 Consider a Raman microscope as shown in Figure 9.4. The laser operates at 488 nm, has a power of 10 mW, and is focused to a 2 µm radius (assume a top-hat intensity profile) on the sample. The microscope objective lens has a numerical aperture of 0.95. The molecule under study has a differential cross section of $(d\sigma/d\Omega) = 5 \times 10^{-30}$ cm^2/Sr/molecule and a Stokes shift of 256 cm^{-1}. Assuming that the microscope images a volume that contains 10^{14} molecules, calculate the power collected by the objective lens. Convert the power to a photon flux per second.

9.6 The differential cross sections of selected molecules measured at 488 nm are listed in Table 9.1. What is the approximate scaling factor for the differential cross section if a 633 nm laser is used? A 10 mW HeNe laser ($\lambda = 633$ nm) is focused to a 20 µm^2 area with a top-hat profile; the monochromator slit width is 100 µm. A 10 cm diameter lens at a distance of 10 cm from the scattering volume optics collects scattered light. The density of O_2 and O_3 molecules in the cell is 2.7×10^{19}/cm^3 and 1.0×10^{19}/cm^3, respectively. Calculate the scattered

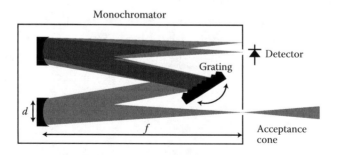

FIGURE 9.15 Schematic diagram of a monochromator.

light power and the number of scattered photons per second for each molecular species.

9.7 Use the classical expression to derive the nonlinear susceptibility expression for the pump wave; the relationship between the polarization and the fields is found in Equation 9.47. Furthermore, show that $\chi_Q^{(3)}(\omega_P; \omega_S, -\omega_S, \omega_P) = \chi_Q^{(3)*}(\omega_S; \omega_P, -\omega_P, \omega_S)$.

9.8 In the limit of high conversion efficiency for a Raman process, the pump experiences depletion, given by Equation 9.49. Derive Equation 9.49 by considering the pump and Stokes fields but ignoring the anti-Stokes.

9.9 Equations 9.74 and 9.75 give the classical expression for the nonlinear polarization when a pump, Stokes, and anti-Stokes fields are present in a medium. Use Equation 8.9 to rewrite these equations in terms of an $\chi^{(3)}$ process. In Equation 8.9, consider the scalar case where we keep the frequency dependence and ignore the Cartesian components. Show that the resulting susceptibilities are given by Equations 9.76 and 9.77.

9.10 Use the classical approach in Section 9.4 to calculate the nonlinear polarization for the pump frequency when a pump, Stokes, and anti-Stokes fields are present in a medium. Show that on resonance (i.e., $\omega_P - \omega_S = \omega_R$ and $\omega_{AS} - \omega_P = \omega_R$) the nonlinear polarization may be written as

$$P^{NL}(\omega_P) = i\frac{3\varepsilon_o \left|\chi_R^{(3)}\right|}{2}\left(|A_S|^2 - |A_{AS}|^2\right)A_P e^{ik_P z}, \tag{9.171}$$

where $\left|\chi_R^{(3)}\right|$ is given by Equation 9.78.

9.11 Use Equations 9.97 and 9.99 to derive an expression for the Stokes and anti-Stokes fields in the undepleted pump approximation for an arbitrary Δk. For the input assume $A_S(0) = A_{S0}$, $A_{AS}(0) = A_{AS0}$, and $A_P(z) = A_P(0)$.

9.12 Use the result of Problem 9.11 to calculate Stokes and anti-Stokes intensities in the undepleted approximation given $|A_S(0)| = 2.7 \times 10^6$ V/m, $|A_{AS}(0)| = 2.7 \times 10^6$ V/m, $|A_P(z)| = 2.7 \times 10^7$ V/m, $\lambda_P = 1.5$ μm, $n = 1.5$, $\left|\chi_R^{(3)}\right| = 2 \times 10^{-22}$ m^2/V^2, and $\delta = 0.03$. Plot $I_S(z)/I_{P0}$ and $I_{AS}(z)/I_{P0}$ from $z = 0$ to $z = 10$ m. Assume that the phase of the Stokes, pump, and anti-Stokes all start at 0.

9.13 Derive an expression for the angle, θ, where the anti-Stokes frequency is noncollinearly phase matched in an isotropic medium in terms of k_{AS}, k_P, and k_S. The geometry is shown in Figure 9.16. What is the angle given that $\omega_R = 2\pi \times 10^5$ GHz, the pump wavelength is 1.064 μm, and the dispersion is given by the following relationship:

$$n^2(\lambda) = A_0 + \frac{A_1}{\lambda^2 - A_2} - A_3\lambda^2, \tag{9.172}$$

where $A_0 = 2.7359$, $A_1 = 0.01878$, $A_2 = 0.01822$, and $A_3 = 0.01354$?

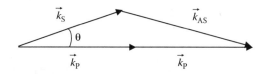

FIGURE 9.16 Noncollinear phase-matching geometry for the Stokes and anti-Stokes waves.

9.14 Numerically integrate Equations 9.97 through 9.99 to obtain the Stokes and anti-Stokes fields. Use the following parameters and plot $I_S(z)/I_{P0}$ and $I_{AS}(z)/I_{P0}$ from $z = 0$ to $z = 50$ m for (a) $\Delta k = 0$ and (b) $\Delta k = 0.3\,\mu m^{-1}$. $|A_S(0)| = 2.7 \times 10^6$ V/m, $|A_{AS}(0)| = 2.7 \times 10^6$ V/m, $|A_P(z)| = 2.7 \times 10^7$ V/m, $\lambda_P = 1.5\,\mu m$, $n = 1.5$, $|\chi_R^{(3)}| = 2 \times 10^{-22}$ m^2/V^2, and $\delta = 0.03$. Assume that the phase of the Stokes, pump, and anti-Stokes all start at 0.

9.15 Show that the angular separation for the Raman-Nath diffracted orders is given by

$$\Delta\theta = \frac{f_{ac}\lambda_{opt}}{nv_{ac}}, \qquad (9.173)$$

where f_{ac} is the acoustic frequency, λ_{opt} is the optical wavelength in vacuum, n is the optical index of refraction, and v_{ac} is the acoustic velocity. Assume low-order diffraction such that $\theta_{diffraction}$ is small.

9.16 In the Raman-Nath process, what is the maximum possible diffraction efficiency for the first-order diffraction?

9.17 Calculate the frequency content of the first-order Raman-Nath diffracted beam if the acoustic wave is a standing wave instead of a traveling wave. Assume that the optical wave is at normal incidence to the acoustic wave.

9.18 Calculate a classical expression for the Brillouin nonlinear susceptibility at the pump frequency following the same procedure as in Section 9.7.3, and show that it is given by Equation 9.147.

9.19 Show that the FWHM for the Brillouin line shape in Equation 9.155 is given by Γ_B.

9.20 Directly integrate the differential equation for the Brillouin intensity given in Equation 9.153 in the undepleted pump approximation, and show that the result is given by Equation 9.156.

9.21 A single-mode fiber has a mode-field diameter of 6 μm. Estimate the maximum fiber length where the Brillouin effects will not be appreciable (i.e., $I_B(0)/I_{P0} < 0.1$). The input pump power is 1 W, $\lambda_P = 1.5\,\mu m$, and $g_B = 3.5 \times 10^{-9}$ cm/W, and assume that the fiber is lossless. Assume that $I_B(L)$ comes from the spontaneous Brillouin process. Estimate $I_B(L)$ using 1 photon/μs scattered into the fiber mode. Treat the intensity in the fiber as a top-hat profile filling the mode-field diameter.

REFERENCES

Agrawal, G. P. 2013. *Nonlinear Fiber Optics* (5th edition). New York: Academic Press.

Brillouin, L. 1922. Diffusion de la lumiere et des rayons X par un corps transparent homogene: Influence de l'agitation thermique. *Annales de Physique (Paris)* 17: 88.

Chen, L. and X. Bao. 1998. Analytical and numerical solutions for steady state stimulated Brillouin scattering in a single-mode fiber. *Optics Communications* 152: 65–72.

Danichkin, S. A., A. A. Eliseev, T. N. Popova, O. V. Ravodina, and V. V. Stenina. 1981. Raman scattering parameters for gas molecules (survey). *Journal of Applied Spectroscopy* 35: 1057–1066.

Demtroder, W. 1996. *Laser Spectroscopy*. New York: Springer-Verlag.

Raman, V. V. and K. S. Krishnan 1928. A new type of secondary radiation. *Nature* 121: 501–502.

Stolen, R. H., E. P. Ippen, and A. R. Tynes. 1972. Raman oscillation in glass optical waveguide. *Applied Physics Letters* 20: 62–64.

Sutherland, R. L. 2003. *Handbook of Nonlinear Optics* (2nd edition). New York: Marcel Dekker, Inc.

NONLINEAR OPTICS INCLUDING DIFFRACTION AND DISPERSION

10.1   INTRODUCTION

Previous chapters developed equations that assume monochromatic beams and largely ignore diffraction effects. Solutions to the wave equation in this simplified limit reveal the basic phenomena that occur in nonlinear media. For example, phase matching, exponential gain, back conversion, self-induced phase modulation processes, and so on are well described by the formalism introduced earlier in this book. However, to have a quantitative understanding of more complex systems needs a more general formalism that includes diffractive and temporal effects. Developing the formalism requires a more complete analysis of the wave equation. As in prior chapters, we introduce the new effects in stages. A large class of nonlinear interactions involves nearly monochromatic sources where temporal effects are negligible. This class of problems extends to nanosecond laser pulses in certain cases. As a general rule of thumb, temporal effects become important when the characteristic interaction length scale becomes comparable to the spatial extent of the laser pulse. For example, a 1-ns laser pulse has spatial extent in air of ~30 cm; hence, we do not expect temporal propagation effects to be important for crystals with interaction lengths of a few centimeters. On the other hand, optical effects in long fibers should include temporal effects such as group-velocity dispersion (GVD), which is defined and discussed below.

In this chapter, we follow a staged approach by first looking at spatial effects following with an analysis of temporal effects. Similar to effects discussed earlier in this book where the linear optical properties of materials play a large role in determining nonlinear effects, the linear propagation effects for more general types of electromagnetic waves also play a critical role in determining the efficiency of nonlinear effects. Specifically, we derive a more general wave equation that first includes diffraction and the Poynting vector walk-off and then temporal effects.

10.2 SPATIAL EFFECTS

This section is devoted to a presentation of diffraction and walk-off effects in the wave equation. Diffraction has been previously introduced in Chapter 2 and used in several discussions, including our discussion of the z-scan method in Section 8.8. Although we devote our attention to spatial effects in this section, we also need to pay attention to the overall temporal pulse shape. Consider an interaction involving nanosecond laser pulses. Here, the variation in intensity of the pulse cannot be ignored because nonlinear effects are strongly intensity dependent. In such cases, the pulse is divided into several time slices of nearly constant intensities; therefore, each segment is considered to be equivalent to a continuous-wave (cw) beam of the same intensity. A separate cw calculation is made for each segment (using that as its average intensity), and then the results from all segments are combined to calculate the net outcome. For the rest of this section we consider the cw case, and note that we can reconstruct the results for long-pulse interactions by using this segmentation procedure.

10.2.1 DIFFRACTION AND THE POYNTING VECTOR WALK-OFF

We start our discussion by recalling the wave equation:

$$\vec{\nabla} \times (\vec{\nabla} \times \vec{E}) = -\mu_0 \frac{\partial^2}{\partial t^2} (\varepsilon_0 \vec{E} + \underline{\vec{P}}^{(1)} + \underline{\vec{P}}^{NL}). \tag{10.1}$$

Here, we assume that the medium is nonmagnetic ($\mu = \mu_0$) and that we can split up the polarization into a linear and nonlinear part. In this situation, the wave equation for one of the beams at frequency ω is

$$\vec{\nabla} \times (\vec{\nabla} \times \vec{E}) = \mu_0 \omega^2 (\varepsilon_0 \vec{E} + \vec{P}^{(1)} + \vec{P}^{NL}). \tag{10.2}$$

The spatial derivatives occur with the terms on the left-hand side of this equation, which is rewritten using the identity

$$\vec{\nabla} \times (\vec{\nabla} \times \vec{E}) = \vec{\nabla}(\vec{\nabla} \cdot \vec{E}) - \nabla^2 \vec{E}. \tag{10.3}$$

In the present analysis, both terms on the right-hand side of this equation must be treated differently than in the plane-wave case. We first look at the Laplacian term. Whereas previously we assumed plane waves, we now consider beams that propagate in a given direction and where diffraction occurs transverse to this direction. The most common convention is for the beam to travel along a laboratory frame of reference called the z-axis. However, when working with beams in crystals, we need to distinguish the propagation direction from the crystallographic axes. Our convention for this chapter is to define the z-axis as the propagation direction of the beam, and when we need to give angles with respect to crystallographic axes, we do so on a case-by-case basis. Under this scenario, the electric field is written as

$$\vec{E} = \frac{1}{2} A(\vec{r}_\perp, z) e^{i(k_0 z - \omega t)} \hat{E} + \text{c.c.}, \tag{10.4}$$

where k_0 is the beam's longitudinal k-vector magnitude and $\vec{r}_\perp$ is the transverse coordinate. In the current discussion, we assume that the beams are monochromatic so that the field's complex amplitude has no time dependence. The Laplacian operator, ∇^2, is broken down into two operators, one for the propagation direction and one for the transverse direction:

$$\nabla^2 = \nabla_\perp^2 + \frac{\partial^2}{\partial z^2}, \tag{10.5}$$

where the transverse Laplacian is defined as

$$\nabla_\perp^2 \equiv \frac{\partial^2}{\partial x^2} + \frac{\partial^2}{\partial y^2}. \tag{10.6}$$

When the Laplacian operates on the field in Equation 10.4, we obtain the same result as with the plane-wave derivation, except that now we have the extra transverse Laplacian term

$$\nabla^2 \vec{E} = \frac{1}{2} \left(\nabla_\perp^2 A - k_0^2 A + 2 i k_0 \frac{\partial A}{\partial z} + \frac{\partial^2 A}{\partial z^2} \right) e^{i(k_0 z - \omega t)} \hat{E} + \text{c.c.} \tag{10.7}$$

In the slowly varying envelope approximation (SVEA), the complex amplitude of the Laplacian term (Equation 10.7) is

$$\left(\nabla_\perp^2 A - k_0^2 A + 2 i k_0 \frac{\partial A}{\partial z} \right) e^{i k_0 z} \tag{10.8}$$

Now we turn our attention to the $\vec{\nabla}(\vec{\nabla}\cdot\vec{E})$ term in Equation 10.3, which takes into account walk-off. In the previous chapters, we treated the Poynting vector walk-off in a somewhat qualitative fashion by calculating walk-off angles and estimating beam overlap. This previous analysis (see Section 2.3.7) shows that the Poynting vector of an e-wave is not parallel to, and the electric field is not perpendicular to, the k-vector direction (see Figure 10.1). In the present discussion, we take a more quantitative approach by including walk-off through the term $\vec{\nabla}(\vec{\nabla}\cdot\vec{E})$. To evaluate the contribution of this expression to the SVEA equations, we assume that an e-wave is propagating in a crystal and that its walk-off direction is in the x–z-plane, as shown in Figure 10.1. This particular convention determines the field direction:

$$\hat{E} = \cos\rho\,\hat{x} + \sin\rho\,\hat{z}. \tag{10.9}$$

Therefore, the divergence of the field given by Equation 10.4 is

$$\vec{\nabla}\cdot\vec{E} = \frac{1}{2}\left(\frac{\partial A}{\partial x}\cos\rho + \left(\frac{\partial A}{\partial z} + ik_0 A\right)\sin\rho\right)e^{i(k_0 z - \omega t)} + \text{c.c.} \tag{10.10}$$

We take the gradient of this equation to obtain

$$\vec{\nabla}(\vec{\nabla}\cdot\vec{E}) \approx \frac{1}{2}\left[ik_0\sin\rho\frac{\partial A}{\partial x}\hat{x} + ik_0\sin\rho\frac{\partial A}{\partial y}\hat{y} + \left[ik_0\frac{\partial A}{\partial x}\cos\rho + 2ik_0\sin\rho\frac{\partial A}{\partial z} - k_0^2\sin\rho A\right]\hat{z}\right]$$
$$\times e^{i(k_0 z - \omega t)} + \text{c.c.}, \tag{10.11}$$

where we make the SVEA (in the transverse direction) and neglect second-order partial derivatives. Only the component of $\vec{\nabla}(\vec{\nabla}\cdot\vec{E})$ in the direction of $\hat{E}$ corresponds to propagation. Rearranging Equation 10.9 gives

$$\hat{x} = \frac{\hat{E}}{\cos\rho} - \tan\rho\,\hat{z}. \tag{10.12}$$

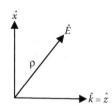

FIGURE 10.1   Illustration of the k-vector and field direction in a birefringent crystal. Note that the walk-off angle, ρ, is exaggerated.

We substitute this result into Equation 10.11 yielding

$$\vec{\nabla}(\vec{\nabla} \cdot \vec{E})|_{\text{along } \hat{E}} \approx \frac{1}{2}\left[2ik_0 \tan\rho \frac{\partial A}{\partial x} e^{i(k_0 z - \omega t)} + \text{c.c.} \right] \tag{10.13}$$

Because walk-off angles are typically less than $3°$, terms on the order of $\sin^2 \rho$ are not included in Equation 10.13, and the small-angle approximation $\tan \rho \approx \rho$ is valid.

Substituting the diffraction term (Equation 10.8) and the walk-off term (Equation 10.13) into Equation 10.2 as well as evaluating the right-hand side of Equation 10.2 gives the following envelope equation:

$$\underset{\underset{\text{diffraction}}{\uparrow}}{\frac{\partial A}{\partial z}} = \frac{i}{2k_0}\nabla_\perp^2 A \; \underset{\underset{\text{walk-off}}{\uparrow}}{+\tan\rho \frac{\partial A}{\partial x}} \; \underset{\underset{\text{loss}}{\uparrow}}{-\frac{\alpha}{2} A} \; \underset{\underset{\text{nonlinearity}}{\uparrow}}{+\frac{i\mu_0\omega^2}{2k_0}P^{\text{NL}}e^{-ik_0 z}} \; . \tag{10.14}$$

The solution to this equation is performed numerically using a split-step technique described in Section 10.2.2.

10.2.2 SPLIT-STEP TECHNIQUE

An analytic solution to Equation 10.14 yielding $A(\vec{r}_\perp, z)$ exists only for special cases. Hence, we turn to numerical approaches to find general solutions. A particularly powerful and useful technique is called the split-step operator approach. This technique subdivides a nonlinear medium into N slices. For each slice, Equation 10.14 is broken down into simpler equations, which are separately solved. Then the solutions for each of the simpler equations are combined to give a net result that is, with reasonable accuracy, close to the exact solutions. As an example, we consider a three-wave mixing process defined by $\omega_1 = \omega_2 + \omega_3$. The nonlinear polarization for the sum frequency is (see Section 3.3)

$$P^{(2)}(\omega_1) = 2\varepsilon_0 d_{\text{eff}} A_2 A_3 e^{i(k_{20} + k_{30})z}, \tag{10.15}$$

where k_{20} and k_{30} are the longitudinal k-vector magnitudes for the fields at ω_2 and ω_3, respectively. Substituting Equation 10.15 into Equation 10.14 gives the envelope equation for A_1 and similarly for A_2 and A_3,

$$\frac{\partial A_1}{\partial z} = \frac{i}{2k_{10}}\nabla_\perp^2 A_1 + \tan\rho_1 \frac{\partial A_1}{\partial x} - \frac{\alpha_1}{2} A_1 + i\frac{\omega_1^2}{k_{10}c^2}d_{\text{eff}}A_2 A_3 e^{-i\Delta k z}, \tag{10.16}$$

$$\frac{\partial A_2}{\partial z} = \frac{i}{2k_{20}} \nabla_\perp^2 A_2 + \tan\rho_2 \frac{\partial A_2}{\partial x} - \frac{\alpha_2}{2} A_2 + i \frac{\omega_2^2}{k_{20}c^2} d_{\text{eff}} A_1 A_3^* e^{i\Delta kz}, \qquad (10.17)$$

$$\frac{\partial A_3}{\partial z} = \frac{i}{2k_{30}} \nabla_\perp^2 A_3 + \tan\rho_3 \frac{\partial A_3}{\partial x} - \frac{\alpha_3}{2} A_3 + i \frac{\omega_3^2}{k_{30}c^2} d_{\text{eff}} A_1 A_2^* e^{i\Delta kz}, \qquad (10.18)$$

where

$$\Delta k = k_{10} - k_{20} - k_{30}. \qquad (10.19)$$

k_{10}, k_{20}, and k_{30} are the z-components of $\vec{k}_1$, $\vec{k}_2$, and $\vec{k}_3$, respectively. For a small step size, the linear and nonlinear parts of the equations are essentially decoupled. With this in mind, our strategy for solving this set of coupled amplitude equations is to split the equations into their linear and nonlinear parts and approximate the exact solution through an iterative approach. We devote the rest of this section to show how to formally implement this algorithm. Although going through the formal process is somewhat cumbersome, keep in mind the end result, which is a justification for splitting the equations.

Formally, we may write Equations 10.16 through 10.18 as the following matrix equation:

$$\frac{\partial \vec{V}}{\partial z} = \begin{bmatrix} O_1 & 0 & 0 \\ 0 & O_2 & 0 \\ 0 & 0 & O_3 \end{bmatrix} \vec{V} + i \frac{d_{\text{eff}}}{c^2} \begin{bmatrix} 0 & \dfrac{\omega_1^2}{k_{10}} A_3 e^{-\Delta kz} & 0 \\ \dfrac{\omega_2^2}{k_{20}} A_3^* e^{i\Delta kz} & 0 & 0 \\ \dfrac{\omega_3^2}{k_{30}} A_2^* e^{i\Delta kz} & 0 & 0 \end{bmatrix} \vec{V}, \qquad (10.20)$$

where

$$\vec{V} = \begin{bmatrix} A_1 \\ A_2 \\ A_3 \end{bmatrix}, \qquad (10.21)$$

and

$$O_1 = \frac{i}{2k_{10}} \nabla_\perp^2 + \tan\rho_1 \frac{\partial}{\partial x} - \frac{\alpha_1}{2}, \qquad (10.22)$$

and similarly for O_2 and O_3. We rewrite Equation 10.20 in a more compact form as

$$\frac{\partial \vec{V}}{\partial z} = (\hat{L} + \hat{N})\vec{V}, \tag{10.23}$$

where $\hat{L}$ and $\hat{N}$ are the linear and nonlinear operators given by the matrices in Equation 10.20. The symbolic solution to the compact equation is

$$\vec{V}(z + dz) = e^{(\hat{L}+\hat{N})dz}\, \vec{V}(z). \tag{10.24}$$

We may approximate this expression by expanding the exponential:

$$e^{(\hat{L}+\hat{N})dz} = 1 + \hat{L}dz + \hat{N}dz + \frac{\hat{L}^2 dz^2}{2} + \frac{\hat{N}^2 dz^2}{2} + \frac{\hat{L}\hat{N}}{2}dz^2 + \frac{\hat{N}\hat{L}}{2}dz^2 + \text{Order}(dz^3). \tag{10.25}$$

Note that in general the order of operators is important and that $\hat{L}\hat{N} \neq \hat{N}\hat{L}$. In other words, $\hat{L}$ and $\hat{N}$ do not commute. The terms in Equation 10.25 involving the product of the two different operators is rewritten as

$$\frac{1}{2}(\hat{L}\hat{N} + \hat{N}\hat{L}) = \hat{L}\hat{N} - \frac{1}{2}[\hat{L}, \hat{N}] \tag{10.26}$$

where $[\hat{L}, \hat{N}]$ denotes the commutation of the two operators and is defined by

$$[\hat{L}, \hat{N}] \equiv \hat{L}\hat{N} - \hat{N}\hat{L}. \tag{10.27}$$

Now note that to the same order,

$$e^{\hat{L}dz} e^{\hat{N}dz} = 1 + \hat{L}dz + \hat{N}dz + \frac{\hat{L}^2 dz^2}{2} + \frac{\hat{N}^2 dz^2}{2} + \hat{L}\hat{N}dz^2 + \text{Order}(dz^3). \tag{10.28}$$

Hence, the exponential operator in Equation 10.24 is approximately

$$e^{(\hat{L}+\hat{N})dz} \approx e^{\hat{L}dz} e^{\hat{N}dz}, \tag{10.29}$$

where the difference between the left-hand side and right-hand side of this equation, to order dz^2, is the commutator term, $(1/2)[\hat{L}, \hat{N}]dz^2$. Therefore, we may write, accurate to order dz,

$$\vec{V}(z + dz) \approx e^{\hat{L}\,dz} e^{\hat{N}\,dz}\vec{V}(z). \tag{10.30}$$

The significance of this result is that $\vec{V}(z+dz)$ is obtained by first operating on $\vec{V}(z)$ with the nonlinear exponential operator alone and then taking this result and operating on it with the linear operator. Note that when we operate with a single exponential operator, it is equivalent to solving the differential equations in the absence of the other operator. For example, $e^{\hat{L}\,dz}\vec{V}(z)$ is the solution to

$$\frac{\partial \vec{V}}{\partial z} = \hat{L}\vec{V}. \tag{10.31}$$

Furthermore, Equation 10.31 is the same set of differential equations as in Equations 10.16 through 10.18 excluding the nonlinear terms. Hence, finding the solution to Equations 10.16 through 10.18 while ignoring the nonlinear terms gives the same solution as the symbolic expression given in Equation 10.31. Similarly, $e^{\hat{N}\,dz}\vec{V}(z)$ is the solution to

$$\frac{\partial \vec{V}}{\partial z} = \hat{N}\vec{V}. \tag{10.32}$$

is symbolic equation has the same solution as that of Equations 10.16 through 10.18 by ignoring the linear terms.

The ability to split the differential equations, accurate to order dz, is exactly the procedure we outlined at the beginning of this section, where we propagate the fields linearly and then solve the nonlinear set of equations. Of course, the order of the operators can be reversed and the same accuracy is achieved. Although we leave the details of the calculation to the reader (see Problem 10.2), we can achieve accuracy to order dz^2 by using the Strang splitting method given by

$$e^{(\hat{L}+\hat{N})dz} \approx e^{(\hat{L}dz/2)}e^{\hat{N}dz}e^{(\hat{L}dz/2)}. \tag{10.33}$$

Following this algorithm, we propagate for 1/2 step linearly (dz/2) and use that result as the input to the nonlinear step, which operates for a full step (dz). Finally, we finish the iteration by 1/2 a linear propagation step. Note that to the same order,

$$e^{(\hat{L}dz/2)}e^{\hat{N}dz}e^{(\hat{L}dz/2)} = e^{(\hat{N}dz/2)}e^{\hat{L}dz}e^{(\hat{N}dz/2)} \tag{10.34}$$

In the numerical approach, we break up the crystal into a number of slices so that the overall iterative solution appears as

$$\left(\frac{L}{2}N\frac{L}{2}\right)\left(\frac{L}{2}N\frac{L}{2}\right)\cdots\left(\frac{L}{2}N\frac{L}{2}\right) = \frac{L}{2}NLNL\ldots LN\frac{L}{2}, \tag{10.35}$$

where L and N represent the linear and nonlinear computational steps. Again, the order of the operators can be reversed and the same second-order accuracy is achieved. In Section 10.2.3, we discuss how the linear and nonlinear steps are carried out.

10.2.3 LINEAR PROPAGATION: BEAM PROPAGATION METHOD

We start with the linear propagation step where we ignore the nonlinear terms in Equations 10.16 through 10.18. Note that by dropping the nonlinear terms we end up with three uncoupled differential equations, and so we solve the propagation of each field independent of the others. An analytic solution to the linear equations does not exist for arbitrary field distributions, and so we take an algorithmic approach. For example, the linear equation for the field A_3 is

$$\frac{\partial A_3}{\partial z} = \frac{i}{2k_0} \nabla_\perp^2 A_3 + \tan\rho \frac{\partial A_3}{\partial x} - \frac{\alpha}{2} A_3. \tag{10.36}$$

The solution to this equation is obtained by first going to the Fourier domain in the transverse coordinates,

$$A_3(x,y,z) = \iint \tilde{A}_3(f_x, f_y, z) e^{i2\pi(f_x x + f_y y)} df_x df_y, \tag{10.37}$$

where f_x and f_y are spatial frequencies in the transverse direction. We use frequencies here instead of angular frequencies to be consistent with numerical FFT routines. Substituting the Fourier transform (Equation 10.37) into Equation 10.36 yields an algebraic equation for the Fourier amplitude,

$$\frac{\partial \tilde{A}_3}{\partial z} = -\left(i \frac{2\pi^2}{k_0} (f_x^2 + f_y^2) + i2\pi f_x \tan\rho + \frac{\alpha}{2} \right) \tilde{A}_3. \tag{10.38}$$

The solution of Equation 10.38 for the Fourier amplitude is

$$\tilde{A}_3(q_x, q_y, z + dz) = \tilde{A}_3(q_x, q_y, z) \exp\{-iQ\, dz\}, \tag{10.39}$$

where

$$Q \equiv \left(\frac{2\pi^2}{k_0} (f_x^2 + f_y^2) + 2\pi f_x \tan\rho - i\frac{\alpha}{2} \right). \tag{10.40}$$

The envelope in real space is the inverse Fourier transform of Equation 10.39,

$$A_3(x,y,z+dz) = \mathfrak{I}^{-1}\tilde{A}_3(f_x, f_y, z+dz), \tag{10.41}$$

where $\mathfrak{I}^{-1}$ represents the inverse Fourier transform operation.

The algorithm for a linear propagation step is represented symbolically as

$$A_3(x,y,z+dz) = \mathfrak{I}^{-1}e^{-iQdz}\mathfrak{I}A_3(x,y,z). \tag{10.42}$$

This expression is accomplished by the following four steps:

1. Define a two-dimensional $N \times N$ matrix whose entries are populated with a digitized version of $A(x,y):A(m\Delta x, n\Delta x)$, where m and n are array indices that run from 1 to N and $\Delta y = \Delta x$. Note that A is a complex variable and so the array should also be complex.
2. Apply a two-dimensional FFT to the matrix, $A(m\Delta x, n\Delta x)$. Note that FFTs are nearly universally available mathematical software. The resultant array is an $N \times N$ matrix with elements $\tilde{A}(f_x, f_y) = \tilde{A}(p\Delta f - 1/(2\Delta x), q\Delta f - 1/(2\Delta x))$, where p and q are the array indices running from 1 to N, and $\Delta f = 1/(N\Delta x)$. Note that the frequency range of the matrix runs from $-1/(2\Delta x)$ to $1/(2\Delta x)$. The specific frequency spacing and frequency range are what is generated by numerical FFTs.
3. Multiply the matrix $\tilde{A}(f_x, f_y)$ by the propagator matrix $e^{-iQ(f_x f_y)dz}$, which is given by[*]

$$e^{-iQ(f_x,f_y)dz}$$
$$= \exp\left\{-\left(i\frac{2\pi^2}{k_0}((p-N/2)^2 + (q-N/2)^2)\Delta f^2 + i(p-N/2)2\pi\Delta f \tan\rho + \frac{\alpha}{2}\right)dz\right\}. \tag{10.43}$$

The propagator matrix is diagonal in the sense that the new value of $\tilde{A}(f_x, f_y)$ is dependent only on its previous complex amplitude. Also the matrix can be defined before the integration steps and applied at each iteration of the algorithm. For a purely linear problem, the size of the step dz is not required to be small.
4. Inverse Fourier transform the result of Step 3 to give $A(x,y,z+dz)$.

The above algorithm is called the beam propagation method (BPM). It is written for a general beam cross section, but in some cases, a simpler routine using only

---

[*] Note that the FFT output is an array with a prescribed frequency ordering. In some cases, the ordering is $-f_{max}, \ldots 0, \ldots +f_{max}$, but in many cases, the ordering is $0, \ldots +f_{max}, -f_{max}, \ldots -\Delta f$. When multiplying by $e^{-iQ(f_x f_y)dz}$, be sure that both have the same frequency ordering.

one dimension is used. One example where a one-dimensional version works is for a plane wave incident on a vertical slit. The one-dimensional version of BPM is a good starting point for learning how to work with the algorithm.

As mentioned above, linear beam propagation over large distances may be made in a single step. To prove this, consider propagation over two dz steps, which is written symbolically as

$$A_3(x,y,z_0 + 2\mathrm{d}z) = (\mathfrak{J}^{-1}e^{-iQ\mathrm{d}z}\mathfrak{J})(\mathfrak{J}^{-1}e^{-iQ\mathrm{d}z}\mathfrak{J})A_3(x,y,z_0).$$

(10.44)

Note that an inverse Fourier transform followed by a forward Fourier transform is equivalent to an identity operation so that Equation 10.44 becomes

$$A_3(x,y,z_0 + 2\mathrm{d}z) = \mathfrak{J}^{-1}e^{-2iQ\mathrm{d}z}\mathfrak{J}A_3(x,y,z_0).$$

(10.45)

This result shows that two linear steps of dz are equivalent to a single step of 2dz. Hence, linear propagations over large distances are possible with a single BPM step. Caution should be exercised when propagating a field with the BPM method over arbitrary distances. At some point, the beam size becomes larger than the numerical grid's bounds, and a larger grid is required. This issue becomes especially important when considering interactions with tightly focused beams where the beam size changes rapidly with propagation. This limitation is due to the periodic boundary conditions inherent in the FFT algorithm. Field amplitudes that pass through one boundary reappear at the opposite boundary and may result in unwanted wraparound effects.

10.2.4 NONLINEAR PROPAGATION FOR THREE-WAVE MIXING

The nonlinear step is the solution to Equations 10.16 through 10.18 when the linear parts are ignored and these equations become

$$\frac{\partial A_1}{\partial z} = i\frac{\omega_1^2}{k_{10}c^2}d_{\mathrm{eff}}A_2A_3e^{-i\Delta kz},$$

(10.46)

$$\frac{\partial A_2}{\partial z} = i\frac{\omega_2^2}{k_{20}c^2}d_{\mathrm{eff}}A_1A_3^*e^{i\Delta kz},$$

(10.47)

$$\frac{\partial A_3}{\partial z} = i\frac{\omega_3^2}{k_{30}c^2}d_{\mathrm{eff}}A_1A_2^*e^{i\Delta kz}.$$

(10.48)

These coupled equations are numerically integrated, typically with the Runge–Kutta (RK) algorithm. The RK integrator is included with most mathematical packages. The most commonly used RK method is good to fourth-order in the step size, dz^4. However, the splitting technique itself is only accurate to order dz^2 (so-called Strang splitting); so a second-order RK method, which is accurate to dz^2, can be used to speed up the calculations. More details on how to set up Equations 10.46 through 10.48 for the RK algorithm are given in Appendix C.

When modeling the nonlinear interaction, we follow the iterative procedure shown in Equation 10.35. For the linear step we use BPM and for the nonlinear step we use the RK integration. This routine has achieved great success and is in close agreement with experimental measurements.

10.3 TEMPORAL EFFECTS

We now incorporate temporal effects into the SVEA equations. These become important for subnanosecond pulse durations and for long interaction lengths (such as in optical fibers). As with the spatial effects introduced in Section 10.2, linear propagation temporal effects play an important role in nonlinear interactions. In this section, we derive an envelope equation that includes both linear and nonlinear effects, and we look at some of the solutions to the envelope equation.

10.3.1 TIME-DEPENDENT FIELD DEFINITIONS

We follow a similar strategy as in Section 10.2, namely, we factor out the rapidly varying part of the field and work with the pulse envelope. To do so, we first define the carrier frequency. We assume a linearly polarized electric field given by

$$\vec{E} = \frac{1}{2}\tilde{E}(t)\hat{E} + \text{c.c.}$$

(10.49)

Our definition of the carrier frequency is

$$\omega_0 = \frac{\displaystyle\int_0^\infty \omega\,|\tilde{E}(\omega)|^2\,d\omega}{\displaystyle\int_0^\infty |\tilde{E}(\omega)|^2\,d\omega},$$

(10.50)

where $\tilde{E}(\omega)$ is the Fourier transform of $\underline{E}(t)$. Figure 10.2 shows an illustration of a spectrum and its carrier frequency. When the optical spectrum is symmetric, the carrier frequency is the central frequency.

With the carrier frequency defined, we rewrite the electric field as

$$\vec{E} = \frac{1}{2}A(t)e^{i(k_0 z - \omega_0 t)}\hat{E} + \text{c.c.},$$

(10.51)

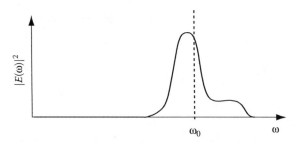

FIGURE 10.2   Illustration of a spectrum and its carrier frequency, ω_0.

where we assume a beam propagating along the z-axis with an on-axis k-vector magnitude of k_0. Our strategy is to Fourier decompose the envelope, $A(t)$, into its spectral components and analyze the interactions in frequency space. We start by considering a monochromatic plane wave. We describe the forward-propagating wave as having an $e^{i(kz-\omega t)}$ dependence. It is natural to treat propagating waves that have a structured envelope as a superposition of forward-propagating plane waves so that we write

$$A(t)e^{-i\omega_0 t} = \frac{1}{2\pi} \int_{-\infty}^{\infty} \tilde{A}(\eta)e^{-i(\omega_0+\eta)t} d\eta. \tag{10.52}$$

This expression is a Fourier transform about the carrier frequency, ω_0. $\tilde{A}(\eta)$ is the bandwidth of the complex amplitude centered on the carrier frequency (see Figure 10.3). The Fourier transform pair to Equation 10.52 is

$$\tilde{A}(\eta) = \int_{-\infty}^{\infty} A(t)e^{i\eta t} dt. \tag{10.53}$$

The above Fourier transform pair is consistent with our picture of describing the fields in terms of forward-propagating waves.[*]

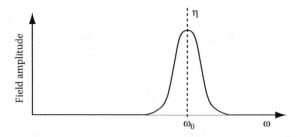

FIGURE 10.3   Frequency-space representation of the field's complex amplitude. In general, the bandwidth is not symmetric about the carrier frequency.

---

[*] Note that standard time-domain signals have a Fourier transform pair with opposite signs as given here. Our choice is motivated by describing forward-propagating waves described by $e^{i(kz-\omega t)}$ instead of the complex conjugate.

10.3.2 TIME-DEPENDENT LINEAR POLARIZATION

The field interacts with the medium via polarization, which can be thought of as a response function to an applied field. We assume that the response splits into a linear and a nonlinear part so that we write the polarization as

$$\underline{P}(t) = \underline{P}^{(1)}(t) + \underline{P}^{NL}(t). \tag{10.54}$$

In the present analysis, we focus only on the temporal effects and so we treat the polarization and fields as scalars. Our approach to including temporal effects is to first look at the linear part of the wave equation and after that we add the nonlinear effects.

For a linear medium and for monochromatic fields, the polarization has the same frequency as the incident field but with a possible phase shift. When the electric field has temporal structure, the relationship between the polarization and the electric field can depend on the past history of the field. That is, the linear polarization for an isotropic medium is written as

$$\underline{P}^{(1)}(t) = \varepsilon_0 \int_{-\infty}^{t} \underline{R}(t - \tau)\underline{E}(\tau)d\tau, \tag{10.55}$$

where R is a linear response function that gives the contribution of a field that occurred previously (at time τ) to the current polarization at time t. A way to visualize the relationship between the field and the polarization is by an analog to the relationship between a mass-spring system and an external force. If an impulsive force is applied to the mass at some time in the past, then the mass still oscillates at a later time. So the mass's response (i.e., its displacement) depends on the past history of the applied force.

The time-domain representation for the polarization is cumbersome to work with; so we switch to a frequency-space representation. We do so by first changing variables to $\tau' = t - \tau$ so that the integral in Equation 10.55 is now

$$\underline{P}^{(1)}(t) = \varepsilon_0 \int_{0}^{\infty} \underline{R}(\tau')\underline{E}(t - \tau')d\tau'. \tag{10.56}$$

We introduce a time-domain susceptibility function,

$$\underline{\chi}(\tau') = u(t')\underline{R}(\tau'), \tag{10.57}$$

where $u(\tau')$ is the unit step function,

$$u(\tau') = \begin{cases} 0 & \tau' < 0 \\ 1 & \tau' > 0. \end{cases} \tag{10.58}$$

The polarization in terms of $\chi(\tau')$ is a convolution:

$$\underline{P}^{(1)}(t) = \varepsilon_0 \int_{-\infty}^{\infty} \underline{\chi}(\tau')\underline{E}(t-\tau')\,d\tau' = \varepsilon_0\underline{\chi}(t) \otimes \underline{E}(t), \qquad (10.59)$$

where $\otimes$ represents the convolution operation. A property of the convolution of two functions is that its Fourier transform is a multiplication of the two functions' Fourier transforms,

$$\underline{P}^{(1)}(t) = \frac{\varepsilon_0}{2\pi} \int_{-\infty}^{\infty} \tilde{\chi}(\omega)\tilde{E}(\omega)e^{-i\omega t}\,d\omega. \qquad (10.60)$$

Note that an *instantaneous* polarization response in the time domain corresponds to an impulsive susceptibility, $\chi(\tau) = \chi_0\delta(t)$. The Fourier transform of an impulse is a flat spectrum, and hence an instantaneous response corresponds to a material with no dispersion. Furthermore, an instantaneous medium is also lossless as can be verified using Poynting's theorem (see Problem 10.7).

The Fourier transform of the electric field is further expanded out using the field definition in Equation 10.51:

$$\tilde{E}(\omega) = \Im(\underline{E}) = \Im\frac{1}{2}(A(t)e^{-i\omega_0 t} + A^*(t)e^{i\omega_0 t}) = \frac{1}{2}(\tilde{A}(\omega-\omega_0) + \tilde{A}^*(\omega+\omega_0)). \quad (10.61)$$

This equation shows that when we factor out the carrier frequency, the Fourier transform is shifted by that frequency. The field in Equation 10.61 is substituted into the linear polarization expression (Equation 10.60):

$$\underline{P}^{(1)}(t) = \frac{\varepsilon_0}{4\pi} \int_{-\infty}^{\infty} \tilde{\chi}(\omega)(\tilde{A}(\omega-\omega_0) + \tilde{A}^*(\omega+\omega_0))e^{-i\omega t}\,d\omega. \qquad (10.62)$$

We make the variable substitutions $\eta = \omega - \omega_0$ and $\eta' = (\omega + \omega_0)$, and rewrite the polarization as two integrals,

$$\underline{P}^{(1)}(t) = \frac{\varepsilon_0}{4\pi}\left(\int_{-\infty}^{\infty} \tilde{\chi}(\omega_0+\eta)\tilde{A}(\eta)e^{-i\eta t}\,d\eta\right)e^{-i\omega_0 t} + \frac{\varepsilon_0}{4\pi}\left(\int_{-\infty}^{\infty} \tilde{\chi}(\eta'-\omega_0)\tilde{A}^*(\eta')e^{-i\eta' t}\,d\eta'\right)e^{i\omega_0 t}.$$

$$(10.63)$$

The linear polarization in this expression is a real quantity; therefore, we may write

$$\underline{P}^{(1)}(t) = \frac{\varepsilon_0}{4\pi} \left(\int_{-\infty}^{\infty} \tilde{\chi}(\omega_0 + \eta)\tilde{A}(\eta)e^{-i\eta t}d\eta \right) e^{-i\omega_0 t} + \text{c.c.} \tag{10.64}$$

This equation is the desired frequency-space representation for the linear polarization, which has the carrier frequency factored out of its complex amplitude. Both the susceptibility and the field reference have the same frequency, since the field envelope is considered to be centered on the carrier frequency.

10.3.3 TIME-DEPENDENT NONLINEAR POLARIZATION

The general nonlinear polarization response also depends on the past history of the incident fields. Such a dependence is especially important for resonantly enhanced nonlinearities and for slow processes, such as a nonlinear process called photorefraction in certain materials, which is treated in books by Banerjee (2004) and Boyd (2008). In Equation 10.54, we assume that the linear and nonlinear polarizations are separable, and in the rest of this chapter, we further assume that the nonlinear polarization may be expanded in a power series in the total field:

$$\underline{P}_i^{\text{NL}}(t) = \varepsilon_0 \int_{-\infty}^{t}\int_{-\infty}^{t} \underline{R}_{ijk}^{(2)}(t-\tau_1; t-\tau_2)\underline{E}_j(\tau_1)\underline{E}_k(\tau_2)d\tau_1 d\tau_2$$

$$+ \varepsilon_0 \int_{-\infty}^{t}\int_{-\infty}^{t} \underline{R}_{ijk\ell}^{(3)}(t-\tau_1; t-\tau_2; t-\tau_3)\underline{E}_j(\tau_1)\underline{E}_k(\tau_2)\underline{E}_\ell(\tau_3)d\tau_1 d\tau_2 d\tau_3 + \cdots \tag{10.65}$$

This expression for the nonlinear polarization shows dependence not only on products of fields, but also on products of fields at different times. In this chapter, we mostly consider nearly instantaneous nonlinearities based on the electronic response of a medium so that we ignore dispersion in the nonlinear polarization (Kleinman symmetry). The instantaneous response assumption gives back the familiar expression

$$\underline{P}_i^{\text{NL}}(t) = \varepsilon_0 \chi_{ijk}^{(2)}\underline{E}_j(t)\underline{E}_k(t) + \varepsilon_0 \chi_{ijk\ell}^{(3)}\underline{E}_j(t)\underline{E}_k(t)\underline{E}_\ell(t) + \cdots \tag{10.66}$$

In the rest of this chapter, we work under the assumption that the nonlinear polarization is an instantaneous response, and we write it as a complex amplitude with the carrier frequency factored out,

---

$$\underline{P}^{\mathrm{NL}}(t) = \frac{1}{2} P^{\mathrm{NL}}(t) e^{-i\omega_0 t} + \text{c.c.}, \tag{10.67}$$

where P^{NL} (without the underscore) is the nonlinear complex amplitude envelope.

10.3.4 WAVE EQUATION FOR FIELDS WITH A TIME-DEPENDENT ENVELOPE

The field and polarizations developed in Sections 10.3.1 through 10.3.3 are substituted sequentially into the wave equation (Equation 10.2). Consider first substituting in the field and the linear polarization:

$$\mu_0 \varepsilon_0 \frac{\partial^2 \underline{E}}{\partial t^2} + \mu_0 \frac{\partial^2}{\partial t^2} \underline{P}^{(1)} = \frac{1}{2c^2} \frac{\partial^2}{\partial t^2} \left(\frac{1}{2\pi} \int_{-\infty}^{\infty} [1 + \tilde{\chi}(\omega_0 + \eta)] \tilde{A}(\eta) e^{-i(\omega_0 + \eta)t} d\eta \right) + \text{c.c.} \tag{10.68}$$

By writing the field and polarization in the frequency domain, we are able to take the time derivative,

$$\mu_0 \varepsilon_0 \frac{\partial^2 \underline{E}}{\partial t^2} + \mu_0 \frac{\partial^2}{\partial t^2} \underline{P}^{(1)} = -\frac{1}{2c^2} \left(\frac{1}{2\pi} \int_{-\infty}^{\infty} (\omega_0 + \eta)^2 [1 + \tilde{\chi}(\omega_0 + \eta)] \tilde{A}(\eta) e^{-i(\omega_0 + \eta)t} d\eta \right) + \text{c.c.}$$

$$\tag{10.69}$$

If we have complete information regarding the susceptibility (or more commonly the permittivity, ε), then we leave Equation 10.69 as it stands. Alternatively, we use the relationship,

$$\frac{[k(\omega_0 + \eta)]^2 c^2}{(\omega_0 + \eta)^2} = 1 + \tilde{\chi}(\omega_0 + \eta). \tag{10.70}$$

Substituting this expression into Equation 10.69 yields

$$\mu_0 \varepsilon_0 \frac{\partial^2 \underline{E}}{\partial t^2} + \mu_0 \frac{\partial^2}{\partial t^2} \underline{P}^{(1)} = -\frac{1}{2} \left(\frac{1}{2\pi} \int_{-\infty}^{\infty} [k(\omega_0 + \eta)]^2 \tilde{A}(\eta) e^{-i(\omega_0 + \eta)t} d\eta \right) + \text{c.c.} \tag{10.71}$$

We expand k in a Taylor series about the carrier frequency,

$$k(\omega_0 + \eta) = k(\omega_0) + \left. \frac{\partial k}{\partial \omega} \right|_{\omega_0} \eta + \frac{1}{2} \left. \frac{\partial^2 k}{\partial \omega^2} \right|_{\omega_0} \eta^2 + \cdots. \tag{10.72}$$

The partial derivatives are related to the group velocity so that Equation 10.72 is written as

$$k(\omega_0 + \eta) = k_0 + \frac{1}{v_g}\bigg|_{\omega_0} \eta - \frac{1}{2v_g^2}\frac{\partial v_g}{\partial \omega}\bigg|_{\omega_0} \eta^2 + \cdots. \tag{10.73}$$

Before substituting this expression back into the integral in Equation 10.71, we note the following Fourier transform relationship:

$$\frac{\partial^n A(t)}{\partial t^n} = \frac{1}{2\pi}\int_{-\infty}^{\infty}(-i\eta)^n \tilde{A}(\eta)e^{-i\eta t}d\eta. \tag{10.74}$$

Therefore, each term in the Taylor expansion in Equation 10.73 when substituted back into Equation 10.71 is an expansion in a series of time derivatives of $A(t)$. By carrying out this substitution, we obtain an expression in terms of the field's envelope,

$$\mu_0 \varepsilon_0 \frac{\partial^2 E}{\partial t^2} + \mu_0 \frac{\partial^2}{\partial t^2}\underline{P}^{(1)} = -\frac{1}{2}\left(k_0^2 A(t) + i\frac{2k_0}{v_g}\frac{\partial A(t)}{\partial t} - \frac{1}{v_g^2}\left(1 - k_0\frac{\partial v_g}{\partial \omega_0}\right)\frac{\partial^2 A}{\partial t^2} + \cdots\right)e^{-i\omega_0 t} + \text{c.c.} \tag{10.75}$$

The group velocity and GVD are evaluated at ω_0. The coefficient of the third term in parentheses is the GVD. Note that the dominant contribution is proportional to k_0 and the GVD is frequently expressed in the form: $((\partial/\partial \omega_0)(1/v_g)) = -((1/v_g^2)(\partial v_g/\partial \omega_0))$.

The GVD is the slope of the inverse of the group velocity. From this analysis, we find that each higher-order derivative corresponds to a higher-order dispersion term. The dominant contributions to the higher-order dispersion coefficients can be expressed as $(\partial^n/\partial \omega_0^n)(1/v_g)$; for $n = 2$, we have the third-order dispersion coefficient, and so on. In the following, we include only GVD, thus truncating the expansion at the second-order time derivative.

Next we add the nonlinear contribution to the wave equation, $((\partial^2 \underline{P}^{\text{NL}}(t))/\partial t^2)$, which we write in terms of its complex envelope (see Equation 10.67). We take the time derivative by writing the nonlinear polarization's envelope as a Fourier transform:

$$\frac{\partial^2 \underline{P}^{\text{NL}}(t)}{\partial t^2} = \frac{1}{2}\frac{\partial^2}{\partial t^2}\left(\int_{-\infty}^{\infty} \tilde{P}^{\text{NL}}(\eta)e^{-(\omega_0 + \eta)t}d\eta\right) + \text{c.c.}$$

$$= \frac{1}{2}\left(-\int_{-\infty}^{\infty}(\omega_0 + \eta)^2 \tilde{P}^{\text{NL}}(\eta)e^{-(\omega_0 + \eta)t}d\eta\right) + \text{c.c.} \tag{10.76}$$

Using the same Fourier approach as in Equation 10.74, we obtain

$$\frac{\partial^2 \underline{P}^{NL}(t)}{\partial t^2} = -\frac{1}{2}\left(\omega_0^2 P^{NL} + 2i\omega_0 \frac{\partial P^{NL}}{\partial t} - \frac{\partial^2 P^{NL}}{\partial t^2}\right)e^{-i\omega_0 t} + \text{c.c.} \tag{10.77}$$

This expression is simplified by recalling that the nonlinear effects are assumed to be small. Therefore, we neglect corrections to the already small perturbation and retain only the leading contribution within the parentheses,

$$\frac{\partial^2 \underline{P}^{NL}(t)}{\partial t^2} \approx -\frac{1}{2}\omega_0^2 P^{NL}(t)e^{-i\omega_0 t} + \text{c.c.} \tag{10.78}$$

We now write down the entire wave equation in terms of the field envelope using the spatial derivatives from Equations 10.7 and 10.13, and the temporal derivatives from Equations 10.67 and 10.75:

$$\frac{\partial A}{\partial z} - \frac{i}{2k_0}\frac{\partial^2 A}{\partial z^2} = i\left[\frac{1}{2k_0}\nabla_T^2 - i\tan\rho\frac{\partial}{\partial x}\right]A \quad \text{Diffraction and walk-off (right-hand side)}$$

$$-\left[\frac{1}{v_g}\frac{\partial}{\partial t} + \frac{i}{2k_0 v_g^2}\left(1 - k_0\frac{\partial v_g}{\partial\omega}\right)\Bigg|_{\omega_0}\frac{\partial^2}{\partial t^2} + \cdots\right]A \quad \text{Dispersion}$$

$$+\frac{i\mu_0\omega_0^2}{2k_0}P^{NL}e^{-ik_0 z} \quad \text{Nonlinearity.} \tag{10.79}$$

This equation shows the pulse evolution along the z-axis in terms of diffraction, dispersion, and nonlinearity. Higher-order spatial and temporal contributions can be included to extend the applicability of the SVEA equations. It describes the change in the field envelope as it propagates along the z-axis (see Figure 10.4a). A more

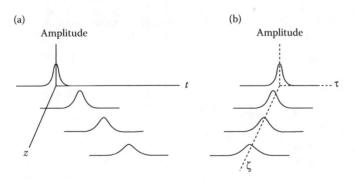

(a) Amplitude

(b) Amplitude

FIGURE 10.4　Pulse propagation as viewed from two perspectives. (a) $A(r = 0, z, t)$ is the center of the beam's amplitude in real space. (b) $A(r = 0, \zeta, \tau)$ is viewed in a coordinate system that moves with the center of the pulse.

convenient description of the envelope is to transform to a frame of reference that moves with the pulse, that is, the copropagating reference frame:

$$\zeta \equiv z,$$

$$\tau \equiv t - \frac{z}{v_g}. \tag{10.80}$$

With this coordinate change, $A(\tau)$ gives the pulse envelope at a given location along the beam's propagation axis. To visualize the difference between coordinate systems, Figure 10.4 shows the center of the beam ($x = 0$, $y = 0$) as a function of z and t, as well as a function of ζ and τ. The transverse coordinates, x and y, are unaffected by the transformation. Writing the envelope in terms of ζ and τ requires that we also transform the derivatives in Equation 10.79:

$$\frac{\partial A(\zeta,\tau)}{\partial z} = \frac{\partial A}{\partial \zeta}\frac{\partial \zeta}{\partial z} + \frac{\partial A}{\partial \tau}\frac{\partial \tau}{\partial z} = \frac{\partial A}{\partial \zeta} - \frac{1}{v_g}\frac{\partial A}{\partial \tau}, \tag{10.81}$$

$$\frac{\partial^2 A(\zeta,\tau)}{\partial z^2} = \frac{\partial^2 A}{\partial \zeta^2} - \frac{2}{v_g}\frac{\partial^2 A}{\partial \tau \partial \zeta} + \frac{1}{v_g^2}\frac{\partial^2 A}{\partial \tau^2}, \tag{10.82}$$

$$\frac{\partial A(\zeta,\tau)}{\partial t} = \frac{\partial A}{\partial \tau}, \tag{10.83}$$

$$\frac{\partial^2 A(\zeta,\tau)}{\partial t^2} = \frac{\partial^2 A}{\partial \tau^2}. \tag{10.84}$$

Substituting the variable changes for the derivatives gives the envelope equation in terms of ζ and τ:

$$\frac{\partial A}{\partial \zeta} - \frac{i}{2k_0}\frac{\partial^2 A}{\partial \zeta^2} + \frac{i}{k_0 v_g}\frac{\partial^2 A}{\partial \tau \partial \zeta} = i\left[\frac{1}{2k_0}\nabla_T^2 - i\tan\rho\,\frac{\partial}{\partial x}\right]A \quad \text{Diffraction and}$$

walk-off (right-hand side)

$$+ i\left(\frac{1}{2v_g^2}\frac{\partial v_g}{\partial \omega}\right)\Bigg|_{\omega_0}\frac{\partial^2}{\partial \tau^2}A + \cdots \quad \text{Dispersion}$$

$$+ \frac{i\mu_0 \omega_0^2}{2k_0}P^{\mathrm{NL}}e^{-ik_0\zeta} \quad \text{Nonlinearity.}$$

$$\tag{10.85}$$

Note that the first-order time derivatives cancelled leaving us with the second-order time derivative GVD contribution. As mentioned above, the leading contributions are retained and a small contribution inversely proportional to k_0 is neglected. In most cases, even with ultrashort-pulse durations, we make two further assumptions to write the final SVEA equation:

$$\left|\frac{\partial A}{\partial \zeta}\right| \gg \left|\frac{1}{2k_0}\frac{\partial^2 A}{\partial \zeta^2}\right| \tag{10.86}$$

and

$$\left|\frac{\partial A}{\partial \zeta}\right| \gg \frac{\partial}{\partial \tau}\left(\frac{1}{k_0 v_g}\frac{\partial A}{\partial \zeta}\right). \tag{10.87}$$

The validity of these approximations can break down under different conditions. To describe ultrashort and ultraintense optical pulses on the femtosecond or attosecond time scales, higher-order linear and nonlinear effects are incorporated into the analysis. Also, resonator structures may be found in photonic band gap materials, so the analysis has to be generalized to include the effects of backward-propagating waves, which invalidates our original assumption that only forward-propagating waves are present in the medium. Under the above conditions in Equations 10.86 and 10.87, the slowly varying envelope equation becomes

$$\frac{\partial A}{\partial \zeta} = i\left[\frac{1}{2k_0}\nabla_T^2 - i\tan\rho\frac{\partial}{\partial x}\right]A + i\left(\frac{1}{2v_g^2}\frac{\partial v_g}{\partial \omega}\right)\bigg|_{\omega_0}\frac{\partial^2}{\partial \tau^2}A + \frac{i\mu_0\omega_0^2}{2k_0}P^{NL}. \tag{10.88}$$

This is the basic scalar SVEA equation that incorporates diffraction, dispersion, and nonlinear effects for ultrashort-pulse propagation in a nonlinear medium. The extension to cover additional effects, such as SRS, pulse polarization dynamics, ultra-intense pulse dynamics, and pulse widths as short as a few optical cycles, are available in the literature.

10.4 DYNAMICAL SOLUTIONS TO THE NONLINEAR ENVELOPE EQUATION

An examination of Equation 10.88 shows that the evolution of the field's envelope is affected by diffraction, dispersion, and the medium's nonlinearity. In linear media, diffraction causes a beam to spread in its transverse dimension while dispersion causes a broadening of the pulse's duration. In the case of a self-induced Kerr nonlinearity, the effect of the nonlinearity can enhance or reduce diffraction and dispersion depending on the sign of the nonlinear coefficient. We illustrate the balance between dispersion and nonlinearity and between diffraction and nonlinearity separately in the following sections.

10.4.1 SELF-PHASE MODULATION

Although the effects of diffraction and walk-off always play a role in pulse propagation, in an optical fiber these effects determine a given fiber mode (see Appendix E). The fiber mode completely determines the transverse beam shape of the pulse. In a fiber optic medium, the transverse effects in the wave equation (Equation 10.88) are managed by the guiding fiber mode, so that the transverse mode profile remains constant and the nonlinearity is weighted by an effective area of the mode in the fiber, S_{eff}. The wave equation for this situation is given by

$$\frac{\partial A}{\partial \zeta} - i \left(\frac{1}{2v_g^2} \frac{\partial v_g}{\partial \omega} \right)\bigg|_{\omega_0} \frac{\partial^2}{\partial \tau^2} A = i \frac{\mu_0 \omega_0^2}{2k_0} P^{\text{NL}}. \tag{10.89}$$

At this point, the solution depends on the specific nonlinearity under consideration. For temporal effects in fibers, the dominant nonlinearities are due to the electronic contribution, Raman and Brillouin processes. To narrow our analysis, we assume that the bandwidth of our pulse is large enough to turn off the Brillouin effects, but small enough to avoid exciting a stimulated Raman contribution. That is, we limit our discussion to the instantaneous electronic response (Kerr effect), giving the nonlinear polarization,

$$P^{(3)}(\tau) = \frac{3\varepsilon_0}{4} \chi^{(3)} |A(\tau)|^2 A(\tau). \tag{10.90}$$

We first investigate the phenomena of self-phase modulation. This effect adds bandwidth to the pulse, which we demonstrate with the following example. Consider a thin medium where GVD is negligible. The envelope equation is then

$$\frac{\partial A}{\partial \zeta} = i \frac{3\omega_0}{8n_0 c} \chi^{(3)} |A(\tau)|^2 A(\tau). \tag{10.91}$$

We rewrite $A(\tau)$ as

$$A(\tau) = u(\tau)e^{i\phi(\tau)}. \tag{10.92}$$

With this substitution and noting that $\chi^{(3)}$ is real for a lossless medium gives

$$\frac{\partial u}{\partial \zeta} = 0, \tag{10.93}$$

$$\frac{\partial \phi}{\partial \zeta} = \frac{3\omega_0}{8n_0 c} \chi^{(3)} u^2(\tau). \tag{10.94}$$

Under the lossless medium approximation, the pulse envelope's amplitude as a function of τ remains the same, but it attains a nonlinear phase shift, which is the integral of Equation 10.94,

$$\phi^{NL} = \frac{3\omega_0}{8n_0 c} \chi^{(3)} u^2(\tau)\Delta\zeta. \tag{10.95}$$

The nonlinear phase term can also be written in terms of the nonlinear index (see Chapter 8):

$$\phi^{NL} = \frac{n_2^I \omega_0}{c} I(\tau)\Delta\zeta. \tag{10.96}$$

This nonlinear phase shift adds bandwidth to the pulse, which becomes clear when we study the instantaneous frequency,

$$\omega_{\text{inst}} = -\frac{\partial \phi^{NL}}{\partial \tau} = -\frac{n_2^I \omega_0}{c} \Delta\zeta \left(\frac{\partial I(\tau)}{\partial \tau} \right). \tag{10.97}$$

The amount of bandwidth added to the pulse is negligible for pulse durations greater than a few picoseconds (see Problem 10.14).

Consider, for example, a Gaussian envelope as shown in Figure 10.5 propagating in a medium with a *negative* $\chi^{(3)}$. The leading edge of the pulse envelope (on the left)

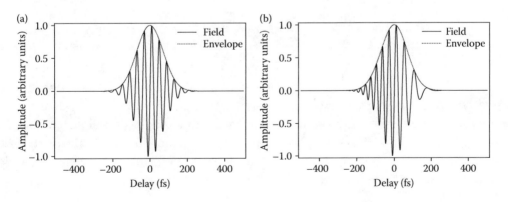

FIGURE 10.5   Electric field and its envelope (a) prior to and (b) after nonlinear interaction. Although the envelope stays the same, excess bandwidth is added to the pulse as seen by the frequency chirp (exaggerated). The nonlinearity $\chi^{(3)}$ or n_2^I is negative.

has a positive slope and hence this part of the pulse has a higher instantaneous frequency (blue-shifted), while the trailing edge of the envelope has a negative slope and is red-shifted. Another term for this phenomenon is *frequency chirping*. Because self-phase modulation is a mechanism to generate excess bandwidth, it is possible, in principle, to utilize dispersion compensation to shorten the pulse width. Now let us consider further the example shown in Figure 10.5b, where the front of the pulse is blue-shifted and the trailing edge of the pulse is red-shifted, but this time dispersion is added. Suppose the pulse propagates through a medium with positive GVD (this is the regime of normal dispersion where blue travels slower than red), then from our previous example with a negative valued nonlinearity the front of the pulse slows down while the back of the pulse speeds up resulting in a self-compressed pulse as it propagates through the medium. The degree of compression depends on how much excess bandwidth is generated and how well the dispersion can compensate for the chirp, since it is not strictly a linear frequency sweep.

It is more common to find nonlinear materials with a positive $\chi^{(3)}$; in this case, negative GVD (also called anomalous dispersion) will compensate the new frequencies. Indeed, as we shall discuss in greater detail below, Equation 10.89 has solutions that balance the contributions of dispersion and nonlinearity to form a pulse that is either propagates without the change of pulse shape or periodically changes its pulse shape. The solutions are called *solitons*, which form an entire class of solutions whose shape repeats periodically with the distance propagated.

10.4.2 NUMERICAL SOLUTIONS WITH PULSES

The discussion on self-phase modulation in Section 10.4.1 is limited to the special case of a thin sample exhibiting no GVD. Such a limitation is lifted by considering a numerical solution to the pulse propagation equation (Equation 10.89). For a self-induced nonlinearity (see Equation 10.90), Equation 10.89 becomes

$$\frac{\partial A}{\partial \zeta} - i \left(\frac{1}{2v_g^2} \frac{\partial v_g}{\partial \omega} \right) \Bigg|_{\omega_0} \frac{\partial^2}{\partial \tau^2} A(\tau) + \cdots = i \frac{3\omega_0}{8n_0 c} \chi^{(3)} |A(\tau)|^2 A(\tau), \qquad (10.98)$$

where the $+\cdots$ corresponds to terms proportional to $(\partial^n/\partial\tau^n)A$. These higher-order derivatives bring in higher-order dispersion terms (see Problem 10.11). If we break the medium down into many small slices in the ζ direction, it is possible to choose an increment where $|A(\tau)|^2$ does not change appreciably from one slice to the next. When this condition is satisfied, we rewrite Equation 10.98 in terms of a linear dispersion operator and a nonlinear operator:

$$\frac{\partial A}{\partial \zeta} = (\hat{D} + \hat{N})A(\tau), \qquad (10.99)$$

where

$$\hat{D} = i\left(\frac{1}{2v_g^2}\frac{\partial v_g}{\partial \omega}\right)\Bigg|_{\omega_0}\frac{\partial^2}{\partial \tau^2} + \cdots, \tag{10.100}$$

$$\hat{N} = i\frac{3\omega_0}{8n_0c}\chi^{(3)}|A(\tau)|^2 A(\tau). \tag{10.101}$$

The dispersion operator is independent of ζ, and the nonlinear operator is approximately independent of ζ (for a small step size) so that we may write the symbolic solution to Equation 10.99:

$$A(\zeta + d\zeta) = A(\zeta)\exp\{(\hat{D} + \hat{N})d\zeta\}. \tag{10.102}$$

We follow the same procedure as in Section 10.2.3 and approximate the solution as

$$A(\zeta + d\zeta) = \exp\left(\frac{\hat{D}}{2}d\zeta\right)\exp(\hat{N} d\zeta)\exp\left(\frac{\hat{D}}{2}d\zeta\right)A(\zeta), \tag{10.103}$$

which is accurate up to order $d\zeta^2$. This symbolic representation indicates that the solution to Equation 10.98 is approximated by propagating the solution for 1/2 step using only dispersion, then propagating for a full step using only the nonlinearity, and finally finishing with another 1/2 dispersion step.

10.4.2.1 DISPERSION STEP

The dispersion step is the solution to

$$\frac{\partial A}{\partial \zeta} = i\left(\frac{1}{2v_g^2}\frac{\partial v_g}{\partial \omega}\right)_{\omega_0}\frac{\partial^2}{\partial \tau^2}A(\tau) + \cdots. \tag{10.104}$$

We Fourier transform this equation yielding

$$\frac{\partial \tilde{A}}{\partial \zeta} = -i\left(\frac{1}{2v_g^2}\frac{\partial v_g}{\partial \omega}\right)\Bigg|_{\omega_0}\omega^2\tilde{A}(\omega) + \cdots. \tag{10.105}$$

In the numerical approach, the higher-order terms are simple to include by making the connection that $(\partial^n/\partial\tau^n) \to (-i\omega)^n$ in the Fourier transform (see Problem 10.11). Equation 10.105 is directly integrated:

$$\tilde{A}(\zeta+d\zeta,\omega) = \tilde{A}(\zeta,\omega)\exp\left\{-i\left(\frac{1}{2v_g^2}\frac{\partial v_g}{\partial\omega}\right)\bigg|_{\omega_0}\omega^2 d\zeta\right\}, \qquad (10.106)$$

where we truncate the dispersion terms at ω^2. This solution shows that the pulse envelope's Fourier transform is multiplied by a phase term quadratic in ω. After performing this operation, we inverse Fourier transform to complete the dispersion step,

$$A(\zeta+d\zeta,\tau) = \Im^{-1}\tilde{A}(\zeta+d\zeta,\omega). \qquad (10.107)$$

Similar to the spatial solution in Section 10.2.3, we may write the dispersion step in a compact symbolic notation,

$$A(\zeta+d\zeta,\tau) = \Im^{-1}e^{-iQ_D d\zeta}\Im A(\zeta,\tau), \qquad (10.108)$$

where the quadratic phase term is

$$Q_D = \left(\frac{1}{2v_g^2}\frac{\partial v_g}{\partial\omega}\right)\bigg|_{\omega_0}\omega^2. \qquad (10.109)$$

The algorithm for propagating the beam by a step, $\Delta\zeta$, using the dispersion step is given by the following:

1. Define a one-dimensional vector array (complex-valued) whose entries are populated with a digitized version of the envelope, $A(n\Delta\tau)$, where n is the array index, which runs from 1 to N.
2. Apply the FFT operation to the array, $A(n\Delta t)$. The resultant array is an N-element array with elements $\tilde{A}(f) = \tilde{A}(p\Delta f - 1/(2\Delta\tau))$, where p is the array index running from 1 to N, and $\Delta f = 1/(N\Delta\tau)$. Note that the frequency range of the matrix runs from $-1/(2\Delta\tau)$ to $1/(2\Delta\tau)$. The specific frequency spacing and frequency range are what is generated by FFTs.
3. Multiply the vector $\tilde{A}(f)$ by the array $e^{-iQ_D(f)\Delta\zeta}$, which is given by[*]

---

[*] See Section 10.2.3 on the frequency ordering of the FFT array.

$$e^{-iQ_D(f)\Delta\zeta} = \exp\left\{-\left(i\left(\frac{2\pi^2}{v_g^2}\frac{\partial v_g}{\partial\omega}\right)\bigg|_{\omega_0}\left(\frac{p-N}{2}\right)^2\Delta f^2\right)\Delta\zeta\right\},$$ (10.110)

where p is the array index.

4. Inverse Fourier transform the result of Step 3 to give $A(\zeta + \Delta\zeta, \tau)$.

10.4.2.2 NONLINEAR STEP

The nonlinear step is the solution to

$$\frac{\partial A}{\partial \zeta} = i\frac{3\omega_0}{8n_0 c}\chi^{(3)}\,|A(\tau)|^2\,A(\tau).$$ (10.111)

This equation is the same as we considered when introducing self-phase modulation. For a thin sample, we assume that $|A(\tau)|^2$ does not change appreciably from one slice to the next so that the nonlinear step's solution is

$$A(\zeta + d\zeta, \tau) = A(\zeta, \tau)\exp\left\{i\frac{3\omega_0}{8n_0 c}\chi^{(3)}\,|A(\tau)|^2\,d\zeta\right\}.$$ (10.112)

An equivalent expression is in terms of the nonlinear index and intensity,

$$A(\zeta + d\zeta, \tau) = A(\zeta, \tau)\exp\left\{i\frac{n_2^I\omega_0 I(\tau)}{c}\,d\zeta\right\}.$$ (10.113)

This equation shows that the nonlinear solution is obtained by multiplying the original envelope by a nonlinear phase exponential. The nonlinear step for propagating $\Delta\zeta$ is performed numerically by the following:

1. Calculate $I(m\Delta\tau)$, where m is the array index, using the relationship

$$I(m\Delta\tau) = \frac{n_0\varepsilon_0 c}{2}\,|A(m\Delta\tau)|^2.$$ (10.114)

2. Multiply the vector, $A(m\Delta\tau)$, by

$$\exp\left\{i\frac{n_2^I\omega_0 I(m\Delta\tau)}{c}\,\Delta\zeta\right\}.$$ (10.115)

A good test to validate the numerical approach is to propagate known solutions to envelope equation; a soliton solution is given in Section 10.4.5.

10.4.3 NONLINEAR SCHRODINGER EQUATION

When higher-order dispersion terms may be neglected in Equation 10.98, the nonlinear Schrodinger equation (NLSE) is

$$\frac{\partial A}{\partial \zeta} + \frac{i}{2} k_2 \frac{\partial^2}{\partial \tau^2} A(\tau) = i\gamma \, |A(\tau)|^2 \, A(\tau), \tag{10.116}$$

where

$$k_2 = -\left(\frac{1}{v_g^2} \frac{\partial v_g}{\partial \omega} \right)\bigg|_{\omega_0} \tag{10.117}$$

and

$$\gamma = \frac{n_0 \varepsilon_0 \omega_0 n_2^I}{2} = \frac{3\omega_0}{8 n_0 c} \chi^{(3)}. \tag{10.118}$$

Equation 10.116 is most commonly expressed in a dimensionless form by making the following substitutions:

$$T \equiv \tau \sqrt{\frac{|\gamma| \, |A_0|^2}{|k_2|}} = \frac{\tau}{T_0}, \tag{10.119}$$

$$Z \equiv |\gamma| \, |A_0|^2 \, \zeta = \frac{\zeta}{L_{\mathrm{NL}}}. \tag{10.120}$$

In addition to giving us dimensionless parameters, these equations introduce a characteristic nonlinear timescale, T_0, and characteristic nonlinear length scale, L_{NL}. Furthermore, we substitute the dimensionless quantity, U, for the field amplitude:

$$U \equiv \frac{A}{|A_0|}. \tag{10.121}$$

With these substitutions, Equation 10.116 becomes

$$\frac{\partial U}{\partial Z} + \mathrm{sign}(k_2) \frac{i}{2} \frac{\partial^2 U}{\partial T^2} = \mathrm{sign}(n_2^I) i |U|^2 U. \tag{10.122}$$

10.4.4 MODULATION INSTABILITY

Before delving into a pulse solution of Equation 10.122, i.e., the NLSE, it is instructive to seek a deeper understanding of the equation by performing a stability analysis of the constant amplitude solution. The modulation instability was discussed and observed in water wave systems and plasma physics, as well as in nonlinear optics. The observation of a constant amplitude wave that spontaneously evolves to a pulse train is the signature observation of the modulation instability phenomenon. The optical version of the modulation instability phenomenon has been a fruitful field of study.

A time-independent solution of Equation 10.122 is

$$U(Z) = U_0 e^{\text{sign}(n_2^I) i U_0^2 Z}, \tag{10.123}$$

where U_0 is a constant related to the beam power and is assumed to be real. The question of the stability of the solution is answered by examining the behavior of small perturbations from the exact solution. If a small perturbation exponentially decays, then the solution is locally stable. On the other hand, if the perturbation grows exponentially, then the solution is unstable. Finally, if the solution is oscillatory and neither grows nor decays, then it is called marginally stable. The stability of the constant solution is found by linearizing the solution. In this instance, we use the perturbative form

$$U(Z) = (U_0 + V(Z,T)) e^{\text{sign}(n_2^I) i |U_0|^2 Z} \tag{10.124}$$

where $V(Z,T)$ is a small, complex amplitude. Insert this into Equation 10.122 and retain terms that are linear in the perturbation amplitude.

$$\frac{\partial V}{\partial Z} + i \, \text{sign}(k_2) \frac{1}{2} \frac{\partial^2 V}{\partial T^2} = i \, \text{sign}(n_2^I) U_0^2 (V + V^*). \tag{10.125}$$

Consider a plane-wave form for the perturbative amplitude; the appearance of the complex conjugate couples conjugate plane waves

$$V(Z,T) = V_{01} e^{i(KZ - \Omega T)} + V_{02} e^{-i(KZ - \Omega T)}; \tag{10.126}$$

we find two linear, homogeneous algebraic equations separated by their common plane-wave factors:

$$KV_{01} - \text{sign}(k_2) \frac{\Omega^2}{2} V_{01} = \text{sign}(n_2^I) U_0^2 \left(V_{01} + V_{02}{}^* \right),$$

$$-KV_{02} - \text{sign}(k_2) \frac{\Omega^2}{2} V_{02} = \text{sign}(n_2^I) U_0^2 \left(V_{02} + V_{01}{}^* \right). \tag{10.127}$$

The coefficients of the two equations (using the complex conjugate of the second equation, i.e., V_{02}^*) satisfy the dispersion relation:

$$K = \pm \frac{|\Omega|}{2}\sqrt{\Omega^2 + 4\,\text{sign}(k_2)\,\text{sign}\left(n_2^I\right)U_0^2}. \tag{10.128}$$

This important result deserves further examination. For the case when the product $\text{sign}(k_2)\text{sign}(n_2^I) = 1$, the value of K is real and the perturbation function does not decay or grow away from the constant solution. In this case, the constant solution is called marginally stable.

The more interesting case where the product is negative, that is, $\text{sign}(k_2)\text{sign}(n_2^I) = -1$, the value of K is imaginary over a range of frequencies: $\Omega \in \{-2U_0, 2U_0\}$. This leads to exponentially growing solutions to Equation 10.125, which means that the constant solution is unstable for a range of frequency perturbations. This is precisely why it is called a *modulation instability*. Using this condition, the linear gain in over the range of frequencies is defined as

$$g = 2\,\text{Im}\{K\} = |\Omega|\sqrt{4U_0^2 - \Omega^2}. \tag{10.129}$$

Figure 10.6 is a plot of the imaginary spatial wave vector for three values of U_0 using Equation 10.129. The perturbations grow and eventually may evolve to a new pulse-shaped solution. One special solution, called a soliton, will be derived and discussed in the following section. The peak gain coefficient favors growth of the modulated amplitude at that frequency. Modulation instability is identified in experiments by the appearance of two side bands in the frequency spectrum at peak gain frequencies that depend on the input intensity.

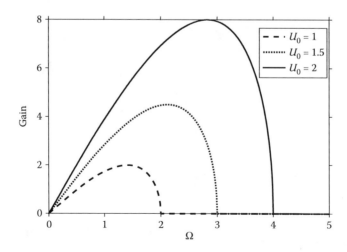

FIGURE 10.6 Modulation instability gain versus (positive) angular frequency for three power values.

A single-mode optical fiber is an archetypal realization of the NLSE, as expressed by Equation 10.122 with some notable differences in the physical interpretation of the wave amplitude. Transverse effects and diffraction that complicate the treatment of waves in free space do not encumber wave propagation description in an optical fiber. In an optical fiber, diffraction effects have been harnessed by confining the transverse dependence of the wave to the form of the fiber mode function. As mentioned above, the mode is characterized by an effective area, S_{eff}, which is defined in Appendix E. At the same time, the fiber can be designed to control the temporal dispersion, such as the group velocity and GVD values, of the wave in the fiber. Fibers can be designed and fabricated to have anomalous, normal, or even zero dispersion at a particular wavelength. Appendix E has a derivation of the NLSE for wave propagation in an optical fiber. For an optical fiber, the wave amplitude squared has the units of power, rather than intensity for the free space derivation in Section 10.4.3. The differences in the interpretation of the NLSE are best seen by evaluating the nonlinear coefficients in each case. The third-order optical nonlinearity of fused silica is reported between $n_2 = 2.2 \times 10^{-20}$ to 3.4×10^{-20} m^2/W, and the area of a standard fiber (so called SMF 28 fiber) mode at the wavelength of 1.55 μm is around $S_{eff} = 85$ μm^2. The nonlinearity in Equation E.14 is between $\gamma = 1.1$–1.6/W-km and the GVD of SMF 28 is $\beta_2 = -23$ ps^2/km. In physical units, the gain in Equation 10.129 is

$$g = |\beta_2| \, |\Omega| \sqrt{4 \frac{\gamma}{|\beta_2|} P_0 - \Omega^2}. \tag{10.130}$$

Using a median value 1.5/W-km for the fiber nonlinearity and wave power $P_0 = 1$ W, the gain band width limits are $\Omega_{extrema} = \pm 0.51$ THz. The limits are proportional to the square root of the power. The maximum gain ($|\Omega_{max}| = 0.36$ THz) for these parameters is $g_{max} = 6.0$/km. This value corresponds to about 400 times increase in the modulation intensity over a 1-km propagation distance. The modulation instability in optical fibers characterized by a growing periodic wave amplitude variation and the appearance of side bands at the modulation frequency in the optical spectrum has been reported in several experiments. The book by Agrawal (2013) has a thorough discussion of the modulation instability in optical fibers.

10.4.5 FUNDAMENTAL SOLITON SOLUTION

We consider the case when the signs of k_2 and n_2 are opposite, such that the nonlinearity offsets the dispersion effects. For our specific case of glass fibers, the sign of the nonlinear index is positive. Hence, we need a negative k_2, which corresponds to operating in the anomalous dispersion region. For this case, Equation 10.122 is

$$\frac{\partial U}{\partial Z} - \frac{i}{2} \frac{\partial^2 U}{\partial T^2} = i |U|^2 U. \tag{10.131}$$

This dimensionless equation is referred to as the NLSE even though our derivation is completely classical. However, the differential equation is in the same form as is found in quantum mechanics (with scaled physical coordinates and their substitutions: $t \to Z$ and $x \to T$); namely, the second term is analogous to the kinetic energy of a quantum particle in one dimension and the right-hand side is analogous to a quantum potential, which oddly is a quadratic function of the wave amplitude $V \to |U|^2$; thus, the word "nonlinear" is placed in the equation's name.

Solutions to the NLSE take many forms, and we examine just one of them called the fundamental soliton. It is a solution where dispersion and nonlinearity are balanced such that the intensity, $I(\tau)$, does not change shape with propagation. Another way to view the balance is that the envelope attains a frequency chirp due to GVD while, at the same time, it obtains a frequency chirp of the opposite sign due to the nonlinearity. When the two contributions balance each other, the pulse envelope remains unchanged.

In the case of the NLSE, an unchanging intensity during propagation corresponds to the envelope shape, $|U(T)|^2$, remaining constant. Our approach is to postulate a form of solution and then plug it into the NLSE. By doing so, we see that the particular solution has restrictions placed on it. The form of soliton solution we work with is given by

$$U(Z,T) = V(T)e^{i\kappa Z}, \tag{10.132}$$

where V is the envelope function and κ is a real constant. The form of Equation 10.124 guarantees that the intensity does not change with Z resulting in the desired soliton property. If such a form is possible, it must satisfy the NLSE (Equation 10.116) and hence we substitute Equation 10.132 into Equation 10.122 yielding

$$\frac{\partial^2 V}{\partial T^2} = 2V(\kappa - V^2). \tag{10.133}$$

We solve this equation by making the substitution

$$\Theta = \frac{\partial V}{\partial T}. \tag{10.134}$$

Note that $(\partial\Theta/\partial T) = (\partial^2 V/\partial T^2)$ and

$$\frac{\partial\Theta}{\partial T} = \frac{\partial\Theta}{\partial V}\frac{\partial V}{\partial T} = \Theta\frac{\partial\Theta}{\partial V}. \tag{10.135}$$

With this substitution we arrive at

$$\Theta d\Theta = 2V(\kappa - V^2)dV. \tag{10.136}$$

Integrating Equation 10.136 gives

$$\left(\frac{\partial V}{\partial T}\right)^2 = 2\kappa V^2 - V^4 + C, \tag{10.137}$$

where C is an integration constant. We expect the envelope, $V(\tau)$, and its derivative to go to zero at infinity; hence, $C = 0$. If we assume that the soliton is symmetric about $\tau = 0$, then at the center of the envelope, $(\partial V/\partial T)|_{\tau=0} = 0$ and $V(0) = 1$ allow us to solve for $\kappa = 1/2$. Substituting $C = 0$ and $\kappa = 1/2$ into Equation 10.137 gives

$$\frac{\partial V}{\partial T} = V(1 - V^2)^{1/2}. \tag{10.138}$$

This equation is integrated by separating variables resulting in (see Problem 10.15)

$$V = \mathrm{sech}(T). \tag{10.139}$$

Therefore, one solution to the dimensionless NLSE given in Equation 10.122 is

$$U = \mathrm{sech}(T)e^{iZ/2}. \tag{10.140}$$

Returning to the original physical variables, the soliton solution is

$$A(\zeta, \tau) = A_0 \mathrm{sech}\left(\sqrt{\frac{n_2^I I_0 \omega_0}{k_2 c}}\tau\right)\exp\left\{i\frac{n_2^I I_0 \omega_0}{c}\zeta\right\}. \tag{10.141}$$

Earlier in our derivation, we chose the signs of k_2 and n_2 as opposite. Had we chosen the same sign, then the solution would diverge. The fundamental soliton solution given in Equation 10.141 is a good test for the split-step numerical approach described by Equation 10.103. The numerical approach should show that the pulse shape does not change as the pulse propagates along Z. The numerical approach is also helpful in understanding what happens when this solution is slightly perturbed.

10.4.6 SPATIAL SOLITONS

When we work with monochromatic fields, the time derivatives in the envelope equation (Equation 10.88) are negligible. Furthermore, we consider an isotropic medium such that the Poynting vector walk-off is not present. Under these conditions, and for a Kerr nonlinearity, the envelope equation is

$$\frac{\partial A}{\partial \zeta} = i\frac{1}{2k_0}\nabla_T^2 A + \frac{i3\omega_0^2}{8k_0c^2}\chi^{(3)}|A|^2 A. \tag{10.142}$$

A close look at this equation shows that it has the same form as the temporal NLSE given in Equation 10.116. We switch to a dimensionless form of Equation 10.135 by introducing the following variable substitutions:

$$Z = \frac{3\omega_0^2}{8k_0c^2}|\chi^{(3)}|\,|A_0|^2\,\varsigma, \tag{10.143}$$

$$(X,Y) = \sqrt{k_0\left(\frac{3\omega_0^2}{8k_0c^2}|\chi^{(3)}|\,|A_0|^2\right)}(x,y), \tag{10.144}$$

$$U = \frac{A}{|A_0|}. \tag{10.145}$$

The dimensionless envelope equation looks nearly identical to the temporal case (see Equation 10.122):

$$\frac{\partial U}{\partial Z} = \frac{i}{2}\frac{\partial^2 U}{\partial X^2} + \frac{i}{2}\frac{\partial^2 U}{\partial Y^2} + i\,\text{sign}(\chi^{(3)})|U|^2 U. \tag{10.146}$$

The dependence on two spatial dimensions is critical for the stability of the beam. In two or more dimensions, including when time dependence is also incorporated in the envelope equation, the nonlinearity will overwhelm the diffraction term and lead to self-focusing and eventually the beam's collapse, as mentioned in the discussion of self-focusing in Chapter 8. A spatial soliton situation occurs for an optical beam that is narrow in one transverse spatial dimension, but is essentially infinite in terms of its extent in the orthogonal direction. Such a situation may be approximated by focusing a laser beam with a cylindrical lens or by confining the beam in a slab waveguide in one dimension. Equation 10.142 then becomes

$$\frac{\partial A}{\partial \zeta} = i\frac{1}{2k_0}\frac{\partial^2 A}{\partial x^2} + \frac{i3\omega_0^2}{8k_0c^2}\chi^{(3)}|A|^2 A. \tag{10.147}$$

A soliton solution forms when the sign of $\chi^{(3)}$ is positive, which gives rise to a positive self-focusing that balances beam spreading due to diffraction. The solution follows the same procedure as in the temporal case and is left as a problem (see Problem 10.16).

10.4.7 DARK AND GRAY SOLITONS

Another dark soliton solution of the NLSE asymptotically approaches a nonzero amplitude. In Equation 10.147, the dark soliton solution exists for dependence on one transverse dimension for the case when the sign of $\chi^{(3)}$ is negative. An assumed solution form using scaled coordinates is

$$U(X,Z) = F(X)e^{-iZ}. \tag{10.148}$$

The differential equation is transformed to

$$\frac{1}{2}\frac{d^2F}{dX^2} + F - F^3 = 0. \tag{10.149}$$

A first integral of the differential equation is

$$\left(\frac{dF}{dX}\right)^2 = C - 2F^2 + F^4. \tag{10.150}$$

The asymptotic solution as $X \to \pm\infty$ with $(dF/dX) \to 0$ and $F \to \pm 1$ determines the integration constant as $C = 1$. Integrating to find the solution for F

$$F = \tanh(X). \tag{10.151}$$

More general dark soliton solutions exist and can be expressed as

$$U(X,Z) = [\cos\phi \tanh(\cos\phi(X - \sin\phi Z)) + i\sin\phi]e^{-iZ}. \tag{10.152}$$

where ϕ is an arbitrary phase. The solution for $\phi = 0$ is the dark soliton, which is zero at $X = 0$. The gray soliton amplitudes $(0 < \phi < \pi)$ have a nonzero minimum value, $|U|_{min} = \sin\phi$. The dark and gray soliton solutions are illustrated in Figure 10.7. It is also instructive to test these solutions by adapting the numerical split-step algorithm to the nonzero boundary conditions.

Beam solutions with two transverse dimensions may be numerically examined by using the split-step BPM. The soliton-like wave solution has been called a Townes soliton to honor early work of the laser pioneer Charles H. Townes. Whereas beam solutions with one transverse coordinate dependence form stable solitons, the beam solutions with two or more (i.e., two spatial and one temporal) coordinates are

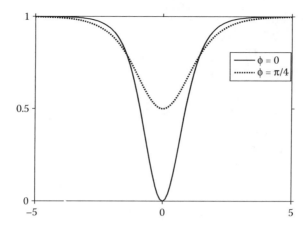

FIGURE 10.7 Spatial dark and gray soliton solutions for three values of the phase $\phi = 0$, $\pi/4$.

unstable and lead to collapse of the beam to a small filament, as discussed under the topic of self-focusing in Chapter 8. To distinguish these solutions from stable solitons, they are often called solitary waves. This is the nature of high-intensity beams propagating in media with the self-focusing nonlinearity.

Due to the negative value of $\chi^{(3)}$, the dark soliton supports a region of lower refractive index around the hole. Therefore, wave at an offset frequency will be guided through the material by the trough formed by the dark soliton. When the dark soliton is off, the offset frequency beam spreads by diffraction in the nonlinear medium. However, by turning the dark soliton on, the offset frequency beam can be guided by cross-phase modulation and beam diffraction is ameliorated.

10.5 DYNAMICAL STIMULATED RAMAN SCATTERING

The description of stimulated Raman scattering (SRS) in Chapter 9 was treated using time-independent field envelopes. Let us reconsider the dynamical evolution of the fields when Raman scattering is present. The vibrational coordinate Q in Equation 9.15 can be written using the complex amplitudes defined in Equation 9.23. We generalize the previous derivation by allowing the complex amplitude to be time dependent, $X(t)$. The equation of motion in the copropagating coordinate frame is

$$\frac{\partial^2 X}{\partial \tau^2} - 2i\omega_\Delta \frac{\partial X}{\partial \tau} - i\omega_\Delta \Gamma \left(X + i\frac{1}{\omega_\Delta}\frac{\partial X}{\partial \tau} \right) + \left(\omega_R^2 - \omega_\Delta^2 \right) X = \frac{1}{4}\frac{d\alpha}{dQ}\bigg|_0 A_p A_s^* e^{i(k_p - k_s)z} \tag{10.153}$$

In the spirit of the SVEA, the first term on the left-hand side is neglected, the second term in parentheses is also a negligibly small term, that is, $\Gamma \ll \omega_\Delta$, and the angular frequency of the driving field is equal to the Raman frequency shift, $\omega_R = \omega_\Delta$. The dynamical equation for the vibrational coordinate is

$$\frac{\partial X}{\partial \tau} + \frac{1}{2}\Gamma X = \frac{i}{8\omega_R}\frac{d\alpha}{dQ}\bigg|_0 A_p A_S^* e^{i(k_p - k_S)\zeta}. \tag{10.154}$$

The equations for the plane-wave pump and Stokes field in the copropagating reference frame are derived from Equation 10.88:

$$\frac{\partial A_p}{\partial \zeta} = \frac{i\mu_0 \omega_p^2}{2k_p} P^{NL}(\omega_p), \tag{10.155}$$

$$\frac{\partial A_S}{\partial \zeta} = \frac{i\mu_0 \omega_S^2}{2k_S} P^{NL}(\omega_S). \tag{10.156}$$

We assume that the group velocities are the same for both fields; this is a reasonable approximation for gases. The nonlinear polarizations are

$$P^{NL}(\omega_p) = N\varepsilon_0 \frac{d\alpha}{dQ}\bigg|_0 A_S X e^{i(k_S - k_p)\zeta}. \tag{10.157}$$

$$P^{NL}(\omega_S) = N\varepsilon_0 \frac{d\alpha}{dQ}\bigg|_0 A_p X^* e^{i(k_p - k_S)\zeta}. \tag{10.158}$$

The equations for the fields are

$$\frac{\partial A_p}{\partial \zeta} = iK(\omega_p) A_S X e^{i(k_S - k_p)\zeta}, \tag{10.159}$$

$$\frac{\partial A_S}{\partial \zeta} = iK(\omega_S) A_p X^* e^{i(k_p - k_S)\zeta}. \tag{10.160}$$

where $K(\omega_p) = (\omega_p/2c)N(d\alpha/dQ)\big|_0$ and $K(\omega_S) = (\omega_S/2c)N(d\alpha/dQ)\big|_0$. From Equations 10.159 and 10.160, it is straightforward to show that

$$\frac{1}{\omega_p}|A_p|^2 + \frac{1}{\omega_S}|A_S|^2 = C(\tau). \tag{10.161}$$

$C(\tau)$ is independent of the spatial coordinate ζ and Equation 10.161 is a Manley–Rowe relation for SRS processes revealing conservation of the total photon flux. Equations 10.154, 10.159, and 10.160 constitute the dynamical equations for the SRS problem.

10.5.1 DYNAMICAL SRS EQUATIONS SOLUTION

When the pulse width is short enough so that $\tau_{\text{pulse}} \ll 1/\Gamma$, the damping term in Equation 10.154 can be neglected. The spatially dependent exponential in the dynamical equations can be eliminated by redefining the variable X as $X \to X e^{i(k_p - k_S)z}$. Furthermore, we can introduce a scaling of the fields, $A_p \to iA_p/\sqrt{\omega_p}$ and $A_S \to A_S/\sqrt{\omega_S}$, and the time and space variables, $\tau \to T = \tau\left(\sqrt{\omega_p \omega_S}/8\omega_R\right)(d\alpha/dQ)\Big|_0$ and $\zeta \to Z = \zeta K(\omega_p)\sqrt{\omega_S}/\sqrt{\omega_p} = \zeta K(\omega_S)\sqrt{\omega_p}/\sqrt{\omega_S}$, to simplify the equations. The three SRS equations are

$$\frac{\partial X}{\partial T} = A_p A_S^*, \tag{10.162}$$

$$\frac{\partial A_p}{\partial Z} = -A_S X, \tag{10.163}$$

$$\frac{\partial A_S}{\partial Z} = A_p X^*. \tag{10.164}$$

Note that the phase mismatch is not relevant for the SRS pump–Stokes wave interaction. The coefficients for both SVEA field equations are nearly equal to unity, since the Stokes shift is small. These equations also admit soliton solutions. To solve the equations, we define the variables $P = 2A_p A_S^*$ and $N = |A_S|^2 - |A_p|^2$ and determine their differential equations

$$\frac{\partial P}{\partial Z} = -2NX, \tag{10.165}$$

$$\frac{\partial N}{\partial Z} = PX^* + P^* X. \tag{10.166}$$

Note that by assuming the functions are real and normalized to unity, that is, $N^2 + P^2 = 1$. Equations 10.162, 10.165, and 10.166 are equivalent to a set of equations called the Maxwell–Bloch equations that describe an electromagnetic field interaction with two-level atoms. Using this analogy, we write $P = \sin\theta$ and $N = \cos\theta$, and Equations 10.165 and 10.166 reduce to a single equation

$$\frac{\partial \theta}{\partial Z} = -2X, \qquad (10.167)$$

Taking the derivative with respect to T of Equation 10.167 and substituting Equation 10.162 yields a single equation

$$\frac{\partial^2 \theta}{\partial Z \partial T} = -\sin\theta. \qquad (10.168)$$

Soliton solutions in the form $\theta(T - Z)$ can be derived using the ordinary differential equation methods we used previously, where the group velocity is unity. Equation 10.168 transforms to an equation of motion identical to that of an inverted mechanical pendulum.

$$\frac{d^2\theta}{dT^2} = \sin\theta. \qquad (10.169)$$

Multiply both sides by $d\theta/dT$ and integrate over time yields

$$\frac{1}{2}\left(\frac{d\theta}{dT}\right)^2 = (1 - \cos\theta). \qquad (10.170)$$

The constant of integration assumes that $d\theta/dT \to 0$ and $\theta \to n2\pi$, $n = 0, 1, 2, \ldots$ as $T \to \pm\infty$.

The solution for θ gives

$$N = \cos\theta = -\tanh(Z - T), \qquad (10.171)$$

$$P = \sin\theta = \operatorname{sech}(Z - T), \qquad (10.172)$$

and

$$X = -\frac{1}{2}\frac{\partial \theta}{\partial Z} = \frac{1}{2}\operatorname{sech}(Z - T). \qquad (10.173)$$

The functional form of the SRS solutions resembles the dark soliton discussed in the previous section. Note that the solutions are related to the pump and Stokes wave scaled intensities by

$$|A_p|^2 = (1 - N)/2 = (1 + \tanh(Z - T))/2,$$
$$|A_S|^2 = (1 + N)/2 = (1 - \tanh(Z - T))/2. \qquad (10.174)$$

These solutions are not pulse solutions, but the pump wave power is transformed entirely into the Stokes wave power via the interaction with the medium. SRS soliton solutions will be examined in Problem 10.21.

PROBLEMS

10.1 Fill in the missing steps to obtain the spatial envelope equation given in Equation 10.14.

10.2 Show that the envelope equation (Equation 10.36) describing linear propagation is simplified to the case without walk-off for the coordinate transformation: $z' = z$ and $x' = x - \tan \rho z$. This is a walk-off centric coordinate system. Plot the solution for several propagation distances using the Gaussian focused beam solution with one transverse dimension using parameters $\tan \rho = 0.01$, $z_R = 10w_0$, $\alpha = 0$, and coordinates scaled to the beam waist at $z = 0$: w_0. Find an expression to determine the propagation distance where the on-axis intensity, that is, at $x = 0$, falls to e^{-2} of its value at $z = 0$.

Note: This is a transcendental equation for z. Solve the equation using the root finding algorithm in Appendix C and compare the result with the limits: zero walk-off and no diffraction.

10.3 Show that (a) $e^{(\hat{L}+\hat{N})dz} \approx e^{\hat{L}\,dz}e^{\hat{N}\,dz}$ to order dz, (b) $e^{(\hat{L}+\hat{N})dz} \approx e^{(\hat{L}\,dz/2)}e^{\hat{N}\,dz}e^{(\hat{L}\,dz/2)}$ to order dz², and (c) $e^{(\hat{L}\,dz/2)}e^{\hat{N}\,dz}e^{(\hat{L}\,dz/2)} = e^{(\hat{N}\,dz/2)}e^{\hat{L}\,dz}e^{(\hat{N}\,dz/2)}$ to order dz^2.

10.4 Write a one-dimensional BPM code using the steps outlined in Section 10.2.3. Validate the code by propagating a one-dimensional Gaussian beam and compare the result with the analytic expressions given in Section 2.6.1 after integrating over the y coordinate. A one-dimensional Gaussian beam can be visualized by considering a plane wave incident on a cylindrical lens. The resultant beam focuses in one dimension while remaining unchanged in the other dimension.

10.5 Write a two-dimensional BPM code using the steps outlined in Section 10.2.3. Validate the code by propagating a Gaussian beam and compare the result with the complex analytic expressions given in Section 2.6.1.

10.6 Numerically integrate Equations 10.46 through 10.48. Use $d_{eff} = 30$ pm/V, the pump and signal wavelengths are 1 and 1.5 µm, respectively, and the nominal refractive index is 3 to calculate the coupling coefficients. The input field amplitudes are $|A_1| = 10^7$ V/m and $|A_2| = 10^6$ V/m. The nonlinear crystal thickness is 25 mm. Solve for several values of Δk, that is, $\Delta kL < 2\pi$ and comparable to and greater than 2π.

10.7 Use Poynting's theorem to show that when a medium has an instantaneous nonlinear response, it is also lossless.

10.8 Verify the Fourier transform relationships given in Equation 10.61.

10.9 Show that the second integral in Equation 10.63 is the complex conjugate of the first.

10.10 Show that the pulse envelope equation (Equation 10.79) becomes Equation 10.85 when we transform to a coordinate system copropagating with the pulse.

10.11 Extend the pulse propagation equation (Equation 10.89) to the next highest order in dispersion (i.e., to $\partial^3/\partial\tau^3$). Ignoring the nonlinearity, solve Equation 10.104 by Fourier transforming to obtain a recipe for a numerical solution.

10.12 Using the perturbation form given by Equation 10.124 show that it leads to the linearized result in Equation 10.125.

10.13 Using the result in Equation 10.129 for the modulation instability gain, find the maximum angular frequency bounding the nonzero gain region and the angular frequency of the maximum gain and the gain at that point.

10.14 Calculate the maximum frequency shift due to self-phase modulation for a pulse that has a Gaussian temporal profile given by

$$I(\tau) = I_0 e^{-2(\tau/T_0)^2},$$

where T_0 is the $1/e^2$ width of the pulse and I_0 is the peak intensity.

10.15 Given that

$$\int \frac{dx}{x\sqrt{(1-x^2)}} = -\ln\left(\frac{1+\sqrt{1-x^2}}{x}\right),$$

show that the soliton solution to Equation 10.138 is as given in Equation 10.139.

10.16 Find the spatial soliton solution, $U(Z,X)$, to Equation 10.147. Convert the solution from dimensionless form into the envelope, $A(\zeta,x)$.

10.17 Use the split-step procedure outlined in Section 10.4.2 to numerically simulate the NLSE (Equation 10.122).

   a. Validate your program by showing that the amplitude of the soliton solution (Equation 10.140) does not change during propagation.
   b. Change the amplitude of the soliton solution by a factor of 2 to observe how the pulse shape changes.
   c. Change the sign of the nonlinearity and propagate the soliton solution used in (a). Show that the pulse width increases with propagation.

10.18 Two soliton-like pulses at $Z = 0$ are separated in time by 2τ. The initial form of the pulses is

$$A(0,T) = A_0 \mathrm{sech}(T + \tau) + A_1 e^{i\phi}\mathrm{sech}(T - \tau).$$

Use the numerical split-step procedure on the scaled NLSE (Equation 10.131) to study the initial wave propagation in the nonlinear medium for the following cases.

   a. Let the two amplitudes equal unity ($A_0 = A_1 = 1$), the delay between them $\tau = 6$, and the phase takes two values $\phi = 0, \pi$. Plot the propagation over a distance of 16π. Comment on the nature of the interaction between the solitons for the two cases.

b. Change the amplitudes $A_0 = 2$, $A_1 = 0$, and for convenience use $\tau = 0$. The wave is no longer constant with propagation distance, but periodically changes, Estimate the propagation distance for a period of the wave shape.

c. Let the two amplitude be equal with $A_0 = A_1 = 1.1$, the delay between them $\tau = 6$, and phase $\phi = 0$. Plot the propagation over a distance of 16π. Comment on the interaction between the two solitons.

10.19 For an optical fiber with physical parameters $\gamma = 1.5/\text{W-km}$ and $\beta_2 = -23 \text{ ps}^2/\text{km}$.

a. Determine the pulse width T_0 so that the nonlinear and dispersion lengths defined in Equation E.15 are identical.

b. Use a sech-shaped pulse with a amplitude $A_0 = \sqrt{P_0}$ and $P_0 = 1 \text{ mW}$. Apply the split-step algorithm to the NLSE (Equation E.14) and show that the soliton propagates without changing.

c. Add linear absorption to your numerical algorithm in (b) and assume its value is $1/L_D$ (the diffusion length, L_D, is defined in Equation E.5). Propagate a distance $4L_D$ and plot the final amplitude normalized to unity and compare the pulse width to the initial solution, which represents the result without absorption. What is the reason for the difference in pulse widths?

10.20 Test the spatial dark soliton solutions in Equation 10.152 by applying the split-step numerical algorithm to Equation 10.147 with a negative nonlinear coefficient. In order to satisfy periodic boundary conditions use two dark soliton solutions separated by $X == 10$ (scaled) units that have a π phase difference. The separation distance is far enough that the two solitons do not interact.

a. Use solutions with $\phi = 0$ and the amplitude is asymptotically unity. Does the amplitude remain unchanged during propagation?

b. Perturb the initial amplitude of the wave and numerically propagate it to observe and note the evolution of the solution. Neglecting the wrap around effect due to periodic boundary conditions. Is the dark soliton solution stable?

10.21 Consider a general SRS solitary wave solution of Equations 10.162 through 10.164 written as

$$A_p = \cos\theta\,\text{sech}(\cos\theta(Z - T))e^{i\sin\theta Z},$$
$$A_S = [\cos\theta\tanh(\cos\theta(Z - T)) - i\sin\theta]e^{-i\sin\theta T},$$
$$X = \cos\theta\,\text{sech}(\cos\theta(Z - T))e^{i\sin\theta(Z + T)}.$$

The angle θ is arbitrary.

a. Show by direct substitution into the equations of motion that these are solutions of the SRS equations of motion (Equations 10.162 through 10.164). This solution represents a Stokes wave that has a constant background and the pump wave is a pulse.

b. Show that the equations of motion satisfy $|A_p|^2 + |A_s|^2 = C(T)$ and that the above solutions satisfy this relation.

10.22 Write a split-step propagation program to study wave propagation using the dimensionless NLSE with two transverse dimensions (Equation 10.146) (see Problem 10.5 for diffraction validation steps). Use an initial Gaussian function

$$E(\vec{r},0) = E_0 e^{-(r/\sigma)^2},$$

where the amplitude E_0 and beam width σ are scaled variables. Fix the value of $\sigma = 1$ and start with small field amplitude values, which are increased until the beam collapses. Compare your values with the critical power given in Equation 8.80 adjusted for our scaled dimensions.

REFERENCES

Agrawal, G. P. 2013. *Nonlinear Fiber Optics* (5th edition). New York, NY: Academic Press.

Banerjee, P. P. 2004. *Nonlinear Optics: Theory, Numerical Modeling, and Applications*. New York, NY: Marcel Dekker, Inc.

Boyd, R. W. 2008. *Nonlinear Optics* (3rd edition). Burlington, MA: Academic Press.

FURTHER READING

Agrawal, G. P. and R. W. Boyd. 1992. *Contemporary Nonlinear Optics*. New York, NY: Academic Press.

Hasegawa, A. and F. Tappert. 1973. Transmission of stationary nonlinear optical pulses in dispersive dielectric fibers. I. Anomalous dispersion. *Applied Physics Letters* 23: 142–144.

Iserles, A. 1996. *A First Course in the Numerical Analysis of Differential Equations*. New York, NY: Cambridge University Press.

Mollenauer, L. F., R. H. Stolen, and J. P. Gordon. 1980. Experimental observation of picosecond pulse narrowing and solitons in optical fibers. *Physical Review Letters* 45: 1095–1098.

New, G. 2014. *Introduction to Nonlinear Optics*. Cambridge, UK: Cambridge University Press.

Smith, A. V. and M. S. Bowers. 1995. Phase distortions in sum- and difference-frequency mixing in crystals. *Journal of the Optical Society of America B* 12: 49–57.

Stegeman, G. I. and R. A. Stegeman. 2012. *Nonlinear Optics: Phenomena, Materials and Devices*. Hoboken, NJ: John Wiley & Sons.

Sutherland, R. L. 2003. *Handbook of Nonlinear Optics* (2nd edition). New York, NY: Marcel Dekker, Inc.

Zakharov, V. E. and L. A. Ostrovsky. 2009. Modulation instability: The beginning. *Physica D: Nonlinear Phenomena* 238: 540–548.

11

QUANTUM NONLINEAR OPTICS

11.1 INTRODUCTION

A signature achievement of twentieth-century science was the development and spectacular success of quantum electrodynamics in explaining the energy difference (called the Lamb shift) between the $2s^{1/2}$ and $2p^{1/2}$ states with the same total angular momentum, and the anomalous intrinsic spin magnetic moment of the electron due to vacuum fluctuations of the electromagnetic field. However, it is not a theory relegated to a few historically celebrated experiments. It generally applies to ordinary nonlinear optical phenomena and lasers, and the applications of quantum electrodynamics are expanding to the nascent fields of quantum information and quantum cryptography. The effects of vacuum fluctuations are ubiquitous in everyday phenomena from the dominance of spontaneous emission from ordinary incoherent light sources to attractive dipole-induced van der Waals forces. The force due to a single dipole is inherently weak; however, in concert, a multitude of dipoles can produce macroscopic effects. One common example is the gravity-defying ability of geckos to climb walls and tread across ceilings. Furthermore, as discovered by Edward Purcell, the vacuum fluctuations can be manipulated by tuning an atomic (or molecular) transition to a cavity resonance that increases the photon density of states, which enhances the atom's coupling to the vacuum field, thereby increasing the spontaneous emission rate (Purcell factor). By extension of the concept, the vacuum fluctuations can also be suppressed or inhibited in a region of space by lowering the photon density of states.

Quantum principles are exploited to study the peculiar nature of light, including a phenomenon referred to as entanglement, to which Einstein objected, calling it "spooky" action at a distance. His objection, first articulated in the famous 1935 Einstein–Podolsky–Rosen (EPR) paper, described quantum entanglement of two particles where the measurement of one particle's state instantaneously (i.e., there is no causal connection between them, which EPR's paper highlighted and was their principal objection) determines the state of the other particle. Additional papers attempted to clarify the issues, culminating in the papers by John Bell who derived inequalities (now called Bell's inequalities) based on the EPR requirements for a more complete theory. Bell's work would offer a path to substantiating the role of local realism in a theory. The word "local" means that the particles are endowed with intrinsic properties, even if those properties are hidden variables, that is, no superluminal effects. The word "realism" means that the particle's properties exist even if not measured. Those properties may be "hidden variables" possessed by each isolated, noninteracting particle. The topic has become especially relevant as technology has advanced to create the requisite entangled particle states needed to conduct experimental tests of EPR. The Bell inequalities were extended by others, enabling the design of experiments to examine violations of the inequalities. The results of the experiments have confirmed the predictions of quantum mechanics, even under "spooky" superluminal conditions where the particles' measurement times and separations preclude communication by the speed of light.

For nearly a century, researchers have confronted the foundation of quantum mechanics, questioning the lack of determinism and the violation of causality inherent in the Copenhagen interpretation of the wave function collapse when a system is measured. It continues to be the subject of deep inquiry. Nevertheless, a growing number of experiments have confirmed the predictions of quantum mechanics, a validation process which continues to engage researchers today. Quantum mechanics has been extended to uncover quantum phenomena in many diverse areas of science and technology; for instance, electronic atomic orbitals have applications to chemical bonding and to solid-state physics that provide insights into very complex multielectronic systems.

High gain systems such as stimulated Raman scattering, optical parametric generation (OPG), optical parametric amplification (OPA), and initiation of coherent laser emission all manifest the characteristics of quantum fluctuations at a macroscopic level. Initially, a feeble signal generated with a single photon is amplified to macroscopic levels, and the quantum statistical (or noise) properties are amplified, as well. By studying high gain systems, we can probe the quantum origins of the emission process through a quantum statistical analysis. Another closely related process with high gain and no feedback is called superfluorescence in the optical regime or superradiance in the microwave regime. In these systems, all the atoms are excited so that the medium polarization is zero; the radiative emission is initiated by vacuum fluctuations of the electromagnetic field.

Two complementary formalisms are applied to solve problems in quantum mechanics; they are called the Schrödinger and Heisenberg pictures. In the former picture, the evolution of the wave functions is described, and averages of measureable quantities are calculated using the time-dependent wave functions. This formalism is commonly taught in books on quantum mechanics. In the latter picture, the evolution of operator variables, which may be decomposed into a set of elementary operators, is followed. The measurable quantities in this case are averages with respect to the initial wave functions. In this chapter, quantum mechanics is developed from the point of view of the Heisenberg picture. Its application connects more simply with the reader because the equations are familiar from our classical treatment of nonlinear optics. The essential complication of quantum mechanics replaces the dynamical variables of classical mechanics by quantum operators.

In Section 6.5, the topic of spontaneous parametric scattering was introduced using a phenomenological treatment. The classical electric field is written as a superposition of plane-wave modes, each with a corresponding field amplitude. The $\vec{E}$ field can be considered as arising from both a scalar potential (charge source) and a vector potential (current source). The scalar potential in a charge-free space is zero and the vector potential written in the form

$$\underline{\vec{A}}(\vec{r},t) = \frac{1}{2}\sum_{\vec{k}} \frac{\mathcal{E}_{\vec{k}}}{\omega_{\vec{k}}} \vec{A}_{\vec{k}}(t)e^{i\vec{k}\cdot\vec{r}} + \text{c.c.} \tag{11.1}$$

where $\mathcal{E}_{\vec{k}}$ is a (real) constant having units of the electric field, $\vec{A}_{\vec{k}}(t)$ is the time-dependent (unitless) vector field amplitude for the plane-wave mode described by the wave vector $\vec{k}$, and $\omega_{\vec{k}}$ is the angular frequency of the mode. The short-hand notation c.c. on the right-hand side denotes the complex conjugate of the first term. The scalar potential is assumed to be zero in a source-free space, and the Coulomb gauge (i.e., $\vec{\nabla} \cdot \underline{\vec{A}}(\vec{r},t) = 0$) ensures that the vector amplitudes are transverse to the wave vector

$$\vec{k} \cdot \vec{A}_{\vec{k}}(t) = 0. \tag{11.2}$$

In free space, where the refractive index $n = 1$, the amplitudes satisfy Maxwell's wave equations; the dispersion relation is $\omega_{\vec{k}}^2 = c^2 k^2$, and the amplitudes satisfy

$$\frac{d^2\vec{A}_{\vec{k}}(t)}{dt^2} + \omega_{\vec{k}}^2 \vec{A}_{\vec{k}}(t) = 0. \tag{11.3}$$

This equation has the same form as the harmonic oscillator equation; solutions are chosen of the form

$$\vec{A}_{\vec{k}}(t) = (\hat{e}_{1,\vec{k}} A_{1,\vec{k}}(t) + \hat{e}_{2,\vec{k}} A_{2,\vec{k}}(t)). \tag{11.4}$$

The two linear polarization unit vectors $\hat{e}_{1,\vec{k}}$ and $\hat{e}_{2,\vec{k}}$ are orthogonal to the wave vector $\vec{k}$. The initial amplitudes $(A_{1,\vec{k}}(t), A_{2,\vec{k}}(t))$ satisfy the scalar version of Equation 11.3 with the solution

$$A_{q,\vec{k}}(t) = a_{q,\vec{k}} e^{-i\omega_{\vec{k}}t} + a^*_{q,-\vec{k}} e^{i\omega_{\vec{k}}t}, \quad q = 1, 2. \tag{11.5}$$

The subscript $-\vec{k}$ on the second amplitude is defined to ensure that the vector potential is real. The corresponding electric field is expressed as

$$\vec{E}(\vec{r}, t) = -\frac{\partial \vec{A}}{\partial t} = \frac{1}{2} \sum_{\vec{k}} \sum_{q=1,2} i\mathcal{E}_{\vec{k}} \hat{e}_{q,\vec{k}} a_{q,\vec{k}} e^{i(\vec{k}\cdot\vec{r} - \omega_{\vec{k}}t)} + \text{c.c.} \tag{11.6}$$

In the quantum version of the electric field, the amplitudes $\left(a_{q,\vec{k}} a^*_{q,\vec{k}}\right)$ are replaced by operators. In the following section, the analogy between the field amplitude dynamics and the classical/quantum harmonic oscillator are utilized to make the transition from classical dynamical variables to quantum mechanical operators in the Heisenberg representation.

11.2 QUANTIZING EQUATIONS OF MOTION

11.2.1 CLASSICAL TO QUANTUM EQUATIONS OF MOTION

In this section, we give the essential elements of quantum mechanics with an emphasis on quantum aspects of nonlinear optics. Quantum mechanics is covered in a nut shell by introducing only those concepts and definitions that will be used in this chapter. The interested reader can refer to one of the many excellent texts on quantum mechanics for more depth. The primary model to describe a linear response of a medium uses a harmonic oscillator, and when applied to atomic systems, it is called the Lorentz model. This model was thoroughly treated in Chapter 3. We are revisiting the model again but ignoring losses for the moment in order to connect it to quantum mechanics.

$$m \frac{d^2 x}{dt^2} + m\omega_0^2 x = -eE. \tag{11.7}$$

The second term is the restoring force of the mass to a neutral position at $x = 0$, and it is linear in the displacement. The electric field on the right-hand side is derived from the time-dependent electromagnetic field. The force can be classical or possess quantum characteristics; the connection between the classical and the quantum harmonic oscillator will be motivated in this section. This familiar equation can be written as two first-order differential equations,

$$m\frac{dx}{dt} = p, \quad \frac{dp}{dt} + m\omega_0^2 x = -eE, \quad (11.8)$$

where p is the particle momentum. In classical mechanics, (p,x) are generalized variables that define energy-conserving dynamical equations for the particle. The energy for the harmonic oscillator is expressed as

$$H = \frac{1}{2m}p^2 + \frac{m\omega_0^2}{2}x^2 + eEx. \quad (11.9)$$

As a function $H(x,p)$, the expression in Equation 11.9 is called the Hamiltonian. Next, we introduce the quantum version of the harmonic oscillator.

In quantum mechanics, these variables are operators representing a system of conjugate variables, and Equation 11.8 is a set of operator equations. *Conjugate variables* in our usage of the term will refer to two operators that are related by Fourier transforms. For the spatial representation of both operators, the momentum has a representation as derivative operator $\hat{p} = (\hbar/i)(\partial/\partial x)$. Planck's constant divided by 2π, denoted as $\hbar$, appears in the numerator on the right-hand side. As a fundamental constant, it plays a central role in quantum mechanics. In the Fourier transformed space, the role of the derivative operator is reversed, that is,

$$(\hat{x},\hat{p}) = \left(x, \frac{\hbar}{i}\frac{\partial}{\partial x}\right) \to \left(-\frac{1}{i}\frac{\partial}{\partial k}, \hbar k\right).$$

As a consequence, the order of the operators is important, and the commutator of the operators is

$$[\hat{p},\hat{x}] = \hat{p}\hat{x} - \hat{x}\hat{p} = \frac{\hbar}{i}. \quad (11.10)$$

Note that the commutator should be manipulated by assuming that it is multiplied by an arbitrary function to the right, that is,

$$[\hat{p},\hat{x}]f(x) = \hat{p}\hat{x}f - \hat{x}\hat{p}f = \frac{\hbar}{i}f + \hat{x}\hat{p}f - \hat{x}\hat{p}f = \frac{\hbar}{i}f. \quad (11.11)$$

In quantum theory, the equations of motion are generated by using the Hamiltonian and the commutator:

$$\frac{d\hat{x}}{dt} = \frac{i}{\hbar}[\hat{H},\hat{x}], \quad \frac{d\hat{p}}{dt} = \frac{i}{\hbar}[\hat{H},\hat{p}]. \quad (11.12)$$

The quantum results for the operators are identical to the classical results in Equation 11.8. The so-called creation and annihilation operators can be defined as linear combinations of the operators $(\hat{p}, \hat{x})$ as follows:

$$a = \frac{1}{\sqrt{2}} \left(i\sqrt{\frac{1}{m\omega_0\hbar}}\hat{p} + \sqrt{\frac{m\omega_0}{\hbar}}\hat{x} \right),$$

$$a^\dagger = \frac{1}{\sqrt{2}} \left(-i\sqrt{\frac{1}{m\omega_0\hbar}}\hat{p} + \sqrt{\frac{m\omega_0}{\hbar}}\hat{x} \right).$$

(11.13)

We refrain from using the arc symbol above these operators to retain simplicity in the notation. The operators are called Hermitian conjugates and they satisfy the commutation relation:

$$[a, a^\dagger] = 1.$$

(11.14)

Using the definitions in Equation 11.13, the equations of motion for the creation and annihilation operators satisfy

$$\frac{da}{dt} = -i\omega_0 a - i\frac{eE}{\sqrt{2m\omega_0\hbar}},$$

$$\frac{da^\dagger}{dt} = i\omega_0 a^\dagger + i\frac{eE}{\sqrt{2m\omega_0\hbar}}.$$

(11.15)

In the absence of an external field, the analytical solutions of Equation 11.15 are

$$a(t) = a(0)e^{-i\omega_0 t} \quad \text{and} \quad a^\dagger(t) = a^\dagger(0)e^{i\omega_0 t}.$$

(11.16)

The Hamiltonian with the applied field equal to zero and expressed using the creation and annihilation operators takes the form

$$H = \hbar\omega_0 a^\dagger a + \frac{\hbar\omega_0}{2}.$$

(11.17)

The added constant energy, $(1/2)\hbar\omega_0$, is called the zero-point energy, which, as we have mentioned in the introduction, is the source of physical phenomena as diverse as spontaneous emission and the Lamb shift. The energy levels in a harmonic oscillator have equidistant separations and form a ladder of levels, labeled by an integer, ideally extending to infinity; the discrete harmonic energy level is shown in Figure 11.1.

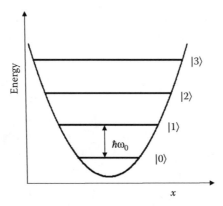

FIGURE 11.1 Parabolic potential energy drawn to demonstrate the step energy dependence on the spatial coordinate. The discrete quantum energy level positions are labeled by kets $\{|n\rangle\}$ with the equidistant offset energy between consecutive levels shown, as well. The ground state is the lowest quantum state, and it is shifted from the bottom of the parabola by $(1/2)\hbar\omega_0$.

Throughout the rest of this chapter, we adopt the bra $(\langle n|)$ and ket $(|n\rangle)$ vector notation, where n is an integer with values $n = 0, 1, 2,...$ to express the wave function. The "bra–ket" notation was introduced by Dirac to describe wave functions and is referred to in the literature as the *Dirac notation*. The set of orthonormal basis states $\{|n\rangle, n = 0, 1, 2,...\}$ forms a complete set that spans the vector space. General wave functions are constructed as a linear combination of *ket* vectors and their dual *bra* vector space. The creation and annihilation operators are also called ladder operators because they add or subtract the excitation number of the harmonic oscillator in the absence of an applied field. The annihilation operator steps the number state $|n\rangle$ down by unity and the creation n operator does the same for the bra state $\langle n|$, that is,

$$a\,|n\rangle = A_n\,|n-1\rangle \quad \text{and} \quad \langle n|a^\dagger = \langle n-1|A_n^*. \tag{11.18}$$

The constants (A_n, A_n^*) are determined below. The *bra* wave function is the conjugate of the *ket* wave function. The creation operator acting on the number *ket* state steps its number up by unity,

$$a^\dagger\,|n\rangle = A_{n+1}\,|n+1\rangle. \tag{11.19}$$

The wave functions for the harmonic oscillator satisfy orthogonality conditions, and the inner product is normalized to unity; the result is written as

$$\langle n\,|\,n'\rangle = \delta_{n,n'}. \tag{11.20}$$

The operator $\hat{N} = a^\dagger a$ represents the excitation number, and eigenstates can be labeled by a nonnegative integer $\{n: n = 0, 1, 2,...\}$,

$$\hat{N} \mid n \rangle = n \mid n \rangle. \tag{11.21}$$

In other words, the normalization of the annihilation operation is $A_n = \sqrt{n}$. The state $n = 0$ is the lowest energy state, also called the *ground state*. From these results, we can use the properties of the creation operator to show that any excited state can be generated from the ground state,

$$\mid n \rangle = \frac{(a^\dagger)^n}{\sqrt{n!}} \mid 0 >. \tag{11.22}$$

The wave function in the coordinate space representation is denoted by $\langle x \mid n \rangle = \Psi_n(x)$. For the ground state, $n = 0$, the wave function is found by using the definition

$$a\langle x \mid 0 \rangle = \frac{1}{\sqrt{2}} \left(i\sqrt{\frac{1}{m\omega_0\hbar}}\, \hat{p} + \sqrt{\frac{m\omega_0}{\hbar}}\, \hat{x} \right) \Psi_0(x) = 0. \tag{11.23}$$

Using the differential form of the momentum operator, the solution of the first-order differential equation (Equation 11.23) is

$$\Psi_0(x) = N \exp(-m\omega_0 x^2/2\hbar). \tag{11.24}$$

This is the well-known Gaussian function representation for the ground-state wave function in the coordinate representation. The normalization coefficient is N, that is, $\int_{-\infty}^{\infty} |\Psi_0(x)|^2 dx = 1$. An arbitrary ket vector state is constructed by a linear combination of all kets that form a complete set spanning the vector space:

$$\mid V \rangle = \sum_{n=0}^{\infty} c_n \mid n \rangle. \tag{11.25}$$

The set of amplitudes $\{c_n: n = 0,...\}$ consists of complex numbers. The bra vector space is spanned by the dual vector to the ket vector:

$$\langle V \mid = \sum_{n=0}^{\infty} c_n^* \langle n \mid. \tag{11.26}$$

c_n^* is the complex conjugates of the complex coefficients in Equation 11.25. The inner product of the vector is denoted as

$$\langle V | V \rangle = \sum_{n=0}^{\infty} \sum_{n'=0}^{\infty} c_n^* c_{n'} \langle n | n' \rangle = \sum_{n=0}^{\infty} |c_n|^2 = 1. \qquad (11.27)$$

The last equality expresses the certainty of finding the particle, and each square amplitude term is

$$P(n) = |c_n|^2, \qquad (11.28)$$

which represents the probability of occupying the number state $|n\rangle$. The states are assumed to form a complete basis for the quantum system, which are also orthonormal.

The average value of an observable defined by the operator A is defined by

$$\langle A \rangle = \langle V | A | V \rangle = \sum_{n=1}^{\infty} \sum_{n'=1}^{\infty} c_n^* c_{n'} \langle n | A | n' \rangle. \qquad (11.29)$$

An observable is generally considered to have a real valued average. This would be the case for position and momentum operators.

The harmonic oscillator in either the classical or quantum domains perpetually remains in an excited state in the form presented here. This is of interest for times that are short enough compared to characteristic times for energy to be exchanged with its environment. In real physical systems, the environment will extract energy from the central oscillator and distribute it among many modes without coherently returning it to the central oscillator. The excited oscillator energy will decay to the lowest available energy state. Classically, the oscillator rests at $x = 0$, but quantum mechanically the oscillator is in the ground state, $|0\rangle$ with dynamics that are described as zero-point motion and probability distribution given by the square amplitude of the positional wave function in Equation 11.24.

11.2.2 HEISENBERG UNCERTAINTY RELATIONS

The Heisenberg uncertainty principle forms a cornerstone of quantum theory with implications on the accuracy with which observable quantities can be measured. In general, two operators, A and B, do not commute with one another. As a consequence, in quantum mechanics, both noncommuting operators corresponding to observable physical quantities may not be *simultaneously* measured to any accuracy. The Heisenberg uncertainty relations give a quantum limit to the accuracy of a pair of measurements, which we shall examine now. The commutator of two observable operators is denoted as

$$C = [A, B] = AB - BA. \tag{11.30}$$

For example, A and B are taken as the position and momentum operators; x and p were already discussed ($C = [x,p] = -\hbar/i$). The averages of A and B with respect to the initial wave function are denoted as $\langle A \rangle$ and $\langle B \rangle$, respectively. We define the fluctuation operators as deviations from their average value:

$$\Delta A = A - \langle A \rangle, \quad \Delta B = B - \langle B \rangle. \tag{11.31}$$

The commutators of ΔA and ΔB are

$$[\Delta A, \Delta B] = [A, B]. \tag{11.32}$$

This property is sufficient to derive the well-known Heisenberg uncertainty relation, the result that the product of the variances is bounded from below. In other words, there is a minimum uncertainty relation between two noncommuting observables:

$$\langle \Delta A^2 \rangle \langle \Delta B^2 \rangle \geq \frac{1}{4} |\langle [A, B] \rangle|^2 . \tag{11.33}$$

The variances are defined as $\langle \Delta A^2 \rangle = \langle V | \Delta A^2 | V \rangle$ and $\langle \Delta B^2 \rangle = \langle V | \Delta B^2 | V \rangle$. To prove this, we form a new state vector:

$$|V'\rangle = \Delta A |V\rangle - \left(\frac{\langle \Delta B \Delta A \rangle}{\langle \Delta B^2 \rangle} \right) \Delta B |V\rangle. \tag{11.34}$$

$|V'\rangle$ is orthogonal to the vector defined by $\Delta B |V\rangle$, that is, $\langle V | \Delta B | V' \rangle = 0$. The inner product of $|V'\rangle$ with itself satisfies the inequality:

$$0 \leq \langle V' | V' \rangle = \langle \Delta A^2 \rangle - \frac{|\langle \Delta B \Delta A \rangle|^2}{\langle \Delta B^2 \rangle}. \tag{11.35}$$

The final step is to rewrite the average in the numerator as

$$\langle \Delta B \Delta A \rangle = \frac{1}{2} \langle [\Delta B, \Delta A] \rangle + \frac{1}{2} [\langle \Delta B \Delta A \rangle + \langle \Delta A \Delta B \rangle]. \tag{11.36}$$

The first term is imaginary (since $[\Delta B, \Delta A]$ is anti-Hermitian, that is, $[\Delta B, \Delta A]^\dagger = [\Delta A, \Delta B]$). The second term is real, and we write

$$|\langle \Delta B \Delta A \rangle|^2 = \frac{1}{4}|\langle [\Delta B, \Delta A] \rangle|^2 + \frac{1}{4}|\langle \Delta B \Delta A \rangle + \langle \Delta A \Delta B \rangle|^2. \qquad (11.37)$$

Neglecting the second (also positive) term in Equation 11.37, the result in Equation 11.35 reduces to Equation 11.33, which is the Heisenberg uncertainty relation. To reiterate, this result plays a central role in the theory of measurement and places a quantum limit on the ultimate sensitivity of simultaneously determining momentum and position of a quantum particle. Measurement schemes devised to circumvent the quantum limits prepare a system in special quantum states, called squeezed states, which reduce the uncertainty of a chosen measured variable at the expense of increasing the uncertainty in a conjugate noncommuting physical variable. The product of the two variances satisfies the inequality given by Equation 11.33. More about this will be discussed later. It is left as a problem to show that the ground state of the harmonic oscillator is a minimum uncertainty state for the position and momentum operators.

11.3   ELECTROMAGNETIC FIELD

Now let us return to the plane-wave electromagnetic vector potential amplitude, which satisfies Equation 11.3, and invoke the same logic as for the transition from the classical to the quantum harmonic oscillator. The plane-wave mode amplitude $a_{q\vec{k}}$ in Equation 11.5 is transformed to an annihilation operator, and its Hermitian conjugate $a_{q\vec{k}}^\dagger$ is the mode's creation operator; the operators satisfy the commutations relations in Equation 11.14. The amplitudes of different plane-wave modes commute with one another. The operator expression for the electric field is given by Equation 11.6, and the magnetic field expression derived from the vector potential is

$$\vec{B}(\vec{r},t) = \vec{\nabla} \times \underline{\vec{A}} = -\frac{1}{2}\sum_{\vec{k}}\sum_{q=1,2} i\frac{\mathcal{E}_{\vec{k}}}{\omega_{\vec{k}}}\vec{k} \times \hat{e}_{q,\vec{k}}\, a_{q,\vec{k}}\, e^{-i\omega_{\vec{k}}t} e^{i\vec{k}\cdot\vec{r}} + \text{h.c.} \qquad (11.38)$$

The second term with h.c. denotes the Hermitian conjugate of the first term. The second term in Equation 11.6 also is changed from c.c. to h.c. The energy of the electromagnetic field using the energy density expression found in Poynting's theorem in Chapter 2 applied to free space is

$$H = \frac{1}{2}\iiint \left(\varepsilon_0 \underline{\vec{E}} \cdot \underline{\vec{E}} + \frac{1}{\mu_0}\underline{\vec{B}} \cdot \underline{\vec{B}} \right) dV. \qquad (11.39)$$

By applying solutions 11.6 and 11.38, the two integrals can be separately evaluated. The volume integrals are

$$\iiint e^{i(\vec{k}\pm\vec{k}')\cdot\vec{r}}\mathrm{d}V = V\delta_{\vec{k},\mp\vec{k}'}. \tag{11.40}$$

For the quantized fields, the complex amplitudes in Equation 11.5 $\{a_{q,\vec{k}}, a_{q,\vec{k}}^*\}$ are transformed to operators $\{a_{q,\vec{k}}, a_{q,\vec{k}}^\dagger\}$ that satisfy the commutation relations,

$$\left[a_{q,\vec{k}}, a_{q',\vec{k}'}^\dagger \right] = \delta_{q,q'}\delta_{\vec{k},\vec{k}'}. \tag{11.41}$$

Using the following relations for the unit vectors,

$$\hat{e}_{q,\vec{k}}\cdot\hat{e}_{q',\vec{k}} = \delta_{q,q'} \quad \text{and} \quad \left(\vec{k}\times\hat{e}_{q,\vec{k}}\right)\cdot\left(\pm\vec{k}\times\hat{e}_{q',\vec{k}}\right) = \pm k^2\delta_{q,q'}, \tag{11.42}$$

the electric portion of the energy is

$$\iiint \left(\varepsilon_0\underline{\vec{E}}\cdot\underline{\vec{E}}\right)\mathrm{d}V = \frac{\varepsilon_0 V}{4}\sum_{\vec{k},q}\mathcal{E}_{\vec{k}}^2\left(a_{q,\vec{k}}a_{q,\vec{k}}^\dagger + a_{q,\vec{k}}^\dagger a_{q,\vec{k}} - a_{q,\vec{k}}a_{q,\vec{k}}e^{-i2\omega_{\vec{k}}t} - a_{q,\vec{k}}^\dagger a_{q,\vec{k}}^\dagger e^{i2\omega_{\vec{k}}t}\right), \tag{11.43}$$

and the magnetic portion of the energy is

$$\iiint \left(\frac{1}{\mu_0}\vec{B}\cdot\vec{B}\right)\mathrm{d}V = \frac{V}{4\mu_0}\sum_{\vec{k},q}\frac{k^2}{\omega^2}\mathcal{E}_{\vec{k}}^2\left(a_{q,\vec{k}}a_{q,\vec{k}}^\dagger + a_{q,\vec{k}}^\dagger a_{q,\vec{k}} + a_{q,\vec{k}}a_{q,\vec{k}}e^{-i2\omega_{\vec{k}}t} + a_{q,\vec{k}}^\dagger a_{q,\vec{k}}^\dagger e^{i2\omega_{\vec{k}}t}\right). \tag{11.44}$$

The field energy is rewritten in an expanded form to emphasize the terms that will not commute in the quantized version,

$$H = \frac{\varepsilon_0 V}{4}\sum_{\vec{k},q}\mathcal{E}_{\vec{k}}^2\left(a_{q,\vec{k}}a_{q,\vec{k}}^\dagger + a_{q,\vec{k}}^\dagger a_{q,\vec{k}}\right). \tag{11.45}$$

Setting the coefficient equal to $\hbar\omega_{\vec{k}}$, the field amplitude is normalized as

$$\mathcal{E}_{\vec{k}}^2 = \frac{2\hbar\omega_{\vec{k}}}{\varepsilon_0 V}. \tag{11.46}$$

In quantized form, the field energy is expressed in the form,

$$H = \sum_{\vec{k},q} \left(\hbar \omega_{\vec{k}} a_{q,\vec{k}}^\dagger a_{q,\vec{k}} + \frac{\hbar \omega_{\vec{k}}}{2} \right). \tag{11.47}$$

It is apparent in this result that the zero-point energy diverges as the number of modes tends to infinity; this enigmatic result has important physical consequences. The vacuum fluctuations couple to excited atoms and induce electronic transitions to a lower state through the ubiquitous process called spontaneous emission, which is the dominant optical process in our lighting technology. The troubling infinity is removed by modern renormalization theory, which extracts only energy differences.

Recall in Section 6.5 that we introduced the spontaneous emission for parametric processes. The zero-point energy in Equation 11.47 now substantiates the assumption of 1/2 photon per mode for the vacuum state. The spontaneous emission is distributed over 4π steradians and all frequencies; however, parametric amplification has gain only over limited angular directions and frequencies chosen by phase-matching conditions. As a result of the spontaneous emission, OPG typically has wide bandwidth (~100 GHz) emission for a ~1-ns pulse width. Seeding the parametric amplifier with a particular signal frequency mode of the nonlinear crystal will overwhelm the spontaneous contributions, and the emission is reduced to a narrow bandwidth ~1 GHz.

We discuss the spontaneous emission phenomenon in greater detail below when we treat the quantum amplifier model. To suppress spontaneous emission, the atom has to be isolated from the electromagnetic environment, which is possible by embedding it in a cavity or three-dimensional periodic dielectric structure. Of course, by controlling the coupling to electromagnetic vacuum modes, spontaneous emission can be enhanced (Purcell effect), as well.

Another prominent vacuum fluctuation effect is the Lamb shift, which is an increase in the $2S_{1/2}$ electronic level relative to the $2P_{1/2}$ level of the hydrogen atom by around 1 GHz. The two levels are otherwise degenerate, even according to the relativistic formulation using the Dirac equations. The agreement between accurate measurements and calculated values of the Lamb shift demonstrated that the vacuum is indeed a real effect that cannot be ignored. Other manifestations of the vacuum modes are discussed in the excellent and thorough book on the subject by Milonni (1994).

11.3.1 COHERENT STATE REPRESENTATION

An important and useful representation of the electromagnetic field is the coherent state, which is an eigenstate of the annihilation operator,

$$a \, | \alpha \rangle = \alpha \, | \alpha \rangle. \tag{11.48}$$

α is a complex number and for $\alpha = 0$ the coherent state is identical to the vacuum state. The coherent state is normalized,

$$\langle \alpha | \alpha \rangle = 1, \tag{11.49}$$

but different coherent states are not orthogonal, as we show below.

The coherent state has the property that it is a minimum uncertainty state. Consider two conjugate observables,

$$Q = \frac{1}{2}(a + a^{\dagger}), \quad P = \frac{1}{2i}(a^{\dagger} - a). \tag{11.50}$$

Their commutator is

$$[Q, P] = \frac{i}{2}. \tag{11.51}$$

The averages and variances are

$$\langle Q \rangle = \langle \alpha | Q | \alpha \rangle = \frac{1}{2}(\alpha + \alpha^{*}), \quad \langle P \rangle = \langle \alpha | P | \alpha \rangle = \frac{1}{2i}(\alpha^{*} - \alpha),$$

$$\langle Q^2 \rangle - \langle Q \rangle^2 = \frac{1}{4}, \quad \langle P^2 \rangle - \langle P \rangle^2 = \frac{1}{4}. \tag{11.52}$$

When these results are inserted into Equation 11.33, the equality is satisfied. This is a somewhat surprising result, since the coherent state is generated from vacuum by a classical applied force, and still it remains a minimum uncertainty state.

The coherent state can be represented as a sum over the number of states:

$$|\alpha\rangle = \sum_{n=0}^{\infty} f(n) | n \rangle. \tag{11.53}$$

Applying Equation 11.48, the amplitudes are related by the recursion relation:

$$f(n) = \alpha f(n-1)/\sqrt{n}. \tag{11.54}$$

The solution is expressed as

$$f(n) = \frac{\alpha^n}{\sqrt{n!}} f(0). \tag{11.55}$$

The coefficient $f(0)$ is determined by the normalization in Equation 11.55.

$$\langle \alpha | \alpha \rangle = \sum_{n=0}^{\infty} | f(n) |^2 = f^2(0) e^{|\alpha|^2} = 1. \tag{11.56}$$

The coherent state using the number state representation is expressed as

$$| \alpha \rangle = e^{-|\alpha|^2/2} \sum_{n=0}^{\infty} \frac{\alpha^n}{\sqrt{n!}} | n \rangle. \tag{11.57}$$

With this result, we can determine the inner product between two coherent states (α, β) as

$$\langle \beta | \alpha \rangle = e^{-|\beta|^2/2} e^{-|\alpha|^2/2} \sum_{n=0}^{\infty} \frac{(\beta^* \alpha)^n}{n!} = e^{-|\beta|^2/2 - |\alpha|^2/2 + \beta^* \alpha}. \tag{11.58}$$

The nonorthogonality of two coherent states is not surprising, since two states differing by only a small complex amplitude are nearly the same.

One of the most common applications we encounter is the transmission and reception of radio frequency radiation. The radiation is generated from an antenna that is described by a current distribution. Finally, the coherent state can be expressed in terms of the ground state and a creation operator power series,

$$| \alpha \rangle = e^{-|\alpha|^2/2} \sum_{n=0,\infty} \frac{1}{n!} (\alpha a^\dagger)^n | 0 \rangle = e^{-|\alpha|^2/2} \exp(\alpha a^\dagger) | 0 \rangle. \tag{11.59}$$

11.3.2 ELECTROMAGNETIC WAVE FUNCTION

The results of the previous section are applied to each plane-wave mode. The electromagnetic field is decomposed into an infinite number of modes. Earlier we decomposed the electromagnetic field into plane-wave modes with each mode having its own set of eigenstates. In the number representation, each state is denoted as $| n_{q,\vec{k}} \rangle$, where $\vec{k}$ is the wave vector of the state, q is its polarization, and $n_{q\vec{k}}$ is the occupation number of the state. A general state of the electromagnetic field is represented as a product over all polarizations and plane waves:

$$| \Psi \{ n_{q,\vec{k}} \} \rangle = \prod_{\{q,\vec{k}\}} | n_{q,\vec{k}} \rangle. \tag{11.60}$$

The values of the occupation number are restricted to nonnegative values, that is, $\{ n_{q,\vec{k}} = 0, 1, 2, \ldots \}$. A general state consists of a superposition of different occupation numbers for each state. Other representations of the photon state can be adopted, such as the coherent state. One special state deserves attention: the vacuum-state wave function of the electromagnetic field is represented as

$$|\Psi_0\rangle = \prod_{\{q\vec{k}\}} |0_{q,\vec{k}}\rangle. \tag{11.61}$$

It is a product of all the plane-wave modes with zero photons in every state. Specific wave functions of the electromagnetic field are generated from the vacuum state by the application of creation operators, as discussed above for the coherent state.

11.3.3 ELECTROMAGNETIC SIGNALS

The field operator is separated into positive and negative frequency components, respectively,

$$\underline{\vec{E}}(\vec{r},t) = \frac{1}{2}\left(\vec{\tilde{E}}^{+}(\vec{r},t) + \vec{\tilde{E}}^{-}(\vec{r},t)\right). \tag{11.62}$$

The positive-frequency field operator contains the annihilation operators, thus reducing the number of photons in each mode, and the negative frequency field operator with creation operators increases the photon number. This fact will be applied in constructing nonlinear equations of motion for nonlinear processes with conserved quantities based on Manley–Rowe relations. Using Equation 11.6, the positive electric field operator is a sum over annihilation operators,

$$\vec{\tilde{E}}^{+}(\vec{r},t) = \sum_{\vec{k}}\sum_{q=1,2}\mathcal{E}_{\vec{k}}\hat{e}_{q,\vec{k}}a_{q,\vec{k}}e^{i(\vec{k}\cdot\vec{r}-\omega_{\vec{k}}t)}, \tag{11.63}$$

and the negative one has creation operators,

$$\vec{\tilde{E}}^{-}(\vec{r},t) = \sum_{\vec{k}}\sum_{q=1,2}\mathcal{E}_{\vec{k}}\hat{e}_{q,\vec{k}}^{*}a_{q,\vec{k}}^{\dagger}e^{-i(\vec{k}\cdot\vec{r}-\omega_{\vec{k}}t)}. \tag{11.64}$$

Before leaving this section, we use the SVEA for a beam propagating along the z-axis. The electric field operator is decomposed into an envelope operator and a rapid spatial and temporal sinusoidal function:

$$\tilde{E}_i^{+}(\vec{r},t) = \frac{1}{2}A_i^{+}(\vec{r},t)e^{i(Kz-\omega t)}. \tag{11.65}$$

The electric field envelope operator, $A_i^+(\vec{r}, t)$, is polarized orthogonal to the wave vector. Since the electric field operator satisfies the wave equation, we invoke the same calculus for the SVEA equation. It has a form identical to its classical counterpart,

$$\frac{\partial A_i^+}{\partial z} + \frac{1}{v}\frac{\partial A_i^+}{\partial t} + \frac{i}{2K}\nabla_T^2 A_i^+ = i\frac{\omega^2}{2K}P_i^{\mathrm{NL}+}. \tag{11.66}$$

The operator $P_i^{\mathrm{NL}+}$ is a defined nonlinear polarization positive frequency contribution, and the group velocity, v, in a gaseous medium is to a good approximation the same as the phase velocity. In dense media, we use the group velocity of the medium.

The extension of the foregoing quantization procedure for electrodynamics in vacuum to dielectric media leads to inconsistencies in the equations of motion. A consistent quantization procedure suitably applied to nonlinear optics may be developed starting with emphasis on the displacement field, since as a plane wave it is orthogonal to the wave vector. That would require further technical details that are not warranted for this introductory presentation. The interested reader can refer to books and papers in the references for the finer details. We shall proceed by applying the quantum properties of electromagnetism, as presented here, to the aspects of nonlinear optics.

11.4 QUANTUM AMPLIFIERS AND ATTENUATORS

Our attention is devoted to nonlinear optics, and we now turn our attention to specific quantum manifestations of nonlinear optics. In Chapter 6, the concept of spontaneous parametric down-conversion (SPDC) was discussed after we heuristically introduced the concept of vacuum fluctuations. In this section, we demonstrate the effects of the bath reservoir vacuum modes on a central system using a quantum-based model approach.

11.4.1 QUANTUM ATTENUATOR MODEL

Losses cannot be introduced into the quantum description by simply inserting a damping contribution in the equations of motion without incurring unphysical consequences. The quantum dynamics and the operator evolution are based on an energy-conserving model, and so the model with damping should preserve energy-conserving characteristics. In this section, we will introduce a simple coupled harmonic oscillator model that has a damping effect on an oscillator while still preserving the conserved dynamics of the system. The illustration of the coupled harmonic oscillator model is shown in Figure 11.2. In the model, a central oscillator with mass m and resonant angular frequency ω_0 is coupled to its environment, which is a bath of oscillators (coupling coefficients $\{\kappa_1, \kappa_2, \kappa_3, \ldots\}$). The bath or reservoir model

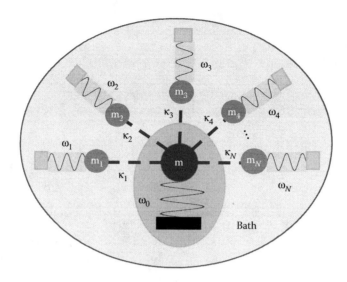

FIGURE 11.2 Illustration of the quantum model for damping. A central oscillator with resonant angular frequency ω_0 is coupled to a bath of harmonic oscillators; each oscillator in the bath is characterized by a different resonant angular frequency.

is composed of many oscillators, each with different masses and resonant angular frequencies. When the environmental coupling is removed, we have only a single harmonic oscillator without damping.

Adding the coupling to the bath of oscillators allows energy to be exchanged between the central oscillator and the bath. While Poincare's classical recurrence theorem states that an initial state will approximately recur, the time required for the recurrence may be orders of magnitude longer than the age of the universe, and so in practical terms the recurrence can be ignored. In other words, the energy being distributed among many oscillators with different oscillation frequencies is coupled back to the central oscillator at different times and with different phases so that with high probability its initial state does not recur. The damping phenomenon is a result of many different oscillators exchanging energy with the central oscillator. Eventually, the bath soaks up the central oscillator energy, leaving it in the ground state.

Based on a model with linear coupling between oscillators (see the illustration in Figure 11.2), the equations of motion for all the harmonic oscillators written with quantum mechanical creation and annihilation operators are

$$\frac{\mathrm{d}a}{\mathrm{d}t} = -\mathrm{i}\omega_0 a - \mathrm{i}\sum_{n=1}^{N} \kappa_n b_n, \tag{11.67}$$

$$\frac{\mathrm{d}b_n}{\mathrm{d}t} = -\mathrm{i}\omega_n b_n - \mathrm{i}\kappa_n a. \tag{11.68}$$

The equations are linear and can be at least formally solved. The second equation for the bath annihilation operators is solved in integral form

$$b_n(t) = b_n(0)e^{-i\omega_n t} - i\kappa_n \int_0^t a(t')e^{-i\omega_n(t-t')}dt'.$$

(11.69)

Substitute this solution for $b_n(t)$ into Equation 11.67,

$$\frac{da(t)}{dt} = -i\omega_0 a(t) - \sum_{n=1}^N \kappa_n^2 \int_0^t e^{-i\omega_0(t-t')}a(t')e^{i(\omega_0-\omega_n)(t-t')}dt' - \sum_{n=1}^N i\kappa_n b_n(0)e^{-i\omega_n t}.$$

(11.70)

The second term is evaluated by extending the sum over oscillators to an integral; in the limit $N \to \infty$, the coupling coefficient is a continuum density, that is, $\kappa_n^2 \to d\omega\rho(\omega)\kappa^2(\omega)$, where $\rho(\omega)$ is the density of the bath oscillator modes. Both $\rho(\omega)$ and $\kappa(\omega)$ are assumed to be smooth and slowly varying function on the scale of the characteristic angular frequency ω_0:

$$\sum_{n=1}^N \kappa_n^2 e^{i(\omega_0-\omega_n)(t-t')} = \int_{-\infty}^\infty d\omega\rho(\omega)\kappa^2(\omega)e^{i(\omega_0-\omega)(t-t')}$$

$$= \rho(\omega_0)\kappa^2(\omega_0)\int_{-\infty}^\infty d\omega e^{i(\omega_0-\omega)(t-t')} = 2\pi\rho(\omega_0)\kappa^2(\omega_0)\delta(t-t').$$

(11.71)

Equation 11.71 is a simplified approximation that neglects a small frequency offset of the oscillator from its isolated resonant angular frequency; the frequency perturbation is called the Bloch–Siegert shift. Using the limit in Equation 11.71, the equation of motion for the central harmonic oscillator (Equation 11.70) is rewritten as

$$\frac{da(t)}{dt} = -i\omega_0 a(t) - \gamma a(t) + F(t).$$

(11.72)

The damping coefficient is related to the coupling coefficient, $\kappa(\omega)$, by

$$\gamma = \pi\rho(\omega_0)\kappa^2(\omega_0).$$

(11.73)

The last term acts as a force contribution; we dub the quantum Langevin force. It is a function of the bath oscillator operators, defined as

$$F(t) = -\sum_{n=1}^N i\kappa_n b_n(0)e^{-i\omega_n t}.$$

(11.74)

The quantum Langevin force term has the commutator

$$[F(t), F^{\dagger}(t')] = \sum_{n=1}^{N} \sum_{n'=1}^{N} \kappa_n \kappa_{n'} [b_n(0), b_{n'}^{\dagger}(0)] e^{-i\omega_n t} e^{i\omega_n t'}. \tag{11.75}$$

It only depends on the initial bath oscillator operators, which obey the commutator relations $[b_n(0), b_{n'}^{\dagger}(0)] = \delta_{n,n'}$. Taking the infinite bath oscillator limit $(N \rightarrow \infty)$ and the same approximations leading to the result in Equation 11.71, the Langevin force commutator is

$$[F(t), F^{\dagger}(t')] = 2\gamma \delta(t - t'). \tag{11.76}$$

The last term is important for consistency of the quantum theory, and its neglect leads to a violation of commutator relations. To show this last statement, the solution of Equation 11.72 is

$$a(t) = e^{-(i\omega_0 + \gamma)t} a(0) + \int_0^t dt' e^{-(i\omega_0 + \gamma)(t-t')} F(t'). \tag{11.77}$$

The commutator of the central oscillator creation and annihilation operators using Equation 11.77 is

$$[a(t), a^{\dagger}(t)] = e^{-2\gamma t} [a(0), a^{\dagger}(0)] + \int_0^t dt'' \int_0^t dt' e^{-(i\omega_0 + \gamma)(t-t')} e^{-(-i\omega_0 + \gamma)(t-t'')} [F(t'), F^{\dagger}(t'')]. \tag{11.78}$$

The first term alone would yield a commutator at time t that exponentially decays to zero, which would violate the Heisenberg uncertainty relations discussed earlier. Evaluating the last term using the commutator identity in Equation 11.75 leads to the invariance of the commutator in time, that is, $[a(t), a^{\dagger}(t)] = 1$.

Given an initial coherent state $|\alpha\rangle$ for the central oscillator, the attenuator decreases the amplitude of the field exponentially:

$$\langle a(t) \rangle = e^{-(i\omega_0 + \gamma)t} \alpha. \tag{11.79}$$

The noise impressed on the signal due to the attenuation process can be calculated by determining the variance of the operators: $O_+ = (1/2)(a + a^{\dagger})$ and $O_- = (1/2i)(a^{\dagger} - a)$.
The average value of O_+ is simply

$$\langle O_+ \rangle = \frac{1}{2} \langle (a + a^{\dagger}) \rangle = \frac{1}{2} (e^{-i\omega_0 t} \alpha + e^{i\omega_0 t} \alpha^*) e^{-\gamma t}. \tag{11.80}$$

Calculating the second moment is a bit more challenging:

$$\langle O_+^2 \rangle = \frac{1}{4}\langle (e^{-i\omega_0 t}a(0)+e^{i\omega_0 t}a^\dagger(0))^2 \rangle e^{-2\gamma t} + \frac{1}{4}\left\langle \left(\int_0^t dt' e^{-(i\omega_0+\gamma)(t-t')}F(t-t')+\text{h.c.}\right)^2 \right\rangle. \quad (11.81)$$

The operator average in the first term in brackets is found by ordering the operators so that the annihilation operators are to the right of the creation operators:

$$\langle (e^{-i\omega_0 t}a(0)+e^{i\omega_0 t}a^\dagger(0))^2 \rangle = \frac{1}{4}\langle (e^{-2i\omega_0 t}a^2(0)+e^{2i\omega_0 t}a^{\dagger 2}(0)+a(0)a^\dagger(0)+a^\dagger(0)a(0)) \rangle$$

$$= \frac{1}{4}(e^{-i\omega_0 t}\alpha+e^{i\omega_0 t}\alpha^*)^2 + \frac{1}{4}. \quad (11.82)$$

The second term with the reservoir operators is reduced to the form,

$$\frac{1}{4}\left\langle \left(\int_0^t dt' e^{-(i\omega_0+\gamma)(t-t')}F(t-t')+\text{h.c.}\right)^2 \right\rangle = \frac{1}{4}\int_0^t dt'\int_0^t dt'' e^{(i\omega_0-\gamma)(t-t')}e^{-(i\omega_0+\gamma)(t-t'')}\langle F^\dagger(t-t')F^\dagger(t-t'')\rangle.$$

$$(11.83)$$

For a reservoir in the vacuum state, $\langle F^\dagger(t-t')F(t-t'')\rangle = 0$. Using this result and average of Equation 11.76, we deduce

$$\langle F(t-t')F^\dagger(t-t'')\rangle = 2\gamma\delta(t'-t''). \quad (11.84)$$

The integrals in Equation 11.83 are evaluated as

$$\frac{1}{4}\left\langle \left(\int_0^t dt' e^{-(i\omega_0+\gamma)(t-t')}F(t-t')+\text{h.c.}\right)^2 \right\rangle = \frac{1}{4}\left(1-e^{-2\gamma t}\right). \quad (11.85)$$

Finally, the operator variance is found to be

$$\sigma_+^2 = \langle O_+^2 \rangle - \langle O_+ \rangle^2 = \frac{1}{4}. \quad (11.86)$$

The same calculation repeated for the O_- variance σ_-^2, which yields the same result. In other words, the attenuator model does not add noise to the measurement of these observables. An oscillator that starts in a minimum uncertainty state remains in that state.

11.4.2 QUANTUM AMPLIFIER MODEL

An amplifier model, analogous to the damping model of the previous section, has elements of the harmonic oscillator model for damping presented in the previous section, but with one significant difference. In the amplification model, the mass and restoring force coefficient of the central oscillator are negative. The parameter inversion has the effect of inverting the energy ladder of the quantum harmonic oscillator. The "ground" state is now the highest energy state in the system, and the excited states have lower energy, as illustrated in Figure 11.3. This model mimics an amplifier without a saturation effect, since there is no lower bound to the energy. The bath of oscillators, on the other hand, remains the same with their coupling coefficients and normal harmonic oscillator parameters.

The Heisenberg equations of motion for the amplifier model are

$$\frac{da}{dt} = i\omega_0 a - i \sum_{n=1}^{N} \kappa_n b_n^\dagger, \tag{11.87}$$

$$\frac{db_n^\dagger}{dt} = i\omega_n b_n^\dagger + i\kappa_n a. \tag{11.88}$$

The sign of the angular frequency term in Equation 11.87 is an important distinction for the model. By following the same solution steps as in the previous section, the reduced dynamical equation of motion is

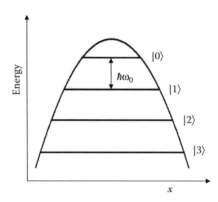

FIGURE 11.3 Amplifier is characterized by an inverted parabolic potential energy that is negative for all values of x. The discrete quantum energy level positions are labeled by kets $\{|n\rangle\}$ with the ground state now being the highest quantum state, and it is shifted from the top of the parabola by $-(1/2)\hbar\omega_0$.

$$\frac{da}{dt} = i\omega_0 a + \gamma a + F(t). \tag{11.89}$$

The expressions for the amplification coefficient, γ, remains unchanged, and results from the attenuator model can be adopted. The quantum Langevin force for the amplifier model is

$$F(t) = -\sum_{n=1}^{N} i\kappa_n b_n^\dagger(0)e^{i\omega_n t}. \tag{11.90}$$

The Langevin force in the amplifier model contains the reservoir creation operators. This has important consequences for measurable quantities as we shall see.

The solution of Equation 11.89 is

$$a(t) = e^{(i\omega_0 + \gamma)t}a(0) + \int_0^t dt' e^{(i\omega_0 + \gamma)(t-t')}F(t'). \tag{11.91}$$

Now we can use the operators O_+ and O_- defined above to calculate the variances. The procedure is the same as followed above, but now the average of the quantum Langevin force is

$$\langle F^\dagger(t-t')F(t-t'')\rangle = 2\gamma\delta(t'-t''). \tag{11.92}$$

As a consequence, the variances exponentially grow,

$$\sigma_+^2 = \sigma_-^2 = \frac{1}{4}e^{2\gamma t}. \tag{11.93}$$

From the growth of the variances, we conclude that the process of amplification always adds noise to the measured observables. Quantum size fluctuations can be magnified to macroscopic levels in high gain systems. Another consequence of this result is that photons cannot be cloned without adding additional uncertainty to the measurement.

11.4.3 QUANTUM INITIATION: OPTICAL PARAMETRIC GENERATOR

The quantum noise initiation as a spontaneous parametric scattering process for OPG was treated in Chapter 6. It differs from OPA because no seed signal/idler field is injected at the input. As a consequence, the emitted bandwidth extends over the gain bandwidth of the parametric amplifier. In the slowly varying envelope approximation, the operator field form of the wave equation is identical to its classical form.

The DFG plane-wave solutions for the nondepleted pump case treated in Chapter 6.2 are generalized for signal and idler input fields and phase mismatch conditions:

$$A_S(z) = A_S(0)\left(-i\frac{\Delta k}{g}\sinh(gz) + \cosh(gz)\right)e^{i\Delta kz/2} + A_I^\dagger(0)i\frac{\kappa_S}{g}\sinh(gz)e^{i\Delta kz/2}, \tag{11.94}$$

$$A_I(z) = A_I(0)\left(-i\frac{\Delta k}{g}\sinh(gz) + \cosh(gz)\right)e^{i\Delta kz/2} + A_S^\dagger(0)i\frac{\kappa_I}{g}\sinh(gz)e^{i\Delta kz/2}. \tag{11.95}$$

The initial conditions are operators, and the parameters are

$$\kappa_{S/I} = \frac{\omega_{S/I}d_{\text{eff}}}{n_{S/I}c}A_P, \quad \Gamma^2 \equiv \kappa_S\kappa_I^* = \frac{2\omega_S\omega_I d_{\text{eff}}^2}{n_P n_S n_I \varepsilon_0 c^3}I_P, \quad g = \sqrt{\Gamma^2 - \left(\frac{\Delta k}{2}\right)^2}. \tag{11.96}$$

Now, spontaneous emission noise is incorporated into the fields. We introduce spontaneous emission into the equations by using the addition of vacuum fluctuations and energy $\hbar\omega/2$ per mode per volume of the crystal interaction region. For a dielectric medium of index n, we replace ε_0 by $n^2\varepsilon_0$ (ignoring birefringence):

$$\left\langle A_I(0)A_I^\dagger(0)\right\rangle = \frac{\hbar\omega_I}{2n^2\varepsilon_0 V}, \quad \left\langle A_S(0)A_S^\dagger(0)\right\rangle = \frac{\hbar\omega_S}{2n^2\varepsilon_0 V}. \tag{11.97}$$

The parameter V is the volume of the beam in the crystal $V = AL$, where L is the crystal length, A is the area of the pump beam, and the refractive index of the medium n was incorporated in these expressions to correspond to the energy of the dielectric medium. The strength of the vacuum fluctuations is proportional to the angular frequency of the field. The solutions for the average intensities of the signal and idler fields are

$$I_S = \frac{n\varepsilon_0 c}{2}\left\langle A_S^\dagger(L)A_S(L)\right\rangle = \frac{c\hbar\omega_I}{2nV}\left(\frac{\kappa_I}{g}\right)^2\sinh^2(gL), \tag{11.98}$$

$$I_I = \frac{n\varepsilon_0 c}{2}\left\langle A_I^\dagger(L)A_I(L)\right\rangle = \frac{c\hbar\omega_S}{2nV}\left(\frac{\kappa_S}{g}\right)^2\sinh^2(gL). \tag{11.99}$$

The bandwidth of the gain is determined by the dispersion of the medium; when the phase mismatch Δk exceeds 2Γ, the fields cease to be amplified. The coefficient of the sech squared function is the noise intensity. For a given beam geometry of cross-sectional area A, multiply both sides by A. The noise power coefficient is calculated with dependence on the crystal length. For phase matching at the degeneracy point, that is, $\omega_s = \omega_i$, the bandwidth is broader due to the parabolic shape of the phase-matching curve, which flattens the wave vector mismatch locally to a quadratic function.

11.4.4 SCHRÖDINGER'S CAT STATES

In response to the EPR paper, Schrödinger devised the famous cat gedanken experiment to highlight issues with the transition from the microscopic world, where quantum superposition of states is manifest, to the macroscopic world, where quantum superposition states are not observed. Cats or any macroscopic objects are not observed in a superposition amplitude state of being simultaneously both "dead" and "alive." It is one or the other outcome. Quantum superposition is a fundamental aspect of the Copenhagen interpretation of quantum mechanics; in other words, the particles do not carry definite properties, and a measurement causes a collapse of the wave function to a definite state. In his argument, Schrodinger envisioned a cat enclosed inside a box containing a radioactive material. The radioactive decay of a nucleus triggers a device, which releases a poisonous gas that kills the cat. Nuclear decay is a microscopic process whose state can be described by a superposition of two states, $|1\rangle$ for the nondecayed nuclear state and $|2\rangle$ for the decayed nuclear state:

$$|\Psi\rangle = c_1 |1\rangle + c_2 |2\rangle. \tag{11.100}$$

On the other hand, the cat represents a macroscopic state. A single decay event cannot itself trigger a poison gas release event without amplification through an amplifier device. In a very lucid article, Glauber (1986) used the amplifier model to grapple with problems of macroscopic quantum states (among other topics) in quantum mechanics.

From the presentation in the previous sections, it is now apparent that an amplifier does not merely magnify the signal amplitude, and it also adds noise during the process. In other words, the macroscopic state includes the stochastic effects of noise, which cannot be ignored in a consistent amplifier theory. In the end, quantum amplifier theory makes statements only about an ensemble average of an observable after many trials of an experiment with identical initial states. Overall, the quantum superposition, as manifest for instance by interference fringes, is smeared by the noise that triggers events caused by vacuum fluctuations.

11.5   QUANTUM DETECTION

11.5.1 DIRECT DETECTION

Experimental tests of the foundations of quantum mechanics require careful analysis of the measurement process. Sensitive measurements of a few photons are

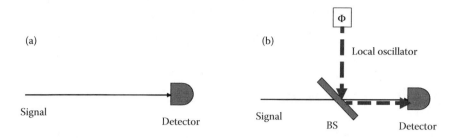

FIGURE 11.4 Signal detection methods: (a) direct detection and (b) coherent detection. BS is a beam splitter.

made possible with modern semiconductor photodetectors. The simplest method of detection is called *direct detection*, which is illustrated in Figure 11.4a. In direct detection, the light intensity is measured and information about the phase of the field is lost. As with any measurement process, there are limitations due to noise. In photodetection, the dominant noise sources include shot noise and thermal noise, which limit the detectability of a feeble quantum signal. The signal current generated from the incident light is expressed as

$$I_S(t) = RP(t).$$ (11.101)

where R is the responsivity of the photodetector and P is the incident power on the photodetection surface. The responsivity is expressed as $R = e\eta/\hbar\omega$, where η is the photodetector efficiency, and e is the elementary electron charge. The optical power may be a pulse of energy leading to a transient photodetected signal current. Noise accompanies the photodetection process and degrades the signal. We introduce two sources of noise called shot noise and thermal noise in this section. Shot noise is due to the random movement of electrons in the circuit. The shot-noise current variance, in the absence of electronic gain in the photodetector, is proportional to the signal current,

$$\sigma_{sn}^2 = 2eI_{sdc}B,$$ (11.102)

where I_{sdc} is the averaged direct current flow through the detector, B is the bandwidth of the measurement, and e is the charge of the electron again. The thermal-noise current variance due to the random scattering of electrons in the detection circuitry is

$$\sigma_{tn}^2 = 4k_BTB/R_L,$$ (11.103)

where k_B is Boltzmann's constant, T is the temperature, and R_L is the load resistance. Thermal noise is also called Johnson or Nyquist noise in the literature. Both sources of noise are independent of one another. The signal to noise ratio (SNR) is defined as

$$S/N = \frac{(RP)^2}{\sigma_{sn}^2 + \sigma_{tn}^2}.$$

(11.104)

Direct detection has been widely adopted in optical communications systems. The advanced development of detectors with low noise and high speed has enabled systems with high data rates. Detectors with gain, called avalanche photodiodes, may provide additional advantages to improve the SNR. A SNR of unity, being an ultimate limit of detectability, is superseded by the practical requirement in communication systems that bit error rates be as small as 1 in 10^9 or 10^{12} bits sent, which demands a larger SNR value.

11.5.2 COHERENT DETECTION

Coherent detection is an alternate avenue to signal detection, which measures both the signal's phase and amplitude. In coherent detection, a typically weak signal is mixed with a strong local oscillator (LO) whose phase can be externally controlled. As illustrated in Figure 11.4b, a beam splitter is used to combine the signal and LO in the detector. The LO phase can be controlled to determine both amplitude and phase of the signal. The properties of the beam splitter need special consideration and are discussed in the following section.

A quantum measurement of one specific observable has an accuracy that is not limited by Heisenberg's uncertainty relation. Of course, the accuracy of one observable does affect the uncertainty of a second (noncommuting) observable determined by a simultaneous measurement. In principle, a single measurement can be noise free. In practice, of course, measurement accuracy is limited by the efficiency of the photodetection system, as discussed above. In the next section, we shall treat a beam splitter as an amplitude divider. Subsequently, we shall see how it is used in homodyne detection to improve measurement accuracy.

11.5.2.1 BEAM SPLITTER

A beam splitter is a simple device that can split and combine the flux of photons from different input ports. Here, we assume that the polarizations are identical and the overlap of the signal and LO on the photodetector surface is coherent. It is a four-port device with two input port amplitudes (a_1, a_2) and two output port amplitudes (a_3, a_4) when viewed with inputs at identical angles to one another displaced from the mirror surface; in Figure 11.5, the angles are 45° with respect to the mirror surface or 90° with respect to each other. The port's function can be inverted so that output ports become input ports and vice versa. The beam splitter satisfies a reciprocity relation, which means that if the direction of the physical process is inverted, the results follow

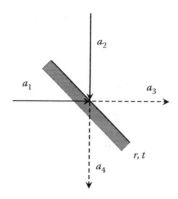

FIGURE 11.5 Beam splitter as a four-port device. The amplitude splitting ratios are r and t for reflection and transmission, respectively.

the splitting as if they were time reversed. In other words, for identically polarized coherent waves, we have the linear relation,

$$\begin{bmatrix} a_3 \\ a_4 \end{bmatrix} = M \begin{bmatrix} a_1 \\ a_2 \end{bmatrix} = \begin{pmatrix} A & B \\ C & D \end{pmatrix} \begin{bmatrix} a_1 \\ a_2 \end{bmatrix}, \tag{11.105}$$

where M is a 2×2 matrix. For an input in Port 1, the output flux in Port 3 is $T = |t|^2$ and the output flux in Port 4 is $R = |r|^2$. Similarly, we can determine the output port fluxes when an input is placed only on Port 2. It is tempting to use real numbers for the coefficients of M, but as we shall see, that would violate reciprocity. For a lossless mirror, conservation of power demands the relation,

$$|a_3|^2 + |a_4|^2 = |a_1|^2 + |a_2|^2 = 1. \tag{11.106}$$

The equality is equivalent to $M^{\mathrm{T}}M = I$, where M^{T} is the transpose of M and I is the unit matrix. This imposes a condition that the mirror is perfect, that is, no light flux is lost. The following results are found for the coefficient M. The diagonal elements satisfy

$$|A|^2 + |C|^2 = 1 \quad \text{and} \quad |B|^2 + |D|^2 = 1, \tag{11.107}$$

and the off-diagonal elements satisfy

$$AB^* + CD^* = 0 \quad \text{and} \quad A^*B + C^*D = 0. \tag{11.108}$$

The amplitudes can be consistently solved by using $A = D = t$ and $C = B = ir$, where it is assumed that t and r are real and $r = \sqrt{1 - t^2}$. This satisfies the above equations imposed by reciprocity. Reciprocity is preserved when the phase of the reflected amplitude is shifted by $\pi/2$. It is left as an exercise for the reader to demonstrate that reciprocity is obeyed.

Classically, we treat the input using a single port without regard to the second-input port. However, quantum mechanically even when the second-input port has no signal, it introduces vacuum fluctuations that affect the measurements by adding noise.

11.5.2.2 COHERENT DETECTION SIGNAL TO NOISE

There are two available output ports in Figure 11.5. For definiteness, let us choose Port 3, which has a field operator,

$$a_3 = \left(ta_1 e^{-i\omega t} + ira_{LO} e^{-i\omega_{LO}t}\right). \tag{11.109}$$

The LO angular frequency ω_{LO} may be different from the signal's angular frequency ω. Consider the case where both signal and LO states are coherent states, that is, the wave function expressed in terms of the input states is $|\Psi\rangle = |\alpha_1\rangle|\alpha_{LO}\rangle$. The photodetector current is proportional to the light intensity

$$i_{pd}(t) = e\eta r_{ph}\langle\Psi|a_3^\dagger a_3|\Psi\rangle = e\eta r_{ph}\left(|t|^2|\alpha_1|^2 + |r|^2|\alpha_{LO}|^2 - 2\operatorname{Re}\{irt\alpha_1\alpha_{LO}^* e^{-i(\omega-\omega_{LO})t}\}\right), \tag{11.110}$$

where r_{ph} is the average photon arrival rate. The common situation is a strong LO, that is, $|\alpha_{LO}| \gg |\alpha_1|$, and the measurements are shot-noise dominated. The dominant contribution is the second term in Equation 11.110, while the first term is negligible. The last term in parentheses contains a linear product of the LO and the input signal. The case $\omega = \omega_{LO}$ defines the process called homodyne detection. The signal current contribution is on top of a large LO-driven dc current. The heterodyne detection technique applies when $\omega \neq \omega_{LO}$, the offset angular frequency is small, that is, $|\omega - \omega_{LO}| \ll \omega, \omega_{LO}$. Picking out the offset frequency from the signal separates it from the LO dc background current.

In the next section, we introduce the concept of balanced homodyne detection to eliminate the dominant LO contribution. The signal amplitude and phase are extracted from the data by changing the LO phase.

11.5.2.3 BALANCED HOMODYNE DETECTION

The problem of homodyne detection is that the LO dominates the dc current. The fields and their uncertainties vary at optical frequencies and cannot be simply measured by a photodetector. The measurement problem is resolved by mixing the electromagnetic wave with an LO and using a *balanced* homodyne detection scheme, which is illustrated in Figure 11.6. This is similar to the method discussed in Section 11.5.2.2, but with an additional photodetector introduced in the second output port. The light

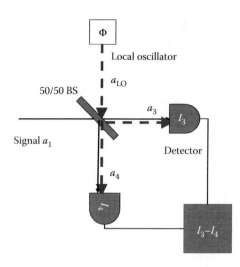

FIGURE 11.6   Balanced homodyne detection scheme.

from both input ports is equally split (50/50 beam splitter) into two output ports. The output current from the two photodetectors is subtracted using correlation electronics. The LO dc current component in Equation 11.110 is cancelled by this method, and the signal contribution is the leading term producing a current due to the reciprocal properties of the beam splitter.

The mirror output field operators are expressed as a linear superposition of the two input port operators:

$$a_3 = \frac{1}{\sqrt{2}}\left(a_1 e^{-i\omega t} + ia_{LO}\right),$$

$$a_4 = \frac{1}{\sqrt{2}}\left(ia_1 e^{-i\omega t} + a_{LO}\right). \tag{11.111}$$

The photodetector current is proportional to the input intensity, which is proportional to the photon number:

$$I_3 = \langle a_3^\dagger a_3 \rangle, \quad I_4 = \langle a_4^\dagger a_4 \rangle. \tag{11.112}$$

The difference between the two currents in the presence of a strong LO is

$$I_3 - I_4 = -i\langle a_{LO}^\dagger a_1 e^{-i\omega t} - a_1^\dagger e^{i\omega t} a_{LO} \rangle. \tag{11.113}$$

The LO in a coherent state, $|\alpha_{LO}e^{-i\omega t}\rangle$, is represented by a complex number of the form $\alpha_{LO} = |\alpha_{LO}|e^{i\Phi}$. The phase Φ may be controlled by displacing the LO path length over a predefined fraction of a wavelength. The measured current is expressed as

$$I_3 - I_4 = |\alpha_{LO}|\langle ia_1^\dagger e^{i\Phi} - ia_1 e^{-i\Phi} \rangle. \tag{11.114}$$

By changing the LO phase, the observable quadrature is changed. For $\Phi = 0$, the balanced homodyne method current is proportional to $\langle O_2 \rangle$, and for $\Phi = \pi/2$, the current is proportional to $\langle O_1 \rangle$. We define the operator $O_{34} = i a_1^\dagger e^{i\Phi} - i a_1 e^{-i\Phi}$. The variance of the output current is proportional to

$$\sigma_{34}^2 = |\alpha_{LO}|^2 \left(\langle (O_{34})^2 \rangle - \langle O_{34} \rangle^2 \right). \tag{11.115}$$

By changing LO phase to measure $\langle O_1 \rangle$ or $\langle O_2 \rangle$, the measured variance corresponds to the variance of the measured observable.

11.6 QUANTUM SQUEEZED LIGHT

11.6.1 SINGLE-MODE SQUEEZED STATES

Quantum mechanical squeezed states are prepared so that an observable has an uncertainty that is below a desired limit, even the so-called standard quantum limit defined by the ground state of the system. This result does not violate the Heisenberg relations because the measurement of one physical observable could have an unlimited accuracy at the expense of adding uncertainty of a second measurement on conjugate observable. Generating squeezed states is conceived as a pathway to beat the quantum-limited uncertainty for measurements on a system. High precision metrology measurements made on the physical quantity in a squeezed state would improve measurement accuracy below the quantum limit.

One challenging application is the detection of gravitational waves. The system requirements are highly demanding since it must be designed to detect the movement of masses less than the size of an atomic nucleus. Recently, the first direct measurements of gravitational waves were reported from the advanced LIGO scientific collaboration, which did not incorporate squeezed light into the system design. However, in future system upgrades, a squeezed-state scheme could be incorporated into the LIGO system. Research results have been reported with squeezed states showing modest noise reduction characteristics; however, even small improvements in measurement could be beneficial. Incorporating squeezed states into gravitational wave detection would be the fruition of more than three decades of research on squeezed states applied to high precision metrology.

Our goal here is simply to analyze how nonlinear optical interactions may be used to generate squeezed states. Squeezed light production is realized by different nonlinear optical phenomena. We first show that squeezed states are a consequence of degenerate parametric down-conversion processes. Squeezed states are also produced using four-wave mixing, a case we will examine later. Using parametric down-conversion, two cases are discussed: single-mode squeezed states and two-mode squeezed states. Two-mode squeezed states also have a connection to EPR tests of quantum mechanics because the same process can produce entangled photon states.

To exemplify the production of squeezed states of the electromagnetic field, we quantize the equations of three-wave processes derived in Chapter 4 (Equations 4.25 through 4.27). The simplest analysis of squeezed-state generation uses the equations for degenerate parametric down-conversion with plane-wave fields. The signal and idler fields have the same frequency, and depending on the type of phase matching, the photon pair may be indistinguishable, possess orthogonal polarizations, or have different propagation directions. Single-mode squeezed states can be generated by second-order and third-order nonlinear processes. We will use Type I three-wave phase-matching conditions for our discussions in this section. More details of the Type I noncollinear phase-matching scheme are found in Section 11.7.1. Potential nonlinear materials include quasi-phase-matched structures such as PPLN or OPGaAs, where the operational wavelengths can be designed. Degenerate noncollinear phase matching in nonlinear crystals, for instance BBO, can be adopted to produce degenerate waves. One weakness of degenerate down-conversion is the broad bandwidth created by noncritical phase matching.

When the generated fields are weak, the nondepleted pump approximation is valid. Since the equations for the signal and idler waves are linear, the quantized version of those equations is identical to the classical case. To transform the classical fields to their quantized counterparts, we use the following identification: the complex, positive-frequency field amplitude is replaced by the annihilation operator, and its complex conjugate with the creation operator, that is, $(\alpha, \alpha^*) \rightarrow (a, a^\dagger)$. The corresponding Heisenberg equations for degenerate down-conversion using the nondepleted pump field are

$$
\frac{da}{dz} = i\kappa a^\dagger,
$$
$$
\frac{da^\dagger}{dz} = -i\kappa^* a,
$$

(11.116)

where $\kappa = (\omega_2/n_2 c)d_{\text{eff}} A_1$ and $\omega_2 = \omega_3 = \omega_1/2$. The pump field is a complex number. The equations are identical in form to the classical equations introduced in Chapter 4. The solutions of linear equations are identical in both classical and quantum domains, and we write the solutions for the creation and annihilation operators as

$$
a(L) = \cosh(|\kappa|L)a(0) + i\frac{\kappa}{|\kappa|}\sinh(|\kappa|L)a^\dagger(0),
$$
$$
a^\dagger(L) = \cosh(|\kappa|L)a^\dagger(0) - i\frac{\kappa^*}{|\kappa|}\sinh(|\kappa|L)a(0).
$$

(11.117)

The corresponding electric field operator for a single, plane-wave mode, after Equation 11.6, is

$$
E = \frac{1}{2}\mathcal{E}_{\vec{k}}\left(-ia^\dagger(L)e^{-i(\vec{k}\cdot\vec{r}-\omega t)} + ia(L)e^{i(\vec{k}\cdot\vec{r}-\omega t)}\right).
$$

(11.118)

It should be noted that the operators always satisfy the commutation relations for a fixed value of L. Two conjugate observables are linear combinations of the two operators:

$$O_1 = \frac{1}{2}\left(a(L) + a^\dagger(L)\right),$$
$$O_2 = \frac{1}{2i}\left(a^\dagger(L) - a(L)\right). \tag{11.119}$$

Using these definitions, the electric field operator is

$$E = \mathcal{E}_{\vec{k}}\left(-O_1 \sin(\vec{k} \cdot \vec{r} - \omega t) + O_2 \cos(\vec{k} \cdot \vec{r} - \omega t)\right). \tag{11.120}$$

The observable operator variances can be measured by the balanced homodyne detection technique presented in Section 11.5.2.3. The two operators are in-phase, $\cos(\omega t)$, and in-quadrature, $\sin(\omega t)$, contributions; in other words, their phases differ by a quarter period. The phase shift can be used to extract independent amplitudes of the signal. Furthermore, the two observable operators satisfy the commutation relation:

$$[O_1, O_2] = \frac{1}{2i}. \tag{11.121}$$

Using the Heisenberg uncertainty relation (Equation 11.33), the two operators satisfy the inequality:

$$\langle \Delta O_1^2 \rangle \langle \Delta O_2^2 \rangle \ge \frac{1}{16}. \tag{11.122}$$

The uncertainty of each operator is calculated defining $\kappa = |\kappa| e^{i\phi}$:

$$\langle \Delta O_1^2 \rangle = \frac{1}{4}|\cosh(|\kappa|L) - ie^{i\phi}\sinh(|\kappa|L)|^2,$$
$$\langle \Delta O_2^2 \rangle = \frac{1}{4}|\cosh(|\kappa|L) + ie^{i\phi}\sinh(|\kappa|L)|^2. \tag{11.123}$$

The phase ϕ includes the pump phase, which determines the orientation of the uncertainty ellipse for the conjugate observables. The maximum and minimum variances are

$$\sigma_{max}^2 = \frac{1}{4}e^{2|\kappa|L},$$
$$\sigma_{min}^2 = \frac{1}{4}e^{-2|\kappa|L}. \tag{11.124}$$

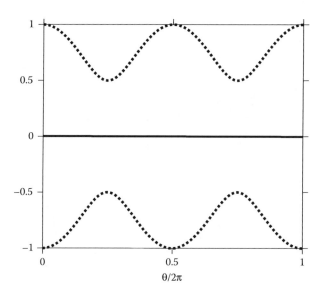

FIGURE 11.7 Standard deviations of one observable for $\sigma_{max} = 1$, $\sigma_{min} = 0.5$. The average field value is zero, that is, vacuum state.

They form an uncertainty ellipse with major and minor axes defined by the standard deviations $(\sigma_{max}, \sigma_{min})$. A measurement of one quadrature for the vacuum state of the electric field in Equation 11.120 has amplitude fluctuations with sinusoidal variations in time. An example is shown in Figure 11.7, defining $\theta = \omega t - \vec{k} \cdot \vec{r}$; the spatial factor $\vec{k} \cdot \vec{r}$ is arbitrary and determined by the detector's position. The product of the major and minor standard deviations satisfies the minimum uncertainty 0.5. In the standard quantum limit, the standard deviations are equal $\sigma_{max} = \sigma_{min} = 1/2$ and the uncertainty ellipse forms a circle. An average field value of zero represents the vacuum state. Squeezing is defined with respect to the variances; hence, the value of squeezing in the example of Figure 11.7 is a reduction by a factor of 2 or 3 dB squeezing.

The uncertainty ellipse accompanies the field when it is shifted to a nonzero average value. The ellipse can be rotated and translated to a nonzero value. In Figures 11.8, the uncertainty ellipse is oriented with its major axis perpendicular to the evolution direction. The ellipse rotates counterclockwise in Figure 11.8a as the field evolves in time; four snapshots are shown in Figure 11.8a. The maximum value of the uncertainty at $\theta = 0$ becomes a minimum uncertainty for $\theta = \pi/2$. The time evolution and the uncertainty in measuring O_1 are plotted in Figure 11.8b. In Figures 11.9, the ellipse is rotated by $\pi/2$. The trajectory in (O_1, O_2) space is drawn in Figure 11.9a and the time evolution of O_1 and its standard deviation (dotted lines) are shown in Figure 11.9b.

11.6.2 SQUEEZED LIGHT EXPERIMENTS

Wu et al. (1986) reported their observation of squeezed light measurements using an optical parametric oscillator with about 5-wt% MgO-doped lithium niobate

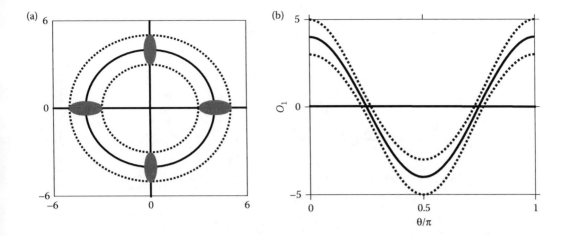

FIGURE 11.8 Time evolution of the electric field amplitude and the uncertainty ellipse. The maximum and minimum standard deviations are denoted by dotted lines. (a) The axes are labeled by the observable O_1 and O_2, and the uncertainty ellipse is drawn at four moments. The evolution of the ellipse is counterclockwise by our definition. (b) The average value (solid line) and standard deviations for the observable O_1. The field amplitude is 4 and the standard deviations are (1, 0.5) for this example.

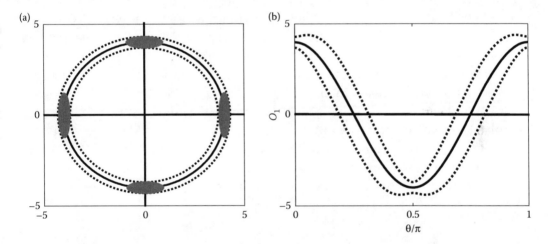

FIGURE 11.9 Two illustrations of field temporal evolution. The ellipse is rotated by $\pi/2$, but the parameters are identical to Figure 11.8.

(LN) crystal in the cavity. MgO:LN has a higher (photorefractive) damage threshold than stoichiometric LN. They achieved around 3 dB quadrature squeezing by this technique. Breitenbach et al. (1995) also used MgO:LN and a monolithically designed OPO cavity with spherically shaped, polished, and coated crystal surfaces for the cavity. The dichroic coating created a relatively high finesse cavity for the degenerate signal/idler pair and a two-pass geometry for the pump to lower the oscillation threshold. Additional electronics, optical feedback, and temperature controls were required to actively stabilize the cavity for longer periods of time.

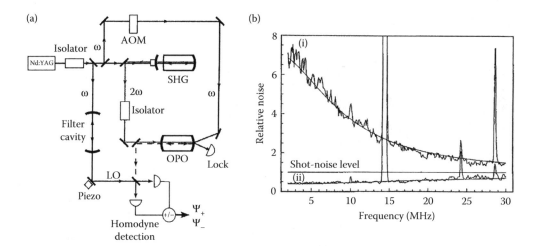

FIGURE 11.10 Squeezed light generation and detection. (a) Experimental setup. For details read the paper by Breitenbach et al. (1995). (b) Spectrum of the relative noise.

The primary coherent radiation source was a diode-pumped CW Nd:YAG laser (1.064 microns); a part of its output was split off and filtered to reduce the laser amplitude fluctuations for the LO. Another part of the laser light was used to inject a frequency-shifted signal into the OPO cavity for cavity stabilization. The OPO was pumped by a frequency doubling crystal outside the laser cavity (532 nm). The laser LO (1.064 microns) was mixed with the OPO output in a balanced homodyne detection apparatus, as previously explained. The spectrum of the relative noise is shown in Figure 11.10b. The shot-noise limit is identified in the figure. The spectrum below the shot-noise level line is the squeezed light noise (labeled ii) and the spectrum above the line is the anti-squeezed light quadrature noise (labeled i). The peaks at 14.5 and 29 MHz are the acousto-optic modulator (AOM) frequency of the signal injected into the OPO cavity and its harmonic used for locking the OPO cavity. The squeezing at lower frequencies ($\langle$5 MHz) exceeds 5 dB.

11.7 MULTIMODE QUANTUM STATES

11.7.1 ENTANGLED QUANTUM STATES

The photons are endowed with properties that allow them to be distinguished in different ways, by polarization, wavelength, or emission direction. In general, a combination of all three is possible. Consider a parametric down-conversion scheme where (signal and idler) photon pairs are generated with vertical (v) or horizontal (h) polarizations but emitted in different directions, which we label 1 and 2. The wave function for the superposition of two distinguishable photons with one photon horizontally and the other vertically polarized is

$$|\Psi(1,2)\rangle = c_1 |1_{1v}\rangle |1_{2h}\rangle + c_2 |1_{1h}\rangle |1_{2v}\rangle. \tag{11.125}$$

For instance, the notation $|1_{2h}\rangle$ means that 1 photon is horizontally polarized in the direction labeled 2; c_1 and c_2 are complex amplitudes satisfying $|c_1|^2 + |c_2|^2 = 1$. States prepared in this way are called *entangled* states. A pure quantum state is a product of the single wave vector states in the form

$$|\Psi(1,2)\rangle = |a_{1x}\rangle |b_{2y}\rangle. \tag{11.126}$$

The states $|a_{1x}\rangle$ and $|b_{2y}\rangle$ each represent an arbitrary superposition of the basis vectors. Equation 11.126 is entangled because it cannot be rewritten to separate the effects of Photon 1 or 2. Entangled states are at the heart of the EPR paradox that describes the essential nonlocality of quantum mechanics. For instance, Equation 11.125 leads to the conclusion that the outcome of a single measurement of one photon *instantaneously* determines the state of the second photon no matter how far the two particles are separated. For example, suppose $c_1 = -c_2 = 1/\sqrt{2}$. Simultaneous polarization measurements can be made using a polarization beam splitter or a Wollaston prism to spatially separate the two polarizations. The separated polarizations of a single photon are measured with two photodetectors. The probability of measuring vertical polarization for Photon 1 without measuring (or comparing) Photon 2 polarization is

$$P(1v) = \left| \frac{(\langle 1_{2v}| + \langle 1_{2h}|)(\langle 1_{1v}|\Psi(1,2)\rangle}{\langle \Psi(1,2)|\Psi(1,2)\rangle} \right|^2 = \frac{1}{2}. \tag{11.127}$$

Measuring the horizontally polarized signal for propagation direction 1 means that the idler photon along 2 is a vertically polarized wave, that is,

$$|\Psi(1v,2)\rangle = \langle 1_{1v}|\Psi(1,2)\rangle = \frac{1}{\sqrt{2}}|1_{2h}\rangle. \tag{11.128}$$

Polarization measurements of only the first or second photon have equal probability. The two-photon initial state represented by the wave function in Equation 11.125 leads to strongly correlated polarization measurements between both photons even though they are no longer interacting with one another. Suppose a vertical polarization of Photon 1 is measured. Then the probability of measuring the horizontal polarization for Photon 2 is certain:

$$P(1v,2h) = \left| \frac{\langle 1_{2h}|\Psi(1v,2)\rangle}{\langle \Psi(1v,2)|\Psi(1v,2)\rangle} \right|^2 = 1. \tag{11.129}$$

Similarly, by measuring the vertically polarized signal in direction 1, the horizontally polarized idler photon is propagating along 2, that is, $|\Psi(1h,2)\rangle = \langle 1_{1h}|\Psi(1,2)\rangle = -(1/\sqrt{2})|1_{2v}\rangle$. The directions of the measured polarizations can be changed by rotating the polarizers by angles $(\theta_{1a}, \theta_{2b})$. The eigenstates of the rotated polarizers for Photon 1 is

$$
\begin{aligned}
|1_{1va}\rangle &= \cos\theta_{1a}|1_{1v}\rangle + \sin\theta_{1a}|1_{1h}\rangle, \\
|1_{1ha}\rangle &= -\sin\theta_{1a}|1_{1v}\rangle + \cos\theta_{1a}|1_{1h}\rangle.
\end{aligned}
\tag{11.130}
$$

The inverse transformation for Photon 1 is

$$
\begin{aligned}
|1_{1v}\rangle &= \cos\theta_{1a}|1_{1va}\rangle - \sin\theta_{1a}|1_{1ha}\rangle, \\
|1_{1h}\rangle &= \sin\theta_{1a}|1_{1va}\rangle + \cos\theta_{1a}|1_{1ha}\rangle.
\end{aligned}
\tag{11.131}
$$

There are similar expressions for Photon 2. The simultaneous measurement of Photon 1 and Photon 2 in the new basis are defined by the probabilities, such as,

$$
P(1va,2hb) = \left| \frac{\langle 1_{2hb}|\langle 1_{1va}|\Psi(1,2)\rangle}{\langle \Psi(1va,2)|\Psi(1va,2)\rangle} \right|^2 = \cos^2(\theta_a - \theta_b).
\tag{11.132}
$$

The notation indicates vertical polarization for Photon 1 in the rotated basis θ_a and horizontal polarization in the basis θ_b. Similarly, the probability of measuring vertical polarization for Photon 1 and vertical polarization for Photon 2 is

$$
P(1va,2vb) = \left| \frac{\langle 1_{2vb}|\langle 1_{1va}|\Psi(1,2)\rangle}{\langle \Psi(1va,2)|\Psi(1va,2)\rangle} \right|^2 = \sin^2(\theta_a - \theta_b).
\tag{11.133}
$$

The results in Equations 11.132 and 11.133 are two examples of predictions that can be made based on quantum mechanics. The entangled twin photons are perfectly correlated when the two polarization apparatus angles are parallel, and the correlation becomes weaker as the two polarization axes become orthogonal. Entangled wave functions are at the heart of experiments designed to demonstrate quantum mechanical violations of local realistic theory assumptions, which are rendered testable by Bell's inequalities.

11.7.2 ENTANGLEMENT VIA SPDC

Two schemes based on phase matching to generate distinguishable or indistinguishable photon pairs are analyzed in this section. The photon entangled states are extracted from specially designed nonlinear crystal geometries applying noncollinear phase matching in SPDC. The scattering geometry is illustrated in

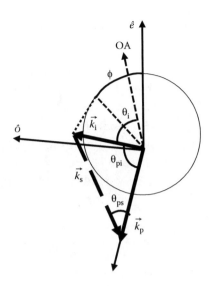

FIGURE 11.11 Noncollinear geometry. The optic axis is labeled as OA.

Figure 11.11. Three orthogonal axes in a uniaxial birefringent crystal are defined by the pump wave vector $\vec{k}_p$, the ordinary $\hat{o}$, and extraordinary $\hat{e}$ polarization directions for the pump. The crystal's design requires analysis before it can be applied to an experiment. The optic axis lies in the $(\vec{k}_p, \hat{e})$ plane. The pump wave is assumed to have extraordinary polarization, so that its wave vector's length depends on its angle with respect to the optic axis θ_p (not labeled in Figure 11.11); hence, the pump wave number may be denoted as $k_p(\theta_p)$. The angles between the idler and signal wave vector and the optic axis are labeled as θ_i and θ_s, respectively. The angle θ_i is indicated in Figure 11.11. The signal and idler wave vectors, in general, do not lie in the plane defined by $(\vec{k}_p, \hat{e})$. Those wave vectors can be rotated about the $\vec{k}_p$ axis by an angle ϕ. The signal and idler wave vectors have fixed magnitudes independent of the angle ϕ.

Spontaneous emitted light forms a cone with its central axis in the direction of the pump wave vector. The spontaneous signal and idler photon emission occurs at any angle ϕ defined by a circle formed in the plane perpendicular to the pump propagation direction. Phase matching of the wave vectors is decomposed into a component along $\vec{k}_p$ and one perpendicular to that direction:

$$k_p(\theta_p) = k_s \cos(\theta_{ps}) + k_i \cos(\theta_{pi}), \tag{11.134}$$

$$k_s \sin(\theta_{ps}) = -k_i \sin(\theta_{pi}). \tag{11.135}$$

The polarization of the signal and idler wave vectors may be ordinary or extraordinary. Two separate cases will be discussed below. The angles θ_{ps} and θ_{pi} are defined in Figure 11.11 as the angle between the pump and signal (s) or idler (i) wave vectors.

11.7.2.1 TYPE I PHASE MATCHING

Type I phase matching of the form $e \to o + o$ can be achieved in negative uniaxial crystals; the process generates photon pairs by spontaneous down-conversion. KDP and BBO are two crystals used for this purpose. The design of the crystals requires analysis before it can be applied to an experiment. The ordinary polarized signal and idler wave vectors have fixed magnitudes, independent of the angle ϕ. The optimum angle for θ_{ps} is chosen so that the angle of the signal/idler is small enough to ignore walk-off away from the pump beam. Figure 11.12 is a plot of the degenerate signal/idler emission angles outside a KDP crystal; the Sellmeier equations in Appendix B were used for all calculations. The external emission angle for the signal wave, θ_{se}, is related to the angle inside the crystal by

$$n_s \sin\theta_{ps} = \sin\theta_{se}. \tag{11.136}$$

This result also holds for nondegenerate signal and idler waves, and the idler wave external angle is analogously defined. The chosen pump wavelength is $\lambda_p = 351.1$ nm. The phase-matched degenerate signal and idler waves are collinear at $\theta_p = 49.9°$ and have an initial steep slope as the angle increases. At larger angles, the walk-off distance from the pump beam is short. Of course, the pump beam's walk-off angle also needs to be considered in the experimental design; it may limit the usable crystal thickness and beam diameter. The choice of θ_p should be close to, but larger than, the critical value $49.9°$.

Figure 11.13a shows the external angles for the signal and idler wavelengths in the KDP crystal for three choices of the pump–optic axis angle θ_p. The correlated signal/idler

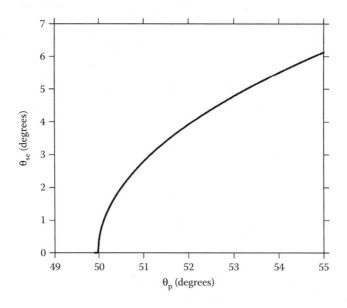

FIGURE 11.12 Degenerate signal/idler phase noncollinear phase-matching angle versus pump–optic axis angle in KDP.

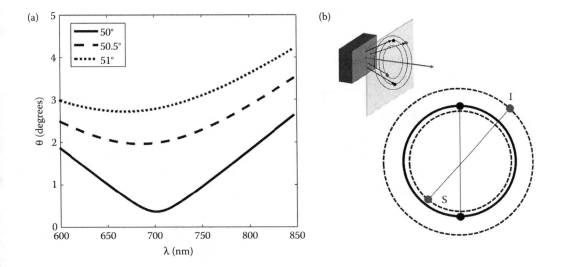

FIGURE 11.13 Type I phase matching in KDP using the spontaneous emission geometry $e \to o + o$. The pump wavelength is $\lambda_p = 351.1$ nm. (a) Spontaneous emission angles outside the KDP crystal. (b) The solid ring represents the case of degenerate signal and idler wavelengths. The dashed line rings are signal and idler wavelengths related by energy conservation.

photon pairs are angularly displaced on their emission cones by 180°. In Figure 11.13b, the circle drawn with solid lines represents the degeneracy wavelength, and one correlated photon pair is indicted by two solid circles and a line connecting them.

For the nondegenerate case, Equations 11.134 and 11.135 define two cones with different emission angles, one for the signal and one for the idler. Consider a screen placed in the path so that the emission cones form circles; the signal circle diameter may be slightly smaller than the circle formed by the degenerate waves, but, as Figure 11.13a shows, the signal wave more likely lies on a larger diameter circle that overlaps with an uncorrelated idler wave.

The idler internal emission angle is found by applying Equation 11.135, and its external angle is determined by applying Snell's law (Equation 11.136) with idler wave parameters. The circles drawn with dashed lines in Figure 11.13b are the signal/idler pair wavelengths; the correlated photon pair is found by connecting a line between two points and passing through the circles' center. In other words, the emitted photon pairs are correlated at the opposite bisectors of their respective circles. Therefore, by placing a screen with holes made at the optimal positions, the properties of photon quantum correlations can be further examined. A filter can eliminate unwanted radiation to select the desired photon wavelengths. Type I phase-matched photon pairs are not entangled, but the two photons are correlated and can become entangled by combining them at a beam splitter, as we shall demonstrate in Section 11.7.5 with an analysis of the Hong–Ou–Mandel (HOM) interferometer.

11.7.2.2 TYPE II PHASE MATCHING

The analysis of SPDC using Type II phase matching is more complex, but the additional effort yields a higher reward because the entangled states are richer in physical content. The creation of entangled photon pair states by this method was first proposed by Kwiat et al. (1995, 1999). For our example, we use a BBO crystal and the Type II phase-matching process: e → o + e; the idler wave has extraordinary polarization. Figure 11.11 is used again for the phase-matching geometry notation. In BBO Type II phase matching, the idler wave vector depends on the angle of the idler wave vector with respect to the optic axis, which is ϕ dependent.

The angle θ_p is fixed, and the angle θ_{pi} is related to θ_{ps} using Equation 11.135. Combining the squares of the wave vector components in Equations 11.134 and 11.135, a function of the angles θ_{pi} and θ_i can be found. Combining the two equations with some algebra, the phase-matching condition becomes

$$k_p^2(\theta_p) - k_s^2 - 2k_p(\theta_p)k_i(\theta_i)\cos(\theta_{pi}) + k_i^2(\theta_i) = 0. \tag{11.137}$$

We need a further connection with the extraordinary phase-matching angle θ_i and (θ_{pi}, ϕ). This is found by expressing the unit vector for the direction of the optic axis:

$$\hat{a}_{OA} = \hat{e}\sin\theta_p + \hat{p}\cos\theta_p, \tag{11.138}$$

where $\hat{p}$ is the unit vector in the direction of the pump wave vector $\vec{k}_p$, and $\hat{e}$ is the extraordinary polarization when the signal and idler wave vectors lie in the $(\hat{e}, \hat{p})$ plane. The idler unit wave vector is

$$\hat{k}_i = \hat{e}\sin\theta_{pi}\cos\phi + \hat{p}\cos\theta_{pi} - \hat{o}\sin\theta_{pi}\sin\phi. \tag{11.139}$$

The angle between the optic axis and idler wave vector is

$$\cos\theta_i = \hat{k}_i \cdot \hat{a}_{OA} = \sin\theta_p\sin\theta_{pi}\cos\phi + \cos\theta_p\cos\theta_{pi}. \tag{11.140}$$

The angle between the idler wave vector and the optic axis is determined from this expression. Solving Equation 11.137 together with Equation 11.140, where the angle θ_i is a function of θ_{pi} and ϕ, the phase-matching angle θ_{pi} can be calculated. The emission out of the plane is a function of an azimuthal geometric angle ϕ around the pump wave vector directions.

To design the SPDC geometry, we first determine the pump angle for collinear phase-matching condition at degeneracy. BBO is a (negative) uniaxial birefringent crystal that is transparent to UV wavelengths. The pump wavelength for our example is $\lambda_p = 325$ nm. The collinear phase-matching angle at degeneracy found by numerically solving Equation 11.139 ($\theta_i = \theta_p$ in this case) using the BBO dispersion formulas in Appendix B is $\theta_p = 54.489°$.

Simultaneously, at degeneracy there is also a noncollinear phase-matching point in the same plane defined by the optic axis and the pump wave vector. These two points are an indication of a noteworthy relationship between the signal and the idler. The azimuthal angle (ϕ) dependence at degeneracy connects those points, and two nearly circular wave vector cones that touch at one on-axis point are emitted; the shape of the cones on the plane perpendicular to $\vec{k}_p$ is shown in Figure 11.14a. The on-axis touching point is indicated by a dot. The degenerate signal and idler with parallel wave vectors are distinguishable by their orthogonal polarizations. Note that the pump and idler are extraordinary polarized, which will cause energy to walk off the center position. Spontaneous parametric processes will initiate the signal and idler fields to radiate out of the plane defined by the optic axis and the pump propagation direction.

Proceeding to simultaneously solving Equations 11.137 and 11.140 for the nondegenerate case, the two cones overlap as shown for two nondegenerate cases in Figure 11.14b and c. Overlap of the two circles indicates positions where the signal and idler photons are emitted in the same direction. For these points, when the signal is emitted along one direction, the idler must be emitted in the complementary direction. The state of the photon pair is entangled:

$$|\Psi(1,2)\rangle = \frac{1}{\sqrt{2}}\left(|1_{v1}\rangle|1_{h2}\rangle - |1_{h1}\rangle|1_{v2}\rangle\right). \tag{11.141}$$

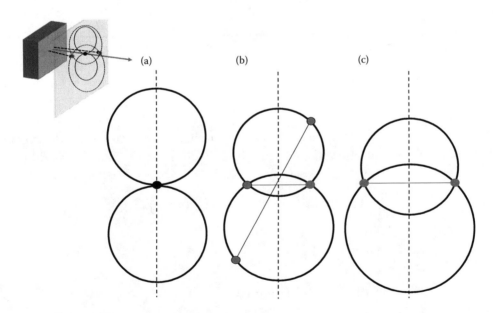

FIGURE 11.14   BBO Type II phase-matching angles with orthogonally polarized signal and idler waves. (a) Wave vectors for signal and idler degeneracy. (b) Wave vectors for the nondegenerate case with dots connected by lines that are correlated signal/idler pairs. (c) Same as case (b) but with the wavelength difference of the nondegenerate pair increased.

This entangled wave function has been adopted to study the foundations of quantum mechanics.

11.7.3 TWO-MODE PARAMETRIC SQUEEZING

Two-mode squeezing is one possible outcome of the creative SPDC process, but other nonlinear processes such as four-wave mixing can be used for photon pair creation. Consider the equations for phase-matched, nondegenerate signal and idler waves in the nondepleted pump approximation, as discussed in Section 6.2. The quantized waves satisfy the operator equations:

$$\frac{da_2}{dz} = i\kappa_2 a_3^\dagger,$$
$$\frac{da_3^\dagger}{dz} = -i\kappa_3^* a_2. \tag{11.142}$$

The coupling coefficients are

$$\kappa_2 = \frac{\omega_2 d_{\text{eff}}}{n_2 c} A_1, \quad \kappa_3 = \frac{\omega_3 d_{\text{eff}}}{n_3 c} A_1.$$

The solution of the equations is written as (define $\kappa = \sqrt{\kappa_2 \kappa_3^*}$)

$$a_2(L) = \cosh(\kappa L) a_2(0) + i \frac{\kappa_2}{\kappa} \sinh(\kappa L) a_3^\dagger(0),$$
$$a_3^\dagger(L) = \cosh(\kappa L) a_3^\dagger(0) - i \frac{\kappa_3^*}{\kappa} \sinh(\kappa L) a_2(0). \tag{11.143}$$

The physical operators are defined as composites of the single-mode operator, analogous to the optical phase conjugation case treated in the previous section:

$$O_1 = \frac{1}{2}\left(a_2^\dagger(L) + a_2(L) + a_3^\dagger(L) + a_3(L)\right),$$
$$O_2 = \frac{i}{2}\left(a_2^\dagger(L) - a_2(L) + a_3^\dagger(L) - a_3(L)\right). \tag{11.144}$$

The commutator for the operators defined above is

$$[O_1, O_2] = i. \tag{11.145}$$

The product of the variances for the operators is

$$\langle \Delta O_1^2 \rangle \langle \Delta O_2^2 \rangle \geq \frac{1}{4}. \tag{11.146}$$

In general, the coupling coefficients can have a complex value from the pump field $\kappa_2 = |\kappa_2| e^{i\phi}, \quad \kappa_3 = |\kappa_3| e^{i\phi}$

$$\langle \Delta O_1^2 \rangle = \frac{1}{2} |\cosh(\kappa L) + ie^{i\phi} \sinh(\kappa L)|^2,$$

$$\langle \Delta O_2^2 \rangle = \frac{1}{2} |\cosh(\kappa L) - ie^{i\phi} \sinh(\kappa L)|^2. \tag{11.147}$$

The product of the two variances reveals that the system is in a minimum uncertainty state. For $\varphi = \pi/2$, the variance $\langle \Delta O_1^2 \rangle$ is squeezed below the quantum limit, and for $\varphi = -\pi/2$, the quadrature variance $\langle \Delta O_2^2 \rangle$ shows that the operator O_2 is in a squeezed state.

11.7.4 QUANTUM OPTICAL PHASE CONJUGATION

To illustrate the quantum treatment of Kerr media, let us reexamine the phenomenon of optical phase conjugation treated in Chapter 8. We solved Equations 8.134 and 8.135 under rather restrictive conditions. The optical phase conjugate mirror (PCM) is a four-port device. As shown in Figure 11.15, in the quantum version the input at $z = L$ cannot be ignored, but, as we have learned, is a port where vacuum fluctuations enter the system. The equations are rewritten here in operator form dropping the prime superscript used in Chapter 8:

$$\frac{dA_{\text{Probe}}}{dz} = i \frac{6\omega\chi^{(3)}}{8nc} |A_P|^2 A_{\text{Signal}}^{\dagger} = i\eta A_{\text{Signal}}^{\dagger}, \tag{11.148}$$

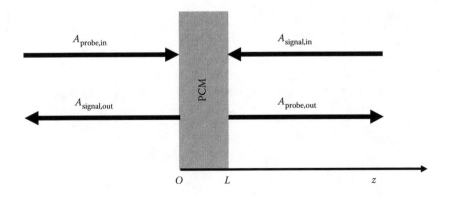

FIGURE 11.15 Quantum analysis of a PCM includes the input signal vacuum fluctuations at $z = L$.

$$\frac{dA_{\text{Signal}}}{dz} = -i\frac{6\omega\chi^{(3)}}{8nc}|A_{\text{P}}|^2 A_{\text{Probe}}^{\dagger} = -i\eta A_{\text{Probe}}^{\dagger}. \tag{11.149}$$

A new field operator $A_{\text{Signal}}(L) = A_{\text{Signal,in}}$ is added to the boundary conditions; the solution has the form,

$$A_{\text{Signal}}(z) = \frac{\cos(\eta z)}{\cos(\eta L)} A_{\text{Signal,in}} + i\frac{\sin(\eta(L-z))}{\cos(\eta L)} A_{\text{Probe,in}}^{\dagger}, \tag{11.150}$$

$$A_{\text{Probe}}(z) = \frac{\cos(\eta(L-z))}{\cos(\eta L)} A_{\text{Probe,in}} + i\frac{\sin(\eta z)}{\cos(\eta L)} A_{\text{Signal,in}}^{\dagger}. \tag{11.151}$$

The output signal operator is expressed as a linear combination of the input creation and annihilation operators:

$$A_{\text{Signal,out}} = A_{\text{Signal}}(0) = \sec(\eta L)A_{\text{Signal,in}} + i\tan(\eta L)A_{\text{Probe,in}}^{\dagger}. \tag{11.152}$$

The output probe solution is similarly found to be

$$A_{\text{Probe,out}} = A_{\text{Probe}}(L) = \sec(\eta L)A_{\text{Probe,in}} + i\tan(\eta L)A_{\text{Signal,in}}^{\dagger}. \tag{11.153}$$

Given the above expanded solutions for the optical phase conjugation geometry, the output field variances are calculated for the conjugate observables defined as

$$O_{\text{P1}} = \frac{1}{2}\left(A_{\text{Probe,out}} + A_{\text{Probe,out}}^{\dagger}\right), \tag{11.154}$$

$$O_{\text{P2}} = \frac{i}{2}\left(A_{\text{Probe,out}} - A_{\text{Probe,out}}^{\dagger}\right). \tag{11.155}$$

The corresponding detection operators are defined for the output signal fields O_{S1} and O_{S2}. Using vacuum states for the input field, $|\Psi\rangle = |0\rangle_{\text{Probe,in}}|0\rangle_{\text{Signal,in}}$. The variance is the same for both cases:

$$\sigma_j^2 = \langle O_{\text{P}j}^{\dagger}O_{\text{P}j}\rangle = \frac{1}{4}\left(\sec^2(\eta L) + \tan^2(\eta L)\right), \quad j = 1, 2. \tag{11.156}$$

The state of each field is no longer that of minimum uncertainty; as the magnitude of the nonlinearity parameter approaches $\eta L = \pi/2$, the variances approach infinity.

This result demonstrates that an active PCM device alters the vacuum state of the input fields through the four-wave mixing process. The nonlinear process endows the field with quantum correlations leading to squeezed states that can be uncovered by mixing the two output fields. Define the linear combination of the two operators as

$$A_{\text{SP}} = \frac{1}{\sqrt{2}}\left(A_{\text{Signal,out}} + A_{\text{Probe,out}}\right)e^{i\theta}. \tag{11.157}$$

The composite, conjugate operator observables are

$$O_{\text{SP1}} = \frac{1}{2}\left(A_{\text{SP}} + A_{\text{SP}}^{\dagger}\right), \quad O_{\text{SP2}} = \frac{i}{2}\left(A_{\text{SP}} - A_{\text{SP}}^{\dagger}\right). \tag{11.158}$$

Using the vacuum initial state, the variances of the combined fields are

$$\sigma_{\text{SP1}}^2 = \langle O_{\text{SP1}}^{\dagger} O_{\text{SP1}} \rangle, \quad \sigma_{\text{SP2}}^2 = \langle O_{\text{SP2}}^{\dagger} O_{\text{SP2}} \rangle, \tag{11.159}$$

$$\sigma_{\text{SP1}}^2 = \frac{1}{4}\left(\sec^2(\eta L) + \tan^2(\eta L) + 2\sec(\eta L)\tan(\eta L)\sin(2\theta)\right)^2, \tag{11.160}$$

$$\sigma_{\text{SP2}}^2 = \frac{1}{4}\left(\sec^2(\eta L) + \tan^2(\eta L) - 2\sec(\eta L)\tan(\eta L)\sin(2\theta)\right)^2. \tag{11.161}$$

For $\theta = \pi/4$, the variances have the form,

$$\sigma_{\text{SP1}}^2 = \frac{1}{4}\left(\sec(\eta L) + \tan(\eta L)\right)^2, \tag{11.162}$$

$$\sigma_{\text{SP2}}^2 = \frac{1}{4}\left(\sec(\eta L) - \tan(\eta L)\right)^2. \tag{11.163}$$

The product of the correlated fields has the form of a minimum uncertainty state. As the variance σ_{SP1}^2 grows, the variance σ_{SP2}^2 correspondingly shrinks, yielding a squeezed state. The variance vanishes as ηL approaches $\pi/2$, a limit that is also the classical oscillation threshold for the PCM. At this point, the analysis presented here becomes invalid, since this limit is characterized by high gain where pump depletion may not be ignored.

11.7.5 HOM INTERFEROMETER

We end our introduction to quantum nonlinear optical phenomena with a treatment of the HOM interferometer. An illustration of the experimental setup is shown in Figure 11.16. Degenerate and distinguishable signal and idler photons are filtered to

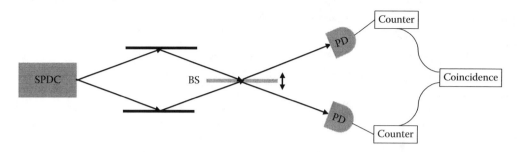

FIGURE 11.16   Sketch of the HOM interferometer experimental apparatus. The beam splitter (BS) is movable, as shown. Spectral filters and aperture stops placed before the photodetectors (PDs) are not shown. More experimental details are found in the paper by Hong et al. (1987).

limit their spectral bandwidth from a Type I SPDC source. The emitted photon pair is correlated in time through the SPDC event. Assuming pure, degenerate frequency photon states, the initial wave function is a pure state represented as

$$|\Psi_{in}\rangle = a_1^\dagger a_2^\dagger |0_1\rangle |0_2\rangle = |1_1\rangle |1_2\rangle. \tag{11.164}$$

The two photons are entangled by a 50/50 beam splitter and collected at two photodetectors. After the beam splitter, the output photon operators are

$$a_3^\dagger = \frac{1}{\sqrt{2}}(a_1^\dagger - ia_2^\dagger), \quad a_4^\dagger = \frac{1}{\sqrt{2}}(-ia_1^\dagger + a_2^\dagger). \tag{11.165}$$

The output state is entangled:

$$|\Psi_{out}\rangle = a_3^\dagger a_4^\dagger |0_1\rangle |0_2\rangle = \frac{1}{2}(a_1^\dagger - ia_2^\dagger)(-ia_1^\dagger + a_2^\dagger)|0_1\rangle |0_2\rangle = \frac{-i}{2}(|2_1\rangle |0_2\rangle + |0_1\rangle |2_2\rangle). \tag{11.166}$$

This nonintuitive result is surprising. Naively, one expects to have one photon split in each arm after the beam splitter or at least have some probability of one photon in each arm. Instead, Equation 11.166 reveals that the output wave function is an entangled state with both photons in one arm or the other arm. The physical explanation for this effect lies in the beam splitter, which destructively superposes the amplitudes of two processes where the photon pair are reflected and transmitted. In other words, no coincidence counts would be measured when the photons arrive at the two photodiodes at the same time.

The interesting feature of the experimental HOM interferometer in Figure 11.16 is the movable beam splitter. As the path length of the two arms is changed, the photon wave packet arrival times are no longer simultaneous, and the counter coincidence reduces to single-photon events. A KDP crystal and the pump wavelength of

351.1 nm (argon ion laser) are used in their experiment. Our prior analysis of Type I phase matching showed it is possible to produce degenerate twin photons with the same polarization; the photons are distinguished by their emission directions.

The emitted photon wavelengths due to spontaneous emission are spread over a broad bandwidth. The single degenerate photon pair discussed above needs to be extended to include this situation. One effect of the wavelength spread is the corresponding coherence time of the two photons. The HOM interferometer experiment was conceived to determine the coherence time of the two photons. In their experiment, spectral filters were placed before the detectors to limit the bandwidth. The coincidence counts on the separate photodetectors are measured for different positions of the beam splitter. As the two path lengths come into coincidence, a dip in the coincident counts is observed. This dip is a clear verification of the expected two-photon entangled state (Equation 11.166) after the beam splitter. The width of the dip is about 100 fs, which corresponds to their spectral filters.

To better understand this result, the wave function in Equation 11.166 is extended to include correlated photon pairs at nondegenerate frequencies. The initial wave function is the vacuum state for all photons, similar to Equation 11.61 but reduced to the relevant two plane-wave modes (1,2) and paired signal/idler frequencies $(\Omega',-\Omega')$:

$$|\Psi_0\rangle = \prod_{\Omega'}|0_{1\Omega'}\rangle|0_{2-\Omega'}\rangle. \tag{11.167}$$

The angular frequency difference from degeneracy is denoted as Ω. The twin photons have frequencies $\omega_0 \pm \Omega$, where ω_0 is the angular frequency at degeneracy. The expanded input wave function is

$$|\Psi_{in}\rangle = \int d\Omega C(\Omega)a_{1,\Omega}^{\dagger}a_{2,-\Omega}^{\dagger}|\Psi_0\rangle = \int d\Omega C(\Omega)|1_{1,\Omega}\rangle|1_{2,-\Omega}\rangle\prod_{\Omega'\neq\Omega}|0_{1\Omega'}\rangle|0_{2-\Omega'}\rangle. \tag{11.168}$$

The amplitude $C(\Omega)$ of each wave function is determined by the spectral filter shape. For definiteness, we assume a Gaussian function,

$$C(\Omega) = \frac{1}{\sqrt{\pi\Delta}}e^{-\Omega^2/(2\Delta^2)}. \tag{11.169}$$

The amplitude is normalized so that $\langle\Psi_{in}|\Psi_{in}\rangle = \int d\Omega|C(\Omega)|^2 = 1$. The positive-frequency, plane-wave electric field operators are

$$E^+(t_\mu,t) = \int d\Omega \mathcal{E}_{\vec{k}_\mu}a_\mu(\Omega)e^{-i(\omega_0+\Omega)(t_\mu+t)}e^{i\vec{k}_\mu\cdot\vec{r}}. \tag{11.170}$$

The subscript $\mu = 1,2$ denotes the two paths of the photon pair. The time t_μ is the arrival time of each photon. The two fields are combined at the beam splitter to yield the fields at the output ports, which are labeled as 3 and 4:

$$
\begin{aligned}
E^+(3,t) &= \sqrt{T}E^+(t_1,t) + i\sqrt{R}E^+(t_2,t), \\
E^+(4,t) &= i\sqrt{R}E^+(t_1,t) + \sqrt{T}E^+(t_2,t).
\end{aligned}
\tag{11.171}
$$

In this expression, the beam splitter may deviate from its ideal 50/50 values. The detection rate at the Arm 3 detector is found by

$$
\begin{aligned}
E^+(3,t)|\Psi_{in}\rangle = \int d\Omega C(\Omega)\mathcal{E}_{\bar{k}}(e^{-i(\omega_0+\Omega)(t_1+t)}\sqrt{T}\,|0_{1,\Omega}\rangle|1_{2,-\Omega}\rangle \\
+ e^{-i(\omega_0-\Omega)(t_1+t)}\sqrt{R}\,|1_{1,\Omega}\rangle|0_{2,-\Omega}\rangle)\prod_{\Omega'\neq\Omega}|0_{1\Omega'}\rangle|0_{2-\Omega'}\rangle.
\end{aligned}
\tag{11.172}
$$

The amplitude $\mathcal{E}_{\bar{k}}$ at the degenerate wavelength replaces the nondegenerate frequency ones, since the filter bandwidth and even angular acceptance limit the field frequencies to be close to degeneracy. The detection rate is proportional to

$$
w(3) = \langle\Psi_{in}|E^-(3)E^+(3)|\Psi_{in}\rangle = \int d\Omega\,|C(\Omega)|^2\,|\mathcal{E}_{\bar{k}}|^2\,(R+T) = |\mathcal{E}_{\bar{k}}|^2\,(R+T).
\tag{11.173}
$$

The same result holds for detection in Arm 4. The rate of two-photon coincidence is calculated from the expression,

$$
\begin{aligned}
E^+(4,t')E^+(3,t)|\Psi_{in}\rangle &= \int d\Omega C(\Omega)\big(\mathcal{E}_{\bar{k}}\big)^2\,(Te^{-i(\omega_0+\Omega)(t_1+t)}e^{-i(\omega_0-\Omega)(t_2+t')} \\
&\quad - Re^{-i(\omega_0+\Omega)(t_1+t')}e^{-i(\omega_0-\Omega)(t_2+t)})\prod_{\Omega'}|0_{1\Omega'}\rangle|0_{2-\Omega'}\rangle \\
&= \sqrt{2}\big(\mathcal{E}_{\bar{k}}\big)^2 e^{-i\omega_0(t_1+t_2+t+t')}(Te^{-\Delta^2(t_1-t_2+t-t')^2/2} \\
&\quad - Re^{-\Delta^2(t_1-t_2-t+t')^2/2})\prod_{\Omega'}|0_{1\Omega'}\rangle|0_{2-\Omega'}\rangle.
\end{aligned}
\tag{11.174}
$$

The coincidence times $\tau = t - t'$ are dictated by the filter bandwidth. Averaging over the time difference τ to arrive at the average photon counting rate, the coincidence rate is proportional to

$$
\begin{aligned}
w(3,4) &= \int d\tau\langle\Psi_{in}|E^-(4,t')E^-(3,t)E^+(4,t')E^+(3,t)|\Psi_{in}\rangle \\
&= \frac{2\sqrt{\pi}}{\Delta}\big(\mathcal{E}_{\bar{k}}\big)^4\,(T^2 + R^2 - 2TRe^{-\Delta^2(t_1-t_2)^2}).
\end{aligned}
\tag{11.175}
$$

This result allows one to explore conditions that are not ideal. For $R = T = 1/2$, the minimum count rate is zero. The reader is invited to explore cases with $R \neq T$, $R + T \langle 1$ in Problem 11.12. Other experimental issues that affect the interference function include detector inefficiencies and partial overlap of the signal and idler waves on the detector.

The HOM experiment has noteworthy features. First, the coincident rate analysis gives femtosecond resolution even though the laser beam is a continuous wave, and the apparatus does not require ultrafast photodetectors. The coincidence is determined by the position of the beam splitter, which can be displaced with (sub) micrometer scale accuracy (i.e., femtosecond accuracy). Second, the photon coincidence is a quantum mechanical nonlinear optics phenomenon, demonstrating that the photons are simultaneously emitted as the pump photon is absorbed. This result is not recovered by classical considerations of superposed coherent states.

The quantum aspects of nonlinear optics and measurements have been explored and elucidated in many publications. The choice of topics in this chapter is severely limited, and there are many further applications to be found by consulting the references. In fact, it is a very dynamic field with new research opportunities and applications made ever more accessible by increasing availability and improvements in photonic components. Experiments such as the HOM interferometer and squeezed state may be simple in conception, but they have a profound effect on our understanding of quantum mechanics. Elegant experiments incorporating entangled photons generated via noncollinear Type II phase-matching schemes first proposed by Kwiat et al. (1995, 1999) were designed to verify quantum mechanics violations of Bell's inequalities among other topics.

Schemes using third-order quantum nonlinear optics are also of note. Readily available low-loss optical fibers provide a long interaction length for accumulating quantum nonlinear effects. For instance, new experiments incorporating third-order nonlinearities have been designed to explore quantum squeezing of solitons in fiber-based devices. We refer the reader to the books by Haus (2000) and Drummond and Hillery (2014) for further discussions.

The quantum description of optical phenomena provides a fertile area of research and also has real-world applications. Pragmatic concepts include high precision metrology, quantum cryptography, quantum teleportation, and quantum information. A rich harvest of new phenomena may be discovered, as well. Quantum cryptography uses entangled photon pairs to securely transmit information because quantum-encoded messages or data are in a fragile state. An eavesdropper cannot make measurements without affecting the state of the entangled pair. The disturbance is detected by the sender and receiver, which alerts them to the intrusion.

PROBLEMS

11.1 The following problem is for the harmonic oscillator with no applied electromagnetic field.

a. Using the momentum operator definition $\hat{p} = (\hbar/i)(\partial/\partial x)$ in Equation 11.23, show that the ground-state wave function in the x space representation is given by Equation 11.24 and determine the normalization factor N.

b. Using the result from (a), calculate the ground-state weighted averages of $[\hat{p}, \hat{x}]$ and $\hat{p}\hat{x} + \hat{x}\hat{p}$ used in the derivation of the Heisenberg uncertainty relation.

c. For the ground state, calculate the variances of $\hat{p}$ and $\hat{x}$. Compare the result with Equation 11.33 to show by direct integration that the ground state is a minimum uncertainty state.

11.2 Operator identities and Heisenberg uncertainty relation.

a. Show that the definition of the output creation and annihilation operators in Equation 11.119 leads to the commutation relation in Equation 11.121 and the variances in Equation 11.122 for vacuum input states.

b. Prove the following identities for the creation and annihilation operators:
$[\hat{x}, \hat{p}^n] = i\hbar n \hat{p}^{n-1}$, $[a^{\dagger}, a^n] = -na^{n-1}$, $[(a^{\dagger})^2, a^2] = -4a^{\dagger}a - 2$.

c. Using the ground-state wave function of the harmonic oscillator, infer the value of the commutator between position and kinetic energy $(1/2m)\hat{p}^2$ of the particle average forming the lower limit in the Heisenberg uncertainty relation.

11.3 Using the following transformation (r real)

$$S(r) = e^{r/2(a^2 - (a^{\dagger})^2)},$$

and the definitions

$$a(r) = S^{\dagger}(r)aS(r), \quad a^{\dagger}(r) = S^{\dagger}(r)a^{\dagger}S(r),$$

prove $a(r) = \cosh(r)a - \sinh(r)a^{\dagger}$, $a^{\dagger}(r) = \cosh(r)a^{\dagger} - \sinh(r)a$.
Note: One way to solve for the operator expressions is by forming the differential equations,

$$\frac{da(r)}{dr}, \quad \frac{da^{\dagger}(r)}{dr};$$

after evaluating and simplifying the operator commutator identities, solve the coupled differential equations.

11.4 The displacement operator is defined as $D(\alpha) = e^{(\alpha a^{\dagger} - \alpha^* a)}$.

a. Prove that $a(\alpha) = D^{\dagger}(\alpha)aD(\alpha) = a + \alpha$ and find the expression or for $a^{\dagger}(\alpha)$. To prove the identities, use $D(\alpha, s) = e^{s(\alpha a^{\dagger} - \alpha^* a)}$ where s is a real variable, and in the first part, solve a differential equation for $a(\alpha, s) = D^{\dagger}(\alpha, s)aD(\alpha, s)$ at $s = 1$.

b. Prove $D(\alpha) = e^{(\alpha a^{\dagger} - \alpha^* a)} = e^{\alpha a^{\dagger}} e^{-\alpha^* a} e^{-|\alpha|^2/2}$. Use the definition $f(s) = e^{s\alpha a^{\dagger}} e^{-s\alpha^* a}$ and solve a differential equation for $f(s)$.

c. $D(\alpha)$ in (b) is called the displacement operator and it is a coherent state generator. Using the previous result, show that the coherent state is

$$|\alpha\rangle = D(\alpha)|0\rangle.$$

11.5 For the amplifier model, calculate the equal-time commutator of the Langevin force and prove that the equal-time commutation relations $[a(t),a^\dagger(t)]$ are preserved.

11.6 Equation 11.89 is an operator equation of motion for a quantum amplifier. Using the averages of the Langevin force in Equation 11.92, calculate the variances of the operators $O_\pm$ when the amplified mode is initially in a coherent state, $|\alpha\rangle$. The variances in Equation 11.93 correspond to vacuum-state averages.

11.7 In a balanced homodyne detector, designate Port 1 for input signal and Port 2 for the LO.
   a. Calculate the difference current between the output ports, $I_3 - I_4$, for the following signals:
      i. Coherent state
      ii. N photon state
   b. The detectors are shot-noise dominated with an integration time of 100 ns; the signal and LO wavelength is $\lambda = 532$ nm. For a 90% detector efficiency, calculate the signal to noise based on the signal photon in a coherent state with $|\alpha_s|^2 = 1$. To calculate the optical power use $P = \hbar\omega|\alpha|^2 B$

11.8 A BBO crystal is pumped at the wavelength $\lambda = 325$ nm.
   a. Determine the pump-optic axis angle for collinear Type I (e → o + o) phase-matching degenerate signal and idler waves.
   b. Make a plot of the pump angle versus external signal wave angle for noncollinear phase-matching degenerate signal and idler waves. (See Figure 11.12 for KDP case.)
   c. Calculate d_{eff} for each pump angle in (b).

11.9 The signal and idler intensities for a parametric amplifier are given by Equations 11.98 and 11.99, respectively. A BBO crystal is pumped with a laser wavelength $\lambda = 325$ nm.
   a. Determine the minimum angle between the pump and optic axis for collinear phase matching at degeneracy (Type II: e → o + e). (Answer: $\theta_p = 54.489°$.)
   b. Calculate d_{eff} for the pump angle in (a) and calculate the gain parameter Γ at phase matching for a pump intensity of 1 GW/cm². Adopt a geometry for the maximum d_{eff} for the fixed angle θ_p.
   c. Calculate the gain bandwidth using the gain in (b). Determine the extremum magnitude of Δk where the gain parameter g vanishes and use that value to find the corresponding signal and idler wavelengths. (Answer: 649.57 and 650.43 nm.)
   d. For a beam of diameter 100 μm and crystal length 1 cm, calculate the noise power coefficient for the degenerate case in Equation 11.99.

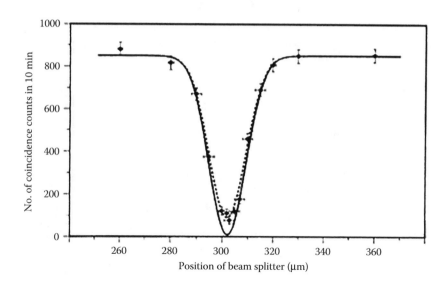

FIGURE 11.17 Coincidence counts versus beam splitter position. (Reprinted with permission from Hong, C. K., Z. Y. Ou, and L. Mandel. *Physical Review Letters* 59: 2044. Copyright 1987 by the American Physical Society.)

11.10 Using Equations 11.134 and 11.135 for Type II phase matching (e → o + e), prove Equation 11.137.

11.11 Show that the solution of Equations 11.116 for two input fields is Equations 11.117.

11.12 In the HOM interferometer, translation of the beam splitter position to the time domain gives an estimate of the simultaneity of signal and idler arrival time. Give a fit of the data in Figure 11.17 by using the ideal $R = T = 0.5$ and subsequently explore unequal values of R and T and $R + T < 1$. Scale the expression to normalize the background counts to unity and to estimate the full width at half maximum of arrival times. To better fit the data HOM introduced a factor that accounts for nonperfect overlap between the signal and idler fields by multiplying the Gaussian function by a constant, $Q \leq 1$. Use the overlap factor to fit the data assuming $T = 0.5$ and $R = 045\ T$.

11.13 A beam splitter is a four-port device with two input ports and two output ports. The ports are labeled in Figure 11.5.

a. Show that the variances of an observable O in output Ports 3 and 4 with statistically independent signals at the input ports are $\langle \Delta O_{13+}^2 \rangle = T \langle \Delta O_{11+}^2 \rangle + R \langle \Delta O_{12-}^2 \rangle$ and $\langle \Delta O_{14+}^2 \rangle = R \langle \Delta O_{11+}^2 \rangle + T \langle \Delta O_{12-}^2 \rangle$, defining $O_{1s+} = (1/2)(a_s + a_s^\dagger)$, $O_{1s-} = (1/2i)(a_s^\dagger - a_s)\ s = 1,2,3,4$.

b. Suppose that the signal at Port 1 is a single-mode squeezed state one observable variance is $\langle \Delta O_{11+}^2 \rangle = (1/4)e^{-2r}$, where r is a nonnegative real number, and the Port 2 input signal is the vacuum state. Plot the variance of the observable in Ports 3 and 4 versus the squeezing parameter and the transmission parameter T (assume $R + T = 1$). A beam splitter represents loss of a signal. What is the effect of loss on the squeezed-state variance?

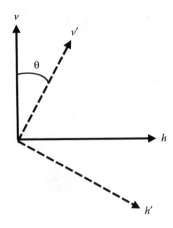

FIGURE 11.18   Rotation of a polarizing beam splitter.

11.14 Bell's inequalities are based on local realistic theories applied to a series of measurements designed to violate quantum mechanical predictions. In this problem, an inequality due to Clauser, Horne, Shimony, and Holt is applied to Type II entangled photon polarization measurements. Define a linear polarization measurement to have values $P = +1$ (v) and $P = -1$ (h). The axes of the polarizing beam splitter can be rotated by an angle θ, as shown in Figure 11.18.

a. Four variables $\{p_1, p_2, p_3, p_4\}$ have values between -1 and $+1$. Show that they satisfy the following inequality: $|B| = |p_1 p_3 + p_2 p_3 + p_1 p_4 - p_2 p_4| \le 2$.

b. Photon 1 is associated with two polarization analyzer orientations; the measurements $\{p_1, p_2\}$ are relative to polarization analyzers oriented at angles (θ_1, θ_2) and Photon 2 is associated with measurements. $\{p_3, p_4\}$ with polarization analyzers oriented at angles (θ_3, θ_4). The correlations of two measurements are defined as

$$C(i, j) = \int d\Lambda \rho(\Lambda) p_i(\Lambda) p_j(\Lambda), \quad i = 1, 2; \quad j = 3, 4,$$

where Λ represents the hidden variables and $\rho(\Lambda) \ge 0$ is the associated probability distribution. The inequality in (a) is $-2 \le B = C(1,3) + C(2,3) + C(1,4) - C(2,4) \le 2$.

Choose the angular differences as follows: $\phi = \theta_3 - \theta_1 = \theta_2 - \theta_3 = \theta_1 - \theta_4$, $3\phi = \theta_2 - \theta_4$. For the entangled state in Equation 11.125 $\left(c_1 = -c_2 = 1/\sqrt{2}\right)$, calculate the quantum correlation values for the four cases (Photon 1 is measured to be vertically polarized and Photon 2 is measured to be horizontally polarized) using the defined angular differences. Plot the function B as a function of ϕ. Over what range of angles is the inequality violated by quantum mechanics and determine the maximum magnitude of the violation?

REFERENCES

Breitenbach, G., T. Müller, S. F. Pereira, J.-Ph. Poizat, S. Schiller, and J. Mlynek. 1995. Squeezed vacuum from a monolithic optical parametric oscillator. *Journal of the Optical Society of America B* 12: 2304.

Drummond, P. D. and M. Hillery. 2014. *The Quantum Theory of Nonlinear Optics.* Cambridge, UK: Cambridge University Press.

Glauber, R. J. 1986. Amplifiers, attenuators, and Schrödinger's cat. *Annals of the New York Academy of Sciences* 480: 336.

Haus, H. A. 2000. *Electromagnetic Noise and Quantum Optical Measurements.* Berlin, Germany: Springer.

Hong, C. K., Z. Y. Ou, and L. Mandel. 1987. Measurement of subpicosecond time intervals between two photons by interference. *Physical Review Letters* 59: 2044.

Kwiat, P. G., K. Mattle, H. Weinfurter, A. Zeilinger, A. V. Sergienko, and Y. Shih. 1995. New high-intensity source of polarization-entangled photon pairs. *Physical Review Letters* 75: 4337.

Kwiat, P. G., E. Waks, A. G. White, I. Appelbaum, and P. H. Eberhard. 1999. Ultrabright source of polarization-entangled photons. *Physical Review A* 60: R773.

Milonni, P. W. 1994. *The Quantum Vacuum: An Introduction to Quantum Electrodynamics.* San Diego, CA: Academic Press.

Wu, L. A., H. J. Kimble, J. L. Hall, and H. Wu. 1986. Generation of squeezed states by parametric down conversion. *Physical Review Letters* 57: 2520.

BOOKS

Gerry, C. L. and P. L. Knight. 2005. *Introductory Quantum Optics.* Cambridge, UK: Cambridge University Press.

Grynberg, G., A. Aspect, and C. Fabre. 2010. *Introduction to Quantum Optics.* Cambridge, UK: Cambridge University Press.

Loudon, R. 2000. *The Quantum Theory of Light.* Oxford, UK: Oxford University Press.

Louisell, W. H. 1973. *Quantum Statistical Properties of Radiation.* New York, NY: Wiley.

Mandel, L. and E. Wolf. 1995. *Optical Coherence and Quantum Optics.* Cambridge, UK: Cambridge University Press.

Scully, M. O. and M. S. Zubairy. 1997. *Quantum Optics.* Cambridge, UK: Cambridge University Press.

Tang, C. L. 2009. *Fundamentals of Quantum Mechanics for Solid State Electronics and Optics.* Cambridge, UK: Cambridge University Press.

SELECTED ARTICLES

Yamamoto, Y. and H. A. Haus. 1986. Preparation, measurement and information capacity of optical quantum states. *Review of Modern Physics* 58: 1001.

AMPLIFIERS AND ATTENUATORS

Fix, A. and R. Wallenstein. 1996. Spectral properties of pulsed nanosecond optical parametric oscillators: Experimental investigation and numerical analysis. *Journal of the Optical Society of America B* 13: 2484.

Guan, Y., J. W. Haus, and P. P. Powers. 2004. Broadband and off-axis optical parametric generation in periodically poled LiNbO₃. *Journal of the Optical Society of America B* 21: 1225.

Guan, Y., J. W. Haus, and P. P. Powers. 2005. Macroscopic quantum fluctuations of pulses nanosecond optical parametric generation in periodically-poled LiNbO₃. *Physical Review A* 71: 023809.

SQUEEZED LIGHT

Caves, C. M. 1981. Quantum mechanical noise in an interferometer. *Physical Review D* 23: 1693.

The LIGO Scientific Collaboration. 2013. Enhanced sensitivity of the LIGO gravitational wave detector by using squeezed states of light. *Nature Photonics* 7: 613.

Grote, H., K. Danzmann, K. L. Dooley, R. Schnabel, J. Slutsky, and H. Vahlbruch. 2013. First long-term application of squeezed states of light in a gravitational-wave observatory. *Physical Review Letters* 110: 181101.

Slusher, R. E., L. W. Hollberg, B. Yurke, J. C. Mertz, and J. F. Valley. 1985. Observation of squeezed states generated by four-wave mixing in an optical cavity. *Physical Review Letters* 55: 2409.

Vahlbruch, H., S. Chelkowski, B. Hage, A. Franzen, K. Danzmann, and R. Schnabel. 2005. Demonstration of a squeezed light enhanced power- and signal-recycled Michelson interferometer. *Physical Review Letters* 95: 211102.

Vahlbruch, H., S. Chelkowski, B. Hage, A. Franzen, K. Danzmann, and R. Schnabel. 2006. Coherent control of vacuum squeezing in the gravitational-wave detection band. *Physical Review Letters* 97: 011101.

EPR AND TESTS OF QUANTUM MECHANICS

Aspect, A., P. Grangier, and G. Roger. 1982. Experimental realization of Einstein-Podolsky-Rosen-Bohm Gedanken experiment: A new violation of Bell's inequalities. *Physical Review Letters* 49: 91.

Bowen, W. P., N. Treps, B. C. Buchler, R. Schnabel, T. C. Ralph, H.-A. Bachor, T. Symul, and P. K. Lam. 2003. Experimental investigation of continuous-variable quantum teleportation. *Physical Review A* 67: 032302.

Einstein, A., B. Podolsky, and N. Rosen. 1935. Can quantum mechanical description of physical reality be considered complete? *Physical Review* 47: 777.

Furusawa, A., J. L. Sørensen, S. L. Braunstein, C. A. Fuchs, H. J. Kimble, and E. S. Polzik. 1998. Unconditional quantum teleportation. *Science* 282: 706.

APPENDIX A: COMPLEX NOTATION

Throughout this book, we use complex variables to represent fields and material properties. This appendix gives a quick overview of complex variables and their terminology.

A.1 RECTANGULAR FORM

We start with the fundamental imaginary number

$$i = \sqrt{-1} \quad \text{or} \quad i^2 = -1. \tag{A.1}$$

A complex number is written as

$$z = a + ib, \tag{A.2}$$

where a and b are real numbers (as opposed to numbers with an i). This notation is called the rectangular form of a complex number. For the complex number, z, given in Equation A.2, the real part is "a" and the imaginary part is "b." It is common to plot z in what is called the complex plane. This is a plane defined with one axis corresponding to the real part and the other axis corresponding to the imaginary part (see Figure A.1). In the complex plane, z is represented by a vector.

The representation in the complex plane logically leads to considering complex numbers as having a magnitude and a phase. The magnitude is the length of the vector and the phase is the angle between the vector and the real axis (θ). The magnitude is

$$\sqrt{a^2 + b^2}, \tag{A.3}$$

and the phase angle is

$$\theta = \tan^{-1}\left(\frac{b}{a}\right). \tag{A.4}$$

Another common term when using complex variables is the complex conjugate. If $z = a + ib$, then its complex conjugate is $z^* = a - ib$. That is, wherever you see an i,

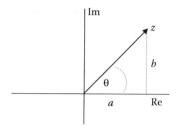

FIGURE A.1 Representation of a complex number, z, in the complex plane.

replace it with $-i$ and vice versa. If we multiply z by its "complex conjugate," then we obtain the magnitude squared:

$$|z|^2 = zz^*. \tag{A.5}$$

A.2 EULER'S FORMULA

The power-series expansion of e^x is

$$e^x = 1 + x + \frac{x^2}{2!} + \frac{x^3}{3!} + \frac{x^4}{4!} + \frac{x^5}{5!} + \cdots. \tag{A.6}$$

Now let $x = i\theta$:

$$e^{i\theta} = 1 + i\theta + \frac{(i\theta)^2}{2!} + \frac{(i\theta)^3}{3!} + \frac{(i\theta)^4}{4!} + \frac{(i\theta)^5}{5!} + \cdots. \tag{A.7}$$

We use the $i^2 = -1$ property to simplify

$$e^{i\theta} = \left[1 - \frac{\theta^2}{2!} + \frac{\theta^4}{4!} + \cdots \right] + i \left[\theta - \frac{\theta^3}{3!} + \frac{\theta^5}{5!} + \cdots \right]. \tag{A.8}$$

Note that the terms in the [] are the power-series expansions for $\cos\theta$ and $\sin\theta$. Hence, we obtain Euler's formula

$$e^{i\theta} = \cos\theta + i\sin\theta. \tag{A.9}$$

Euler's formula also allows us to write

$$\cos\theta = \frac{e^{i\theta} + e^{-i\theta}}{2}, \tag{A.10}$$

$$\sin\theta = \frac{e^{i\theta} - e^{-i\theta}}{2i}. \tag{A.11}$$

Another common way to write Equations A.10 and A.11 is

$$\cos\theta = \text{Re}\{e^{i\theta}\} = \text{Re}\{e^{-i\theta}\}, \tag{A.12}$$

$$\sin\theta = \text{Re}\{-ie^{i\theta}\} = \text{Re}\{ie^{-i\theta}\}. \tag{A.13}$$

A.3 POLAR FORM

z is written in the polar form by identifying from Figure A.1

$$a = |z|\cos\theta, \tag{A.14}$$

$$b = |z|\sin\theta. \tag{A.15}$$

Hence, z is written as

$$z = R(\cos\theta + i\sin\theta), \tag{A.16}$$

where $R = |z|$. With the help of Euler's formula (Equation A.9), the polar form of z is

$$z = Re^{i\theta}. \tag{A.17}$$

Note that the complex conjugate of Equation A.17 is $Re^{-i\theta}$. So, we have the same rule as with the rectangular form when finding the complex conjugate; we replace i with −i and vice versa. The polar form gives a compact way of identifying the magnitude and phase angle of the complex variable.

A.4 COMPLEX AMPLITUDES

Throughout this book, we work with sinusoidal time-varying quantities such as the scalar electric field,

$$\underline{E} = \frac{1}{2}A^{i(kz-\omega t)} + \text{c.c.} \tag{A.18}$$

where A is the complex amplitude of $\underline{E}$. As an another example, we take the derivative of $\underline{E}$,

$$\frac{\mathrm{d}E}{\mathrm{d}t} = \frac{-i\omega}{2} A e^{i(kz-\omega t)} + \text{c.c.} \tag{A.19}$$

The complex amplitude of Equation A.19 is $-i\omega A$. We see that the complex amplitude is the factor multiplying $1/2e^{i(kz-\omega t)}$. In some cases, the e^{ikz} is factored into the complex amplitude.

APPENDIX B: SELLMEIER EQUATIONS

This appendix gives the dispersion equations for crystals that appear in the problems discussed in this book. For a more complete tabulation of crystals, see Dmitriev et al. (1997).

B.1 β-BaB₂O₄, BETA BARIUM BORATE, BBO

Uniaxial crystal: $n_x = n_y = n_o$
Point group: 3m
Transparency range: 185–2500 nm
$d_{22} = -2.2$ pm/V (Eckardt et al., 1990)
$d_{31} = 0.08$ pm/V

Sellmeier equations from Kato (1986):

$$n_o^2 = 2.7359 + \frac{0.01878}{\lambda^2 - 0.01822} - 0.01354\lambda^2, \tag{B.1}$$

$$n_z^2 = 2.3753 + \frac{0.01224}{\lambda^2 - 0.01677} - 0.01516\lambda^2, \tag{B.2}$$

where λ is entered in microns.

B.2 LiTaO₃, LITHIUM TANTALATE (CONGRUENT)

Uniaxial crystal: $n_x = n_y = n_o$
Point group: 3m
Transparency range: 280–5500 nm
$d_{33} = -13.8$ pm/V (Shoji et al., 1997)

Sellmeier equations from Meyn and Fejer (1997):
Note that lithium tantalate has a very small birefringence, and we only list n_z for quasi-phase matching.

$$n_z^2 = 4.5284 + \frac{7.2449 \times 10^{-3} + b(T)}{\lambda^2 - [0.2453 + c(T)]^2} + \frac{7.7690 \times 10^{-2}}{\lambda^2 - [0.1838]^2} - 2.3670 \times 10^{-2}\lambda^2, \tag{B.3}$$

where λ is measured in microns. $b(T)$ and $c(T)$ give the temperature dependence:

$$b(T) = 2.6794 \times 10^{-8}(T + 273.15)^2, \tag{B.4}$$

$$c(T) = 1.6234 \times 10^{-8}(T + 273.15)^2, \tag{B.5}$$

where T is measured in °C.

B.3   LiNbO₃, LITHIUM NIOBATE (CONGRUENT)

Uniaxial crystal: $n_x = n_y = n_o$
Point group: 3m
Transparency range: 330–5500 nm
$d_{22} = 2.1$ pm/V (Miller et al., 1971)
$d_{31} = -4.35$ pm/V
$d_{33} = -27.2$ pm/V

Use Sellmeier equations: Equations B.6 and B.7 (Edwards and Lawrence, 1984) for birefringent phase matching and Equation B.9 (Jundt, 1997) for quasi-phase matching.

$$n_o^2 = 4.9048 + \frac{0.11775 + 2.2314 \times 10^{-8} F}{\lambda^2 - (0.21802 - 2.9671 \times 10^{-8} F)^2} + 2.1429 \times 10^{-8} F - 0.027153\lambda^2, \tag{B.6}$$

$$n_z^2 = 4.5820 + \frac{0.09921 + 5.2716 \times 10^{-8} F}{\lambda^2 - (0.21090 - 4.9143 \times 10^{-8} F)^2} + 2.2971 \times 10^{-7} F - 0.021940\lambda^2, \tag{B.7}$$

where λ is entered in microns. F gives the temperature dependence:

$$F = (T - T_o)(T + T_o + 546), \tag{B.8}$$

where $T_o = 24.5$°C and T is entered in °C.

For quasi-phase matching problems, the updated Sellmeier equation is used (Jundt, 1997):

$$n_z^2 = 5.35583 + 4.629 \times 10^{-7} F + \frac{0.100473 + 3.862 \times 10^{-8} F}{\lambda^2 - (0.20692 - 0.89 \times 10^{-8} F)^2}$$
$$+ \frac{100 + 2.657 \times 10^{-5} F}{\lambda^2 - 128.806} - 1.5334 \times 10^{-2}\lambda^2. \tag{B.9}$$

B.4 LiB₃O₅, LITHIUM TRIBORATE, LBO

Class: mm2
Transparency range: 160–2600 nm
$d_{31} = -0.67$ pm/V (Dmitriev and Nikogosyan, 1993)
$d_{32} = 0.85$ pm/V
$d_{33} = 0.04$ pm/V

Sellmeier equations from Kato (1994):

$$n_x^2 = 2.4542 + \frac{0.01125}{\lambda^2 - 0.01135} - 0.01388\lambda^2, \tag{B.10}$$

$$n_y^2 = 2.5390 + \frac{0.01277}{\lambda^2 - 0.01189} - 0.01849\lambda^2 + 4.3025 \times 10^{-5}\lambda^4 - 2.9131 \times 10^{-5}\lambda^6, \tag{B.11}$$

$$n_z^2 = 2.5865 + \frac{0.01310}{\lambda^2 - 0.01223} - 0.01862\lambda^2 + 4.5778 \times 10^{-5}\lambda^4 - 3.2526 \times 10^{-5}\lambda^6, \tag{B.12}$$

where λ is entered in microns.

B.5 KH₂PO₄, POTASSIUM DIHYDROGEN PHOSPHATE, KDP

Uniaxial crystal: $n_x = n_y = n_o$
Point group: $\bar{4}2m$
Transparency range: 174 nm–1.57 μm
d_{36} (1064 nm) = 0.39 pm/V (Roberts, 1992)
d_{36} (351 nm) = 0.5 pm/V

Sellmeier equations ($T = 20°C$) from Barnes et al. (1982):

$$n_o^2 = 2.259276 + \frac{13.00522\lambda^2}{\lambda^2 - 400} + \frac{0.01008956}{\lambda^2 - (77.26408)^{-1}}, \tag{B.13}$$

$$n_z^2 = 2.132668 + \frac{3.2279924\lambda^2}{\lambda^2 - 400} + \frac{0.008637494}{\lambda^2 - (81.42631)^{-1}}, \tag{B.14}$$

where λ is entered in microns.

B.6 KTiOPO₄, POTASSIUM TITANYL PHOSPHATE, KTP

Point group: mm2
Transparency range: 350 nm–4 μm

$d_{31} = 1.95$ pm/V (Pack et al., 2004)
$d_{32} = 3.9$ pm/V
$d_{33} = 15.3$ pm/V

Sellmeier equations from Vanherzeele et al. (1988):

$$n_x^2 = 2.1146 + \frac{0.89188\lambda^2}{\lambda^2 - 0.0435181} - 0.01320\lambda^2, \qquad (B.15)$$

$$n_y^2 = 2.1518 + \frac{0.87862\lambda^2}{\lambda^2 - 0.0475284} - 0.01327\lambda^2, \qquad (B.16)$$

$$n_z^2 = 2.3136 + \frac{1.00012\lambda^2}{\lambda^2 - 0.0567917} - 0.01679\lambda^2, \qquad (B.17)$$

where λ is entered in microns.

B.7   ZnGeP$_2$, ZINC GERMANIUM PHOSPHIDE, ZGP

Uniaxial crystal: $n_x = n_y = n_o$
Point group: $\overline{4}2m$
Transparency range: 740 nm–11 μm
$d_{14} = 48$ pm/V (Boyd et al., 1971)

Sellmeier equations from Zelmon et al. (2001):

$$n_o^2 = 8.0409 + \frac{1.68625\lambda^2}{\lambda^2 - 0.40824} + \frac{1.2880}{\lambda^2 - 611.05}, \qquad (B.18)$$

$$n_z^2 = 8.0929 + \frac{1.8649\lambda^2}{\lambda^2 - 0.41468} + \frac{0.84052\lambda^2}{\lambda^2 - 452.05}, \qquad (B.19)$$

where λ is entered in microns.

B.8   GaAs, GALLIUM ARSENIDE, GaAs

Nonbirefringent: $n_x = n_y = n_z$
Point group: $\overline{4}3m$
Transparency range: 0.9–17 μm
$d_{14} = 94$ pm/V (Skauli et al., 2002)

Sellmeier equations from Skauli et al. (2003):

$$n^2 = 5.372514 + \frac{27.83972}{1/\lambda_1^2 - 1/\lambda^2} + \frac{0.031764 + F}{1/\lambda_2^2 - 1/\lambda^2} + \frac{0.00143636}{1/\lambda_3^2 - 1/\lambda^2}, \tag{B.20}$$

where λ is entered in microns and

$$F = 4.350 \times 10^{-5}\Delta T + 4.664 \times 10^{-7}\Delta T^2, \tag{B.21}$$

$$\lambda_1 = 0.44313071 + 5.0564 \times 10^{-5}\Delta T, \tag{B.22}$$

$$\lambda_2 = 0.8746453 + 1.913 \times 10^{-4}\Delta T - 4.882 \times 10^{-7}\Delta T^2, \tag{B.23}$$

$$\lambda_3 = 36.9166 - 0.011622\,\Delta T, \tag{B.24}$$

where $\Delta T = T - 22$ and T is entered in °C.

B.9   SCHOTT GLASS SF10

Transparency range: 380–2600 nm
Sellmeier equation from Schott (2015):

$$n^2 = 1 + \frac{1.62153902\lambda^2}{(\lambda^2 - 0.0122241457)} + \frac{0.256287842\lambda^2}{(\lambda^2 - 0.0595736775)} + \frac{1.64447552\lambda^2}{(\lambda^2 - 147.468793)}, \tag{B.25}$$

where λ is entered in microns.

B.10   FUSED SILICA

Sellmeier equation from Corning Incorporated (2003):

$$n^2 = 1 + \frac{0.683740494\lambda^2}{\lambda^2 - 0.00460352869} + \frac{0.420323613\lambda^2}{\lambda^2 - 0.01339688560} + \frac{0.58502748\lambda^2}{\lambda^2 - 64.4932732}, \tag{B.26}$$

where λ is entered in microns.

REFERENCES

Barnes, N. P., D. J. Gettemy, and R. S. Adhav. 1982. Variation of the refractive index with temperature and the tuning rate for KDP isomorphs. *Journal of the Optical Society of America* 72: 895–898.

Boyd, G. D., E. Buehler, and F. G. Storz. 1971. Linear and nonlinear optical properties of ZnGeP$_2$ and CdSe. *Applied Physics Letters* 18: 301–304.

Corning Incorporated. 2003. HPFS® fused silica standard grade. http://www.corning.com/media/worldwide/csm/documents/5bf092438c5546dfa9b08e423348317b.pdf. Accessed December 20, 2016.

Dmitriev, V. G., G. G. Gurzadyan, and D. N. Nikogosyan. 1997. *Handbook of Nonlinear Optical Crystals*. Heidelberg, Germany: Springer-Verlag.

Dmitriev, V. G. and D. N. Nikogosyan. 1993. Effective nonlinearity coefficients for three-wave interactions in biaxial crystal of mm2 point group symmetry. *Optics Communications* 95: 173–182.

Eckardt, R. C., H. Masuda, Y. X. Fan, and R. L. Byer. 1990. Absolute and relative nonlinear optical coefficient of KDP, KD*P, BaB$_2$O$_4$, LiIO$_3$, MgO:LiNbO$_3$, and KTP measured by phase-matched second-harmonic generation. *IEEE Journal of Quantum Electronics* 26: 922–933.

Edwards, G. J. and M. Lawrence. 1984. A temperature-dependent dispersion equation for congruently grown lithium niobate. *Optical and Quantum Electronics* 16: 373–374.

Jundt, D. H. 1997. Temperature dependent Sellmeier equation for the index of refraction, NE, in congruent lithium niobate. *Optics Letters* 22: 1553–1555.

Kato, K. 1986. Second-harmonic generation to 2048 Å in β-BaB$_2$O$_4$. *IEEE Journal of Quantum Electronics* 22: 1013–1014.

Kato, K. 1994. Temperature-tuned 90 degree phase-matching properties of LiB$_3$O$_5$. *IEEE Journal of Quantum Electronics* 30: 2950–2952.

Meyn, J.-P. and M. M. Fejer. 1997. Tunable ultraviolet radiation by second-harmonic generation in periodically poled lithium tantalate. *Optics Letters* 22: 1215–1217.

Miller, R. C., W. A. Nordland, and P. M. Bridenbaugh. 1971. Dependence of second-harmonic-generation coefficients of LiNbO$_3$ on melt composition. *Journal of Applied Physics* 42: 4145–4147.

Pack, M. V., D. J. Armstrong, and A. V. Smith. 2004. Measurement of the $\chi^{(2)}$ tensors of KTiOPO$_4$, KTiOAsO$_4$, RbTiOPO$_4$, and RbTiOAsO$_4$ crystals. *Applied Optics* 43: 3319–3323.

Roberts, D. A. 1992. Simplified characterization of uniaxial and biaxial nonlinear optical crystals: A plea for standardization of nomenclature and conventions. *IEEE Journal of Quantum Electronics* 28: 2057–2074.

Schott. 2015. Optical glass collection data sheets. http://www.us.schott.com/d/advanced_optics/102fefee-c1cb-4772-a784-1ef2e328eb4c/1.0/schott-optical-glass-collection-data-sheets-july-2015-us.pdf. Accessed December 20, 2016.

Shoji, I., T. Kondo, A. Kitamoto, M. Shirane, and R. Ito. 1997. Absolute scale of second-order nonlinear-optical coefficients. *Journal of the Optical Society of America B* 14: 2268–2294.

Skauli, T., P. S. Kuo, K. L. Vodopyanov, T. J. Pinguet, O. Levi, L. A. Eyres, J. S. Harris, and M. M. Fejer. 2003. Improved dispersion relations for GaAs and applications to nonlinear optics. *Journal of Applied Physics* 94: 6447–6455.

Skauli, T., K. L. Vodopyanov, T. J. Pinguet, A. Schober, O. Levi, L. A. Eyres, M. M. Fejer, J. S. Harris, B. Gerard, L. Becouarn et al. 2002. Measurement of the nonlinear coefficient of orientation-patterned GaAs and demonstration of highly efficient second-harmonic generation. *Optics Letters* 27: 628–630.

Vanherzeele, H., J. D. Bierlein, and F. C. Zumsteg. 1988. Index of refraction measurements and parametric generation in hydrothermally grown KTiOPO$_4$. *Applied Optics* 27: 3314–3316.

Zelmon, D. E., E. A. Hanning, and P. G. Schunemann. 2001. Refractive-index measurements and Sellmeier coefficients for zinc germanium phosphide from 2 to 9 μm with implications for phase matching in optical frequency-conversion devices. *Journal of the Optical Society of America B* 18: 1307–1310.

APPENDIX C: PROGRAMMING TECHNIQUES

C.1 PROGRAMMING Δk

As an example, consider Δk for a Type I difference-frequency interaction ($o \rightarrow e + e$) in a positive uniaxial crystal

$$\Delta k\left(\lambda_1, \lambda_2, T, \theta\right) = 2\pi\left(\frac{n_o(\lambda_1)}{\lambda_1} - \frac{n_e(\lambda_2, \theta)}{\lambda_2} - \frac{n_e(\lambda_3, \theta)}{\lambda_3}\right), \tag{C.1}$$

where λ_3 is determined from energy conservation. The pseudocode for Δk appears as follows:

```
Function Deltak(lambda1,lambda2,temp,theta)
    lambda3=1/(1/lambda1-1/lambda2)
    Deltak=2*pi*(no(lambda1)/lambda1-ne(lambda2,theta)/lambda2-
ne(lambda3,theta)/lambda3)
    return Deltak
EndFunction
```

Note that the program makes two calls to subfunctions no(lambda) and ne(lambda,theta). no(lambda) is a function that returns the index of refraction based on the Sellmeier equation for the ordinary index. ne(lambda,theta) returns the extraordinary index based on the equation,

$$n_e(\lambda, \theta) = \left(\frac{\cos^2\theta}{n_o^2(\lambda)} + \frac{\sin^2\theta}{n_z^2(\lambda)}\right)^{-1/2}. \tag{C.2}$$

The pseudocode for the extraordinary index is

```
Function ne(lambda,theta)
    ne=1/sqrt((cos(theta)/no(lambda))^2 + (sin(theta)/nz(lambda))^2)
    return ne
EndFunction
```

nz(lambda) is obtained from the Sellmeier equation, similar to no(lambda). With a function programmed for Δk, we are now free to plot it against any of its independent variables over arbitrary ranges. Similarly, we can write functions that depend on Δk such as $\mathrm{sinc}^2(\Delta k L/2)$ and plot them against any of the independent variables of Δk.

C.2 ROOT FINDING BY BISECTION

As an example, we look at finding roots of Equation C.1 (i.e., $\Delta k = 0$). Instead of plotting Δk over a given range, we use the bisection technique to quickly zoom in on the zero crossing. Bisection is illustrated in Figure C.1. The routine successively finds the value of Δk at the midpoint of a given range.

The key to the routine is in how it chooses the correct interval. It does so by looking at the sign of the function at the endpoints of a given interval. If the function passes through zero in the interval, then the signs of the function at the endpoints are opposite. If the function does not pass through zero in the interval, then the signs at the endpoints are the same.

As an example, we look at the pseudocode for a bisection program that finds the phase-matching angle when λ_1, λ_2, and the temperature are fixed. The program runs until $|\Delta k|$ becomes equal to zero within some tolerance.

```
Function GetTheta(lambda1,lambda2,temp,tolerance)
// Comment: tolerance is how close to we want to get to zero
    thetastart=0
    thetaend=p1/2
// Comment: first determine that deltak passes through zero
    somewhere between 0 and 90 degrees
If (sign(Deltak(lambda1,lambda2,temp,thetastart)*sign(Deltak(lambda1,
lambda2, temp,thetaend))<0)
//sign is a function that returns +1 if the argument is positive and
    -1 if it is negative
```

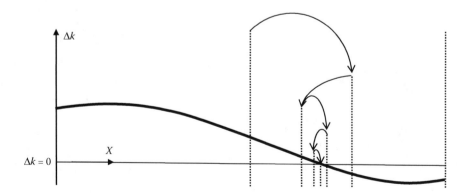

FIGURE C.1   Illustration of the bisection technique. Δk is plotted as a function of X, which can be any one of its dependencies. Successive iterations seek the value of Δk at the midpoint of an interval. By choosing the correct interval, the routine homes in on the zero crossing.

```
    Return NAN // NAN="not a number," which is recognized by most
programming languages
    Break
    EndIf
    Loop
        thetamid=(thetaend-thetastart)/2
        deltak_start=Deltak(lambda1,lambda2,temp,thetastart)
        deltak_mid=Deltak(lambda1,lambda2,temp,thetamid)
        If (ABS(deltak_mid)<tolerance)//ABS is absolute value
            return thetamid
            Break
        Else
            If (sign(deltakmid)*sign(deltakstart)=1)// if true the zero
is between mid and end
                thetastart=thetamid
        Else
                thetaend=thetamid
        Endif
    Endif
  EndLoop
EndFunction
```

The coupled amplitude equations are accurately integrated numerically using the Runge–Kutta (RK) technique. This technique starts on a known boundary and then uses the coupled differential equations to propagate to the next position. RK comes in different accuracies, typically second order and fourth order. The second-order approach is accurate to the second order in a Taylor series expansion. As an example, consider the following coupled equations:

$$\frac{dA_1}{dz} = i\kappa_1 A_2 A_3 e^{-i\Delta kz}, \tag{C.3}$$

$$\frac{dA_2}{dz} = i\kappa_2 A_1 A_3^* e^{i\Delta kz}, \tag{C.4}$$

$$\frac{dA_3}{dz} = i\kappa_3 A_1 A_2^* e^{i\Delta kz}. \tag{C.5}$$

The coefficients κ_1, κ_2, κ_3, and Δk are known parameters; the values of the complex amplitudes A_1, A_2, and A_3 are known at $z = 0$. Before performing the RK integration, we write separate functions for each of the equations. The pseudocode for Equation C.4 is

```
Function dA2dz(kappa2,Deltak,z,A1,A2,A3)
// We include A2 to make the next program clearer
      return i kappa2*A1*Conjugate(A3)*exp(i*Deltak*z)
EndFunction
```

Similarly, we can write functions dA1dz(kappa1,Deltak,z,A1,A2,A3) and dA3dz (kappa3,Deltak,z,A1,A2,A3). The pseudocode for integrating Equations C.3 through C.5 with a fourth-order RK method is

```
Function Integratefields(A1,A2,A3,Dz,iterations)
// We enter the input field amplitudes, A1, A2, and A3
// Enter the parameter values, kappa1, kappa2, kappa3, Deltak
// Dz is the stepsize, iterations is the number of steps to take
// counter, k1, k2, k3, k4, L1, L2, L3, L4, m1, m2, m3, m4 are
    variables defined
// locally within the function

counter=0
Loop
   k1=Dz*dA1dz(kappa1,Deltak,z,A1,A2,A3)
   L1=Dz*dA2dz(kappa2,Deltak,z,A1,A2,A3)
   m1=Dz*dA3dz(kappa3,Deltak,z,A1,A2,A3)

   k2=Dz*dA1dz(kappa1,Deltak,z+Dz/2,A1+k1/2,A2+L1/2,A3+m1/2)
   L2=Dz*dA2dz(kappa2,Deltak,z+Dz/2,A1+k1/2,A2+L1/2,A3+m1/2)
   m2=Dz*dA3dz(kappa3,Deltak,z+Dz/2,A1+k1/2,A2+L1/2,A3+m1/2)

   k3=Dz*dA1dz(kappa1,Deltak,z+Dz/2,A1+k2/2,A2+L2/2,A3+m2/2)
   L3=Dz*dA2dz(kappa2,Deltak,z+Dz/2,A1+k2/2,A2+L2/2,A3+m2/2)
   m3=Dz*dA3dz(kappa3,Deltak,z+Dz/2,A1+k2/2,A2+L2/2,A3+m2/2)

   k4=Dz*dA1dz(kappa1,Deltak,z+Dz,A1+k3,A2+L3,A3+m3)
   L4=Dz*dA2dz(kappa2,Deltak,z+Dz,A1+k3,A2+L3,A3+m3)
   m4=Dz*dA3dz(kappa3,Deltak,z+Dz,A1+k3,A2+L3,A3+m3)

   A1=A1+(k1+2*k2+2*k3+k4)/6
   A2=A2+(L1+2*L2+2*L3+L4)/6
   A3=A3+(m1+2*m2+2*m3+m4)/6
   Z=z+Dz

   counter=counter+1
   LoopUntil (counter>iterations)
   Return A1, A2, A3
EndFunction
```

Perhaps, the most difficult part of the above program is accounting for the fact that the amplitudes A_1, A_2, and A_3 are complex. Most programming languages permit working with complex-valued functions, but they may require a specific declaration that the variable is complex. Without a proper accounting for the complex values, the program will clearly return erroneous results.

FURTHER READING

Cheney, W. and D. Kincaid. 1999. *Numerical Mathematics and Computing*. Pacific Grove, CA: Brooks/Cole Publishing Company.

Iserles, A. 1996. *A First Course in the Numerical Analysis of Differential Equations*. Cambridge, UK: Cambridge University Press.

Press, W. H., S. A. Teukolsky, W. T. Vetterling, and B. P. Flannery. 2007. *Numerical Recipes, the Art of Scientific Computing* (3rd edition). New York, NY: Cambridge University Press.

APPENDIX D: EXACT SOLUTIONS TO THE COUPLED AMPLITUDE EQUATIONS

Exact solutions to the coupled amplitude equations for three monochromatic plane waves were first presented by Armstrong et al. (1962). The following parallels their development. The coupled amplitude equations from Section 4.1.4 are repeated for convenience:

$$\frac{dA_1}{dz} = i \frac{\omega_1}{n_1 c} d_{eff} A_3 A_2 e^{-i\Delta kz},$$ (D.1)

$$\frac{dA_2}{dz} = i \frac{\omega_2}{n_2 c} d_{eff} A_1 A_3^* e^{i\Delta kz},$$ (D.2)

$$\frac{dA_3}{dz} = i \frac{\omega_3}{n_3 c} d_{eff} A_1 A_2^* e^{i\Delta kz}.$$ (D.3)

Instead of carrying various constants through the calculation, we start by writing the coupled amplitude equations in "normalized" form. The amplitudes are normalized by rewriting them in magnitude and phase form,

$$A = \alpha u e^{i\phi},$$ (D.4)

where u and ϕ are real quantities and α is a real normalizing term. Note that from the definition of irradiance, we have

$$|A| = \sqrt{\frac{2I}{\varepsilon_0 nc}}.$$ (D.5)

The normalizing factor is slightly different, with an extra factor of ω under the square root so that, for instance,

$$A_1 = \sqrt{\frac{2I\omega_1}{\varepsilon_0 n_1 c}} u_1 e^{i\varphi_1},$$ (D.6)

the Manley–Rowe relation, $I = I_1 + I_2 + I_3 =$ constant is used in Equation D.6. Substituting Equation D.6 and analogous expressions for A_2 and A_3 into the coupled amplitude equations yields

$$\frac{d}{dz}\left(u_1 e^{i\varphi_1}\right) = id_{\text{eff}} \left(\frac{2I\omega_1\omega_2\omega_3}{\varepsilon_0 c^3 n_1 n_2 n_3}\right)^{1/2} u_2 u_3 e^{i(\varphi_3 + \varphi_2 - \Delta kz)}, \tag{D.7}$$

$$\frac{d}{dz}\left(u_2 e^{i\varphi_2}\right) = id_{\text{eff}} \left(\frac{2I\omega_1\omega_2\omega_3}{\varepsilon_0 c^3 n_1 n_2 n_3}\right)^{1/2} u_1 u_3 e^{i(\varphi_1 - \varphi_3 + \Delta kz)}, \tag{D.8}$$

$$\frac{d}{dz}\left(u_3 e^{i\varphi_3}\right) = id_{\text{eff}} \left(\frac{2I\omega_1\omega_2\omega_3}{\varepsilon_0 c^3 n_1 n_2 n_3}\right)^{1/2} u_1 u_2 e^{i(\varphi_1 - \varphi_2 + \Delta kz)}. \tag{D.9}$$

Now, we see that the constants group together in the same way in each of the coupled equations. We define $\zeta \equiv d_{\text{eff}}\sqrt{2I\omega_1\omega_2\omega_3/(\varepsilon_0 c^3 n_1 n_2 n_3)}z$ so that after taking the derivatives we obtain

$$\left(\frac{du_1}{d\zeta} + iu_1 \frac{d\varphi_1}{d\zeta}\right) = iu_2 u_3 e^{i(\varphi_3 + \varphi_2 - \varphi_1 - \Delta kz)}, \tag{D.10}$$

$$\left(\frac{du_2}{d\zeta} + iu_2 \frac{d\varphi_2}{d\zeta}\right) = iu_1 u_3 e^{i(\varphi_1 - \varphi_2 - \varphi_3 + \Delta kz)}, \tag{D.11}$$

$$\left(\frac{du_3}{d\zeta} + iu_3 \frac{d\varphi_3}{d\zeta}\right) = iu_1 u_2 e^{i(\varphi_1 - \varphi_2 - \varphi_3 + \Delta kz)}. \tag{D.12}$$

Recall that the u's and ϕ's are real quantities, so we equate the real and imaginary parts of Equations D.10 through D.12,

$$\frac{du_1}{d\zeta} = -u_2 u_3 \sin\theta \quad \text{and} \quad \frac{d\varphi_1}{d\zeta} = +\frac{u_2 u_3}{u_1}\cos\theta, \tag{D.13}$$

$$\frac{du_2}{d\zeta} = +u_1 u_3 \sin\theta \quad \text{and} \quad \frac{d\varphi_2}{d\zeta} = +\frac{u_1 u_3}{u_2}\cos\theta, \tag{D.14}$$

$$\frac{du_3}{d\zeta} = +u_1 u_2 \sin\theta \quad \text{and} \quad \frac{d\varphi_3}{d\zeta} = +\frac{u_1 u_2}{u_3}\cos\theta, \tag{D.15}$$

where

$$\theta \equiv \varphi_3 + \varphi_2 - \varphi_1 - \Delta kz. \tag{D.16}$$

Therefore,

$$\frac{d\theta}{d\zeta} = \frac{d}{d\zeta}(\varphi_3 + \varphi_2 - \varphi_1 - \Delta kz). \tag{D.17}$$

Substituting the phase derivatives from Equations D.13 through D.15 yields

$$\frac{d\theta}{d\zeta} = \left(\frac{u_1 u_2}{u_3} + \frac{u_1 u_3}{u_2} - \frac{u_2 u_3}{u_1}\right)\cos\theta - \frac{d(\Delta kz)}{d\zeta}; \tag{D.18}$$

a substitution for $u_1 u_2 / u_3$ is made by using the equations derived from the real part of Equations D.13 through D.15:

$$\frac{d\theta}{d\zeta} = \left(\frac{du_1}{d\zeta}\frac{1}{u_1 \sin\theta} + \frac{du_2}{d\zeta}\frac{1}{u_2 \sin\theta} + \frac{du_3}{d\zeta}\frac{1}{u_3 \sin\theta}\right)\cos\theta - \Delta S, \tag{D.19}$$

where

$$\Delta S = \Delta k \frac{dz}{d\zeta} = \frac{\Delta k}{d_{\text{eff}}}\left(\frac{2I\omega_1\omega_2\omega_3}{\varepsilon_0 c^3 n_1 n_2 n_3}\right)^{-1/2}. \tag{D.20}$$

Note that Equation D.19 is equivalent to

$$\frac{d\theta}{d\zeta} = \frac{d(\ln[u_1 u_2 u_3])}{d\zeta}\cot\theta - \Delta S. \tag{D.21}$$

Finally, we obtain the coupled equations in the normalized variables as

$$\frac{du_1}{d\zeta} = -u_2 u_3 \sin\theta, \tag{D.22}$$

$$\frac{du_2}{d\zeta} = +u_1 u_3 \sin\theta, \tag{D.23}$$

$$\frac{du_3}{d\zeta} = +u_1 u_2 \sin\theta, \tag{D.24}$$

$$\frac{d\theta}{d\zeta} = \frac{d(\ln[u_1 u_2 u_3])}{d\zeta} \cot\theta - \Delta S. \tag{D.25}$$

These equations are solved exactly by first integrating Equation D.25. The integration is facilitated by multiplying both sides of Equation D.25 by $u_1 u_2 u_3 \sin\theta$:

$$u_1 u_2 u_3 \sin\theta \frac{d\theta}{d\zeta} = u_1 u_2 u_3 \sin\theta \frac{d(\ln[u_1 u_2 u_3])}{d\zeta} \cot\theta - \Delta S u_1 u_2 u_3 \sin\theta. \tag{D.26}$$

This equation is rewritten by expanding the derivative of the ln term:

$$u_1 u_2 u_3 \sin\theta \frac{d\theta}{d\zeta} - \frac{d(u_1 u_2 u_3)}{d\zeta} \cos\theta = -\Delta S u_1 u_2 u_3 \sin\theta. \tag{D.27}$$

Equation D.27 is equivalent to

$$-\frac{d(u_1 u_2 u_3 \cos\theta)}{d\zeta} = -\Delta S u_1 u_2 u_3 \sin\theta. \tag{D.28}$$

Substituting $\sin\theta$ from Equation D.24 yields

$$-\frac{d(u_1 u_2 u_3 \cos\theta)}{d\zeta} = \Delta S u_1 \frac{du_1}{d\zeta} = \frac{1}{2}\Delta S \frac{d(u_1)^2}{d\zeta}. \tag{D.29}$$

Equation D.29 is directly integrated:

$$-\int d(u_1 u_2 u_3 \cos\theta) = \frac{1}{2}\Delta S \int d(u_1)^2 + \text{const.} \tag{D.30}$$

In what follows we define the integration constant as $-\Gamma$, and note that it is constant with respect to ζ, that is, $d\Gamma/d\zeta = 0$.

$$\Gamma = +u_1 u_2 u_3 \cos\theta + \frac{1}{2}\Delta S u_1^2. \tag{D.31}$$

Equation D.31 serves as a substitution for $\cos\theta$. The goal is now to rewrite Equation D.24 in terms of u_1 only so that it can be integrated:

$$\frac{du_1}{d\zeta} = -u_2 u_3 \sin\theta = -u_2 u_3 \sqrt{1 - \cos^2\theta} = -u_2 u_3 \sqrt{1 - \frac{(\Gamma - \frac{1}{2}\Delta S u_1^2)^2}{(u_1 u_2 u_3)^2}}. \tag{D.32}$$

Multiplying both sides by u_1 yields

$$\frac{1}{2}\frac{d\left(u_1^2\right)}{d\zeta} = -\sqrt{(u_1 u_2 u_3)^2 - \left(\Gamma - \frac{1}{2}\Delta S u_1^2\right)^2}. \tag{D.33}$$

The integral of Equation D.33 is aided by substituting in various "constants of motion" from the Manley–Rowe relations. Rewriting $I = I_1 + I_2 + I_3$ in terms of normalized variables gives

$$\omega_1 u_1^2 + \omega_2 u_2^2 + \omega_3 u_3^2 = 1. \tag{D.34}$$

Three other constants that result from the Manley–Rowe relations are

$$m_1 = u_2^2 - u_3^2, \tag{D.35}$$

$$m_2 = u_1^2 + u_3^2, \tag{D.36}$$

$$m_3 = u_1^2 + u_2^2. \tag{D.37}$$

The constants of motion allow us to solve for just one of the field amplitudes, for example, $u_1(\zeta)$, and then obtain the other two using the boundary conditions and Equations D.35 through D.37. In the following, we solve for $u_1(\zeta)$.

In terms of the constants of motion, we write Equation D.33 purely in terms of u_1 and ζ:

$$\frac{d\left(u_1^2\right)}{d\zeta} = 2\sqrt{u_1^2\left(m_3 - u_1^2\right)\left(m_2 - u_1^2\right) - \left(\Gamma - \frac{1}{2}\Delta S u_1^2\right)^2}. \tag{D.38}$$

Rearranging and integrating give

$$\zeta = \int_{u_1^2(0)}^{u_1^2(\zeta)} \frac{d\left(u_1^2\right)}{2\sqrt{\left(m_2 - u_1^2\right)\left(m_3 - u_1^2\right)u_1^2 - \left(\Gamma - \frac{1}{2}\Delta S u_1^2\right)^2}}. \tag{D.39}$$

This equation is in the form of an elliptic integral as we now show. The equation,

$$\left(m_2 - u_1^2\right)\left(m_3 - u_1^2\right)u_1^2 - \left(\Gamma - \frac{1}{2}\Delta S u_1^2\right)^2 = 0, \tag{D.40}$$

has three roots, which with magnitudes ordered as $u_{1c}^2 \geq u_{1b}^2 \geq u_{1a}^2$, that is,

$$\left(u_1^2 - u_{1a}^2\right)\left(u_1^2 - u_{1b}^2\right)\left(u_1^2 - u_{1c}^2\right) = \left(m_2 - u_1^2\right)\left(m_3 - u_1^2\right)u_1^2 - \left(\Gamma - \tfrac{1}{2}\Delta S u_1^2\right)^2. \tag{D.41}$$

In terms of the roots, Equation D.39 is written as

$$\zeta = \int_{u_1^2(0)}^{u_1^2(\zeta)} \frac{d\left(u_1^2\right)}{2\sqrt{\left(u_1^2 - u_{1a}^2\right)\left(u_1^2 - u_{1b}^2\right)\left(u_1^2 - u_{1c}^2\right)}}. \tag{D.42}$$

The following variable substitutions put the integral into a simpler form:

$$y^2 \equiv \frac{u_1^2 - u_{1a}^2}{u_{1b}^2 - u_{1a}^2}, \tag{D.43}$$

$$\gamma = \frac{u_{1b}^2 - u_{1a}^2}{u_{1c}^2 - u_{1a}^2}. \tag{D.44}$$

With these variable substitutions, Equation D.42 becomes

$$\zeta = \frac{1}{\sqrt{\left(u_{1c}^2 - u_{1a}^2\right)}} \int_{y(0)}^{y(\zeta)} \frac{dy}{\sqrt{(1 - y^2)(1 - \gamma y^2)}}, \tag{D.45}$$

where

$$y(\zeta) = \sqrt{\frac{u_1^2(\zeta) - u_{1a}^2}{u_{1b}^2 - u_{1a}^2}}. \tag{D.46}$$

Equation D.45 is close to a Jacobi elliptic integral, except that the lower limit is not zero, and so the integral is rewritten as

$$\zeta = \frac{1}{\sqrt{\left(u_{1c}^2 - u_{1a}^2\right)}} \left(\int_0^{y(\zeta)} \frac{dy}{\sqrt{(1 - y^2)(1 - \gamma y^2)}} - \int_0^{y(0)} \frac{dy}{\sqrt{(1 - y^2)(1 - \gamma y^2)}} \right). \tag{D.47}$$

The second integral gives a constant contribution, denoted as ζ_0, since the limits of integration are independent of ζ. We bring ζ_0 to the left-hand side of the equation.

Note that the constant, ζ_0, is also written in the form of an elliptic integral. The final integral is written as

$$\zeta + \zeta_0 = \frac{1}{\sqrt{\left(u_{1c}^2 - u_{1a}^2\right)}} \int_0^{y(\zeta)} \frac{dy}{\sqrt{(1-y^2)(1-\gamma y^2)}}.$$ (D.48)

This integral has a solution in terms of the elliptic function, sn,

$$y(\zeta) = \text{sn}\left[(\zeta + \zeta_0)\sqrt{u_{1c}^2 - u_{1a}^2}, \gamma \right].$$ (D.49)

The value of ζ_0 is found by evaluating

$$\zeta_0 = \frac{1}{\sqrt{\left(u_{1c}^2 - u_{1a}^2\right)}} \int_0^{y(0)} \frac{dy}{\sqrt{(1-y^2)(1-\gamma y^2)}} = \frac{1}{\sqrt{\left(u_{1c}^2 - u_{1a}^2\right)}} \text{sn}^{-1}(y(0)).$$ (D.50)

The normalized field amplitude $u_1(\zeta)$ is obtained from Equation D.43 to give

$$u_1^2(\zeta) = u_{1a}^2 + \left(u_{1b}^2 - u_{1a}^2\right)\text{sn}^2\left[(\zeta + \zeta_0)\sqrt{u_{1c}^2 - u_{1a}^2}, \gamma \right].$$ (D.51)

From the boundary conditions and Equations D.35 through D.37,

$$u_2^2(\zeta) = -u_1^2(\zeta) + m_3 = -u_1^2(\zeta) + u_1^2(0) + u_2^2(0).$$ (D.52)

Similarly for u_3,

$$u_3^2(\zeta) = -u_1^2(\zeta) + u_1^2(0) + u_3^2(0).$$ (D.53)

In final form, the exact solutions to the coupled wave equations for collinear plane waves are

$$u_1^2(\zeta) = u_{1a}^2 + \left(u_{1b}^2 - u_{1a}^2\right)\text{sn}^2\left[(\zeta + \zeta_0)\sqrt{u_{1c}^2 - u_{1a}^2}, \gamma \right],$$ (D.54)

$$u_2^2(\zeta) = u_2^2(0) + u_1^2(0) - u_{1a}^2 - \left(u_{1b}^2 - u_{1a}^2\right)\text{sn}^2\left[(\zeta + \zeta_0)\sqrt{u_{1c}^2 - u_{1a}^2}, \gamma \right],$$ (D.55)

$$u_3^2(\zeta) = u_3^2(0) + u_1^2(0) - u_{1a}^2 - \left(u_{1b}^2 - u_{1a}^2\right)\mathrm{sn}^2\left[(\zeta + \zeta_0)\sqrt{u_{1c}^2 - u_{1a}^2}, \gamma\right]. \qquad \text{(D.56)}$$

The solutions give us an insight into the general behavior of three-wave interactions. The most general trend is that the solutions are periodic. This means that even for a phase-matched interaction, energy flow from the inputs to a desired output changes its direction periodically (see Chapter 5 for plots of the solutions).

REFERENCE

Armstrong, J. A., N. Bloembergen, J. Ducuing, and P. S. Pershan. 1962. Interactions between light waves in a nonlinear dielectric. *Physical Review* 127: 1918–1939.

APPENDIX E: OPTICAL FIBERS— SLOWLY VARYING ENVELOPE EQUATIONS

An optical fiber in its simplest form consists of a core and cladding glass, as illustrated in Figure E.1. The core glass has a refractive index that is higher than the cladding index so as to confine the electromagnetic wave near the core region. The electric field in an optical fiber expressed as an angular frequency-dependent function expanded around the signal angular frequency, ω_0, is written as

$$E(\vec{r},\omega) = F(\vec{r}_\perp,\omega)\tilde{A}(z,\omega-\omega_0)e^{i\beta_{0z}}, \tag{E.1}$$

where $F(\vec{r}_\perp,\omega)$ is the transverse field profile, which describes the shape of the fiber mode and it has a SI unit as will become apparent below. $\tilde{A}$ is the envelope of the signal, and the exponential factor governs the spatial phase shift with $\beta_0 = \beta(\omega_0) = \omega_0 n/c$, where n is the effective index of the wave in the optical fiber. The nature of the propagation constant at the carrier angular frequency will be discussed below. The scalar wave equation is solved by the assumed solution as (Agrawal 2007)

$$\tilde{A}\nabla_\perp^2 Fe^{i\beta_{0z}} + \mu_0\varepsilon(\vec{r}_\perp,\omega)\omega^2\tilde{A}Fe^{i\beta_{0z}} + F\left[\frac{\partial^2 A}{\partial z^2} + i2\beta_0\frac{\partial A}{\partial z} - \beta_0{}^2 A\right]e^{i\beta_{0z}} = -\mu_0\omega^2 P^{(NL)}. \tag{E.2}$$

The optical fiber's dielectric function $\varepsilon(\vec{r}_\perp,\omega)$ is a function of the transverse coordinates that are often locally constant with separate core and cladding regions. By the separation of variables method, the fiber mode function satisfies the wave equation in the following form:

$$\nabla_\perp^2 F + \left(\mu_0\varepsilon(\vec{r}_\perp,\omega)\omega^2 - \beta^2(\omega)\right)F = 0. \tag{E.3}$$

The solution of Equation E.3 for an optical fiber geometry can be described as propagation modes in the optical fiber and the dispersive eigenvalues are the propagation constant. The propagation constant $\beta^2(\omega)$ determined by solving the wave equation (Equation E.3) is dispersion with both material and waveguide

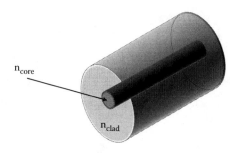

FIGURE E.1   Illustration of a simple optical fiber with a core and cladding.

contributions to its value. It defines the effective index of the wave, and its value at the carrier frequency is β_0^2. Equation E.3 can be reorganized as follows:

$$i2\beta_0 F \frac{\partial \tilde{A}}{\partial z} + \tilde{A}\left[\nabla_\perp^2 F e^{i\beta_0 z} + \mu_0 \varepsilon(\vec{r}_\perp, \omega)\omega^2 F\right] - \beta_0^2 F \tilde{A} + F \frac{\partial^2 \tilde{A}}{\partial z^2} = -\mu_0 \omega^2 P^{(\mathrm{NL})} e^{-i\beta_0 z}. \quad \text{(E.4)}$$

In keeping with the slowly varying envelope approximation (SVEA), the term $\partial^2 A / \partial z^2$ is neglected and the second term in square brackets is replaced with $\beta^2(\omega)F$ using Equation E.3. The envelope equation is expressed as

$$i2\beta_0 F \frac{\partial \tilde{A}}{\partial z} + \left[\beta(\omega)^2 - \beta_0^2\right] F \tilde{A} = -\mu_0 \omega^2 P^{(\mathrm{NL})} e^{-i\beta_0 z}. \quad \text{(E.5)}$$

The fiber's mode function may be normalized, which is expressed as

$$\iint |F|^2 \, d\vec{r}_\perp = 1. \quad \text{(E.6)}$$

As mentioned before, the mode functions have the units of an inverse square root of an area. As a result, the electric field amplitude $\tilde{A}$ has units that may be scaled to the wave's power in the fiber, rather than to the intensity of the wave. This is an important distinction between the wave amplitude in free space and its amplitude in a fiber. The third-order nonlinear polarization in vector form may be expressed as

$$P_i^{(3)}(\omega) = \left(\frac{\varepsilon_0}{8} \chi_{ijkl}^{(3)} A_j F A_k F (A_l F)^*\right) e^{i\beta_0 z}, \quad \text{(E.7)}$$

where the fiber mode function is assumed to be degenerate for both polarizations. The mode function is removed from the wave equation (Equation E.5) by multiplying by F^* and integrating over the transverse coordinates. Assuming an x polarized field,

$$2\beta_0 i \frac{\partial \tilde{A}}{\partial z} + \left[\beta(\omega)^2 - \beta_0^2\right] \tilde{A} = -\omega^2 \frac{3\mu_0 \varepsilon_0}{4} \chi_{xxxx}^{(3)} |\tilde{A}|^2 \tilde{A} \frac{1}{S_{\mathrm{eff}}}. \quad \text{(E.8)}$$

The factor of three in the nonlinear term is due to permutations of the cubic field function. S_{eff} is an effective mode area defined by

$$S_{\text{eff}} = \frac{\iint |F|^2\, d\vec{r}_\perp}{\iint |F|^4\, d\vec{r}_\perp} = \frac{1}{\iint |F|^4\, d\vec{r}_\perp}. \tag{E.9}$$

The dispersion is small and we expand the propagation constant $\beta(\omega)$ in the same fashion as done for the longitudinal wave vector in Section 10.3.4:

$$\beta(\omega) = \beta_0 + \left.\frac{\partial \beta}{\partial \omega}\right|_{\omega_0} \eta + \frac{1}{2}\left.\frac{\partial^2 \beta}{\partial \omega^2}\right|_{\omega_0} \eta^2 + \cdots. \tag{E.10}$$

The second term is $(\partial\beta/\partial\omega)|_{\omega_0} = \beta_1 = 1/v_g$ and the third term is the group velocity dispersion, $(\partial^2\beta/\partial\omega^2)|_{\omega_0} = \beta_2$.

$$\beta(\omega)^2 - \beta_0{}^2 \approx 2\beta_0(\beta(\omega) - \beta_0) = 2\beta_0\left(\beta_1\eta + \frac{1}{2}\beta_2\eta^2 + \cdots\right). \tag{E.11}$$

Equation E.9 is transformed to time space and has the form

$$\frac{\partial A}{\partial z} + \beta_1 \frac{\partial A}{\partial t} + i\frac{1}{2}\beta_2 \frac{\partial^2 A}{\partial t^2} + \frac{\alpha}{2}A = i\gamma\,|A|^2\,A. \tag{E.12}$$

The absorption coefficient is denoted by α; to retain simplicity in the above derivation, it was not explicitly treated in the above derivation, but could be added by considering an imaginary contribution to the fiber's dielectric function. The nonlinear coefficient is defined as

$$\gamma = \frac{3\omega_0}{4n^2 S_{\text{eff}}}\mu_0\chi^{(3)}_{xxxx} = \frac{\omega_0}{c S_{\text{eff}}}n_2, \tag{E.13}$$

and it has the units of inverse power length product. The refractive index is denoted by n. Using the copropagating coordinate transformation, the group velocity is removed from consideration. The result is the nonlinear Schrodinger equation with linear absorption as an additional contribution:

$$\frac{\partial A}{\partial z} + i\frac{1}{2}\beta_2 \frac{\partial^2 A}{\partial t^2} + \frac{\alpha}{2}A = i\gamma\,|A|^2\,A. \tag{E.14}$$

The soliton solutions of Equation E.14 can be characterized by a pulse width T_0 and a peak power P_0. The scaling time and the field amplitude are accordingly $T = t/T_0$ and $|\bar{A}|^2 = |A|^2/P_0$. Two length scales can be defined for the coefficients: the dispersion and nonlinear lengths

$$L_\mathrm{D} = \frac{T_0^2}{\beta_2} \quad \text{and} \quad L_\mathrm{NL} = \frac{1}{(\gamma P_0)}. \tag{E.15}$$

The dispersion length, L_D, is the distance over which an initial Gaussian pulse's width broadens by $\sqrt{2}$, and the nonlinear length, L_NL, is the distance over which the phase of the wave changes by 1 radian due to the nonlinearity. When two lengths are comparable in size, the pulse may form soliton pulses.

REFERENCE

Agrawal, G. P. 2007. *Nonlinear Fiber Optics* (5th edition). Amsterdam, Netherlands: Elsevier.

INDEX

Exponential growth, 188, 192, 305
Extraordinary waves, 34, 38

F

Fabrication, 172
    problem, 172–173
    technique, 175–176
Fabry–Perot-type cavity, 217
Fabry–Perot etalon, 259
Faraday's laws, 11
Fast Fourier transform (FFT), 170, 345–346
    algorithms, 170
Femtosecond accuracy, 431
Femtosecond laser, 4, 103–105
Ferroelectric crystals, 174–175
FFT, *see* Fast Fourier transform
Fiber amplifier, 316
Field amplitudes, 166, 169, 201, 220
"Fingerprint" scattered frequency signature, 4–5
First-order coupled differential equations, 184
First-order differential equation, 123
First-order susceptibility, 76
Fluorescence signal, 298
Formatting beam for SHG, 52–55
Forward-propagating waves, 349, 357
Four-wave mixing parametric amplifier, 263
Fourier approach, 172
Fourier optics, 31
Fourier transform, 21, 349
    treatment of QPM, 168–172
Fourier transformed space, 385
Free current density, 189–190
Frequency-converting crystal, 207
Frequency-resolved optical gate (FROG), 233
Frequency chirping method, 360
Frequency shift, 292
    Raman frequency shifts, 293
Fresnel's equation, 37
FROG, *see* Frequency-resolved optical gate
Full permutation symmetry in lossless media, 81
Full-width-at-half-maximum (FWHM), 142
    temperature-tuning bandwidth, 143
Fused silica, 447
FWHM, *see* Full-width-at-half-maximum

G

Gallium arsenide (GaAs), 176, 446–447
Gaussian beams, 31, 45–47, 256–257, 278
    down-collimating telescope, 49–51
    lens, 47–48
    with nonlinear polarization, 208–210
    OPO with, 221–223
    optical system, 48–49
    optimizing Gaussian beam interactions, 213–215
    parametric interactions with focused, 211–212

propagation using q-parameter, 47
    threshold, 226
    transforms, 228
Gaussian DFG interaction, 212
Gaussian function, 388, 429
Gaussian laser beam, 211
Gaussian units, 11
Geometrical function, 168–169
Gray solitons, 371–372
Ground state, 388
Group-velocity dispersion (GVD), 231, 337, 354

H

Hamiltonian equation, 385
Harmonic oscillator
    equation, 71, 311
    equation of motion for, 300
Heisenberg equations
    for degenerate down-conversion, 412
    of motion, 402
Heisenberg pictures, 382–383
Heisenberg uncertainty relations, 389–391
Hermitian conjugates, 386
Hermitian property, 34
Heterodyne detection technique, 409
High-order harmonic waves, 4
Higher-order effects, 65–66
HOM interferometer, *see* Hong–Ou–Mandel interferometer
Homodyne detection, 409
Homogeneous isotropic crystal, 41
Homogeneous isotropic materials, wave equation in, 30–32
Hong–Ou–Mandel interferometer (HOM interferometer), 421, 427–431
Hooke's law, 71
Hyperbolic sine, 187
Hysteresis curve, 261

I

Impermeability tensor (B_{ij}), 99
Incident beam, 292
Index ellipsoid, 43–45
Index of refraction, 32–33, 41, 141–142, 163
Inelastic scattering process, 291
Infrared region (IR region), 33
Instantaneous polarization, 351
Intensity, 19–20, 70, 127, 187–188, 193, 211
Interference, 15
Inverse Fourier transform, 21, 346–347
Irradiance, 10, 19
IR region, *see* Infrared region
Isotropic medium, 120, 131

J

Johnson noise, *see* Thermal noise